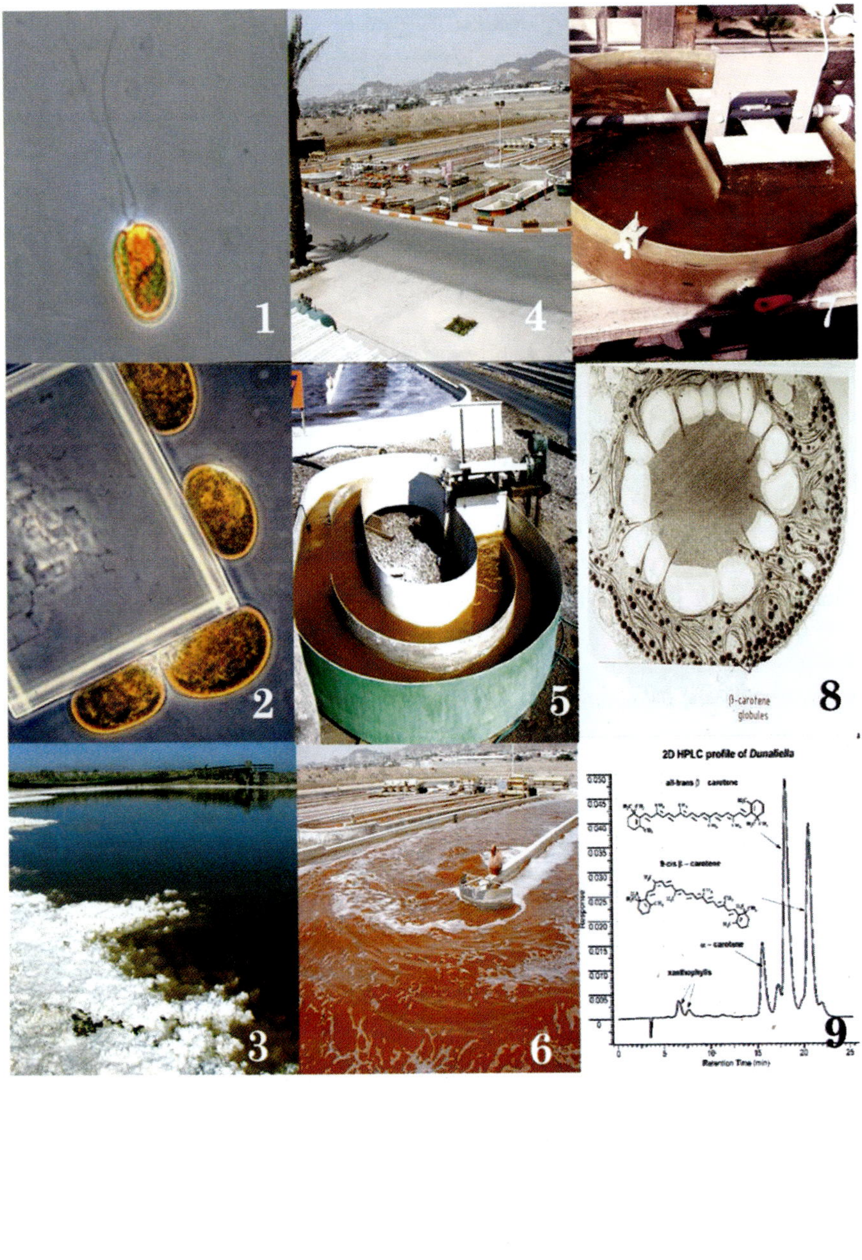
1
4
7
2
5
β-carotene
globules
8
3
6
2D HPLC profile of Dunaliella
all-trans β carotene
9-cis β – carotene
α – carotene
xanthophylls
Response
Retention Time (min)
9

Front Cover shows: **Cells of *Dunaliella bardawil* loaded with β catotene.**

Frontipiece 1

Figure 1: Stressed cells of *Dunaliella bardawil.*

Figure 2: Cells of *Dunaliella bardawil* next to a salt crystal within a concentrated brine solution.

Figure 3: Photograph of a typical hypersaline habitat in which *Dunaliella* algae are found in the San Francisco Bay, USA.

Figure 4: Production facility of NBT with high-rate algae ponds in Eilat, Israel.

Figure 5: Small experimental raceway pond containing *Dunaliella bardawil.*

Figure 6: Close-up view of a large raceway pond with *Dunaliella bardawil* at NBT in Eilat, Israel.

Figure 7: Small 5m^2 raceway pond with *Dunaliella bardawil* at NBT in Eilat, Israel.

Figure 8: Electron micrograph showing the posterior part of an orange, stressed cell of *Dunaliella bardawil.* Visible are the pyrenoid, b-carotene globules, and thylakoid membranes.

Figure 9: Shown is a two-dimensional HPLC profile obtained from orange cells of *Dunaliella bardawil.*

Photo credits: # 3 Dr. J. Polle; #1,2,4,5,6,7,8, and 9 Dr. Ami Ben-Amotz

Back Plate

Figure 1: A *Dunaliella salina* cell from a salt lake in Australia.

Figure 2: Stressed cells of *Dunaliella bardawil.*

Figure 3: Cells of *Dunaliella bardawil* next to a salt crystal within a concentrated brine solution.

Figure 4: Non-stressed green cells of *Dunaliella bardawil.*

Figure 5: Small experimental raceway pond containing *Dunaliella bardawil.*

Figure 6: 100 m^2 raceway pond with *Dunaliella bardawil* at NBT in Eilat, Israel.

Figure 7: Production facility of NBT with high-rate algae ponds in Eilat, Israel.

Figure 8: Photograph of a typical hyper-saline habitat in which *Dunaliella* thrive in the Great Salt Lake in Utah, USA.

Photo credits: #1 Dr. M. Dyall-Smith; #2,3,4,5,6, and 7 Dr. Ami Ben-Amotz; #8 Dr. J. Polle.

The Alga *Dunaliella*

Biodiversity, Physiology, Genomics and Biotechnology

Editors

Ami Ben-Amotz

Marine Biology, The National Institute of Oceanography

Haifa, Israel

Jürgen E.W. Polle

Department of Biology, Brooklyn College of CUNY

Brooklyn, NY, USA

D.V. Subba Rao

Bedford Institute of Oceanography, Dartmouth, NS

Canada

Science Publishers

Enfield (NH) Jersey Plymouth

Science Publishers
234 May Street
Post Office Box 699
Enfield, New Hampshire 03748
United States of America

www.scipub.net

General enquiries : *info@scipub.net*
Editorial enquiries : *editor@scipub.net*
Sales enquiries : *sales@scipub.net*

Published by Science Publishers, Enfield, NH, USA
An imprint of Edenbridge Ltd., British Channel Islands
Printed in India

ISBN 978-1-57808-545-3 (hardcover)

Library of Congress Cataloging-in-Publication Data

The alga Dunaliella : biodiversity, physiology, genomics and biotechnology/ editors, Ami Ben-Amotz, Jürgen E.W. Polle, D.V. Subba Rao. — 1st ed.
p. cm.
Includes bibliographical references and index.
ISBN 978-1-57808-545-3 (hardcover)
1. Dunaliella. I. Ben-Amotz, Ami, 1943- II. Polle, Jürgen E. W. III. Subba Rao, D. V.

QK569.P65A44 2008 579.8'32--dc22 2008048393

To

M. Avron, R.W. Butcher, A. Labbé, W. Lerche, N.P. Massjuk,
F.S. Milko, V.E. Semenenko, and E.C. Teodoresco
for their pioneering contributions on Dunaliella.

Preface

First described by Teodoresco in 1905, the unicellular green flagellate of the genus *Dunaliella* has been one of the most studied members of Chlorophyceae and is represented by 27 species, of which 23 are from salt water. Unique to *Dunaliella* cells is the absence of rigid cell walls, which characterize other unicellular green algae. Species of *Dunaliella* occur in freshwater, euryhaline habitats of all continents, oceans including the Dead Sea and even the salt lakes of the Antarctic. These extremophiles thrive in habitats with a wide range of salinity, pH, light intensity and temperature. The green vegetative cells of *Dunaliella salina* and *D. bardawil* under stress and light turn red due to over-accumulation of α and β-carotene, a feature that has biotechnological applications. Due to these unique features, interest in the fundamental physiological and biochemical research of *Dunaliella* has been increasing. Perhaps no other alga is studied as intensively as *Dunaliella* as a model for osmoregulation, pigment production and for commercial mass cultures.

The 21 chapters in this volume present a state-of-the art research in selected fields of *Dunaliella* including research in biochemistry, molecular biology and medical application. A glossary of specialized terms is appended. Each chapter is contributed by an expert or group of experts dedicated to increase our understanding of *Dunaliella*. All the chapters were reviewed internally by their colleagues, editors and external reviewers; this was followed by a final revision. Due to the range of subject matter, the 21 chapters contributed by 41 experts from 13 nations may vary in their format and style. Nevertheless, it is hoped this book provides a balanced multi-disciplinary communication and contributes to our understanding of this unique alga. We are most grateful to each of the contributors for their understanding, high level of professional and scholarly efforts, and for offering cordial and prompt cooperation in the preparation of this book. To our external reviewers we express our gratitude for their help with constructive comments in improving the content.

This book is addressed to postgraduate students and scientists as a summary of current thoughts on *Dunaliella*. We hope this book will serve as a stimulus and catalyst for further research on *Dunaliella*; if so, as a contribution it will have served a useful purpose.

Subba Rao expresses special thanks to his wife Bala T. Durvasula for her infinite patience, and computer skills while keeping track of the correspondence between the three editors, widely-scattered contributors, and reviewers, and for excellent support in formatting the chapters and art work.

Acknowledgements

For Color figures/Photographs

We greatly appreciate the kind support of the Nikken Sohonsha Company and its late President Y. Tanaka, which made it possible to include color figures and photographs in this publication.

To Reviewers

The Editors reviewed the manuscripts first and submitted them to external reviewers. They extend their grateful thanks to the following conscientious reviewers who have selflessly contributed their time by providing constructive comments on the manuscripts and thus helped in producing this book. *Denotes reviewed more than one manuscript.

Dr. John Beardall, Monash University, Clayton, VIC, Australia

Dr. Lesley Clementson, CMAR, Hobart, Tasmania, Australia

Dr. Ravi V. Durvasula*, University of New Mexico, NM 87131, USA

Dr. John Erdman, Nutrition Research Chair, 451 Bevier Hall, University of Illinois, 905 South Goodwin Ave,Urbana, IL 61801 USA

Dr. Mike Fulton, National Ocean Service, Center for Coastal Environmental Health and Biomolecular Research Charleston Laboratory, 219 Fort Johnson Rd., Charleston, SC 29412 USA

Dr. Willem Goedkoop, Swedish University of Agricultural Sciences, Uppsala, Sweden

Dr. Govindjee, Professor Emeritus of Biochemistry, Biophysics and Plant Biology, University of Illinois at Urbana-Champaign,USA

Dr. Glen W. Harrison, Bedford Inst. Oceanography, Dartmouth, Canada

Dr. L. Jahnke, University of New Hampshire, USA

Dr. Carlos Jiminez, Department of Ecology, Faculty of Sciences, University of Málaga, E-29071 Málaga, Spain

Dr. Angray Kang, Dept. Biosciences University of Westminster School of Biosciences, London, UK

Dr. Anastasios Melis, Department of Plant and Microbial Biology University of California, Berkeley, USA

Dr. Marta Gomez-Chiarri, University of Rhode Island, Kingston, RI, USA

Dr. Martin Lohr, Johannes Gutenberg-Universitat, Mainz, Germany

Dr. Stephen Mayfield, Department of Cell Biology and The Skaggs Institute for Chemical Biology, The Scripps Research Institute 10550 N. Torrey Pines Rd. La Jolla, Calif., 92037, USA

Dr. James E. Stewart, Bedford Inst. Oceanography, Dartmouth, NS, Canada

Dr. Thomas Pröschold, Dunstaffnage Marine Laboratory, Argyll, UK

Dr. Suzanne Roy, Universite du Quebec a Rimouski, Rimouski, Canada

Dr. Erica B. Young, University of Wisconsin-Milwaukee, Milwaukee, USA

Contents

List of Contributors

Maria Barbosa

Wageningen University, Bioprocess Engineering Group, P.O. Box 8128, 6700, EV Wageningen, The Netherlands

John Beardall

School of Biological Sciences, Monash University, Clayton, Australia
email: john.beardall@sci.monash.edu.au

Ami Ben-Amotz

Marine Biology, The National Institute of Oceanography, Israel Oceanographic & Limnological Research, POB 8030, Haifa 31080, Israel
email: ami@ocean.org.il

John R. Benemann

Benemann Associates, 3434 Tice Creek Dr. No.1, Walnut Creek, CA 94595, USA
email: jbenemann@aol.com

Hofit Cohen

The Bert W. Strassburger Lipid Center, The Chaim Sheba Medical Center, Tel-Hashomer 52621 Israel *and* Sackler Faculty of Medicine, Tel Aviv University Israel
email: Hofit.Cohen@sheba.health.gov.il

Pedro Duarte

CIAGEB-University Fernando Pessoa, Prąca 9 de Abril, 349, 4249-004, Portugal
pduarte@ufp.edu.pt

Marie E. DeLorenzo

NOAA, National Ocean Service, 219 Fort Johnson Rd., Charleston, SC 29412, USA
email: Marie.Delorenzo@noaa.gov

Ravi Durvasula

Division of Infectious Diseases, Department of Internal Medicine, School of Medicine, MSC 10 5550, School of Medicine, 1 University of New Mexico, NM 87131
email: Ravi.Durvasula@med.va.gov

Einar Skarstad Egeland

Faculty of Biosciences and Aquaculture, Bodø University College, Main building, Mørkved, NO-8049 Bodø, Norway
email: einar.skarstad.egeland@hibo.no

Mario Giordano

Department of Marine Sciences, Università Politecnica Delle Marche, 60121 Ancona, Italy

Mariela A. González

Departamento de Botánica, Facultad de Ciencias Naturales y Oceanográficas, Universidad de Concepción, Casilla 160-C, Concepción, Chile
email: mgonzale@udec.cl

Patricia I. Gómez

Departamento de Botánica, Facultad de Ciencias Naturales y Oceanográficas, Universidad de Concepción, Casilla 160-C, Concepción, Chile

Ayelet Harari

Sheba Medical Center, Tel Hashomer and Sackler Faculty of Medicine, Tel Aviv University, Tel Hashomer 52621, Israel

Dror Harats

Sheba Medical Center, Tel Hashomer and Sackler Faculty of Medicine, Tel Aviv University, Tel Hashomer 52621, Israel

Michael H. Huesemann

Pacific Northwest National Laboratory (PNNL), Marine Sciences Laboratory, 1529 West Sequim Bay Road, Sequim, Washington 98382, USA
email: michael.huesemann@pnl.gov

Ivy Hurwitz

Division of Infectious Diseases, Department of Internal Medicine, MSC 10 5550, School of Medicine, 1 University of New Mexico, NM 87131. USA

Leland S. Jahnke

Department of Biological Sciences, Rudman Hall, University of New Hampshire, Durham, NH 03824, USA
email: lsj@cisunix.unh.edu

Marcel Janssen

Wageningen University, Bioprocess Engineering Group, P.O. Box 8128, 6700 EV Wageningen, The Netherlands
email: Marcel.Janssen@wur.nl

Shirley W. Jeffrey

CSIRO Division of Marine and Atmospheric Research, GPO Box 1538, Hobart, Tasmania, Australia 7001
email: Shirley.Jeffrey@csiro.au

EonSeon Jin

Department of Life Science, College of Natural Science, Hanyang University, 133-791, Seoul, South Korea
email: esjin@hanyang.ac.kr

Yehuda Kamari

Sheba Medical Center, Tel Hashomer and Sackler Faculty of Medicine, Tel Aviv University, Tel Hashomer 52621, Israel

Adriana Katz

Dept. of Biological Chemistry, Weizmann Institute of Science, Rehovot, 76100, Israel

Scott Matthews

Division of Infectious Diseases, Department of Internal Medicine, School of Medicine, University of New Mexico, 1 University of New Mexico, NM 87131, USA
email: skm4150@yahoo.com

Denis P. Maxwell

Department of Biology, University of Western Ontario, North Campus Building, Room 406, London, Ontario, N6A 5B7, Canada
email: dmaxwell@uwo.ca

Anastasios Melis

Department of Plant and Microbial Biology, 111 Koshland Hall, University of California, Berkeley, CA 94720-3102, USA
email: melis@nature.berkeley.edu

Liki von Oppen-Bezalel

IBR, Israeli Biotechnology Research, Ltd., Rehovot, 76100, Israel
email: Likiob@yahoo.com

Yakov Paz

Dept. of Biological Chemistry, Weizmann Institute of Science, Rehovot, 76100, Israel

Uri Pick

Dept. of Biological Chemistry, Weizmann institute of science, Rehovot, 76100, Israel
email: uri.pick@weizmann.ac.il

Jürgen E.W. Polle

Department of Biology, Brooklyn College of CUNY, 2900 Bedford Ave. 200NE, Brooklyn, NY 11210, USA
email:jpolle@brooklyn.cuny.edu

Song Qin

Institute of Oceanology, Chinese Academy of Sciences, Qingdao 266071, China
email: sqin@ms.qdio.ac.cn

John A. Raven

Division of Plant Sciences, University of Dundee at SCRI, Scottish Crop Research Institute, Invergowrie, Dundee DD2 5DA, UK
email j.a.raven@dundee.ac.uk

Priya Sampath-Wiley

Department of Biological Sciences, Rudman Hall, University of New Hampshire, Durham, NH 03824, USA

Gustav Schonfeld

6602 Wohl Hospital, 660 S. Euclid Ave., Campus Box 8046, St. Louis, MO 63110, USA
e mail:gschonfe@wustl.edu

Aviv Shaish

The Bert W. Strassburger Lipid Center, The Chaim Sheba Medical Center, Tel-Hashomer, 52621, Israel
email: avivsha@netvision.net.il

D.V. Subba Rao

12 The Horseshoe, Manor Park, Dartmouth, NS, B2Y 4E5 Canada
email:seshu35@gmail.com

Bobban Subhadra

Division of Infectious Diseases, Department of Internal Medicine, School of Medicine, University of New Mexico, 1 University of New Mexico, NM 87131, USA
email:bsubhadra@salud.unm.edu

Duc Tran

Department of Biology, Brooklyn College of CUNY, 2900 Bedford Ave. 200NE,-Brooklyn, NY 11210, USA

Andrea L. White

Department of Biological Sciences, Rudman Hall, University of New Hampshire, Durham, NH 03824, USA

René H. Wijffels

Wageningen University, Bioprocess Engineering Group, P.O. Box 8128, 6700 EV Wageningen, Netherlands
email: Rene.wijffels@wur.nl

Kittisak Yokthongwattana

Department of Biochemistry and Center for Excellence in Protein Structure and Function, Faculty of Science, Mahidol University, Rama 6 Rd., Bangkok 10400, Thailand
email: tekyw@mahidol.ac.th

Ada Zamir

Department of Biological Chemistry, Weizmann Institute of Science, Rehovot 76100, Israel
email: ada.zamir@weizmann.ac.il

1

History, Distribution, and Habitats of Algae of the Genus *Dunaliella* Teodoresco (Chlorophyceae)

Jürgen E. W. Polle[1], Duc Tran[1], and Ami Ben-Amotz[2]*

Abstract

Unicellular green algae of the genus *Dunaliella* TEOD. (Chlorophyta) have been used extensively as model organisms for various research areas and for mass culture. Since the early 19th century researchers have investigated the origin of the orange, purple, or pink color of water from salt lakes and salterns. Earlier it already was noticed that some kind of unicellular alga was causing the orange color of these hypersaline water bodies. This chapter reviews the history of the genus *Dunaliella* beginning with the earliest traceable records. In addition, worldwide distribution and habitats of algae of the genus *Dunaliella* are covered. As a specific example the Great Salt Lake in Utah, USA is described as a habitat of three recognized *Dunaliella* species.

Introduction

Unicellular green algae of the genus *Dunaliella* Teodoresco (Teodoresco 1905) were originally grouped in the class of Chlorophyceae, the order of *Volvocales*, and the family of *Polyblepharidacae*. Currently, order and family are still under debate. According to the National Center of Biotechnology Information of the National Institute of Health, algae of the genus *Dunaliella* belong to the class of *Chlorophyceae*, the order of *Chlamydomonadales*, and into the family of *Dunaliellaceae*. Chapter 2 by González et al. of this book discusses classification of the genus *Dunaliella* in more detail.

[1] Department of Biology, Brooklyn College of CUNY, 2900 Bedford Ave. 200NE, Brooklyn, NY 11210, USA.

[2] The National Institute of Oceanography, Israel Oceanographic and Limnological Research, Haifa 31080, Israel.

* Author for correspondence.

Dunaliella species are mostly radially symmetrical, sometimes bilaterally symmetrical, flattened, dorsoventrally curved or slightly asymmetrical. The cell shape varies from ellipsoidal, ovoid, cylindrical, pyriform, or fusiform to almost spherical. Figure 1 shows a representative cell for *Dunaliella*. *Dunaliella* cells lack rigid cell walls, a feature distinguishing them from other unicellular green algae such as the genus *Chlamydomonas*. Nevertheless, *in vivo* cells of *Dunaliella* contain an almost invisible outer cell coat of variable thickness (Hamburger 1905, Teodoresco 1905, Figure 1). Already Teodoresco (1905) compared the cell coat as an envelope 'de nature peut-être protéïque' similar to the 'Hautschicht' of *Euglena*. Due to lack of a cell wall containing cellulose (Hamburger 1905), *Dunaliella* cells were often described as being 'naked' flagellates. Hamburger (1905) called the cell coat a 'Gallerthülle' (=mucilaginous coat) and stained it by use of DELAFIELD Hämatoxylin or visualized it with Chinese ink. Later Labbé (1925) addressed the *Dunaliella* cell coat as a 'chlamyde' which could be stained by use of "blue-polychrome-orcéine". Subsequent analysis (Oliveira et al. 1980, see also Ginzburg 1987) revealed that the cell coat could be stained by use of cationic dyes (Ruthenium Red, Alcian Blue) and it could be digested by trypsin or pronase. These results indicated that the mucilaginous cell coat of *Dunaliella* contains glycoproteins. The proposed makeup of the cell coat by glycoproteins is consistent with findings by Sadka et al. (1991) who described a high molecular weight glycoprotein of about 150 KDa existing in the outer layer surrounding *Dunaliella* cells. As the mucilaginous *Dunaliella* cell coat appears to be made up of glycoproteins, it is often referred to as a 'glycocalyx' (Borowitzka and Borowitzka 1988).

In general, vegetative cells of *Dunaliella* have two flagella of equal length of at least one body-length. If present, both flagella originate at the anterior end of cells.

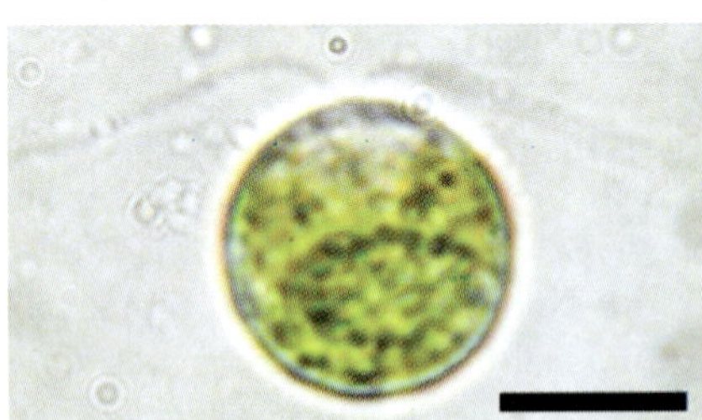

Figure 1: Photograph of an exemplary cell for algae of the genus *Dunaliella* (Magnification 360x). Both flagella originate at the anterior of the cell. The chloroplast fills the posterior region of the cell. The glycocalyx surrounding the cell is clearly visible as a white ring around the cell. The bar shown represents 10 μm.

Further, cells contain one single cup-shaped chloroplast with one central pyrenoid (Figure 2) surrounded by the storage product starch. Typical other organelles are anterior eyespots, anterior nucleus with nucleolus. Golgi bodies, and vacuoles. Cell size and shape may change within a given species depending on different environmental conditions varying in length between 2 to 28 μm and in width between 1 to 15 μm. Under non-optimal salt concentration, cell morphology may change to asexual sub-spherical, thick-walled cysts with bumpy surfaces that are often referred to as aplanospores (Borowitzka and Huisman 1993, Margulis et al. 1980). Nevertheless, the existence of aplanospores was questioned by Lerche (1937), who claimed that aplanospores are actually zygotes. Many species are extremely halotolerant and thrive on a wide range of salinities from seawater to salterns and evaporation ponds even in saturated salt solutions like the Dead Sea, Israel, and the Great Salt Lake, USA. Figure 2 shows exemplary cells of *D. viridis* and *D. salina* growing in concentrated brine solution. In addition, Figure 3 shows a cell of *Dunaliella spec.* that thrives in a concentrated brine solution containing visible salt crystals.

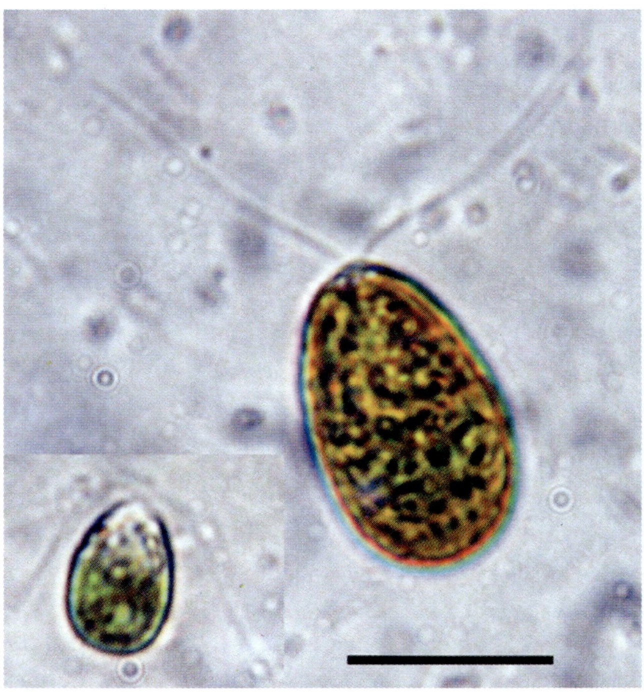

Figure 2: Shown are exemplary cells of *D. viridis* Teodoresco (lower left) and of *D. salina* Teodoresco (right side) grown in saturated brine solution (Magnification 360x). The bar represents 10 μm.

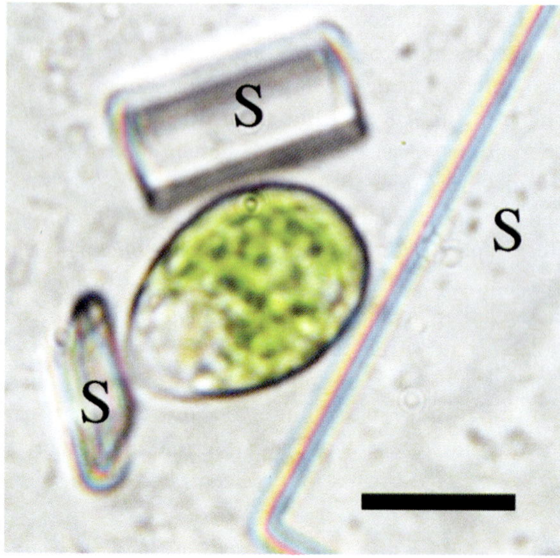

Figure 3: Photograph of a cell of *Dunaliella sp.* that is thriving in saturated brine solution at a magnification of 360x. S = salt crystals.

In past decades the species of *D. salina* TEODORESCO and *D. tertiolecta* BUTCHER were widely used as model organisms for different research areas such as osmoregulation, carotenoid production, and photosynthesis under extreme conditions (Oren 2005). Currently, numerous strains of *Dunaliella* are deposited in different culture collections (http://www.dunaliella.org/dunabase/strains/strains.php), many with poor classification or erroneous characterization. In addition, various strains of *D. salina* are grown commercially in mass cultures for the production of natural β-carotene (Borowitzka and Borowitzka 1988, Ben-Amotz 2003).

History of the Genus *Dunaliella*

For centuries people wondered about orange-red lakes and red snow. Often the red color of waters reminded people of blood: "To the Moabites across the way, the water looked red—like blood" (Bible, 2 Kings 3.22). However, it was not until the microscope was discovered in the late 16th century that scientists could begin to reveal the mystery of red-colored waters and snow. A very recent review of the history of the species *Dunaliella* was presented by Oren (2005).

The earliest currently traceable scientific publication is from Turpin (1836, 1839) who recognized that the reddish coloration of salt lakes is caused by a microscopic alga, which he named *Globularia kermesina*. At about the same time Dunal (1838) investigated the origin of the reddish color of salterns and described two microalgal species *Protococcus salinus* DUNAL and *Haematococcus salinus* DUNAL. Table 1 shows that following Dunal (1838) several scientists renamed the alga found in salt lakes and marine habitats multiple times. Finally, Teodoresco (1905) created the genus *Dunaliella* TEODORESCO for the unicellular, halotolerant alga and recognized the type species *D. salina* TEODORESCO (vegetative cells capable to turning red) (Teodoresco 1905, Hamburger 1905). Figure 4 shows cells of the type species *D. salina*. Vegetative cells are green under optimal environmental conditions and are 5.0 to 30.0 μm long and 2.5 to 21 μm wide. Depending on cell size and shape of the cell, the type species was divided into sub-species (Lerche 1937, Massjuk 1972, Preisig 1992). When exposed to environmental stress such as high salinity, vegetative cells turn orange (Figure 4B, 4C) due to over-accumulation of the pigment β-carotene (Loeblich 1982). Two decades later Labbé (1925) renamed the species *D. salina* TEOD. into *D. kermesina* TURPIN, because he recognized that Turpin had named this organism first. Labbé (1925) was not sure if an even earlier description of the species *Lepraria kermesina* by Wrangel from the year 1823 was also a synonym for *D. salina*. Unfortunately, the publication of Wrangel from 1823 cited by Labbé (1925) cannot be traced nor the accuracy verified, because of an incomplete citation by Labbé. Nevertheless, subsequent researchers did not accept the classification of the alga as *D. kermesina* TURPIN and continued to use the name *D. salina* TEODORESCO. Later, Ben-Amotz and Avron (1980a, b) isolated a new strain from the Bardawil Lagoon and named it *D. bardawil* BEN-AMOTZ et AVRON. However, *D. bardawil* BEN-AMOTZ et AVRON seems to fit the species characteristics of *D. salina* TEODORESCO (Borowitzka and Borowitzka 1988, Ginzburg 1987, Gonzalez et al. 2001). At this stage classification of *D. bardawil* BEN-AMOTZ et AVRON awaits reassessment through methods of modern molecular biology.

Table 1: Listed in chronological order by year of identification, names of species, and the author generating the taxon.

1 = *Dunaliella salina* variety according to species description of Teodoresco (1905)
2 = "*Chlamydomonas* I" Atkins, W.R. and Parke, M., 1951, J. Mar. Biol. Ass. U.K., 29, 609

Year	Species	Author
1836/39	*Globularia kermesina*	Turpin
1838	*Haematococcus salinus/ Protococcus salinus*	Dunal
1840	*Monas dunalii*	Joly
1841	*Diselmis dunalii*	Dujardin

(Table 1 Contd.)

(Table 1 Contd.)

1865	*Chlamydomonas dunalii*	Cohn
1872	*Protococcus salinus*	Geleznow
1886	*Sphaerella lacustris* var. *dunalii*	Hansgirg
1891	*Chlamydomonas dunalii*	Blanchard
1905	*Dunaliella salina*	Teodoresco
1906	*Dunaliella viridis*	Teodoresco
1925	*D. kermesina*[1]	Labbé
1935	*D. peircei*	Nicolai and Baas-Becking
1937	1. *D. parva* 2. *D. media* 3. *D. euchlora* 4. *D. minuta*	Lerche
1938	*D. spec.* 1 *D. spec.* 2 *D. spec.* 3 *D. spec.* 4	Ruinen
1956 1959a	*D. bioculata*	Eddy Butcher
1959b	1. *D. tertiolecta*[2] 2. *D. primolecta* 3. *D. polymorpha* 4. *D. quartolecta*	Butcher
1969	*D. turcomanica*	Massjuk
1971	*D. asymmetrica*	Massjuk
1973a	1. *D. maritima* 2. *D. granulata*	Massjuk
1973b	1. *D. terricola* 2. *D. gracilis* 3. *D. ruineniana* 4. *D. baas-beckingii* 5. *D. minutissima* 6. *D. carpatica* 7. *D. jacobae*	Massjuk
1973	*D. pseudosalina*	Massjuk and Radcnenko
1980a,b	*D. bardawil*[1]	Avron and Ben-Amotz
1980	*D. marina*	Kombrink and Wöber

Soon after creation of the genus with the type species *D. salina*, Teodoresco recognized the second species *D. viridis* TEODORESCO (vegetative cells always

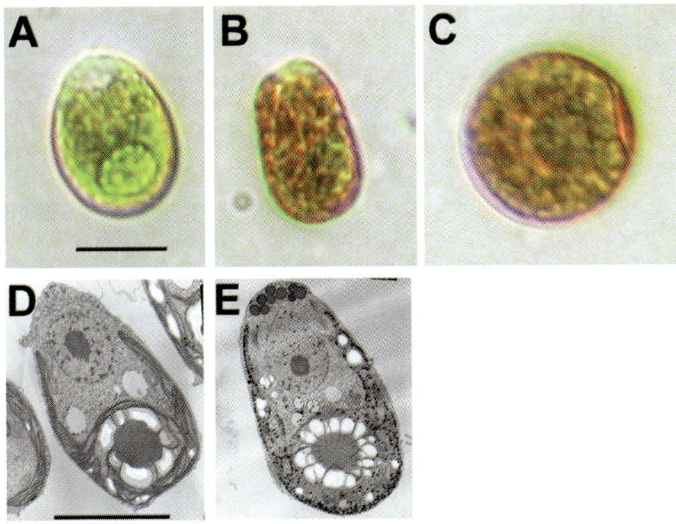

Figure 4: Photographs of cells of the type species *D. salina* Teodoresco at a magnification of 360x. A) Vegetative, motile, green cell grown at 1.5 M NaCl - the bar represents 10μm; B) Vegetative, motile, orange cell grown at 4.5 M NaCl; C) Orange cell grown at 4.5 M NaCl that is non-motile, either an aplanospore or a zygote; D) Electron micrograph of *Dunaliella bardawil, D. salina* variety according to species description of Teodoresco (1905), a green cell - The bar represents 10 μm; E) Electron micrograph of an orange cell. Photographs D & E reproduced by Dr. A. Ben-Amotz.

green) (Teodoresco 1906). Following generation of the genus *Dunaliella* and recognition of the two halotolerant species *D. salina* and *D. viridis*, many species were recognized during the last century, and characterized based on morphological and physiological features. Table 1 lists the year of identification, name of the species, authors, and provides references. Unfortunately, some of the species were only described once and not in great detail. Also, some of the *Dunaliella* species are not available from culture collections, or only one isolate exists among multiple collections.

Besides Teodoresco, major contributors to classification of species within the genus of *Dunaliella* were Lerche (1937), Butcher (1959a, b), and Massjuk (1969, 1971, 1973a, b, c, Massjuk and Radcnenko 1973). Following Labbé (1925), Lerche (1937) recognized great diversity within the species of *D. viridis*. Therefore, Lerche (1937) divided the species of *D. viridis* into several new species: *D. parva* Lerche, *D. peircei* Nicolai, *D. media* Lerche, *D. euchlora* Lerche, *D. minuta* Lerche. Later, Massjuk alone isolated and classified almost half of all currently known *Dunaliella* species. Massjuk's monograph contains very detailed information on the morphology and physiology of various *Dunaliella* species (Massjuk 1973c). Unfortunately, this monograph is still only available in Russian and not well known. Overall, since Teodoresco (1905), 26 saltwater species have been described for the genus *Dunaliella*. Most of those species were grouped by Massjuk (1972) into four sections: *Tertiolectae* MASSJUK, *Dunaliella* MASSJUK, *Virides* MASSJUK, and *Peirceinae* MASSJUK.

In addition to saltwater species, several freshwater species were grouped in the genus of *Dunaliella* (Kalina 1965, Massjuk 1972, Pascher 1930, 1932, Pascher and Jehoda 1928, Skvortzov 1968). The freshwater species are placed into the subgenus *Pascheria*, whereas all saltwater species are placed into the subgenus *Dunaliella* (Massjuk 1972, Preisig 1992). Only five freshwater species are currently still classified as belonging to the genus *Dunaliella* (Preisig 1992). All five species are rare and only two of the species are available from culture collections. One of them is *D. acidophila* (KALINA) MASSJUK (Melkonian and Preisig 1984) that was used specifically for investigation of pH homeostasis for several decades (Pick 1999). Classification of freshwater species to the genus *Dunaliella* is still questionable (Melkonian and Preisig 1984; Preisig 1992). For example, recent phylogenetic analysis revealed that the species of *Dunaliella lateralis* PASCHER et JAHODA (Pascher and Jahoda 1928) should no longer be classified as belonging to the genus of *Dunaliella* (Gonzalez et al. 1999, Gonzalez et al. 2001).

Distribution and Habitats

Chlorophytes (green algae) of the genus *Dunaliella* can be found in freshwater and euryhaline waters of all continents and oceans (Lerche 1937, Massjuk 1972, Ginzburg 1987, Borowitzka and Borowitzka 1988, Preisig 1992). In addition, one species is found in acidic waters (*D. acidophila* MASSJUK).

Reports of occurrence of euryhaline *Dunaliella* (*D. tertiolecta, D. primolecta, D. polymorpha*, and *D. quartolecta*) include marine saline waters in Europe and America. However, in marine phytoplankton *Dunaliella* species seem to play only a minor role, and their ecology is not well studied. In contrast, hypersaline species of *Dunaliella* (*D. viridis, D. parva, D. pseudosalina, D. salina*, and *D. spp.*) are commonly found from the Antarctic salt lakes to the salt lakes of Africa, America, Asia, Australia, and Europe (Table 2). In addition, hypersaline species of *Dunaliella* are present in evaporation ponds of salterns and salt marshes around the world. In hypersaline environments, species such as *D. salina, D. viridis*, and *D. parva* may play a major role in the ecosystems as primary photosynthetic producers of biomass. Often *Dunaliella* species give rise to blooms in salt lakes such as the Dead Sea (Oren and Shilo 1982, Oren 2005) or the Great Salt Lake (Post 1975). Typical examples of salt lakes and evaporation ponds with high brine concentrations are shown for North America in Figure 5. Although *Dunaliella* species were found in numerous hypersaline environments, they do not exist in all hypersaline habitats (Ginzburg 1987). It appears that water chemistry plays a role in determining if *Dunaliella* species will be present in a hypersaline environment. According to Ginzburg (1987) the major limiting factors among the nutrients in salt lakes are nitrogen and phosphate.

Table 2: Habitats and distribution of some of the species of the genus *Dunaliella*.

Species	Habitats	Distribution
D. salina TEOD.	Salt Lakes, Evaporation Ponds of Salterns, Hypersaline Salt Marshes or Lagoons	Africa, Americas, Asia, Australia, Europe
D. viridis TEOD.	Salt Lakes, Salterns	Africa, Americas, Asia, Australia, Europe
D. parva LERCHE	Salt Lakes	Asia, Europe
D. pseudosalina MASSJUK	Salt Lakes, Hypersaline Salt Marshes or Lagoons	South America, Europe
D. tertiolecta BUTCHER	Brackish or Marine Waters	America, Europe
D. species	Salt Lakes	Antarctica

One example for a typical habitat for *Dunaliella* species is the Great Salt Lake which is located in the northern part of the state of Utah, USA. Pictures of the Lake are shown in Figure 5. It is one of the largest terminal lakes on earth. As the lake is relatively shallow, its size and salinity fluctuate depending on cycles of evaporation and surface freshwater inflow. For a long period the hydrogeochemistry of the Great Salt Lake was investigated (Spencer et al. 1985a, Spencer et al. 1985b, Stephens 1990). Since 1957 the Great Salt Lake has been divided by a causeway into the northern and southern basins. In general, the southern basin is less saline with 5-15% in comparison to the northern basin -also called Gunnison Bay- which has higher salinities of 15-28%. The phytoplankton in the Great Salt Lake has been the subject of various studies (Kirkpatrik 1934, Brock 1975, Post 1977, Post 1981, Stephens and Gillespie 1976, Van Auken and McNulty 1973). Among other microbes, at least three species of *Dunaliella* occur in the Great Salt Lake in significant numbers. *D. salina* exists mainly in the northern basin where the salinity is high (Brock 1975; Post 1981). In summer the entire northern basin may appear orange or pinkish in color due to blooms of *D. salina* and halobacteria (Figure 5). The reddish forms of *D. salina* contribute to an orange-ochre color wherease halobacteria are responsible for pink coloration. *D. viridis* exists in various parts of the lake, whereas a third unclassified *Dunaliella* species was found around Antelope Island (Brock 1975, Van Auken and McNulty 1973). The third as of yet unclassified species appears to belong to the section

Tertiolectae (Polle, unpublished results). Distribution of *Dunaliella* species within the lake is determined by salinity with *D. salina* being predominant at salinities around 25%. With an optimal salinity for growth of *D. salina* around 12% (Lerche 1937), its abundance at salinities of 25% is probably due to lack of competition by other algae and predation (Brock 1975). In contrast, the species *D. viridis* with an optimal salinity for growth of about 6% (Lerche 1937; Borowitzka et al. 1977) and the unidentified *Dunaliella* species are predominant at lower salinities in the southern basin. This example of the Great Salt Lake demonstrates that multiple species of *Dunaliella* co-exist in the same habitat.

Figure 5: Photographs showing views of the Great Salt Lake in Utah, USA. Green color of water is due to presence of a mixture of unicellular green algae including green forms of *Dunaliella* species. The pink color of water is caused by an abundance of halobacteria. A more orange water color is due to the unicellular green alga *Dunaliella salina*.

Summary

Unicellular green algae of the genus *Dunaliella* were studied since the early 19th century and numerous species were characterized and classified since then. Classification and ecology will be discussed in more detail in following chapters. In general, *Dunaliella* species are ubiquitous in saline environments and often multiple species occur in the same habitat. This wide distribution of species of the genus *Dunaliella* may be contributed to their tolerance to a wide range of salinities, light intensities, and temperatures as well as their ability to survive for many years among salt crystals (Ginzburg 1987).

References

Avron, M. and A. Ben-Amotz. 1980a. United States Patent #4,199,895.

Avron, M. and A. Ben-Amotz. 1980b. United States Patent #PP4,511.

Ben-Amotz, A. 2003. Industrial production of microalgal cell-mass and secondary products – major industrial species. pp.273-280. In Handbook of Microalgal Cultures, Biotechnology and Applied Phycology. A. Richmond (ed.) Blackwell, UK.

Blanchard, R. 1891. Resultats d'une excursion zoologique en Algerie. Mem. de la Soc. Zool. de France. p. 243.

Borowitzka, L.J., D.S. Kessly and A.D. Brown. 1977. The salt relations of *Dunaliella*. Further observations on glycerol production and its regulation. Arch. Microbio. 113: 131-138.

Borowitzka, M.A. and L.J. Borowitzka. 1988. *Dunaliella*. pp. 27-58. In Microalgal Biotechnology. M.A. Borowitzka and L.J. Borowitzka (eds) Cambridge University Press, Cambridge, UK.

Borowitzka, M.A. and J.M. Huisman. 1993. The ecology of *Dunaliella salina* (Chlorophyceae, Volvocales): Effect to environmental conditions on aplanospore formation. Bot. Mar. 36 (3): 233-243.

Brock, T.D. 1975. Salinity and the ecology of *Dunaliella* from the Great Salt Lake. J. Gen. Microbiol. 89: 285-92.

Butcher, R.W. 1959a. An undescribed species of *Dunaliella* from the Cambridge collection of algae. Hydrobiol. 12: 249-250.

Butcher, R.W. 1959b. An introductory account of the smaller algae of British coastal waters. I. Introduction and Chlorophyceae. Fish. Invest. London Ser. VI(1):1.

Cohn, F. 1865. *Chlamydomonas marina*. Hedwegia (Notizblatt für kryptogamische Studien, nebst Repertorium für kryptog. Literatur) 7: 97-99.

Dujardin, F. 1841. Histoire Naturelle des Zoophytes (Infusoires). Paris, p. 344.

Dunal, F. 1838. Les algues qui colorent en rouge certaines eaux des marais salants mediterraneens. Ann. D. sc. Nat. Bot. 2e Serie. T.IX p. 172.

Eddy, O.B.P. 1956. The suitability of some algae for mass cultivation for food, with special reference to *Dunaliella bioculata*. J. Exp. Bot. 7: 327.

Geleznow, N. 1872. Über die Ursache der Färbung des Salzwassers im See Sak der Krim. Bull. De l'Acad. Imp. De St. Petersbourg T.17 p. 557.

Ginzburg, M. 1987. *Dunaliella*: A green alga adapted to salt. Adv. Bot. Res. 14: 93-183.

González, M.A., P.I. Gómez and R. Montoya. 1999. Comparison of PCR-RFLP analysis of the ITS region with morphological criteria of various strains of *Dunaliella*. J. Appl. Phycol. 10: 573-580.

González, M.A., A.W. Coleman, P.I. Gomez and R. Montoya. 2001. Phylogenetic relationship among various strains of *Dunaliella* (Chlorophyceae) based on nuclear ITS rDNA Sequences. J. Phycol. 37: 604-11.

Hamburger, C. 1905. Zur Kenntnis der *Dunaliella salina* und einer Amöbe aus Salinenwasser von Cagliari. Arch. f. Protistenkd. 6: 111-130.

Hansgirg, A. 1886. Prodromus der Algenflora von Böhmen. *T.I.* p. 106.

Kalina, T. 1965. Zur Morphologie und Taxonomie der Gattung *Spermatozopsis* Korschikow (Volvocales). *Spermatozopsis acidophila* sp. n., Preslia (Prague) 37: 9.

Kirkpatrick, R. 1934. The life of Great Salt Lake with special reference to the algae. M.S. Thesis Univ. Utah.USA .

Joly, N. 1840. Histoire d'un petit Crustacé (*Artemia salina*) auquel on a faussement attribué la coloration en rouge des marais salants méditerranéens, suivi de recherches sur la cause réelle de cette coloration. Ann. d. sc. Nat. Zool. 2e Série. T13: 225-289.

Labbé, A. 1925. 1er Mémoire: Les cycles biologiques des *Dunaliella. Arch. d'Anat. Microsc.* T. XXI: 314-399.

Lerche, W. 1937. Untersuchungen über Entwicklung und Fortpflanzung in der Gattung *Dunaliella*. Arch. Protistenkd. (88): 236-268.

Loeblich, L.A. 1982. Photosynthesis and pigments influenced by light intensity and salinity in the halophile *Dunaliella salina* (Chlorophyta). J. Mar. Biol. Assoc. UK 62: 493-508.

Margulis L., E.S. Barghoorn, D. Ashendorf, S. Banerjee, D. Chase, S. Francis, S. Giovannoni and J. Stolz. 1980. The microbial community in the layered sediments at Laguna Figueroa, Baja California, Mexico: Does it have Precambrian analogues? Precamb. Res. 11(2): 93-123.

Massjuk, N.P. 1969. A new species of the genus *Dunaliella* Teod. Ukr. Bot. Zh. (26): 87-90.

Massjuk, N.P. 1971. New species of *Dunaliella* with asymmetric cells. Ukr. Bot. Zh. (28): 148-152.

Massjuk, N.P. 1972. On phylogeny and taxonomy of the genus *Dunaliella* Teod. Ukr. Bot. Zh. 29: 744-750.

Massjuk, N.P. 1973a. New taxons from the genus *Dunaliella* Teod., I. Ukr. Bot. Zh. (30): 175-183.

Massjuk, N.P. 1973b. New taxons from the genus *Dunaliella* Teod., II. Ukr. Bot. Zh. (30): 345-354.

Massjuk, N.P. 1973c. Morphology, systematic, ecology, geographic distribution of the genus *Dunaliella* Teod. including perspectives of the practical application (translation of Russian title). АКАДЕМИЯ НАYК УКРАИНСКОЙ ССР ИНСТИТУТ БОТАНИКИ НМ. Н .Г . ХОЛОДНОГО, ИЭДАТЕЛЪСТВО «НАУКВА ДУМКА», КИЕВ.

Massjuk, N.P and M.I. Radcnenco. 1973. New taxons from the genus *Dunaliella* Teod., III. Ukr. Bot. Zh. (30): 468-447.

Melkonian, M. and H.R. Preisig. 1984. Ultrastructure of the flagellar apparatus in the green flagellate *Spermatozopsis similis*. Pl. Syst. Evol. 146: 145-162.

Nicolai, E. and L.G.M. Baas-Becking. 1935. Einige Notizen über Salzflagellaten. Arch. Protistenkd. (85): 319-328.

Oliveira, L., T. Bisalputra and J.A. Naval. 1980. Ultrastructural observation of the surface coat of *Dunaliella tertiolecta* from staining with cationic dyes and enzyme treatments. New Phytologist 85(3): 385-392.
Oren, A. 2005. A hundred years of *Dunaliella* research: 1905-2005. Sal. Systems 1: 2
Oren, A. and M. Shilo. 1982. Population dynamics of *Dunaliella parva* in the Dead Sea. Limnol Oceanogr. 27: 201-211.
Pascher, A. 1930. Neue Volvocalen (Polyblepharidinen-Chlamydomonadinen). Arch. Protistenkd. 69: 103.
Pascher, A. 1932. Zur Kenntnis der einzelligen Volvocalen. Arch. Protistenkd. 76: 1.
Pascher, A. and R. Jahoda. 1928. Neue Polyblepharrideen und Chlamydomonadinen aus den Alttûmpeln um Lunz. Arch. Protistenkd. 76.
Post, F.J. 1977. The microbial ecology of the Great Salt Lake. Microb Ecol. (3): 143-165.
Post, F.J. 1981. Microbiology of the Great Salt Lake north arm. Hydrobiologia 81: 59-69.
Preisig, H.R. 1992. Morphology and Taxonomy. pp.1-15. In *Dunaliella*: Physiology, Biochemistry, and Biotechnology. Avron and Ben-Amotz (eds) CRC Press Boca Raton, USA.
Ruinen, J. 1938. Notizen über Salzflagellaten. II Über die Verbreitung der Salzflagellaten. Arch. Protistenkd. (90): 210-258.
Sadka, A., S. Himmelhoch and A. Zamir. 1991. A 150 Kilodalton cell surface protein is induced by salt in the halotolerant green alga *Dunaliella salina*. Plant Physiol. 95: 822-831.
Skvortzov, B.V. 1968. New genera of green flagellatae recorded in N.-O. China. Rev. Algol. 9: 121.
Spencer, R.J., H.P. Eugster and B.F. Jones. 1985b. Geochemistry of Great Salt Lake, Utah II: Pleistocene-holocene evolution. Geochim. Cosmochim. Acta. 49(3): 739-747.
Spencer, R.J., H.P. Eugster, B.F. Jones and S.L. Rettig. 1985a. Geochemistry of Great Salt Lake, Utah I: Hydrochemistry Since 1850. Geochim. Cosmochim. Acta 49(3): 727-737.
Stephens, D.W. 1990. Changes in lake levels, salinity and the biological community of Great Salt Lake (Utah, USA), 1847–1987. Hydrobiologia 197(1): 139-146
Stephens, D.W. and D.M. Gillespie. 1976. Phytoplankton production in the Great Salt Lake, Utah, and a laboratory study on algal response to enrichment. Limnol. Oceanography 21(10):74-87.
Teodoresco, E.C. 1905. Organization et developpement du *Dunaliella*, nouveau genre de Volvocacee-Polyblepharidee. Beih. Bot. Zentralblatt. Bd.18 Abt.1: 215-232.
Teodoresco, E.C. 1906. Observations morphologiques et biologiquessur le genre *Dunaliella*. Rev. Gén. d. Bot. T18: 353-371.
Turpin, P.J.F. 1836. Examen de la substance rouge observée par M. Payen sur la surface des marbres blancs de la carrière de Serraveza. C. R. Ac. Sc., III, p. 766.
Turpin, P.J.F. 1839. Quelques observations nouvelles sur les *Protococcus*, qui colorent en rouge les eaux des marais salants. Comp. Rend. Acad. Sci. nov. p. 626.
Van Auken, O.W. and I.B. McNulty. 1973. The effect of environmental factors on the growth of a halophylic species of algae. Biol. Bull. 145: 210-22.

2

Taxonomy and Phylogeny of the Genus *Dunaliella*

Mariela A. González[1], Patricia I. Gómez[1] and Jürgen E. W. Polle[2]*

Abstract

In this chapter, the taxonomic position of the genus *Dunaliella* within the *Chlorophyta* is reviewed briefly, and some of the taxonomic difficulties arising from the use of morphological features at species level, evidenced by molecular studies, are analyzed. The genus *Dunaliella* comprises 28 recognized species separated in 2 subgenera (*Pascheria* and *Dunaliella*) with the subgenus *Dunaliella* being divided into 4 sections: *Tertiolectae*, *Dunaliella*, *Virids* and *Peirceinae*. All species belonging to the subgenus *Pascheria* were found in freshwater and there exist some cytological and molecular arguments suggesting that their taxa do not belong to the genus *Dunaliella* at all. Within the subgenus *Dunaliella,* sections are characterized by specific biochemical and physiological attributes while the species have been defined primarily by the morphological criteria. Consequently, the taxonomic validity of some of the 23 morphospecies traditionally described can be questioned. Further, the enormous intraspecific physiological and molecular variability of the hypersaline species *Dunaliella salina* Teodoresco is highlighted. Finally, some aspects on the major evolutionary trends and phylogeny within the subgenus *Dunaliella* are discussed.

[1] Departamento de Botánica, Facultad de Ciencias Naturales y Oceanográficas, Universidad de Concepción, Casilla 160-C, Concepción, Chile.

[2] Department of Biology, Brooklyn College of CUNY, 2900 Bedford Ave. 200NE, Brooklyn, NY 11210, USA.

* Author for correspondence. Email: mgonzale@udec.cl

Taxonomic Position of the Genus *Dunaliella* within the Chlorophytes

When Teodoresco described the genus *Dunaliella* in 1905 he located it in the order *Volvocales* within the family *Polyblepharidaceae* (Teodoresco 1905), a position that was maintained by Lerche (1937) based on cell morphology, that is, flagellated unicells devoid of a cell wall. Much later Massjuk (1973) included the genus in the same order but in a new family, *Dunaliellaceae* erected by Christensen (1967) which in the original description includes "*single monad cells, with no membrane or lorica clothing them, known by their displaying the special nude flagellum of the Chlorophyceae*". Later, Ettl (1981) separated *Dunaliella* from the walled flagellates (i.e., order *Chlamydomonadales*, class *Chlamydophyceae*) into a newly erected order, *Dunaliellales* within the class *Chlorophyceae* (Ettl 1981, 1983). However, the order *Dunaliellales* sensu Ettl was polyphyletic (Nakayama et al. 1996). In consequence, three different views about the taxonomic position of *Dunaliella* currently exist which differ only in the order into which the family *Dunaliellaceae* sensu Christensen is placed:

	Alternative 1	Alternative 2	Alternative 3
Class:	*Chlorophyceae*		
Order:	*Chlamydomonadales*	*Dunaliellales*(sensu Melkonian 1990)	*Volvocales*
Family: *Dunaliellaceae*			
Genus: *Dunaliella*			

Alternative 1 is currently adopted by the National Center for Biotechnology Information (USA) and appears to be adopted by most researchers. Nevertheless, even though it is clear now that *Dunaliella* does not belong to the family *Polyblepharidaceae* - since most of those genera, if not all of them, have been transferred to the Class *Prasinophyceae* within the order *Polyblepharidales*, sensu Ettl 1983 - there is still confusion about the taxonomic position of *Dunaliella* at the ordinal and/or class level. An example of this is Oren (2005) who in his recent review paper '*A hundred years of Dunaliella research: 1905-2005*' still maintained the classical genus location (*Volvocales* and *Polyblepharidaceae*).

The ambiguity about the taxonomic position of the genus relates not only to the absence of a cell wall made up of cellulose, but presence of a 'Hautschicht' (Lerche 1937) consisting of a distinctive, probably glycoprotein cell cover (Ginzburg 1987, Massjuk 1973, Oliveira et al. 1980 in *D. tertiolecta*; Melkonian and Preisig 1984 in *D. salina*). It has been recognized by Melkonian and Preisig (1984) that *Dunaliella* is not just a wall-less equivalent of *Chlamydomonas*. The authors mention that there are ultrastructural cell differences (i.e., the parabasal position of the dictyosomes, the system I fibres underlying two stranded microtubular roots, the presence of a prominent system II fibre [rhizoplast] and its association with mitochondria). On the other hand, Floyd (1978) and Mattox and Stewart (1984) cited similarities between both taxa: i.e., cell division – phycoplast and flagellar apparatus architecture - clockwise (CW) basal body configuration, respectively, and Chappell et al. (1989) suggested that *Dunaliella* lost the capacity to form a one-piece cell wall but retained the remnants of a phycoplast, and based on that presume a flagellated cell-walled ancestor for *Dunaliella* .

The advent of the molecular era and the relative simplicity in obtaining molecular data today together with the set of ultrastructural attributes available, will undoubtedly help to find the closest relatives to *Dunaliella* within the class *Chlorophyceae*. This aspect will be treated in the discussion of phylogeny.

How Many Species are Recognized within the Genus?

Since the original description of the type species *D. salina* (Teodoresco 1905) and later of *D. viridis* (Teodoresco 1906), many species from a wide range of habitats have been described (mostly by Lerche 1937, Butcher 1959a, b, and Massjuk 1973) (Table 1). The first monograph dedicated to *Dunaliella* was done by Labbé (1925) who focused on the species *D. salina*, which he renamed to *D. kermesina*, and *D. viridis*. A following monograph by Lerche (1937) realized that several different algae were combined under the name of *D. viridis* given by Teodoresco (1906). Consequently, Lerche (1937) erected several new species and eliminated *D. viridis* Teod. However, the species *D. viridis* was then again erected by Massjuk (1973). Lerche (1937) also focused on the life cycle and sexuality exhibited by some selected species (carotenogenic and non-carotenogenic taxa). She successfully described isogamy among five of the six species investigated by her (*D. salina*, *D. parva*, *D. peircei*, *D. euchlora*, and *D. minuta*); most of the species were homothallic. She found *D. salina* to be heterothallic with a weak tendency to homothallism. We have found repeatedly in different strains of *D. salina* that the species is homothallic (unpublished results). Lerche presented many valuable data on the physiology and biology of these organisms. She made the first attempt to use physiological (i.e. NaCl concentration for optimal growth, rate of cell division) and biological parameters (particularly, sexual reproduction) for species characterization. Later, Butcher (1959a) conducted taxonomic work based on cell morphology describing four new species (*D. tertiolecta*, *D. primolecta*, *D. quartolecta*, *D. polymorpha*) from the Bristish sea-coast. He used only cytological attributes such as pyrenoid shape, and number and location of granules within the cytoplasm, all criteria highly variable in green unicells.

Most of the above taxonomic studies were based solely on the morphological criteria (i.e., cell size and shape, cell symmetry, chloroplast shape, presence/absence

Table 1: List of the subgenera, sections and species of *Dunaliella* with their authorities. (*) Species included in this paper.

Subgenus *Pascheria*	
	1.- *D. acidophila* (KALINA 1965) MASSJUK 1971 2.- *D. flagellata* SKVORTZOV 1968 3.- **D. lateralis* PASCHER & JAHODA 1928 4.- *D. obliqua* (PASCHER 1930) MASSJUK 1973 5.- *D. paupera* PASCHER 1932
Subgenus *Dunaliella*	
Section *Tertiolectae* (oligo-euhaline)	
	6.- **D. maritima* MASSJUK 1973 7.- **D. polymorpha* BUTCHER 1959 8.- **D. primolecta* BUTCHER 1959 9.- **D. quartolecta* BUTCHER 1959 10.- **D. tertiolecta* BUTCHER 1959
Section *Dunaliella* (hyperhaline, acumulate carotenes)	
	11.- **D. parva* LERCHE 1937 12.- **D. pseudosalina* MASSJUK & RADCENKO 1973 13.- **D. salina* TEODORESCO 1905
Section *Virides* (hyperhaline, do not acumulate carotenes; cells radially symmetrical)	
	14.- *D. baas-beckingii* MASSJUK 1973 15.- **D. bioculata* BUTCHER 1959 16.- *D. carpatica* MASSJUK 1973 17.- *D. gracilis* MASSJUK 1973 18.- *D. granulata* MASSJUK 1973 19.- *D. media* LERCHE 1937 20.- **D. minuta* LERCHE 1937 21.- *D. minutissima* MASSJUK 1973 22.- *D. ruineniana* MASSJUK 1973 23.- **D. terricola* MASSJUK 1973 24.- **D. viridis* TEODORESCO 1906
Section *Peirceinae* (hyperhaline, do not accumulate carotenes; cells bilaterally symmetrical)	
	25.- *D. asymmetrica* MASSJUK 1973 26.- *D. jacobae* MASSJUK 1973 27.- **D. peircei* NICOLAI & BAAS BECKING 1935 28.- *D. turcomanica* MASSJUK 1969

of pyrenoid and eyespot, presence/absence of refractile granules). But it is widely recognized that cell shape is highly variable due to the absence of a rigid cell cover (Oliveira et al. 1980), and it is known that the cell size as well as the cell shape depend on external factors such as nutrients, pH, salt concentration, temperature, and irradiation (Brown and Borowitzka 1979, Riisgard 1981, Ginzburg 1987). The presence/absence of eyespot and of refractile granules in the cells was also questioned by Preisig (1992). In contrast to the macro-morphological characteristics of numerous species which were studied by light microscopy, ultrastructural studies (Oliveira et al. 1980, Melkonian and Preisig 1984, Chardard 1987, Watanabe and Floyd 1989) were carried out on only 6 out of the 28 species of the genus. These ultrastructural studies described aspects of the flagellar root, cellular cover, chloroplast, pyrenoid, Golgi apparatus, and location of mitochondria within the cell. Most of these attributes appear useful at the genus level. Further, they appear to be very stable at the species level, at least in the studied taxa.

The most comprehensive work done on *Dunaliella* was that of Massjuk in 1973. She combined morphological and structural features (species level) with some physiological and biochemical attributes (section level) to divide the genus *Dunaliella* into two subgenera, *Pascheria* and *Dunaliella*.

The Subgenus *Pascheria*

The subgenus *Pascheria* (Table 1) comprises five species (*D. acidophila*, *D. flagellata*, *D. lateralis*, *D. obliqua*, and *D. paupera*), all of which occur in freshwater habitats. According to Melkonian and Preisig (1984) their inclusion within the genus *Dunaliella* is uncertain. They argued that *D. paupera*, *D. obliqua*, and *D. flagellata* lack a pyrenoid which is a characteristic attribute of the marine species. Moreover, both *D. paupera* and *D. obliqua* have unusual cell division, often producing unequal daughter cells. The fourth species, *D. acidophila* was originally described as *Spermatozopsis acidophila* Kalina (a name still in use by some scientists) from highly acidic habitats. The fifth freshwater species, *D. lateralis* is characterized by a distinctly lateral position of the pyrenoid, a feature not found in any of the marine species of *Dunaliella*. Apart from the distinctive features discussed above, all the freshwater species of the genus exhibit contractile vacuoles not found in their marine counterparts.

The results of ITS sequence data (González et al. 2001) agreed with the traditional view (based on cytological attributes) that *D. lateralis* is only distantly related to species included in the subgenus *Dunaliella*. To the best of our knowledge, there are no ITS or 18S rDNA sequences of the other species of the subgenus *Pascheria* available in the Genbank for comparative studies.

The Subgenus *Dunaliella*

Massjuk (1973) utilized physiological and biochemical criteria to group species in sections within the subgenus *Dunaliella*. She used the ability of various taxa to grow optimally under different salt concentration ranges to separate them into two groups: Oligo-euhaline (2-4% NaCl) and hyperhaline (6-12% NaCl). Then, within the hyperhaline taxa she distinguished those that were capable of turning their cells yellow, brown, or brick-red in color (carotenoid accumulation) under extreme conditions from those taxa that kept their cells green.

Many authors have recognized the extraordinary capacity of taxa of *Dunaliella* to grow under a wide range of salt concentrations (e.g., Borowitzka et al. 1977, Ginzburg and Ginzburg 1981, Ginzburg 1987). Further, it was realized that this ability is related to the intrinsic characteristics of the strain and to its culture history (e.g., Brown and Borowitzka 1979, Latorella and Vadas 1973, Borowitzka et al. 1977). It was also documented that the oligo-euhaline taxa (sensu Massjuk 1973) were halotolerant (=grow better at low salinities of 0.4-4 % NaCl, but can tolerate up to 34% NaCl) whereas the hyperhaline taxa were halophilic (=grow better at higher salinities 6-12% NaCl). However, according to the above authors, the way in which halophilic species like *D. viridis* and halotolerant species like *D. tertiolecta* respond to different concentrations of salt is far from being understood. Although a physiological characteristic such as growth at specific salinities appears to be a good measure to delineate species, Ginzburg and Ginzburg (1981) showed the difficulty encountered in establishing optimal growth conditions for different halotolerant and halophilic species. This is due to interference between the environmental factors salinity, temperature, CO_2 level, and light. Further, these abiotic factors interact with others such as general ionic and nutrient composition of the medium and/or the natural environment (i.e., higher concentrations of NaCl can be tolerated under higher irradiances and higher concentrations of CO_2). In spite of such difficulties, Cifuentes et al. (2001) experimentally reaffirmed the validity of the physiological and biochemical attributes used by Massjuk (1973) to discriminate sections among the subgenus *Dunaliella*. In their work, Cifuentes et al. (2001) included two oligo-euhaline strains of *D. tertiolecta* (section *Tertiolectae*), and seven hyperhaline strains belonging to sections *Dunaliella*, *Virides*, and *Peircei*. They studied the effect of a wide range of salinities (1 to 30% NaCl w/v) on growth and pigment content in isolates that were acclimated or not acclimated to specific experimental conditions. The results revealed, as it was expected, two groups of strains: oligo-euhaline and hyperhaline. However, unexpectedly some strains of supposedly hyperhaline species (i.e., *D. parva* CCMP 362, *D. parva* CCAP 19/9, and *D. peircei* UTEX 2192) fell within the oligo-euhaline group (Figure 1 B, C, D). It was also found that even though *D. parva* UTEX 1983 fell within the hyperhaline group, it responded physiologically like *D. viridis* CONC-002 (high growth rate and optimal growth rates at intermediate salinities) (Figure 1 A, G). Based on these results, it was concluded that all the strains mentioned above (*D. parva* CCMP 362, *D. parva* CCAP 19/9, *D. peircei* UTEX 2192, as well as *D. parva* UTEX 1983) were probably misnamed when they were incorporated into the culture collections.

From the hyperhaline group, *D. pseudosalina* CONC-010 was the only strain whose culture turned light orange at late stationary phase of growth

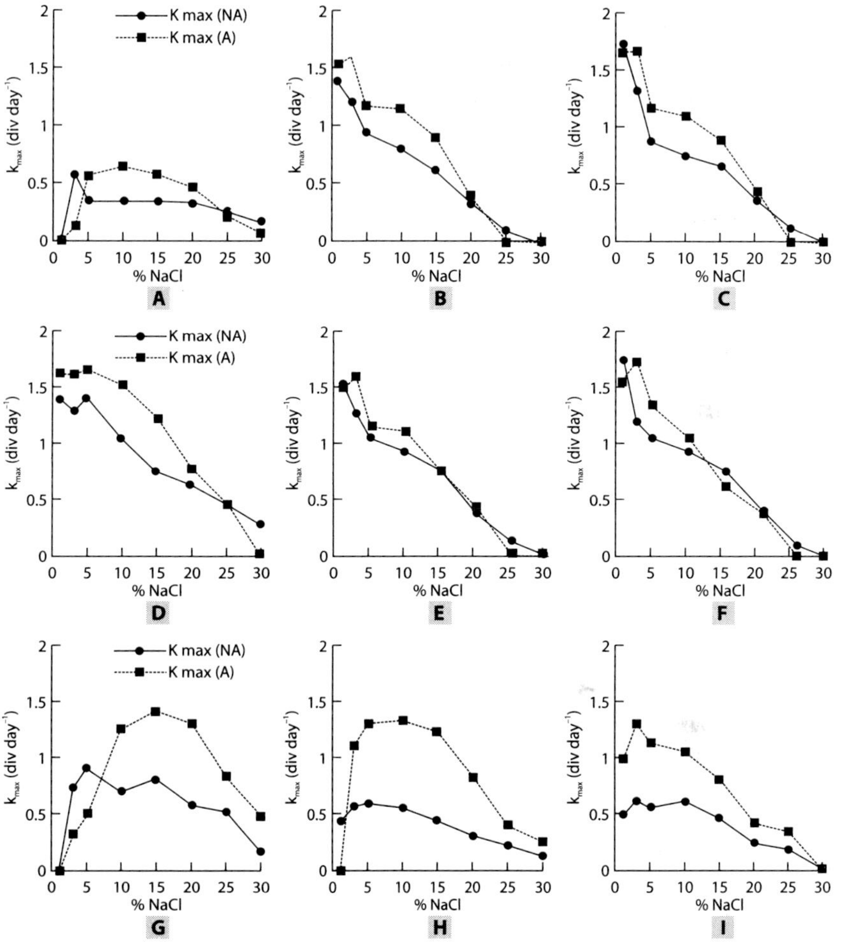

Figure 1: Exponential growth rates (k_{max}, div day^{-1}) in nonacclimated (NA) and acclimated (A) strains of various species of *Dunaliella* at each salt concentration. A-C: *D. parva* strains UTEX 1983 (A), CCMP 362 (B), CCAP 19/9 (C). D: *D. peircei* UTEX 2192. E-F: *D. tertiolecta* UTEX 999 and CCMP 1320, respectively. G: *D. viridis* CONC 002. H: *D. salina* UTEX 200. I: *D. pseudosalina* CONC 010 (extracted from Fig. 4-6 of Cifuentes et al. 2001, J. Phycol.37:339-341, with authors and journal permission).

(*ca.* three months), due to the formation of cysts. This species was isolated from the 'Salar de Atacama' (Chile) and identified by Dr. O. Parra. It will be of great interest to compare it with the type strain from Russia that is deposited in the Microalgae Culture Collection of Ukraine (IBASU-A) (Borisova and Tsarenko 2004).

The authors further concluded that even though all the strains have been maintained in culture for months, or years in some cases, under the same salinity (15% NaCl), they showed a distinct growth rate pattern immediately after being transferred to the experimental salinity concentrations (3, 5, 10, 15, 20, 25, 30% NaCl), reaffirming the theory that these algae do not adapt to a determined salt concentration, but rather tolerate a wide range of salinity (see Figure 1). Moreover, the differences found between oligo/euhaline and hyperhaline strains were not only evidenced by their different salt growth optima, but also by their growth rates: Oligo/euhaline grew faster than hyperhaline.

To date, physiological attributes have been studied only in a limited number of taxa within the genus. Further, physiological studies focused mainly on two species: *D. tertiolecta* which is widely used in hatcheries as a live feed (McLachlan 1960, Grant 1967, Franz and Wegmann 1974, Borowitska and Brown 1974, Borowitska et al. 1977, Jones and Galloway 1979) and *D. salina*, a beta-carotene producer (Loeblich 1982, Borowitska et al. 1977, Ben-Amotz et al. 1991, Orset and Young 1999).

Even though the species of the subgenus *Dunaliella* possess qualitatively the same pigments (chlorophyll *a* and *b*, lutein, violaxanthin, neoxanthin, α- and β-carotene), Massjuk (1973) stated that pigment content values (absolute and relative content ratio between the major groups) may be used to characterize various *Dunaliella* taxa at inter- and intraspecific level when they are cultivated under the same optimal growth conditions. Further, she claimed that the taxonomic value of carotenoid profiles as markers increases when the algae are placed in extreme conditions, which contribute to an increase in the magnitude of differences between separate taxa.

In her monograph of the genus, Massjuk (1973) mentioned some of the results she and Radchenko obtained from various species: For example, the xanthophyll/carotene ratio decreased under extreme conditions in *D. salina* and *D. granulata* and increased in *D. maritima* and *D. minuta* (in some experiments). *D. tertiolecta* was characterized by having a high absolute violaxanthin content, thus differing from the hyperhalobiontic *D. viridis* and *D. minuta*. On the other hand, *D. viridis* exhibited the highest relative content of lutein and antheraxanthin, and *D. tertiolecta* and *D. maritima* the lowest. Despite Massjuk's results, more recently pigments were only used taxonomically to distinguish *D. salina* from the other species of the subgenus. Only *D. salina* is able to store significant carotene quantities (*ca.* 30 to 90% relative carotene content), while for all other *Dunaliella* species, carotenes do not exceed 20% of the total pigment content (Loeblich 1969, Massjuk 1973).

Molecular attributes have only recently been added to understand the genetic and phylogenetic relationship among taxa within the genus *Dunaliella*. These studies have included mainly two regions of the genome: The Internal Transcribed Spacer regions (ITS1 + ITS2 with or without the 5.8S rDNA) (González et al. 1999, 2001) and the 18S rDNA (Wilcox et al. 1992, Olmos et al. 2000, 2002). All studies were based on restriction fragment length polymorphism (RFLP) and sequence comparison analyses. Further, more recently many partial sequences of the 26S rRNA, 18S rRNA, rbcS, and rcbL genes of various species of

Dunaliella have been incorporated in the GenBank. In spite of that abundance of sequences, only a few papers are available yet that use those sequences to help understand the phylogenetic relationship within the genus.

Wilcox et al (1992) described the presence of introns in the 18S rDNA of two species of *Dunaliella*: two introns in *D. parva* (UTEX 1983 = *D. viridis*) and one intron in *D. salina* (M 84320 from Chile which corresponds to CONC 001). Olmos et al. (2000) based on the work of Wilcox et al. (1992) went further and designed a set of conserved and specific oligonucleotide primers which together with RFLP analysis of the 18S rDNA were capable of identifying *D. salina* (M 84320), *D. parva* (LB 1983= UTEX 1983) and *D. bardawil* (LB 2538 = ATCC 30861) as three different species, containing one, two, and two different introns, respectively. In addition, they found that *D. salina* (LB1644= UTEX 1644) and two strains of *D. tertiolecta* (LB 999, CCMP 1320) form a group of strains without introns inside the 18S rRNA gene. However the number of introns does not seem to be a good character for species discrimination in *Dunaliella*. The gain and loss of introns in the 18S rRNA gene could be either an ancestral or a recent event; besides, it could be happening just in some sublines of evolution. Overall it is concluded that 18S rDNA sequence data from many more taxa need to be analyzed before one can infer the phylogenetic impact that the gain/lost of introns may have within the genus *Dunaliella*.

Here, we include a more comprehensive view on the phylogenetic relationship within the species of the genus *Dunaliella*, adding more taxa to the ITS sequence molecular analysis (Table 2) already done by González et al. (2001). Overall, results of phylogenetic analysis using the ITS sequences confirm the existence of three sections of the genus *Dunaliella* as erected by Massjuk (1973) mainly based on morphological and physiological attributes: *Tertiolectae*, *Dunaliella*, and *Virides*. Further, phylogenetic analysis of ITS sequences often confirmed what was suspected previously based on physiological analysis. For example, based on ITS sequences (González et al. 2001) and on physiological attributes (Cifuentes et al. 2001) it was found that several strains deposited in culture collections were inadequately identified. Thus, *D. parva* strains CCMP 362 and *D. parva* strain CCAP 19/9, which group physiologically and molecularly with *D. tertiolecta*, should be renamed as *D. tertiolecta*. Similarly, *D. parva* strain UTEX 1983 should be treated from now on as *D. viridis*.

The Section *Tertiolectae*

All the ITS-1 and ITS-2 sequences from strains of species of the section *Tertiolectae* (*D. maritima*, *D. quartolecta*, *D. primolecta*, *D. polymorpha*, and *D. tertiolecta*) fall basically into the same clade (Figures 2 and 3). This result suggests that either the ITS as markers do not give sufficient resolution to discriminate among species of this section, or that all investigated strains are very closely related and probably

Table 2: List of the species and strains of the genus *Dunaliella* used in the phylogenetic analysis.

Species/Isolate	**Culture Collection/source**	**Origin/Collection date**	**GenBank accession number (ITS-1, ITS-2)**
Chlamydomonas reinhardtii	"CC620"		U66954
Dunaliella bardawil (1) *D. bardawil* (2)	ATCC 30861 "hd7"	Bardawil Lagoon, north Sinai, Israel, 1978 Unknown	AF313430, AF313431 DQ116744 (only ITS-2)
D. bioculata	UTEX 199	Salt Lake, Russia, 1952	DQ182330, DQ157433
D. lateralis	R.Lewin	Nepal	AF313444, AF313445
D. maritima	SAG 42.89	Marine, USSR, 1989	AY582086 (only ITS-2)
D. minuta	SAG 23.86	Unknown	AY582085 (only ITS-2)
D. parva	CCMP 362 CCAP 19/9 UTEX 1983 "hd9"	Unknown Saltmarsh, NortheyIsland, Essex, England Dead Sea, 1973 China	AF313436, AF313437 AF313438, AF313439 AF313440, AF313441 DQ116746 (only ITS- 2)
D. polymorpha	CCAP 19/7	Brackish, River Crouch, Essex, England, 1953	DQ157050, DQ157053
D. primolecta	UTEX 1000 "hd8"	English Channel, England, 1936 China	DQ157052, AY582942 DQ116745 (only ITS-2)
D. pseudosalina	CONC 010	Salar Atacama, Chile, 1990	AF313420, AF313421
D. quartolecta	CCAP19/8	Marine, Southampton, Hampshire, England, 1953	DQ157051, DQ157054
D. salina	UTEX 1644 CONC 006 CONC 007 UTEX 200	Point Colorado, Salinas, Baja California, Mexico, 1967 Salar Atacama , Chile, 1990 Salar Atacama , Chile, 1990 Russia	AF313428, AF313429 AF313424, AF313425 AF313426, AF313427 AF313422, AF313423

D. tertiolecta	CCMP 1320	Unknown	AF313432, AF313433
	A2	Unknown	U66956
	UTEX 999	Oslo fjord, Norway, 1938	AF313434, AF313435
	CCAP 19/6B	Oslo fjord, Norway, 1938	AY572957 (only ITS-2)
	CCAP 19/27	Unknown	AY654300 (only ITS-2)
	DCCBC 5	Great Lake, Utah, USA, 2004	AY686683, AY686683
	DCCBC26	Urmia Salt Lake, Iran	DQ224338 (only ITS-2)
D. terricola	SAG 43.89	Russia, 1989.	ITS-2 (Polle, J.)
D. viridis	CONC 002	Salar Atacama, Chile, 1990	AF313418, AF313419
	SHU	China	AY878699, AY878700
	DCCBC4	Great Sat Lake, Utah USA	AY686682, AY686685
	DCCBC3	Great Sat Lake, Utah USA	AY828227 (ITS-2)
	SAG 44.89	USSR, 1989	AY828228 (ITS-2)
D. peircei	UTEX 2192	Lake Marina CA, USA, 1931	AF313442, AF313443

correspond to only one species. Moreover, no compensatory base pair change (CBC) is observed in the conserved regions of helix II and III in the ITS-2 secondary structure among these morphological species, indicating that all the strains could fall into the same mating group (Coleman et al. 1998, Coleman 2003).

Considering that most of the morphological criteria used by Butcher (1959) to delineate species within the section *Tertiolectae* are known to be unreliable as markers, the results of the above phylogenetic analysis were not surprising.

The Section *Virides*

This section includes 11 species (Table 1) which appear to be very diverse. Unfortunately, the majority are currently not available for phylogenetic analysis. These unavailable taxa include *D. ruineniana*, *D. gracilis*, *D. granulata*, *D. carpatica*, *D. baas-beckingii*, *D. media*, and *D. minutissima*.

Of the four available species of section *Virides*, the location of two strains is intriguing. *D. minuta* (strain SAG 23.86) and *D. bioculata* (strain UTEX 199) fall in the same clade with taxa of section *Tertiolectae* (see Figures 2 and 3). *D. bioculata* was originally described by Butcher (1959b) on the basis of a type culture from the Cambridge Collection (strain 19/4). According to the Culture Collection of Algae and Protozoa (CCAP; originally, Cambridge Collection) the alga was isolated by F. Mainx from a sample of a brackish salt lake in the Soviet Union. On the other hand, according to Butcher (1959b) and Massjuk (1973) the isolate (strain 19/4) used in the original description was collected by V. Czurda (unknown locality) and sent to the collection by Pascher (Prague). Moreover, Massjuk (1973) stated that *D. bioculata* was unknown in the USSR (= Soviet Union). However, irrespective of its origin, the two distinguishing features of the species - two elongated stigmas and the almost invariable cell shape with a pointed anterior- were questioned by Massjuk (1973), who argued that with additional data on its morphological variability and physiological traits, *D. bioculata* may be identified as a form of *D. viridis*. On the other hand, the strain of *D. minuta* used in this study is of unknown origin, and according to Massjuk (1973) differed from *D. viridis* in its cell shape (cylindrical in *D. minuta*, versus elliptical or pear-shaped with wide posterior end in *D. viridis*). Due to the uncertainty of strain origin and lack of more reference strains, for unambiguous transfer of the species *D. minuta* and *D. bioculata* into the section *Tertiolectae* new strains would need to be isolated and analyzed.

This leaves *D. viridis* and *D. terricola*, plus one misidentified isolate of *D. parva* (UTEX1983). If we add to the ITS sequence analysis of González et al. (2001) four more strains of *D. viridis* from locations such as China, Russia, and USA, all these strains fall in the same clade originally occupied by *D. viridis* CONC 002 and *D. parva* UTEX 1983. The bootstrap support differs when ITS-1 + ITS-2 sequences (91% in Figure 2) or only ITS-2 (77% in Figure 3) are included in the cladistic analysis. In the present phylogenetic analysis based on ITS

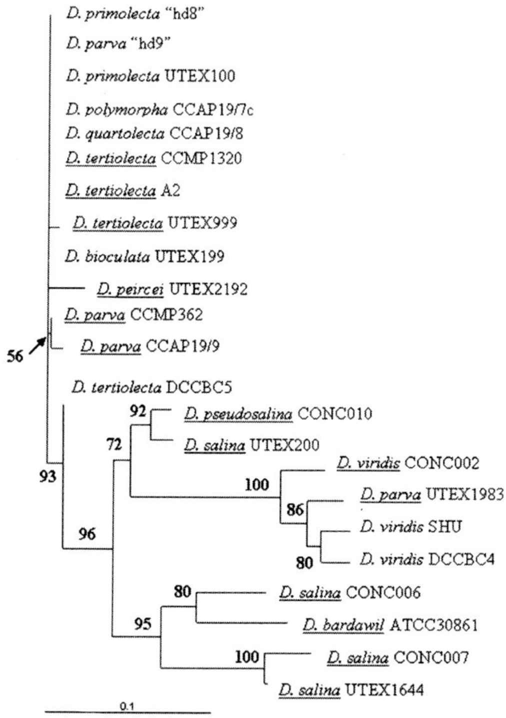

Figure 2: Phylogenetic relationships of 25 *Dunaliella* taxa inferred by ITS-1 + ITS-2 sequence comparison using 512 aligned positions. The tree shown resulted from a maximum likelihood analysis using PAUP (version 4.0b10; Swofford, 2002), the model GTR (SYM+G) with estimated gamma distribution of 0.6718 (calculated as the best model by Modeltest 3.04 (Posada and Crandall 1998). ML analyses used a heuristic search with a TBR algorithm; 10 replicates with random order of sequence addition (RAS). –LogL = 2572.47406. Bootstrap values ≥ 50% are shown above branches (100 replicates). Taxa underlined are the ones used in González et al.´s paper (2001).

(Figure 2 and Figure 3) all strains of *D. viridis* group within the same clade, but the clade shows great diversity. Even different isolates from the same water sample display this biodiversity within their ITS sequences (Polle, unpublished data).

Dunaliella terricola (strain SAG 43.79), apparently endemic from the USSR, formed its own clade with an undescribed species of *Dunaliella* from the Antarctic (Figure 3). Even though Massjuk (1973) found *D. terricola* very closely related to *D. minuta* (based on morphological and ecological grounds), *D. terricola* molecularly is closer to the *D. viridis* group as seen in the tree (Figure 3) and not to

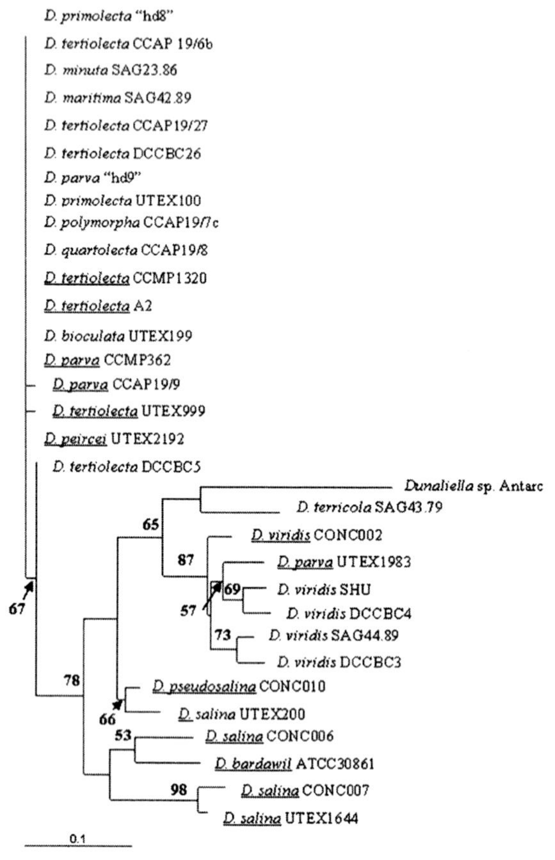

Figure 3: Phylogenetic relationships of 34 *Dunaliella* taxa inferred by ITS-2 sequence comparison using 271 aligned positions. The tree shown resulted from a maximum likelihood analysis using PAUP (version 4.0b10; Swofford, 2002), the model Tamura Nei (Trn+G) with estimated gamma distribution of 0.4392 (calculated as the best model by Modeltest 3.04 (Posada and Crandall 1998). ML analyses used a heuristic search with a TBR algorithm; 10 replicates with random order of sequence addition (RAS). –LogL = 1558.42071. Bootstrap values ≥ 50% are shown above branches (100 replicates). Taxa underlined are the ones used in González et al.´s paper (2001).

D. minuta. Massjuk´s information derived from observations of *D. minuta* collected by her in different areas of the USSR, while the ITS-2 sequence comes from the strain SAG 23.86 from an unknown origin. Only when the ITS sequences from the Massjuk strain (deposited in the Microalgae Culture Collection of Ukraine

[IBASU-A]) can be compared with the SAG 23.86 strain will we be in a better position to understand this result.

In light of this information, we are still far from solving the taxonomic problem of some of the above species. Some of the existing problems in proper analysis of phylogenetic data of taxa originate from the uncertainty about original type material from culture collections. In order to gain better knowledge on the phylogenetic relationship among the taxa within the genus, which should lead to a better definition of the species, more focused work is required on biochemical (pigment content analysis), physiological (growth rates under different salt concentration), and on molecular attributes. Specifically, sequence analyses from other regions of the genome may help to delineate clearly taxa within the section *Virides* from each other.

The Section *Peirceinae*

To our knowledge, only one species representative of section *Peirceinae* is available from culture collections, *D. peircei*. The sole strain originated from a salt lake in California, USA. According to the phylogenetic analysis based on ITS sequences shown in Figure 3 and based on physiological attributes, the strain of *D. peircei* UTEX 2192 is closely related to taxa in the section *Tertiolectae*. Consequently, it should be treated as *D. tertiolecta*. This result is not unexpected since Starr and Zeikus (1993) affirmed that this strain does not closely resemble the published description (George 1976) in spite of being the reputed type material.

Unfortunately, no strains for other taxa of the section *Peirceinae* are available for evaluation at this time. For a more conclusive phylogenetic analysis of taxa belonging to the section *Peirceinae*, first new strains need to be isolated from natural habitats.

The Section *Dunaliella*

The section *Dunaliella* includes three described species: *D. parva*, *D. pseudosalina*, and *D. salina*. The grouping of *D. parva* into the section *Dunaliella* appears questionable (see above). For clarification of its grouping, more new isolates of *D. parva* need to be analyzed. Regarding the species *D. pseudosalina*, our ITS data (Figure 2 and Figure 3) tend to group it closer to *D. viridis* than to *D. salina*, but there exists only the one isolate.

Based on morphological and physiological markers the type species *D. salina* was originally already very well characterized by Teodoresco (1905) and later by Lerche (1937) and Massjuk (1973). Nevertheless, several isolates existing in culture collections are still misidentified. Among these strains is for example *D. salina* UTEX

200, a strain that does not turn orange (physiological marker) and yet it groups with *D. pseudosalina* when analyzed based on ITS sequences (molecular marker). Another strain that does not turn orange and that based on ITS data groups closely with *D. salina UTEX 200* is the strain *D. salina* SAG 184.80 (Polle, unpublished data). In consequence, both strains (UTEX 200 and SAG 184.80) should no longer be considered as belonging to the taxon *D. salina*. This previous misidentification of isolates as *D. salina* strains resulted in some confusion in literature and in the erection of a new species *D. bardawil* Ben-Amotz et Avron (strain ATCC 30861). However, the species of *D. bardawil* fits the original species description of Teodoresco (1905). For this reason and because it does not have an original Latin description (*nomen nudum*) some researchers and culture collections consider it as *D. salina* (Borowitzka and Borowitzka 1988, Sammlung von Algenkulturen [SAG], Goettingen, Germany) or as a variety of *D. salina*. Regarding the strain *D. bardawil* 2 (unknown provenance), Figure 4 shows that it apparently does not correspond to *D. bardawil* nor to *D. salina*, since it is genetically more closely related to *D. tertiolecta* (proportion of sites differing: 3.046/440= 0.0069) than to any *D. salina* (proportion of sites differing ranging from 41.69/441 to 50.87/441) or *D. bardawil* (ATCC 30861) (proportion of sites differing: 45.85/444= 0.1033).

The isolates of *D. salina* appear to be very diverse and previously at least two subspecies and three forms were distinguished (Massjuk 1973, Preisig 1992). Consequently, biodiversity of the species *D. salina* is described in more detail below.

Intraspecific Variability within *Dunaliella Salina*

Comparative data on biochemical and physiological traits among strains of microalgal species are scarce but, when they exist, are generally indicative of tremendous intraspecific variability. Significant variability has been found every time comparisons were performed at this level (reviewed in Wood and Leathan 1992).

Wood and Leathan (1992) established that without an estimate of intra-species variation, it is not possible to determine whether or not the differences observed between species are any greater than might be observed in a random sample of different members of the same species. In spite of that, it is standard practice to characterize the phenotype of a microalgal species using only a single isolated clone (Behra et al. 1999). In the past, intra-species variation has caused many problems in identification and characterization of *Dunaliella* species.

In commercially important microalgal species, intraspecific variability has advantages and disadvantages. A clear advantage is that it permits genetic improvement by means of simple selection of new strains from nature; meanwhile it could lead to erroneous assumptions when information coming from one strain (or clone) is extrapolated to the whole species (Gómez and González 2004).

It is known that genotype and environmental interactions are crucial in determining intraspecific differences in adaptive physiological responses

(Behra et al. 1999). Gallagher (1980, 1982) demonstrated that changes in environmental conditions correlate to changes in the genetic composition of a phytoplankton population. Nevertheless, it is important to keep in mind that environmental changes do not always occur on the same spatial and temporal scale. Closed systems like temporary ponds that some microalgae inhabit may constitute a variety of unique microhabitats with very specific and distinguishable characteristics. Such variety of microhabitats is specifically found in hypersaline environments such as salt lakes in which *Dunaliella* species thrive.

Comparison of physiological attributes of different strains within a microalgal species is rather difficult since the between-strains data usually come from different physiological studies, most of them performed under different experimental conditions. In spite of that methodological problem, physiological variability within the carotenogenic green microalga *D. salina* was demonstrated in various papers (Araneda et al. 1992a, b, Cifuentes et al. 1992, 1996a, b, Markovitz et al. 1993, Gómez et al. 2003, Aguilar et al. 2004, Gómez and González 2005). Comparisons on growth as well as quantity and quality of accumulated carotenoids were carried out in several strains of this species coming from different regions around the world, with some of them exhibiting striking physiological variability. These between-strain differences were found under the same environmental (= culture) conditions, strongly suggesting that these differences have a genetic basis.

Cifuentes et al. (1992, 1996a, b), compared growth responses and total carotenoid accumulation among eight strains of *D. salina* isolated mostly from different salt ponds at 'Salar de Atacama' , a hypersaline environment of 3000 km^2 located at an altitude of 2340 m in the Atacama Desert, north of Chile. The strain CONC 006 was significantly different from its conspecifics, having the largest cells, the lowest maximum cell density and being one of the two most carotenogenic strains described so far; all of these attributes persisted under various different culture conditions. It is important to stress that the morpho-physiological attributes of these strains have not changed since their isolation in 1990, which demonstrates that their characteristics are fixed in their genotype. These Chilean strains were also genetically characterized by the fingerprinting technique Random Amplified polymorphic DNA (RAPD) (Gómez and González 2001). Great genetic variability was found among the strains in spite of the proximity between their geographic origins. The strain CONC 006 was the most genetically divergent among the strains from the 'Salar de Atacama' which was coincident with its morpho-physiological peculiarities.

Another study compared the physiological behavior of strains coming from different countries under different temperature and irradiance regimes (Gómez and González 2005). Statistical analysis revealed that strain identity was the most significant source of variation in total carotenoids accumulation. Some general tendencies were observed among the strains. For instance, independent of their country of origin, coastal strains exhibited higher growth rates and maximum cell densities than the two isolated from Andean environments (CONC 006 and CONC 007, both Chilean) under any culture condition. On the other hand, Chilean strains accumulated α-carotene (15% - 40%) at both assayed temperatures (15°C and 26°C), while the non-Chilean strains accumulated it in detectable amounts (<15%) only during growth at the lowest temperature (see Table 3). Random amplified polymorphic DNA (RAPD) band patterns and nuclear ribosomal DNA internal transcribed spacer (ITS-1 and ITS-2) sequences were used to genotypically characterize these strains (Gómez and González 2004). Although the data and the type of analysis used by both approaches were completely different, they were consistent in resolving the same genetic relatedness among the strains.

As we had observed previously for some strains of *D. salina* (Gómez and González 2001, 2004), when more strains were included in the present genetic distance analysis, using ribosomal ITS sequences, they did not cluster according to their geographic origin (Figure 4). Especially interesting is the case of the Chilean strains CONC 006 and CONC 007 which were isolated from neighboring ponds in the same geographic area. These strains are not closely related genetically as expected according to the proximity of their geographic origins; in fact they did not even fit in the same cluster of the genetic distance tree (Gómez and González 2001, 2004, and Figure 4). Therefore, it is not correct to assume, a priori, that strains originating from different geographic regions are always different genotypes and, on the contrary, that those strains originating from the same geographic location are genetically identical.

As physiological traits are influenced by the genotypic make-up of a strain (Behra et al. 1999) and, in *D. salina* a close correlation does exist between physiological attributes and genetic relatedness, information on the genetic diversity of this species can be used for selection of new strains usable in mass culture.

One of the remaining questions that one may ask is: Are these physiologically diverse strains of *Dunaliella salina* the same biological species? Or should some of them be considered intraspecific categories (i.e., subspecies, varieties, or forma)? Based on geographical, physiological, and morphological grounds, Massjuk (1973) recognized two subspecies (*D. salina* ssp. *salina* and *D. salina* ssp. *sibirica* Massjuk et Radch.) and three forma (*D. salina* ssp. *salina* f. *salina*, *D. salina* ssp. *salina* f. *oblonga* Lerche, and *D. salina* ssp. *salina* f. *magna* Lerche) within the species. One way to test for the biological species concept in *D. salina* will be to perform crossing experiments among the strains. This task would be difficult, since as explained earlier most strains are homothallic. Another way to approach the problem is to look at the genome. Molecular data have revealed great diversity within this single taxonomic species. Genetic data have shown that traditional species can be polyphyletic or consist of highly divergent lineages (Fabry et al. 1999, Coleman 1999,2000). According to Coleman (2000), in some taxa of *Volvocales* (where mating experiments have been performed), by the time a double-sided compensatory base pair change (CBC) has appeared in certain conserved regions of helix II and III in the ITS-2 secondary structure of a group of related organisms, their mating genes will also have diverged to the point where the two subgroups no longer interact at the gamete level. Moreover, Coleman found for *Gonium pectorale*, that if a strain differs by even a one-sided CBC among the relative conserved pairing positions,

Table 3: Chlorophyll *a* (Chl *a*) and *b* (Chl *b*), total carotenoid content (Car) per volume (mg L^{-1}) and per cell (pg $cell^{-1}$), Car/Chl *a*, relative content of α and β-carotene (α-car %; β-car %), and 9-cis/all-trans β-carotene ratio (given as mean value ± standard deviation) in different strains of *D. salina* (estimated at stationary phase of growth), grown at the following conditions of temperature and irradiance: 15°C/110 μmol photons m^{-2} s^{-1} and 26°C/110 μmol photons m^{-2} s^{-1}. (extracted from Gómez and González 2005, Biol. Res.38:151-162, with authors and journal permission)

15°C/110 μmol photons $m^{-2}s^{-1}$

Strain	Ch1 *a* (μgL^{-1})	Ch1 *b* (μgL^{-1})	Car (μgL^{-1})	Car/Ch1 *a*	Car (pg $cell^{-1}$)	α–car (%)	β–car (%)	9–cis/ alltrans βcar
CONC–001	1207±100	452±54	12355±306	10.3±0.6	11.1±0.3	32.7±0.4	67.3±0.4	1.2±0.06
CONC–006	494±24	276±23	10010±32	20.3±1.0	61.1±1.6	36.3±0.4	63.7±0.4	2.0±0.1
CONC–007	644±48	326±12	21789±207	33.9±2.8	90.7±0.8	39.9±1.6	60.1±1.6	2.2±0.1
Mexican	1924±34	1144±137	31857±1016	16.6±0.5	18.7±0.6	12.9±1.0	87.1±1.0	1.8±0.2
Chinese	1547±45	832±31	27315±1263	17.7±0.3	21.7±1.0	11.6±1.0	88.4±1.0	2.0±0.06
Australian	1729±24	599±7	40665±1051	23.3±1.2	33.9±1.3	10.4±0.2	89.6±0.2	1.2±0.1
D. bardawil	1660±38	689±49	25812±754	15.5±0.4	21.2±0.4	13.0±1.0	87.0±1.0	1.9±0.1

(Table 3 Contd.)

(Table 3 Contd.)

26°C/110 μmol photons $m^{-2}s^{-1}$

Strain	Ch1 *a* (μgL^{-1})	Ch1 *b* (μgL^{-1})	Car (μgL^{-1})	Car/Ch1 *a*	Car (pg cell^{-1})	α–car (%)	β–car (%)	9–cis/ alltrans βcar
CONC–001	1338±32	473±32	10292±298	7.7±0.2	9.0±0.3	17.4±0.4	82.6±0.4	1.1±0.0
CONC–006	431±12	250±19	5421±48	12.6±0.2	44.3±0.4	30.6±1.7	69.4±1.7	1.8±0.2
CONC–007	694±15	375±58	16000±952	23.2±1.0	72.1±1.1	24.5±0.8	75.5±0.8	2.1±0.0
Mexican	1935±29	672±10	22765±477	11.8±0.1	17.7±0.3	ND	100±0.0	1.9±0.3
Chinese	1585±63	544±11	21092±402	13.3±0.2	11.2±0.2	ND	100±0.0	1.8±0.1
Australian	1807±13	530±5	24580±19	13.6±0.1	13.2±0.1	ND	100±0.0	1.4±0.1
D. bardawil	2702±88	924±11	20504±526	7.6±0.05	16.9±0.4	ND	100±0.0	2.1±0.2

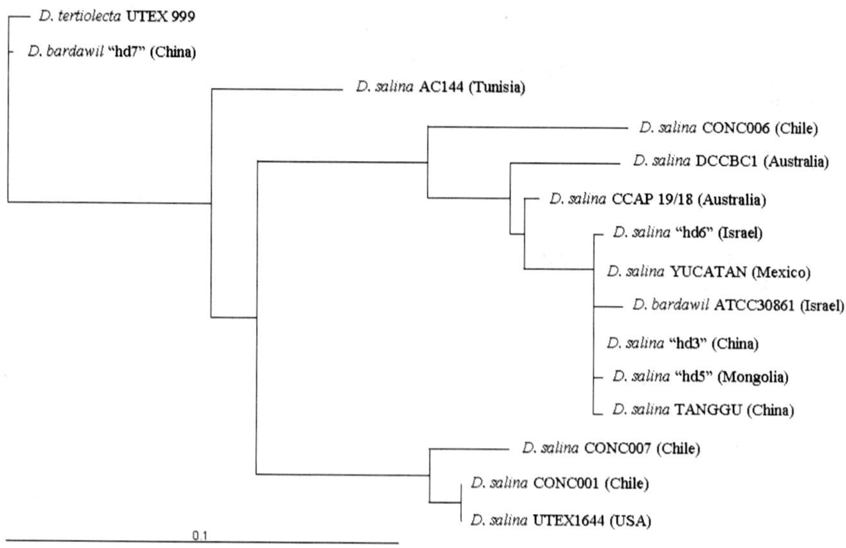

Figure 4: Genetic distance tree among 14 *D. salina* strains inferred by ITS 1+ ITS 2 sequence comparison using 452 aligned positions. The tree shown resulted from a Neighbor Joining analysis using the model after Tamura and Nei (1993) with proportions of invariable sites (= 0.5752) and equal rates for all sites (TrNef+I), calculated as the best model by Modeltest 3.04 (Posada and Crandall 1998). The outgroup was *D. tertiolecta* UTEX999.

its gametes will probably not interact at all with the other strains, and even if they do, no viable products will result. If we look at the ITS-2 secondary structure of the 15 strains of *D. salina* (including *D. bardawil*) analyzed, the region on the 5' side of helix III is identical for the 21 terminal nucleotides. If we then focus carefully on the conserved 12 nucleotides starting from the base of helix II (Figure 5), we can observe the presence of one double-sided CBC in the 11th nucleotide pair position (G-C changed to A-U) in a group of strains that happen to cluster together in the genetic distance tree (see Figure 4); this group of strains may be reproductively isolated from the others. On the other hand, the strains CONC 006 ('Salar de Atacama', Chile) and DCCBC1 (Lake Tyrell, Australia) that stand by themselves in the tree, bear a one-sided CBC in the fifth (G-C changed to G-U) and in the tenth (A-U changed the U with a G) nucleotide position of helix II, respectively. As outlined above, these data may indicate that at least two biological species exist within the clade of *D. salina*. However, we feel that is too premature to propose any hierarchical taxonomical change until we have a more

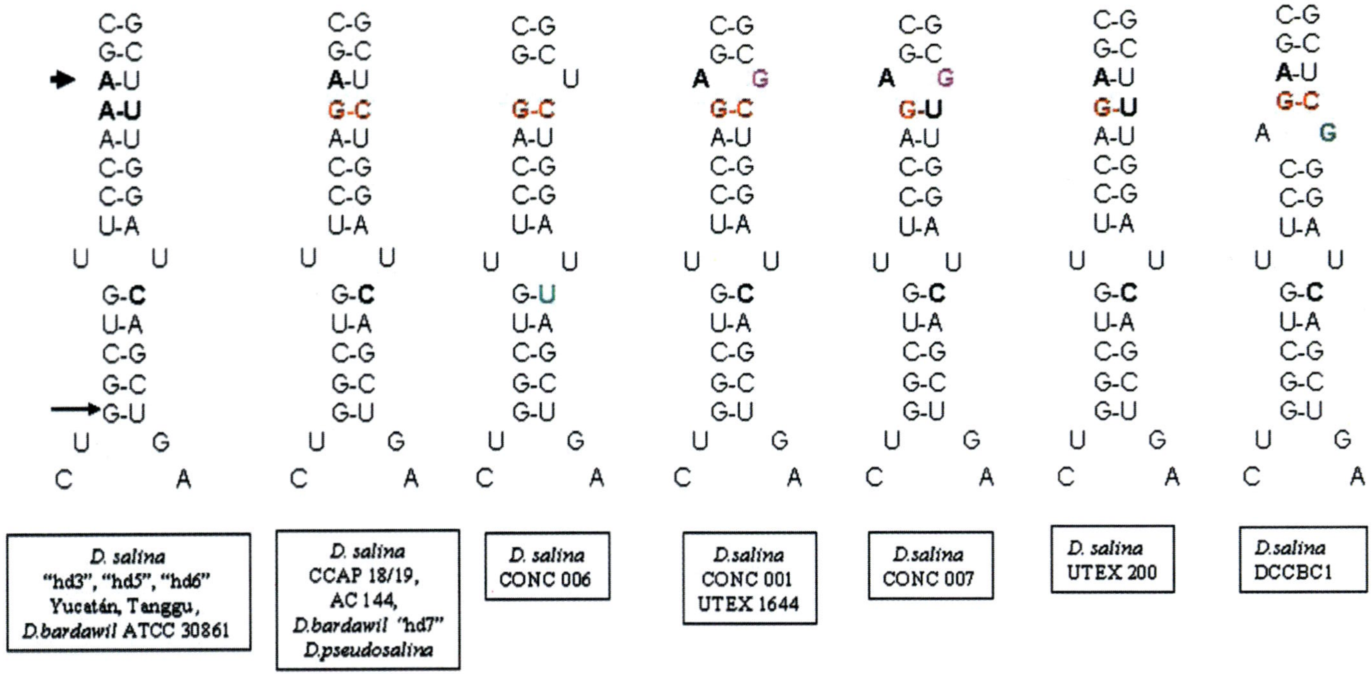

Figure 5: Structure of the basal part of helix II from the secondary structure of ITS-2 of various strains of *D. salina*, showing the basal 12 nucleotides of the helix II and their pairings. Notice the presence of a double-sided CBC at the 11th nucleotide pair position (G-C changed to A-U) in the first group of strains with respect to the others; one can also see various one-sided CBC at the fifth, tenth, eleventh, and twelfth nucleotide pair positions in different strains of *D. salina* which may help to understand the genetic similarity or dissimilarity among the strains described in Figure 4.

Table 4: List of different *D. salina* and of *D. bardawil* isolates used in the molecular analysis, with their geographical origin and GenBank accession number.

Species/ Isolate	Geographic Origin	ITS-1	ITS-2
D. salina CCAP19/18	Hutt Lagoon, Australia	AF546097	AF546098
D. salina	Tanggu, China	AF546095	AF546096
D. salina	Yucatan, Mexico	AF546093	AF546094
D. salina CONC-001	Laguna La Rinconada, Chile	AF546091	AF546092
D. salina CONC-006	Salar de Atacama,Chile	AF313424	AF313425
D. salina CONC-007	Salar de Atacama,Chile	AF313426	AF313427
D. salina UTEX LB1644	Baja California, Mexico	AF313428	AF313429
D. salina AC144	Tunisia, North Africa	AY545542	AY49441
D. salina DCCBC1	Lake Tyrell, Australia	AY545543	AY549442
D. salina "hd6"	Israel	DQ116743	DQ116743
D. salina "hd3"	China	DQ116740	DQ116740
D. salina "hd5"	Inner-Mongolia	DQ116742	DQ116742
D. bardawil "hd7"	China	DQ116744	DQ116744
D. bardawil ATCC 30861	Near Bardawil Lagoon, North Sinai, Israel	AF313430	AF313431

comprehensive set of data coming from other regions of the genome (i.e., 18S rDNA and the Rubisco genes). But, at this point we can make one clear statement, at least for the seven strains in which we have both traditional (morphological, biochemical, and physiological) and molecular data (ITS and RAPDs): the biochemical and physiological attributes are in good agreement with the genetic ones (Figure 6).

Major Evolutionary Trends and Phylogeny

According to Massjuk (1973) any hypothetical phylogeny within the genus should be based solely on comparative attributes of modern *Dunaliella* species, due to the absence of fossils. She proposed a freshwater origin for the genus, and based on the clearly defined taxonomic limits of modern freshwater species (presence/absence pyrenoid, eyespot, cell shape, and symmetry) and their limited geographical distribution (only one known location in central Europe for *D. lateralis*, *D. paupera*, *D. cordata*, *D. acidophila*, or in western China for *D. flagellata*), she argued that they apparently are the remnants of a large variable and formerly well-developed freshwater *Dunaliella* group, most of which disappeared because of competition with other algae which have dense cell walls. In spite of this, today

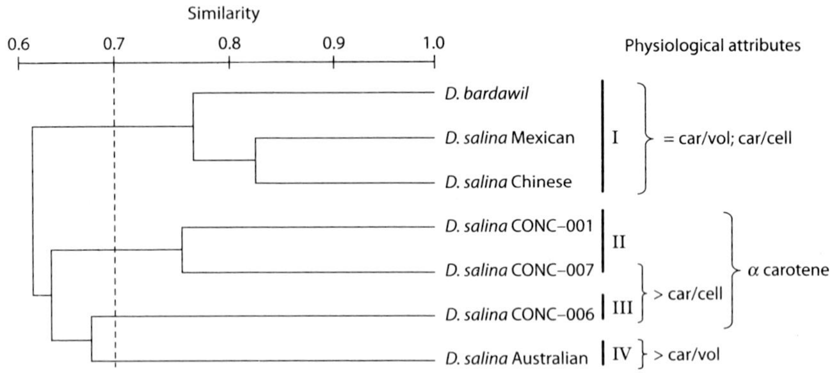

Figure 6: UPGMA cluster analysis for seven strains of *D. salina* derived from RAPD data (Fig 2 extracted from Gómez and González 2004, Aquaculture 233:155, with authors and journal permission) with their physiological attributes.

there are sufficient arguments to suggest that the freshwater species of *Dunaliella* may not belong to the genus at all (see above). More DNA data (18S rDNA, ITS, rbcS, rbcL) are needed from all species of the subgenus *Pascheria* in order to get a clear- cut answer on this issue.

What is clear from the molecular data, is that *Dunaliella* is phylogenetically related to walled chlamydomonadalean flagellates, not to wall-less ones, even though only one to three species have been included in the analysis. Nakayama et al (1996) in their study on the phylogeny of wall-less green flagellates inferred from 18S rDNA sequence data argue that *Dunaliella* and *Asteromonas* evolved from walled chlamydomonadalean algae and that the wall-less nature of this taxon is not symplesiomorphic but an advanced character (secondarily lost). Nakayama et al. (1996) also indicated that the *Dunaliellales* sensu Ettl (1981) is polyphyletic and their 18S rDNA analysis supports the monophyly of the *Dunaliellales* sensu Melkonian (1990) which restricted the order to the genera *Dunaliella* and *Asteromonas*. However, the authors argue that since the phylogenetic trees show that *Chlamydomonas applanata* is more closely related to the *Dunaliellales* sensu Melkonian than to other members of the CW- Chlorophyceae group, it is not necessary to separate them at the ordinal level.

Buchheim et al. (1996) in their phylogeny of *Chlamydomonadales* based on nuclear (18S rDNA) and chloroplast (LSU 23S rDNA) gene sequences, included *D. parva* with 29 species of the genera *Chlamydomonas*, *Chlorococcum*, *Chlorogonium*, *Haematococcus*, *Stephanosphaera*, and *Carteria* and found that in any tree, *D. parva* grouped together with *Chlamydomonas humicola*, *Chlamydomonas agloeformis*, *Chlorogonium elongatum*, and *Stephanosphaera pluvialis* with different branch support (48 to 86%). Only in the MP and ML tree from LSU rDNA data did *H. pluvialis* appear within the above group, and as a sister clade of *D. parva* (72% and 49% support, respectively).

Later, Chapman et al. (1998) and Pröschold et al. (2001) included two (*D. parva*, *D. salina*) or three (the previous two plus *D. bardawil*) species of *Dunaliella* in their analyses. Both analyses gave very similar results: A *Dunaliella*-clade (with *Asteromonas gracilis* in it) sister to a *Polytoma*-clade (i.e., *P. uvella*, *P. obtusum*, *Chlamydomonas applanata*), but according to Pröschold et al. (2001), the phylogenetic relationship between the two clades was not completely resolved due to moderate bootstrap support (58%).

The presence of the genus *Asteromonas* within the *Dunaliella* clade makes sense if we agree with Melkonian and Preisig (1984) and Preisig (1992) who argue that *Dunaliella* appears to be more closely related to the genus *Asteromonas* than to any other green flagellates, based on the morphological (i.e., overall cell shape, presence of surface coat) and ultrastructural (a similar event occurring during mitosis and cytokinesis, similar flagellar apparatus, dictyosomes in the anterior part of the cell associate with the nucleus) studies done by Peterfi and Manton (1968) and Floyd (1978).

Massjuk (1973) hypothesized that the major evolutionary direction of the genus was towards adjustment to environments with high salinity, developing a versatile metabolism, and rapid morphological reactions facilitated by the absence of thick cell wall, followed by the loss of their contractile vacuoles. She argued that the presence of hexokinase in marine *Dunaliella* (i.e., *D. tertiolecta*), without the ability to utilize glucose (as can the majority of other marine algae), also may be regarded as indirect proof of the freshwater origin of the genus. Moreover, because the majority of hyperhaline reservoirs inhabited by *Dunaliella* species have a clearly marine origin, Massjuk suggested that the hyperhalobiontic species of the genus originated from euhalobiontic marine ancestors. The similarities of the two ecological groups both morphologically and physiologically support this assumption. Also, there appears to be a connection between intraspecies diversity and either marine or hypersaline habitat. Based on the ITS marker, species living in marine environments seem to have lower genetic diversity than species from hypersaline environments.

The next evolutionary stage within this genus may have been adaptation to hyperhaline environments, which followed at least two different paths: Carotenogenic (capacity for massive carotene accumulation in unfavorable environments) and non-carotenogenic species groups. Massjuk argued that *D. salina* is one of the most ancient and specialized among the hyperhalobiontic species. It has a wide geographical distribution, with high plasticity and adaptability to the environment. *D. pseudosalina* accumulates mainly canthaxanthin in extreme conditions; it has a separate geographical distribution and may be phylogenetically younger than *D. salina*, having common ancestors with it. *D. parva* occupies an intermediate position between section *Dunaliella* and section *Tertiolectae*, based on its limited ability to form carotenoids.

According to Massjuk (1973), the section *Virides* may comprise the phylogenetically younger and more progressive *Dunaliella* forms, characterized by slight variability and the absence of clear gaps between separate taxa, most of which have continuous geographical distribution. *D. viridis* appears to be the most ancient species, having the widest geographical distribution. Within section *Virides*, there is a transition to soil habitats (*D. terricola*, *D. minuta*).

The section *Peirceinae* is connected with section *Virides* in its origin, but comprised of morphologically highly specialized non-carotenogenous *Dunaliella* species, with asymmetrical, dorsiventral, or flat cell forms and with endemic geographical distribution.

The phylogenetic trees of Figure 3 and 4 agree with Massjuk's hypothesis in showing all taxa from section *Tertiolectae* (oligohaline/euhaline organisms) more basal in the tree, from which all the hyperhaline taxa supposedly evolved (= section *Virides* and *Dunaliella*).

Acknowledgements

We would like to thank Annette W. Coleman (Brown University) for valuable suggestions to improve the sequence alignments and the manuscript and Oscar Toro (University of Concepcion) for helping with the maximum likelihood analyses. We also thank the Sammlung von Algenkulturen at the Georg-August Universitaet in Goettingen, Germany for providing strains for analysis.

References

Aguilar, C.P., M.A. González, A.S. Cifuentes and M. Silva. 2004. Growth and accumulation of total carotenoids in two strains of *Dunaliella salina* Teod. (Chlorophyceae) from the Northern and Central coast of Peru. J. Chil. Chem. Soc. 49 (1): 69-74.

Araneda, P., I. Tapia, B. Gómez-Silva and L. Tapia. 1992a. Microalgae from Northern Chile. II. Cultivo en medios de bajo costo de dos cepas de *Dunaliella salina* (Teodoresco 1905) nativas del desierto de Atacama. Estud. Oceanol. 11: 53-59.

Araneda, P., C. Jiménez and B. Gómez-Silva. 1992b. Microalgae from Northern Chile. III. Growth and β-carotene content of three isolates of *Dunaliella salina* from the Atacama Desert. Rev. Biol. Mar. Valparaíso 27: 157-162.

Behra, R., G.P. Genoni and A.L. Joseph. 1999. Effect of atrazine on growth, photosynthesis, and between-strain variability in *Scenedesmus subspicatus* (Chlorophyceae). Arch. Environ. Contam. Toxicol. 37: 36-41.

Ben-Amotz , A., A. Shaish and M. Avron. 1991. The biotechnology of cultivating *Dunaliella* for production of β-carotene rich algae. Biores. Tech. 38: 233-235.

Borisova, E.V. and P.M.Tsarenko. 2004. Microalgae Culture Collection of Ukraine (IBASU-A). Nova Hedwigia 79(1-2): 127-134.

Borowitzka, L.J. and A.D. Brown. 1974. The salt relations of marine and halophilic species of the unicellular green alga, *Dunaliella*. The role of glycerol as a compatible solute. Arch. Microbiol. 96: 37-52.

Borowitzka, L.J., D.S. Kessly and A.D. Brown. 1977. The salt relations of *Dunaliella*. Further observations on glycerol production and its regulation. Arch. Microbiol. 113: 131-138.

Borowitzka, M.A. and L.J. Borowitzka.1988. *Dunaliella*. pp. 27–58. In Microalgal Biotechnology. MA. Borowitzka and L.J. Borowitzka, [eds.] Cambridge University Press, Cambridge, UK.

Brown, A.D. and L.J. Borowitzka. 1979. Halotolerance of *Dunaliella*. pp.139-190. In Biochemistry and Physiology of Protozoa. M. Levandowsky and S.H. Hunter [eds.] Academia Press, New York, USA.

Buchheim, M.A., C. Lemieux, C. Otis, R.R. Gutell, R.L. Chapman and M. Turmel. 1996. Phylogeny of the Chlamydomonadales (Chlorophyceae): A comparison of ribosomal RNA gene sequences from the nucleus and the chloroplast. Mol. Phylogenet. Evol. 5(2): 391-402.

Butcher, R.W. 1959a. An introductory account of the smaller algae of Bristish coastal waters. I. Introduction to Chlorophyceae. Fish. Invest. 4: 1-74.

Butcher, R.W. 1959b. An undescribed species of *Dunaliella* from the Cambridge Collection of Algae. Hydrobiologia 12: 249-250.

Cifuentes, A.S., M.A. González, I. Inostroza and A. Aguilera. 2001. Reappraisal of physiological attributes of nine strains of *Dunaliella* (Chlorophyceae): growth and pigment content across a salinity gradient. J. Phycol. 37: 334-344.

Cifuentes, A.S., M.A. González, M. Conejeros, V. Dellarossa and O.O. Parra. 1992. Growth and carotenogenesis in eight strains of *Dunaliella salina* Teodoresco from Chile. J. Appl. Phycol. 4: 111-118.

Cifuentes, A.S., M.A. González and O.O. Parra. 1996a. The effect of salinity on the growth and carotenogenesis in two Chilean strains of *Dunaliella salina* Teodoresco. Biol. Res. 29: 227-236.

Cifuentes, A.S., M.A. González, O.O. Parra and M. Zúñiga. 1996b. Cultivo de cepas de *Dunaliella salina* (Teodoresco 1905) en diferentes medios bajo condiciones de laboratorio. Rev. Chilena Hist. Nat. 69: 105-112.

Chardard, R. 1987. L'infrastructure du plasmalemme de *Dunaliella bioculata* (Algue verte) mise en évidence d'un cell-coat; essai de localisation des charges négatives. Cryptogamie Algologie 8(3): 173-189.

Chappell, D.F., H.J. Hoops and G.L. Floyd. 1989. Strip-like covering revealed on "wall less" *Asteromonas* (Chlorophyceae). J. Phycol. 25: 197-199.

Chapman, R.L., M.A. Buschheim, C.F. Delwiche, T. Friedl, V.A.R. Huss, K.G. Karol, L.A. Lewis, J. Manhart, R.M. McCourt, J.L. Olsen and D.A. Waters. 1998. Molecular Systematics of the Green Algae. pp. 508-540. In Molecular Systematics of Plants II. D.E. Soltis, P.S. Soltis and J.J. Doyle [eds.]. Kluwer Academic Publishers, Boston, USA.

Christensen, T. 1967. Two new families and some new names and combinations in the algae. Blumea 15(1): 91-94.

Coleman, A.W. 1999. Phylogenetic analysis of "Volvocaceae" for comparative genetic studies. Proc. Natl. Acad. Sci. USA 96: 13892-13897.

Coleman, A.W. 2000. The significance of a coincidence between evolutionary landmarks found in mating affinity and a DNA sequence. Protist 151: 1-9.

Ettl, H. 1981. Die neue Klasse Chlamydophyceae, eine naturliche Gruppe der Grunalgen (Chlorophyta). Plant Syst. Evol. 137: 107-126.

Ettl, H. 1983. Suesswasserflora von Mitteleuropa. Band 9: Chlorophyta I. Phytomonadina. pp. 95-720. Gustav Fisher Verlag, Stuttgart, New York.

Fabry, S., A. Kohler and A.W. Coleman. 1999. Intraspecies analysis: comparison of ITS sequence data and gene intron sequence data with breeding data for a worldwide collection of *Gonium pectorale*. J. Mol. Evol. 48: 94-101.

Floyd, G. L. 1978. Mitosis and cytokinesis in *Asteromonas gracilis*, a wall-less green monad. J. Phycol. 14: 440-445.

Franz, G. and K. Wegmann. 1974. Physiology and biochemistry of glycerol biosynthesis in *Dunaliella*. Biol. Zbl. 93: 707-723.

Gallagher, J. C. 1980. Population genetics of *Skeletonema costatum* (Bacillariophyceae) in Narragansett Bay. J. Phycol. 16: 464-474.

Gallagher, J. C. 1982. Physiological variation and electrophoretic banding patterns of genetically different seasonal populations of *Skeletonema costatum* (Bacillariophyceae). J. Phycol. 18: 148-162.

George, E. A. 1976. Culture Centre of Algae and Protozoa List of Strains 1976, 3rd ed. Institute of Terrestrial Ecology, Natural Environment Research Council, Cambridge, England. 120 p.

Ginzburg, M. 1987. *Dunaliella*: a green alga adapted to salt. Adv. Bot. Res. 14: 95-183.

Ginzburg, M. and B.Z. Ginzburg. 1981. Interrelationships of light, temperature, sodium chloride and carbon sources in growth of halotolerant and halophylic strains of *Dunaliella*. Br. Phycol. J. 16: 313-324.

Gómez, P. and M.A. González. 2001. Genetic polymorphism in eight Chilean strains of the carotenogenic microalga *Dunaliella salina* Teodoresco (Chlorophyta). Biol. Res. 34: 23-30.

Gómez, P. and M.A. González. 2004. Genetic variation among seven strains of *Dunaliella salina* (Chlorophyta) with industrial potential, based on RAPD banding patterns and on nuclear ITS rDNA sequences. Aquaculture 233:149-162.

Gómez, P. and M.A. González. 2005. The effect of temperature and irrandiance on the growth and carotenogenic capacity of seven strains of *Dunaliella salina* (Chlorophyta) cultivated under laboratory conditions. Biol. Res. 38: 151- 162.

Gómez , P., A. Barriga, A.S. Cifuentes and M.A. González. 2003. Effect of salinity on the quantity and quality of carotenoids accumulated by *Dunaliella salina* (strain CONC-007) and *Dunaliella bardawil* (strain ATCC 30861). Biol. Res. 36: 185- 192.

González, M.A., P. Gómez and R. Montoya. 1999. Comparison of PCR-RFLP analysis of the ITS region with morphological criteria of various strains of *Dunaliella*. J. Appl. Phycol. 10: 573-580.

González, M.A., A.W. Coleman, P. Gómez and R. Montoya. 2001. Phylogenetic relationship among various strains of *Dunaliella* (Chlorophyceae) based on nuclear ITS rDNA sequences. J. Phycol. 37: 604-611.

Grant, B.R. 1967. The action of light on nitrate and nitrite assimilation by the marine chlorophyte, *Dunaliella tertiolecta*. J. Gen. Microbiol. 48: 379-389.

Jones, T. and R. Galloway. 1979. Effect of light quality and intensity on glycerol content in *Dunaliella tertiolecta* (Chlorophyceae) and the relationship to cell growth/ osmoregulation. J. Phycol. 15: 101-106.

Labbé, A. 1925. 1er Mémoire: Les cycles biologiques des *Dunaliella*. *Arch. d'Anat. Microsc.* T. XXI: 314-399.

Latorella, A.H. and R.L. Vadas. 1973. Salinity adaptation by *Dunaliella tertiolecta*. I. Increases in carbonic anhydrase activity and evidence for a light-dependent Na^+/H^+ exchange. J. Phycol. 9: 273-277.

Lerche, W. 1937. Untersuchungen ueber Entwicklung und Fortpflanzung in der Gattung *Dunaliella*. Arch. f. Protistenkd. 88: 236-268.

Loeblich, L. A. 1969. Aplanospores of Dunaliella salina (Chlorophyta). J. Protozool. 16 (suppl.).

Loeblich, L.A. 1982. Photosynthesis and pigments influenced by light intensity and salinity in the halophyte *Dunaliella salina* (Chlorophyta). J. Mar. Biol. Ass. UK 62: 493-508.

McLachlan, J. 1960. The culture of *Dunaliella tertiolecta* Butcher – a euryhaline organism. Can. J. Microbiol. 6: 367-379.

Markovits, A., M.P. Gianelli, R. Conejeros and S. Erazo. 1993. Strain selection for b-carotene production by *Dunaliella*. World Journal of Microbiology and Biotechnology 9: 534-537.

Massjuk, N.P. 1973. Morphology, Taxonomy, Ecology and Geographic Distribution of the Genus *Dunaliella* Teod. And Prospects for Its Potential Utilization. Naukova Dumka, Kiev. [original in Russian].

Mattox, K.R. and K.D. Stewart.1984. Classification of the green algae: a concept based on comparative cytology. pp. 29-72. In Systematics of the Green Algae. D.E.G. Irvine and D.M. John [eds.]. Academic Press, London. UK

Melkonian, M.1990. Phylum Chlorophyta, class Chlorophyceae. pp. 600-607. In Handbook of *Protoctista*. L. Margulis, J.Corliss, M.Melkonian and D.J. Chapman(eds.) Jones and Bartlett, Boston. USA

Melkonian, M. and H.R. Preisig. 1984. An ultrastructural comparison between *Spermatozopsis* and *Dunaliella* (Chlorophyceae). Plant Syst. Evol. 164: 31-46.

Nakayama, T., S. Watanabe and I. Inouye. 1996. Phylogeny of wall-less green flagellates inferred from 18SrDNA sequence data. Phycol. Res. 44: 151-161.

Oliveira, L., T. Bisalputra and N.J. Antia. 1980. Ultrastructural observation of the surface coat of *Dunaliella tertiolecta* from staining with cationic dyes and enzyme treatments. New Phytol. 85: 385-392.

Olmos, J., J. Paniagua and R. Contreras. 2000. Molecular identification of *Dunaliella* sp. utilizing the 18S rDNA gene. Lett. Appl. Microbiol. 30: 80-84.

Olmos, J., J. Paniagua, R. Contreras and L. Trujillo. 2002. Identification of β-carotene hyper producer strains of *Dunaliella* from saline environments using species-specific oligonucleotides. Biotech. Letters 24: 365-369.

Oren, A. 2005. A hundred years of *Dunaliella* research: 1905-2005. Review Saline Systems 1(2): 1-14.

Orset, S. and A.J. Young. 1999. Low-temperature-induced synthesis of β-carotene in the microalga *Dunaliella salina* (Chlorophyta). J. Phycol. 35: 520-527.

Peterfi, L.S. and I. Manton. 1968. Observations with the electron microscope on *Asteromonas gracilis* emend. (*Stephanoptera gracilis* (Atari) Wisl.), with some comparative observations on *Dunaliella* sp. Br. Phycol. Bull. 3(3): 423-440.

Posada, D. and K.A. Crandall. 1998. Modeltest: testing the model of DNA substitution. Bioinformatics 14: 817-818.

Preisig, H. R.1992. Morphology and taxonomy. pp. 1-15. In *Dunaliella*: Physiology, Biochemistry and Biotechnology .A. Ben-Amotz and M. Avron [eds.]. CRC Press, Boca Raton Press, Inc. USA.

Pröschold, T., B. Marin, U.G. Schlosser and M. Melkonian. 2001. Molecular phylogeny and taxonomic revision of *Chlamydomonas* (Chlorophyta). I. Emendation of *Chlamydomonas* Ehreneberg and *Chloromonas* Gobi, and description of *Oogamochlamys* gen. nov. and *Lobochlamys* gen. nov. Protist 152: 265-300.

Riisgard, H.U. 1981. Cell-volume responses in the naked marine flagellate *Dunaliella marina* transferred from darkness to light of different intensities. Bot. Mar. 24: 657-659.

Starr, R. C. and J. A. Zeikus. 1993. UTEX- The Culture Collection of Algae at the University of Texas at Austin. J. Phycol. 29 (suppl.): 39.

Swofford, D.L. 2002. PAUP*, phylogenetic analysis using parsimony and other methods, version 4.0b10. Sinauer Associates, Sunderland, MA, USA

Tamura, K. and M. Nei. 1993. Estimation of the number of nucleotide substitutions in the control region of mitochondrial DNA in humans and chimpanzees. Mol. Biol. Evol. 10: 512-526.

Teodoresco, E.C. 1905. Organization et development du *Dunaliella*, nouveau genre de Volovocacée-Polyblépharidée. Bot. Centralbl. Beih. 18: 215-232.

Teodoresco, E.C. 1906. Observations morphologiques et biologiquessur le genre Dunaliella. Rev. Gén. d. Bot. T18: 353-371.

Watanabe, S. and G.L. Floyd. 1989. Variation in the ultrastructure of the biflagellate motile cells of six unicellular genera of the Chlamydomonadales and Chlorococcales (Chlorophyceae), with emphasis on the flagellar apparatus. Amer. J. Bot. 76(2): 307-317.

Wilcox, L.W., L.A. Lewis, P.A. Fuerst and G.L. Floyd. 1992. Group I introns within nuclear-encoded small-subunit rRNA gene of three green algae. Mol. Biol. Evol. 9(6): 1103-1118.

Wood, A.M. and L. Leathan. 1992. The species concept in phytoplankton ecology. J. Phycol. 28: 723-729.

3 Cultivation, Growth Media, Divison Rates and Applications of *Dunaliella* Species

D.V. Subba Rao

Abstract

This chapter focuses on the culturing techniques of *Dunaliella* in media that support both autotrophic and heterotrophic growth. The simplest medium is seawater enriched with nitrate, phosphate, thiamine plus soil extract, and the most complicated is the ATCC-1194 DV medium with 29 enrichments. *Dunaliella* species seem to tolerate a latitude of salinity conditions, and grow well under a wide range of nutrient concentrations. Procedures for scaling up cultures to large volumes are outlined. A comprehensive discussion is presented on the growth of *Dunaliella* in batch, continuous, pH stat, cyclostat, turbidostat, chemostat forms, growth under several environmental variables such as pH, temperature, salinity, light and nutrients such as phosphorus, nitrogen, iron and other metals. Further, applications of *Dunaliella* as live feed in mariculture, for harvesting commercial sources of bioactive compounds, or in wastewater treatment and production of biofuel are pointed.

Introduction

Dunaliella is a remarkable unicellular chlorophyte, represented by extremophile species that can grow in a much wider range or extreme environments. It can adapt to subzero temperatures, tolerate extreme salinities, pH, and have a high degree of photoadaptability which places it in a league of its own. More than a scientific curiosity, species of *Dunaliella* are a source of bioactive compounds and so are of special interest. As a live feed organism in mariculture and as a source of ß carotene, isolates of *Dunaliella* have been cultured since 1925 (Labbe 1925, Støttrup and McEvoy 2003, Oren 2005).

D.V. Subba Rao, MESD, Bedford Inst. Oceanography, PO Box 1006, Dartmouth, Nova Scotia, Canada, B2Y 4A2. email: seshu35@gmail.com

Dunaliella has been relatively well researched as evident by nearly 574 publications based on culture studies. Lack of a rigid cell wall in *Dunaliella* provides an excellent model to study stress signaling besides applying principles of plant physiology namely growth response to environmental conditions, pollutants and toxicants. Besides their value in studying biochemical and physiological processes, *Dunaliella* cultures find a variety of biotechnological applications such as production of the bioactive compounds, carotene, and glycerol. Some of the unique features may be linked to their unique physiological and genetic properties. Riesing (2005) estimated that micro-algae could produce 5000-15000 gallons of oil per acre per year, 8-25 times more than palm oil and 40-120 times that of rapeseed. It is plausible to manipulate this alga genetically to utilize CO_2 and wastewater from municipalities and agricultural lands and to grow at a higher rate to produce more methane or hydrogen fuel (Polle et al. 2002).

The aim of this chapter is firstly to provide details on autotrophic cultivation studies that focus on the division rates of *Dunaliella* in various undefined, semidefined and defined media; secondly, to describe heterotrophic cultivation of *Dunaliella* without artificial lights which has implications in rearing this alga especially under commercial settings; thirdly to show that *Dunaliella* can be grown in batch, continuous, pH stat, cyclostat, turbidostat, chemostat forms depending on its applications such as utilizing the cells for physiological ecology studies, as live feed in mariculture, for harvesting bioactive compounds, or in wastewater treatment and production of biofuel, and finally to discuss the extraordinary adaptability of *Dunaliella* to environmental variables such as pH, temperature, salinity, light and nutrients such as phosphorus, nitrogen, iron and other metals which can be manipulated to optimize growth and for production of bioactive compounds.

Cultivation

Details of media preparation, isolation, maintenance, and ecology studies can be obtained from Andersen (2005), and Subba Rao (2006). For successful cultivation of any alga a suitable medium can be prepared by enrichment of either natural seawater or artificial seawater with micronutrients, trace metals and vitamins (Andersen 2005 and Subba Rao 2006). To ensure water quality, it is best to collect offshore seawater from a clean source, filter it through 165 μm nitex to remove larger particles and let it age in plastic barrels with tight lids.

As the nutrient levels in natural seawater vary in space and time, enrichment of artificial sea water is preferred for a medium with defined nutrient and trace element concentrations. This will facilitate inland investigators with no access to the sea. Usually for enrichment stock solutions (1000 x) are prepared and frozen. From the stock solutions working solutions are prepared and refrigerated. Any pH adjustments are made before autoclaving the medium. Culture media are usually dispensed into tubes, flasks and carboys and are autoclaved for 15-45 min at 120^{0}C and 20psi or sterilized at 80^{0}C for 1-2 h. As autoclaving expels most of the carbon dioxide and also results in precipitation of some elements it is best to rapidly cool the medium for 24 h and bring it up to the required temperature for culturing. After autoclaving, to the sterile enriched seawater vitamins and other organics are added aseptically.

Media for Autotrophic Growth of *Dunaliella*

Dunaliella species are cultured in a variety of media (Table 1) some include only minor modifications to suit individual needs. For example *D. tertiolecta* was grown in media: ASP with no buffer, ASP 5mM, ASP2, ASP6, ASP7, ASP12, ASW-III, ASP plus ASW-III (McLachlan 1964). These media can be graded as (1) undefined, (2) semi-defined and (3) defined. The simplest medium, the medium belongs to the undefined category is seawater enriched with nitrate, phosphate, thiamine plus soil extract (Madgwick 1966) and the most complicated is the ATCC-1194 DV medium with 29 enrichments. The General Purpose Medium (GPM) is essentially Erd Schrieber medium supplemented with trace elements and vitamins (Loeblich 1975). Medium ES (Cifuentes et al. 1996) is based on enrichment of 2-4M NaCl with nitrate, phosphate, manganese and soil extract. As the constituents of the soil extract vary depending on the origin of the soil, the ingredients remain a 'mystery' (McLachlan 1973) and thus the medium is undefined. The second category encompasses media which use natural seawater as a base for known enrichments. Although the enrichments are precisely known, the constituents of the basic seawater remain unknown and so these media are semi-defined. The most complex and precisely defined broad spectrum medium is the artificial seawater medium (ESAW) (Harrison et al. 1980, Berges et al. 2001) that is based on enrichment of artificial seawater. Daynao Sea Salt, Instant Ocean, Rilla Marine Mix, Sea Salt Mix and Utility Seven-Seas Mix #156, Tropic Marine Sea Salt, Ultramarine Synthetica, Sigma-Aldrich dry sea salt mixture (S9883) are some of the commercial preparations of artificial seawater (Guillard 1975, Harrison and Berges 2004) of which Instant Ocean (Glaude and Maxey 1994) and Tropic Marine Sea Salt, Ultramarine Synthetica have used. Common fertilizers are used for enrichment but growth is not as good as that obtained with defined media (Herrero et al. 1991a).

Currently 29 strains probably belonging to 17 species are cultured today. That *Dunaliella* grows in such a wide range of media demonstrates its adaptability to grow under quite different nutrient conditions. For example good autotrophic growth of *D. tertiolecta* was observed in a variety of media such as Walne medium (Hererero et al. 1991), Erd Schreiber medium (Madjwick 1966, Cifuentes et al.

1996), and various modifications of medium F (Eppley and Coatsworth 1966, Subba Rao 1981), ASP-2 (Eppley and Sloan 1965), McLaughlin's medium (1960), Johnson J1 medium and in Giardano medium (Table 1). In *D. tertiolecta* the maximum cell density (10^6 ml^{-1}) and the proximate composition (pg cell^{-1}) differed with the medium used for cultivation (Herrero et al. 1991a). In Walne, ES, f/2 and Algal-1 media the range for cell density was 4.04 to 8.45, for protein 13.26 to 18.82, for carbohydrate 13.32-17.91 and for lipids 18.18 to 23.94 (Herrero et al. 1991a and b). Using cultures of *D. primolecta* grown in f/2 medium Uriarte et al. (1993) reported values (pg cell^{-1}) ranging from 9.72 to 24.90 for protein, 18.04 to 22.22 for lipids and 10.60 to 26.74 for carbohydrates. Phosphorus deficiency reduced cell division rate in *D. tertiolecta* but resulted in an increase in cellular total fatty acid particularly the oleic acid (Siron et al. 1989).

The enrichments of selected major nutrients in the media ranged between 50 for silica and 20000x for phosphate respectively (Table 2) which shows that in respect of a nutrient (P,Si,N) *Dunaliella* can grow in a wide range. However, there is no evidence to support that silicon is an essential nutrient for *Dunaliella* growth (Personal communication –Dr. Ben-Amotz). Media with trace metal (Fe, Cu, Zn, Co, Mn and Mo) concentrations 10 times more than those in medium F, inhibited *D. tertiolecta* growth (Subba Rao 1981).

That the halotolerant *D. salina* grows in J/1 (Munoz and Gonzalez 2001), ES (Gomez et al. 2003), PES (Cifuentes et al. 1996) media (Table 1) with a wide range of nutrients shows the adaptability of this species. For special investigations such as the the ion fluxes Pick et al. (1986) used a medium with two times more phosphorus and iron. Using Pick's medium (1986) with the salinity reduced to 0.5M NaCl, Polle (personal communication) cultivated several strains of *Dunaliella* on sterile agar slants made by addition of 1.5% agarose to the medium before it is autoclaved. For determination of carbonic anhydrase activity Latorella and Vadas (1973) used a specially modified medium (Table 1) with extremely low trace element concentrations (Johnson et al. 1968).

In most media (Table 1), vitamins are added routinely (3.69pM-7.4nM VitaminB_{12}, 0.3nM- 3mM Thiamine hydrochloride and 3.27nM- 0.2μM Biotin). However, Vorst (1995) using modified Bold's Basal medium without any vitamins obtained maximal specific growth rates of 1.2 d^{-1} and 1.7 d^{-1} for *D. bardawil* and *D. salina* respectively at 28^{0}C and 1M NaCl. It is of interest to note several strains of *Dunaliella* grow in the absence of vitamins demonstrating that vitamins are not essential or to support growth. Strains of *D. salina, D. viridis, D. tertiolecta, D. granulata, D. terricola, D. parva, D. primolecta, D. bardawil* and *D. acidophila* are grown in the absence of vitamins in the laboratories of Dr. Ben-Amotz (Personal communication –Dr. Ben-Amotz).

The choice of the medium depends on the utility of cultures. *Dunaliella* species were autotrophically grown at temperatures ranging between 10^{0}C and 30^{0}C in various media (Table 3). The maximum cell density varied between 0.3x10^6 ml^{-1} and 24x10^6 ml^{-1} between the species and so did the day of attaining

Table 2: Range of selected major nutrients in media used in *Dunaliella* cultures.

Enrichment	***Range***
Phosphorus	0.5μM-10 mM
Nitrogen	1 μM-9.9mM
Silica	10μM-500μM
Iron	1.0 μM-1.8mM
Copper	0.24nM-0.063mM
Cobalt	0.063nM-0.04mM
Zinc	0.3nM-76.5μM
Manganese	0.207nM-1.4mM
Vitamin B_{12}	3.69pM -7.4nM
Thiamine hydrochloride	0.3nM -3mM
Biotin	3.27nM-0.2μM

their maximum (4 to 35 d). However, for large scale cultures sea water enriched with Algal-1 (Herrero et al. 1991a) or with commercial nutrient Algal-1 can be used (Herrero et al. 1991a, b). Artificial seawater enriched with 150 mg $NaNO_3$, 10mg $NaH_2PO_4.2H_20$, 15 mg Fe-EDTA and 360 μg $MnCl_2.4H_2O$ per liter supported better algal growth than did medium F (Okauchi and Kawamura 1997); its suitability for *Dunaliella* needs be evaluated.

Heterotrophic Cultivation

Heterotrophic cultivation of algae facilitates growth of large quantities of algae in the dark and has several adavantages. It allows low cost cultivation without artificial lights thereby reducing an energy source that is becoming increasingly expensive. Such heterotrophic cultures will not be light limited unlike those grown photosynthetically. A few studies exist on the possible heterotrophic cultivation of *Dunaliella* species (Table 4). Attempts to grow *D. viridis* and *D. salina* in dark over three wk in artificial seawater enriched with 50mg KNO_3, 10mg KH_2PO_4 and Hutner's trace elements mixture and glucose, acetate, yeast glucose, yeast acetate, brain heart infusion glucose and brain-heart infusion acetate were unsuccessful (Gibor 1956).

Gladue and Maxey (1994) tested 121 strains of algae including 16 strains of *Dunaliella* for heterotrophic cultivation in SK1A medium containing high levels as glucose, tryptone, yeast extract and urea as sources of carbon (100-250mM) and with complex nutrients. *Dunaliella salina* strain 3024, *D. tertiolecta* strain 3028 and

Table 3: Maximum cell density of *Dunaliella* species in culture.

Species	Medium	Temp °C	Inoculum density 10^6 ml^{-1}	Maximum cell density 10^6 ml^{-1}	Days to attain maximm	Reference
D. tertiolecta	F	10	0.2	12	12	Subba Rao 1981
tertiolecta	Algal-1 Commercial medium	18	0.5	8.45		Herrero et al. 1991a
tertiolecta UTEX	ES	?	?	12	35	Cifuentes et al. 1996
tertiolecta CCMP	ES	?	?	19.5		Cifuentes et al. 1996
tertiolecta	Commercial medium	18	0.5	12	10	Fábregas et al. 1994
bardawil	J1	22	0.25	6.7	10	Sánchez-Saavedra et al. 1996
parva CCMP	J1	20	NA	14.5	30	Cifuentes et al. 2001
parva CCAP	ES			14.5	30	Cifuentes et al. 1996
pseudosalina CONC 010	ES			1.8	30	Cifuentes et al. 1996
piercei UTEX 2192	ES			9.2	30	Cifuentes et al. 1996
salina UTEX	ES			1.95	30	Cifuentes et al. 1996
salina	J1	30	0.005	3	30	Cifuentes et al. 1996
salina	f/2	24	0.0164	0.3	7	Marin et al. 1998
viridis CONC 002	ES			24	35	Cifuentes et al. 1996b
viridis	ES	20		18.5	4	Jimenez and Niell 1991a and b
	ES	25		17.3	4	Jimenez and Niell 1991a and b
	ES	30		13.6	4	Jimenez and Niell 1991a and b

Table 4: Molar concentrations of enrichments in culture media tested for heterotrophic growth of *Dunaliella* species.

Medium ID ▸	*SK (1A)*	*Formulated*
Reference	Glaude and Maxey 1994	Hard and Gilmour 1996
Dunaliella cultivated	*D. salina, D. tertiolecta*	*D. parva*
Enrichment ▾		
Na_2SO_4		48mM SO_4
Na NO_3	8.82 μM	5mM NO_3
Na_2HPO_4. $2H_2O$	108.7μM	0.1mMPO_4
$MgSO_4$. $7H_2O$		10mM Mg
$CaSO_4$		10mM Ca
$NaHCO_3$	10-25; 100-250 mM C	120μM
Tris hydrochloride		20mM
Fe EDTA		1.5μM
H_3BO_3	0.283mM	185μM
$CuSO_4$. $5H_2O$	0.063μM	0.02nM Cu
$CoCl_2$. $6H_2O$	16.7μM	0.002μM Co
$ZnCl_2$	22μM	0.8μM Zn
$MnCl_2$. $4H_2O$	23.9 μM	7μM Mn
$MgSO_4$. $7H_2O$		44mM Mg
Na_2MoO_4. $2H_2O$	0.103 μM	
NiSO4.6H_2O	1.02 μM	
Glucose	52.6mM	2.92mM
Urea	4.995mM	
NH_4Cl	0.467mM	
Glycerol		7.57mM
Sodium acetate		12.19mM
Tryptone	0.5g	
Made up to 1 liter	Instant Ocean	0.4 to 1.5M NaCl

3029 were cultivated heterotrophically with qualified success. The generation time was ~72h; this growth was too slow and needs to be improved. *D. bardawil* could not be grown heterotrophically. *D. parva* could not be grown heterotrophically (Hard and Gilmour 1996) in a medium with 0.4 to 1.5M NaCl with glucose, glycerol and sodium acetate as carbon sources (Table 3). Compared to Glaude and Maxey's medium which had >57.59mM C, the medium tested by Hard and Gilmour had only 26.8mM C; levels of boron, Cu, Co, Zn, Mn were also low.

Different species of *Dunaliella* were investigated for heterotrophic growth in the dark and there is no convincing evidence in support of heterotrophic growth Dr. Ben-Amotz concluded that *Dunaliella* is on obligate autotroph (Dr. Ben-Amotz -Personal communication -) which is corroborated by findings of Dr. Polle (Personal communication).

Scaling up Cultures from Laboratory to Outdoor Tanks

The several steps involved from the isolation of single cells and scaling up to large volume cultures of *Dunaliella* are the same as those for other microalgal cultures. Using standard microbiological methods (see Andersen et al. 2005) algal cells from natural brine samples (PLATE,1) can be streaked on seawater agar plates (PLATE, 2) and incubated in light. Colonies are isolated and purified to asceptic conditions by streaking the cells repeatedly, and purified cells are transferred to test tubes containing culture medium (PLATE, 3). Single cells can be isolated by micropipetting and by serial dilution method and are used for establishing cultures (PLATE, 4-9) and for scaling up cultures (PLATE, 10-12).

Periodically cells are tested for bacterial contamination by plating on bacteriological media followed by treatment with suitable antibiotics. One ml of culture is transferred to four flasks (#1-4), each containing 50 ml sterile medium. Flask one serves as a control and the other flasks receive 0.25, 0.50 and 1.0 ml antiobiotic mix. Cultures are tested for their sterility and to determine what level of antibiotics are non-lethal to algal growth. Kviderova and Henley (2005) found a mix of 25 mg l^{-1} streptomycin and 50mg l^{-1} ampicillin suitable for maintenance of bacteria free stock cultures of *Dunaliella* species for physiological experiments. It is of interest to note that out of 84 marine algae *D. primolecta* exhibited the highest antibiotic activity and strongly inhibited the growth of *Staphylococcus aureus, Bacillus cereus, Bacillus subtilis* and *Enterobacter aerogenes* (Chang et al. 1993). The simple elegant agar plating technique developed by Lakeman and Cattolico (2007) to isolate clones of algae using high-metal stress as a selective screen for assessing genetic variability could be probably applied to *Dunaliella*.

Usually, diluted culture medium (10, 20 or 50x) is used in the initial stages to establish cultures. Starter cultures are maintained in 20-50 ml screw cap polystyrene sterile culture tubes or flasks (PLATE,10) and are scaled up using either125, 250, 500, 1000 ml Erlenmeyer flasks(PLATE,11) or 2.8 l Fernbach flasks finally to 3.5 to 5 gallon carboys (PLATE,12,13). The ratio of inoclum to fresh medium is 1:9 for 125ml, 1:5 for 250ml and 50: 50 for larger volume flasks (PLATE,14). A

Plate 1: Various stages (1-12) in the isolation, purification and scaling up of *Dunaliella* cultures. For explanation see text.

volume head space of 25mm for 12 ml tubes, 4cm for 125ml flasks, 15cm for 2l flasks and 23cm for Fernbach flasks is desirable (Guillard 1975).

From a commercial grower's point of view, it is important to lower the growth time and to increase production. Use of large exponentially growing culture as an inoculum reduces the lag phase and yields high density cultures as shown below:

Assuming that there are no physical or nutritional constraints on the algae, the final yield of cells depends on the cell division rate (Table 5). In the formula F=A*2^n after n doublings:

Table 5: Calculated cell densities of *Dunaliella* cell densities based on division rates, size of inoculum.

Divisions per day	Start Day 0	Cell numbers (x1000 ml^{-1})					
		Day 1	Day 2	Day 3	Day 4	Day 5	Day 6
0	10	10	10	10	10	10	10
0.5	10	14	20	28	40	57	80
1.0	10	20	40	80	160	320	640
1.5	10	28	80	226	640	1,810	5,120
2.0	10	40	160	640	2,560	10,240	40,960
2.5	10	57	320	1.810	10,240	57,926	327,680

Let A=Initial cell number, D = number of days, r = number of divisions per day, F = final cell number, then F = A*2^(D*r), and A = F/2^(D*r)

(where * denotes multiplication, /denotes division, and ^ denotes exponentiation)

Note that n=D*r is the number of divisions that have taken place in D days, so this is just the simple exponential growth formula F=A*2^n after n doublings.

For example with one divison per day, when growth is initiated with 10 cells per ml, the final concentration by 6th d will be 640 x 10^3 cells ml^{-1}; for a culture with 2.5 divions a d $^{-1}$ the corresponding cell density will be 327,680 x 10^3 cells ml^{-1}. By changing the cell densities of the inoculum the time required for attaining a peak density can be suitably altered. For example to have 30,000 x 10^3 cells ml^{-1} on d 4, when the division rate is 1, 1.5, 2 and 2.5 d $^{-1}$, corresponding cells initially required on day 1 are 1875.00, 468.75, 117.19 and 29.30 x 10^3 cells ml^{-1}.

For scaling up, cultures raised in 3.5 and 5 gallon carboys (Nalgene and sterilizable) are used with a 15 cm air space (Guillard 1975). Aeration of *D. tertiolecta* cultures at a flow rate of 6.51 liters min-1 liter of culture-1 improved the growth rate and yielded 12.46 x 10^6 cells ml^{-1} compared to 7 x 10^6 cells ml^{-1} without aeration (Fábregas et al. 1994).The source of illumination was at the rear.

For larger volumes such as 200 liter indoor cultures, free standing 240cm x 30-35cm diameter fibreglass transparent towers have been used. The towers were washed with a hot power jet, soaked in 200ml 12% bleach, and then rinsed with power jet. At the Bedford Institute of Oceanography, Dartmouth, NS, Canada seawater from Bedford Basin from a distance of 300m from the shore is pumped through a sand filter, and fiber filter bags (Filter Innovation bags PE 25 P2S, to remove particles >25 ìm). Progressive filtration of seawater (5, 3 and 0.45 ìm filtration cartridges) reduces bacterial contamination by 60%; alternatively UV exposure of 212, 424, 636 and 848 J m^{-2} yielded bacterial reduction factors of 99.86, 99.969, 99.997 and 100%, respectively (Torrentera et al. 1994). At the Bedford Institute of Oceanography we routinely use International Water –Guard Industries IWG-1-15 single long lamp with 18.4 watts UV output at 254nm for disinfection of sea water. The sequential steps in processing large volumes of seawater involve: a) filtration through a 1 ìm nominal pore size cartridge b) UV radiation c) heating to 85^{0}C d) chilling e) enrichment and f) seeding. Alternatively seawater is filtered through 1 ìm nominal pore size, acidified with hydrochloric acid to pH 3, and after 24h neutralized with sodium carbonate. Biota in the seawater can be also killed by chlorination (1-2 mg l^{-1}, incubation for 24 h, followed by aeration to remove the chlorine. Any residual chlorine can be removed by addition of sodium thiosulphate.

Various designs for out door mass culture of large volumes are described elsewhere in this book. Distilled, demineralised tapwater can be enriched with commercially prepared salt mixtures supplemented with nitrogen sources as well as vitamin B_{12} and thiamine (Ukeles 1965). A 280 liter fiber glass tank can be placed in a larger fiber glass tank and cooled by pumping coastal seawater in the outer tank. Air pumped through the inner tank will eliminate dead spots in the culture tank and prevent bacterial growth and clumping of algal cells. These culture techniques offer various advantages and disadvantages. Axenic, indoor closed batch cultures are stable, predictable, but expensive unlike the open, non-axenic outdoor cultures. Continuous culturing can be efficient, but usually provides a continuous supply of limited quantities of algae. However, the Cawthorne Institute, New Zealand has overcome many problems by developing a cost effective continuous culture system (2000 liters per day) for micro algae. For pilot scale studies ≤125,000 liter ponds are also used (Borowitzka et al. 1984).

In mariculture operations that utilize *Dunaliella* as live feed, it is essential to ensure a steady supply of healthy algae. Standard operating procedures (SOPs) exist and for hatcheries protocols were developed for Hazard Analysis Critical Control Point (HACCP). HPACCP principle and practices are discussed by (Pierson and Cortlett Jr. 1992) and in particular for applied phycology industry by Grobbelaar (2003). To prevent die-offs of algal cultures it is essential to: 1. Maintain a temperature and light controlled clean culture facility with effective security 2. Supply of high quality aged oceanic seawater, free from pathogens 3. Achieve good biosecurity through training and practice, 4. Separation of chemicals, enrichments and clean cultureware exclusively for algal cultures, 5. Usage of standard sterization procedures for isolation, transfers and maintenance of cultures 6. Maintenance of a log book with details of media prepared, transfers of cultures and monitoring algal growth, and 7. Save backup cultures to prevent, eliminate or reduce 'crashes' to cultures.

Production Costs of Marine Microalgae

The largest variable cost items for culturing marine microalgae are labour and supplies. As light is critical for photosynthetic growth, usage of natural solar energy in the outdoor tanks in particular, can reduce costs of production substantially.

Commercially available nutrient stock solutions will reduce the preparation labour costs. Culture media kits can be purchased from The Provasoli Guillard National Center for Cultures of Marine Phytoplankton (CCMP). For example stock solutions for medium f/2 in 20liter and 50liter quantities are priced $25 and US $50 whereas cost of actual f/2 medium 1, 2, 4, 10 liters is US $50, 80, 120 and 250 respectively. CCMP also supplies cells harvested (from 15-18 liter carboy), frozen, pelletized and stored over liquid nitrogen vapors for $500.00. For diversified microalgal biotechnological applications large-scale extensive systems are used with enrichments with commercially available laboratory grade fertilizers such as Clewat-32 (Ronquillo et al., 1997) or others mentioned earlier.

Due to its commercial implications, firm costs for production of *Dunaliella* are not readily available but the data for cultures of *T-iso, Skeletonema* sp. *Pavlova lutheri, Nannochloropsis* sp. are indicative. Cultures raised in tanks can show a substantial reduction in production costs (Table 6) depends upon actual practice. For example what used to cost $1000 per kg^{-1} dry weight of culture earlier now costs a fraction when grown in waste-water (De Pauw et al.1984).

Dunaliella Strains

Strains of *Dunaliella* isolated and studied by various investigators include:

D. *acidophila* (Kalina) Massjuk 1971
D. *baas-beckingii* Massjuk 1973
D. *bioculata* Butcher 1959
D. *carpatica* Massjuk 1973
D. *flagellata* Skvortzov 1968
D. *gracilis* Massjuk 1973
D. *granulata* Massjuk 1973
D. *jacobae* Massjuk 1973
D. *lateralis* Pascher et Jahoda 1928
D. *maritima* Massjuk 1973
D. *media* Lerche 1937
D. *minuta* Lerche 1937
D. *minutissima* Massjuk 1973
D. *obliqua* (Pascher1930) Massjuk 1973
D. *parva* Lerche 1937
D. *paupera* Pascher 1932
D. *peircei* Nicolai et Baas-Becking 1935
D. *polymorpha* Butcher 1959
D. *primolecta* Butcher 1959
D. *pseudosalina* Massjuk et Radcenko
D. *quartolecta* Butcher 1959

D. ruineniana Massjuk 1973
D. salina Teodoresco 1905
D. terricola Massjuk 1973
D. tertiolecta Butcher 1959
D. turcomanica Massjuk 1969
D. virides Teodoresco 1906

Table 6: Production costs of marine microalgae.

Taxa	*Nature of culture*	*Production cost US$ per kg-¹ dry weight*	*Reference*
T-iso, Skeletonema sp. *Pavlova lutheri, Nannochloropsis* sp.	Tanks	1000	Bennemann 1992
Tetraselmis suecica	Batch	300	Coutteau and Sorgeloos 1992
Various diatoms	Continuous flow cultures 240 m^3	167	Walsh et al. 1987
Nannochloropsis sp	Photobioreactors	100	Chini Zittelli et al. 1999
Monospecific algal culture	Indoors or in a green house	120-200	De Pauw et al. 1984
	Outdoor culture	4-20	De Pauw and Persoone 1988
	Tank culture 450 m^3		Donaldson 1991
Algal biomass (Autotrophic)	Photobioreactors and Fermentors	11.22	Behrens 2005
Algal biomass (Heterotrophic)	Photobioreactors and Fermentors	2.01	Behrens 2005
Tetraselmis suecica *Cyclotella cryptica* *Nitzschia alba* *Chlorella* sp. *Cyclotella* *Chlorella* sp. *Crypthecodium cohnii* *Schizochytrium* sp	Fermentors	10 170 12 160 600	Day et al. 1991 Gladue and Maxey 1994 " " " " " " Barclay et al. 1994 De Swaaf et al. 1999
Induced blooms of marine hytoplankton species		4-23	De Pauw et al. 1984
Wastewater- grown microalgae		0.17 – 0.29	De Pauw et al. 1984

There are 23 salt-water species of *Dunaliella* (Preisig 1992) which are divided into four sections: Dunaliella, Peirceinae, Tertiolectae and Virtidis of which only members of *Dunaliella* accumulate carotenoids. Ben-Amotz and Avron isolated *Dunaliella* from a salt pond near Bardawil Lagoon in the North Sinai in 1976 and called it *D. bardawil.* Several investigators used this name although its characteristics suggest that it is in fact *D. salina* (Ben-Amotz et al. 1982) but in this presentation to avoid confusion *D. bardawil* is retained. Several strains of *D. salina Teod.* were isolated; for example from Geroiskoe Lake,Cherson, from Krasnoe Lake, Crimea, from Sasyk Lake, Crimea, strains IBASU-A D-11, and IBSS. Species of *Dunaliella* that are not brought into culture include: *Dunaliella baas-beckingii, D. carpatica, D. cordata D. euchlora, D. granulata, D. jacobae, D. asymmetrica, D. media,, D. ruineniana* and *D. turcomanica.*

Details on obtaining some of these strains are given in Appendix 1.

Of the cultured *Dunaliella* species, about 10 are widely cultivated for studies on their growth, physiology, biochemistry, mariculture and as sources of bioactive compounds such as glycerol and carotenoids. Details follow (Table 7).

Depending on their utility *Dunaliella* cultures can be raised as simple laboratory cultures in batch, continuous, pH stat, cyclostat, turbidostat, chemostat culture forms (Table 8). They also can be cultivated in large volume cultures using photobioreactors, enclosed out-door photobioreactors and fermentors.

Division Rates

Growth of the cultures can be monitored regularly by determining the changes in algal biomass expressed as cell numbers, chlorophyll *a*, cell- carbon, cell volume, optical density and dry weight of the culture. From a growth curve of *Dunalella tertiolecta* culture raised at 10°C and 42.5μmol m^{-2} s^{-1} (Figure 1) three phases can be recognized: 1) initial lag phase, soon after the inoculation of a culture, 2) exponential log phase, when the cells grow at an exponential rate, 3) a light limited growth phase and a stationary phase, when the cells attain their maximum volume and do not divide. In aged cultures (~40d not shown here) due to limitation of light and nutrients usually a senescent phase characterized by a decline in cell density follows the stationary phase.

When the cells are doubling in an exponential rate their growth follows:

$$dn/dt = rN \qquad (1)$$

Table 7: Cultures of *Dunaliella* species, applications and growth media.

Species	*Applications*	*ID Medium*	*Reference*
tertiolecta	Biomass yield	Walne	Concepcion et al. 1991
tertiolecta	chlorophylls	Erd-Schreiber	Madgwick 1966
tertiolecta	growth	F 0.01 to F10	Subba Rao 1981
tertiolecta	Cell division periodicity	f/2	Nelson and Brand 1979
tertiolecta	Iron, phosphorus and nitrogen deficiency, protein induction	ASW + f/2 nutrients and trace metals	La Roche et al. 1993
tertiolecta	UV radiation effects	Various combinations	Heraud et al. 2005
tertiolecta	Fish diet	f/2	Fabergas and Herrero, 1986
tertiolecta	Glycerol content, osmoregulation	McLachlan	Jones and Galloway 1979
tertiolecta	Vitamin E	ASW	Carballo-Cárdenas et al. 2003
	Temperature and biochemical composition	Modified ESWA	Thompson et al. 1995
tertiolecta	Uptake of organic selenides	Modified f/2	Baines et al. 2001
tertiolecta	Production of volatile organohalogens	L1 medium modified F	Murphy et al. 2000
tertiolecta	Carotene destruction due to herbivory	f/2	McLeroy-Theridge and McManus 1999
tertiolecta	Absorption of copper	f/2	Gonzalez-Davila et al. 1995
tertiolecta	Photoadaptation and package effect	f/2	Berner et al. 1989
tertiolecta	Photosynthesis: Respiration	F	Humphrey 1975, 1979
tertiolecta	Quantum yield - temperature	General Purpose Medium	Sosik and Mitchell 1994
tertiolecta	Herbicide sensitivity	f/2	DeLorenzo et al. 2004
tertiolecta	Carbon:Nitrogen: Phosphorus - growth rate	Modified f	Goldman et al. 1979
tertiolecta	Carbon-balance rates	ASP-2	Eppley and Sloan, 1965
tertiolecta	Photoperiod - growth	F	Eppley and Coatsworth 1966

(Table 7 Contd.)

(Table 7 Contd.)

tertiolecta	Heterotrophic growth	Sk (1A)	Glaude & Maxey 1994
tertiolecta, parva, pseudosalina, virides, peirceinae, peircei	Physiological attributes and pigment content	Johnsons J/1	Cifuentes et al. 2001
parva	Biotransformation of aromatic aldehydes	ASW	Hook, et al. 1999
parva, pseudosalina, tertiolecta, virides, peirceinae, peircei	Growth, Photosynthesis irradiance, biochemistry	Giardano	Giardano 1997
tertiolecta	Carbon, nitrogen metabolism	1/10th enriched	Laws and Wong 1978
primolecta	Production of anti-herpes simplex virus substances	BG11	Ohta et al. 1998
bardawil	Phytoene and carotene accumulation	Ben-Amotz et al. medium	Ben-Amotz et al. 1987
bardawil	Far-red light, carotenoids	Johnsons J/1	Sanchez-Saavedra et al. 1996
tertiolecta	Cell cycle proteins Cyto biochemistry	Aquil	Lin et al. 1995 Lin et al 2001
	Biochemical composition	Provosoli ESAW modified (Bold and Wynee 1978)	Thompson et al. 1992.
bardawil	Far-red light, carotenoids	Johnsons J/1	Sanchez-Saavedra et al. 1996
salina, bardawil	Carotenoid accumulation	Provosoli PES	Gómez et al. 2003
salina, bardawil	Carotenoid accumulation	Provosoli ART	
salina, bardawil	Carotenoid accumulation	Provosoli PES	
salina	Gene expression	ASP-7	Wada et al. 1996
salina	Lipid metabolism		Muradyan et al. 2004
acidophila	Various applications	Dun Ac	SAG
Dunaliella sp.	Sponge cultivation	ASW +f/2	Osinga et al. 1998

Table 8: Types of culture systems of *Dunaliella*.

Type	*Taxa*	*Application*	*Reference*
Microalgal cultures	*D. tertiolecta*	General culture studies	Rech et al. 2003
Small volume	*D. tertiolecta*	Cytology, biochemistry, Fe stress	Lin et al. 2001
Batch cultures	*D. tertiolecta*	Nutrient uptake	Subba Rao 1981
		Feeding brine shrimp	Marques et al. 2005
Batch and steady state cultures	*D. tertiolecta*	*Physiological responses*	Graziano et al. 1996
Continuous cultures	*D. tertiolecta*	Nutrient kinetics	Goldman and Peavey 1979
	D. salina, D. viridis D. bardawil	ECOTOX bioassy	Millán et al. 2006
Chemostat	*D. salina, D. bardawil*	Carotene production	Vorst et al. 2004 1994
	D. tertiolecta	Nutrient kinetics	Geider et al.1998
		carotene production	Vorst et al.1994
pH stat	*D. tertiolecta*	cell harvest	Horiuchi et al. 2003
	D. tertiolecta	Production of liquid fuel	Tsukahara and Sawayama 2005
	D. tertiolecta	Genome studies	Walker et al. 2005
	D. tertiolecta	Vitamin E production	Carballo-Cárdens et al 2003
	D. tertiolecta	Dynamic cultivation	Barbosa et al. 2003
Cyclostat	*D. tertiolecta*	Production	Fabregas et al. 1995
Mass cultures	*D. euchlora*	Oyster larval feed	Ukeles 1980
	D. extract	Shrimp feed	Supamattaya et al. 2005
	D. tertiolecta	Scallop larval feed	Nevejan et al. 2003.
Mass culture	*D. salina*	Hydrogen production	Polle et al 2002 Park et al. 1998
Vertical tubular reactors	*Dunaliella* sp.	Mass production	Tredici 2004

(Table 8 Contd.)

(Table 8 Contd.)

Stirred –tank photobioreactors	*D. salina*	Optimal growth model	Li et al.2003
Bioreactors	*D. tertiolecta*	Removal of NOx	Nagase et al. 1997
Bubble column and air-lift bioreactors	*D. tertiolecta*	CO_2 fixation	Suzuki et al. 1995
Tubular photobioreactors	*D. salina*	carotene and luetin production	Garcia-Gonzalez et al. 2005
Flat paner bioreactor with A- stat	*D. tertiolecta*	optimization of biomass	Barbosa et al. 2005
Two-phase bioreactors	*D. salina*	Milking carotene	Hejazi et al. 2004
Enclosed out-door photobioreactors	*D. tertiolecta*	High efficiency growth	Janssen et al. 2003
Fermentors	*Dunaliella* sp.	Waste treatment and ethanol production	Shirai et al. 1998

Where N is the population size, t is the time and r is the intrinsic rate of increase.

Their growth constant (the number of logarithm-to-base –e units of increase per day can be expressed either as 'logarithm-to-base 10' units of increase per day K_{10} or as 'logarithm-to-base-2' units of increase per day k Generation or doubling time in days or also referred as growth coefficient) as follows:

$$K_{10}= \log_{10} (n_2/n_1)/ (t_2-t_1) \quad (2)$$

where n_1 and n_2 represent cell numbers at times t_1 and t_2 or as:

$$k= \log_2 (n_2/n_1)/(t_2-t_1) \quad (3)$$

In most algal culture studies it is convenient to use logarithm-to-base-2. By using log 2, and selecting t as 1 day, the growth constant k, becomes equivalent to the number of doublings per day.

If logarithm-to-base 10 is used, multiplying the K_{10} values by 3.322 (K_e/ ln 2= 2.3026/0.6931= 3.322) gives k generation or doubling time day $^{-1}$. Some use the notation μ to express k. As in the bacterial cultures that divide rapidly if time t is in hours 24/k gives generation time hours per division. Then doubling can be calculated as Doubling G= (0.301/μ) x 24.

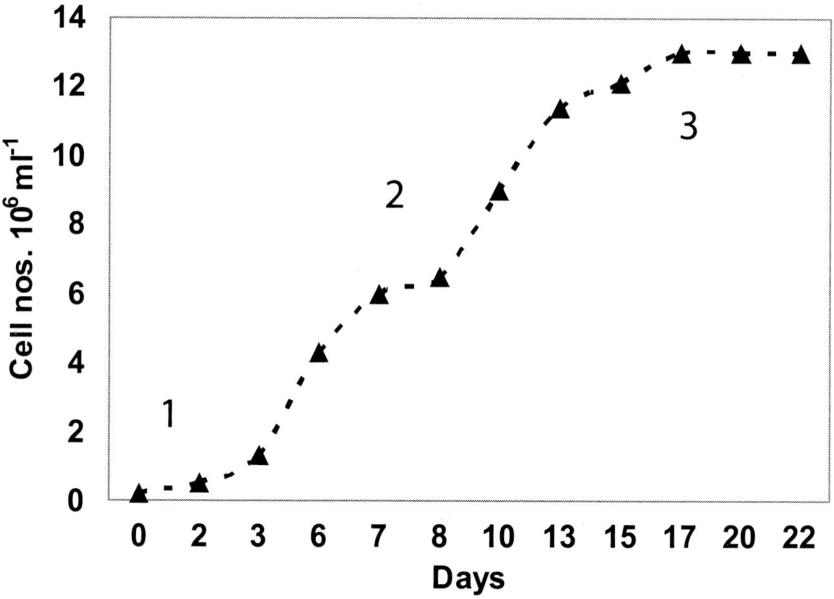

Figure 1: Growth curve showing various phases in *Dunaliella* tertiolecta culture.

Growth rates are a measure of the response of cells to nutrients and environmental conditions. I have assembled growth rates of *Dunaliella* (Table 9) grown under a variety of culture conditions which indicate a range of values. The maximum growth rates reflect those at optimal conditions. Between the species the growth rates differed. The maximum growth rate for *D. teriolecta* is 3.0 based on biovolume (Sciandra et al. 1997), 1.7 for *D. parva* (Cifuentis et al. 2001), 1.5 for *D. viridis* (Cifuentis et al. 2001), 1.4 for *D. salina* (Giardano 1997), 1.3 for *D. pseudosalina* (Cifuentis et al. 2001) and 1.0 for *D. peircei* (Cifuentis et al. 2001). For *D. viridis* division rates are available at different temperatures which indicate their increase (Cifuentis et al. 2001). Further, divison rates were higher for *D. tertiolecta* that received light 24 h a day and those that were raised at higher light levels < 430 μ mol photons m^{-2} s^{-1}(Cifuentis et al. 2001).

The growth rate and final cell yield of *D. tertiolecta* can be improved by increased aeration; cell density was 12.46 x 10^6 cells ml^{-1} under air flow rate of 6.51 l min^{-1} liter of culture (Fábregas et al. 1994). This aeration rate, however, caused evaporation and consequently an increase in the salinity. When cultures were supplied with CO_2 but not aerated the yield was 7 x 10^6 cells ml^{-1}. Besides acceleration in growth by the media and aeration, growth of *Dunaliella* can be influenced by light, temperature, pH, salinity, and nutrients (such as phosphorus, nitrogen, iron and metals).

Table 9. Growth rates of *Dunaliella*.

Species	*Medium*	*Culture*	*⁰C*	**Light μ mol photons $m^{-2} s^{-1}$*	*Lt:dk hours*	*@NaCl M*	*Div. day $^{-1}$*	*Method*	*Reference*
parva	J/1	Batch	20	50	14:10	0.17	1.4-1.7	fluorescence	Cifuentis et al. 2001
salina						2..57	~1.4		
pseudo salina						0.51 1.71	1.3 0.6		
viridis							~1.5		
peircei						0.17-1.71	1.4 -1.0		
						1.71	1.6-1.5		
salina	Giardano		NA	175	NA	1.5	0.3-0.5	numbers	Giordano 1997
salina	Bold's basal medium	Batch	19	150	24		1.7		Vorst 1995
baradawil	Bold's basal medium	Batch	19	150	24		1.2		Vorst 1995
viridis	Jimenez and Niell	Batch	20 25 30	150			0.46-0.90 0.57-1.11 0.80-1.40	numbers	Jimenez and Niell 1991a
viridis	NA	NA	NA	NA	NA	NA	0.74-1.27	numbers	Borowitzka and Borowitzka 1988
tertiolecta	f/2	Batch	25	86	24	0.34	0.75	numbers	DeLorenzo et al. 2004
tertiolecta	J/1		20	50	14:10	0.17	~1.7	numbers	Cifuentis et al. 2001
tertiolecta		Batch	12 28	165	24	NA	0.5 2.2	fluorescence	Sosik and Mitchell 1994

tertiolecta	ASW +f/2	Batch		250	24	~3.08	1.3	numbers	LaRoche et al.
tertiolecta	Aquil		20	85	12:12	~0.42	0.31 to 0.69 0.15 to 0.50	PCNA Stain	Lin et al. 2001
tertiolecta	f/2	chemostat	18	75-430	12:12	NA	0.5 to 3.0	biovolume	Sciandra et al. 1997
tertiolecta	F	Batch	19-21	80	24 12:12	NA	1.1 0.8	numbers	Humphrey 1979
tertiolecta	F	Batch	20	175	3:21 18:6	NA	0.68-2.64 0.8	numbers	Eppley and Coatsworth 1966
tertiolecta	F10 –F0.01	Batch	10	65	24	0.58	0.53 to 1.98	numbers	Subba Rao 1981
tertiolecta	ASW+f/2 nutrients	Batch	NA	250	24	NA	1.2	numbers	Roche et al. 1993
tertiolecta	McLachlan	Batch	30	150 33	24	0.25 to 1.00	2.07 0.93- 1.42	numbers	Jones and Galloway 1979
tertiolecta	Special medium	Continuous	20	270	12:12	NA	0.12-1.44	numbers	Laws and Wong 1978
tertiolecta	f	Batch	25	871	16:8	29ppt	0.8	numbers	Jitts et al. 1964
tertiolecta	f/2	Batch	2015	85	12:12	NA	0.77	PCNA Stain	Lin et al. 1995
tertiolecta	Modified ESWA	Exponential	10-25	220	24	0.363	0.29 to 1.7	Fluorescence or Coulter counts	Thompson et al. 1992
tertiolecta	Chemostat and batch	NH_3 limited medium	25	5.8 tg cal cm^{-2} h^{-1}	24	NA	0.216-1.68	numbers	Caperon and Meyer 1972

*light units and @ salinity are based on conversions; NA= not available

Light

Response of *Dunalilela* to light differs with the various species although growth was very slow at high light intensities. For *D. salina* cultivated in two phase reactors, the cells remained viable for >47 d (Hezari et al. 2004). *D. salina* growth increased in the range 50-800μmol photons m^{-2} s^{-1}, attained a maximuum at 800μmol photons m^{-2} s^{-1}, remained constant up to 1500 μmol photons m^{-2} s^{-1} and at higher irradiances declined (Baroli and Melis, 1996). Although exposure to full sun simulator light caused a marked drop in photochemical efficiency, irradiation with Photosynthetically Active Radiation (PAR) and Ultra Violet Radiation (UV-A) radiation or with PAR alone caused only a slight inhibition of photosynthetic activity (Hermann et al. 1996). Ghetti et al. (1999) observed inhibition of photosynthesis at all levels of simulated solar radiation and it increased when the cut-off was shifted towards shorter wavelengths.

With *D. bardawil,* however, when growth PAR is held constant, but supplemented with UV-A (320-400 nm), carotenoid levels increased and were proportional to the intensity of UV-A (Jahnke 1999). The accompanying result was an 80-310% increase in carotenoid: chlorophyll ratios. *D. salina* grown either at 13°C/150 μmol m^{-2} s^{-1} or 30°C and 2,500 μmol m^{-2} s^{-1} accumulated comparable levels of carotenoids and the zeaxanthin-binding protein, Cbr (Krol et al. 1997). These authors attributed the carotenoid and one Cbr abundance as a response to changes in Photo System II (PSII) 'excitation pressure' rather than to higher light levels per se. For *D. Parva* a halotolerant and a thermoresistant species, photosynthesis was inhibited at 3 M compared to that at 1M and 2 M of NaCl at 31 °C and 600 mol $m^{-2}s^{-1}$ irradiance, but the cells were photosynthetically active even at 42°C (Jiménez et al. 1990). With this species the generation time decreased between the 5 and 200 mol $m^{-2}s^{-1}$ irradiance (Ginzburg and Ginzburg 1993); at 5 mol $m^{-2}s^{-1}$ irradiance growth was possible and the generation time was about 43 to 46 h. This is in contrast to generation time of 15 to 21 h at 200 mol $m^{-2}s^{-1}$ irradiance. In *D. bardawil,* spectral quality of light influenced accumulation of carotene; with photon irradiance of white light provided by fluorescent lights cell density and pigment content increased and carotene accumulation was upto10% of their dry weight, much higher than those supplied with white light +far red light (Sanchez-Saavedra et al. 1996). Far red light induced growth rate, cell volume and total carotenoid content byabout 50% on a per cell basis.

Biomass of *D. tertiolecta* cultivated at a sub-optimal temperature (20 °C) increased more at low photon flux densities (57-370 μmol m^{-2} s^{-1}) than it did at 600 μmol m^{-2} s^{-1} (Janssen et al. 2000). Under nitrogen limitation biovolume and growth rates decreased as the irradiance decreased and increased with increase in light (Sciandra et al. 1997). Although fluctuating light did not effect photosynthesis in *D. tertiolecta* (Stramski et al. 1993), a transition of *D. tertiolecta* cells from 700 to 70 μmol m^{-2} s^{-1} reduced division rates from 1.1 to 0.4 per day within 24 h .

Accompanying changes include a slower accumulation of photosynthetic pigments, light harvesting antenna complexes, Photosystem II reaction centers and structural lipids that constitute the thylakoid membranes which seem to optimize light harvesting capabilities as cells adapt to low photon flux densities (Sukenik et al. 1990).

D. tertiolecta cultivated in photobioreactors under 94/94 light/dark ms cycles yielded higher biomass than when under 440-455 μmol m^{-2} s^{-1} continuous light (Janssen et al. 2001). *D. tertiolecta* cultures, alpha-tocopherol (α-T- vitamin E) increased with increasing cell density. Due to the commercial applications it is of interest to note that unlike in the freshwater alga *Euglena gracilis*, production of this antioxidant is not limited when high density *D. tertiolecta* cultures were grown under diminished light (Carballo-Cardenas et al. 2003).

Temperature

The habitats of *Dunaliella* species are diverse ranging from the Antarctic to the tropical lagoons indicating its wide tolerance to temperatures from -35 °C (Teodoresco 1906) to about 40°C (Wegmann et al. 1980). The optimum temperature range varies with the species i.e. for *D. salina* 20-40 °C (Borowitzka 1981), for *D. viridis* 14-30 °C, for *D. bioculata* 25°C, *D. primolecta* 29 °C and for *D. tertiolecta* 20 °C (see Borowitzka and Borowitzka 1988). Considering their natural habitat the upper limit of this range (30° to 40 °C) is not unreasonable.

Dunaliella sp exhibited progressively more heat tolerance with increasing salinity (Henley et al. 2002). The photosynthetic capacity of cells cultured at 23°-25°C decreased to 15 and 40% at 100 and 50 psu respectively when they were maintained for 2 h at 41.5 °C – the upper temperature limit. Their PSI:PSII fluorescence emission ratio also increased. However when returned to 23°C they recovered within 30 min (Henley et al. 2002).

In *D. viridis* the cell division rate depended systematically on the temperature, salinity and initial nitrogen; over the temperature range from 20 to 30 °C generation time decreased (Jiménez and Neill 1991a and b). Cell motility and resistance to high temperature (35-60 °C) increased in *D.salina, D. parva* and *D. psuedosalina* with increasing NaCl in the range 0.1 to 4M (Shariati 2003). Photosynthesis steadily increased between 15 and 31 °C , reached its peak followed by a decrease upto 42 °C in *D. parva* ; its halotolerance and thermo-resistance are evidenced by its net photosynthesis and positive oxygen evolution at 42 °C at 1-3M NaCl (Jiménez et al. 1990). Borowitzka and Borowitzka (1988) attribute this thermo-resistance to the protective effects of intracellular glycerol.

Cultures of *D. salina* grown at 4 °C accumulated intracellular proline that allowed some cells to withstand freezing without the addition of a cryoprotectant.

Moreover a period of proline treatment before freezing enhanced survival and division rates (Helliot and Mortain-Bertrand 1999). In this species when the temperature was decreased from 34 to 17°C a 7.5 fold increase in the levels of β-carotene was observed whilst Beta carotene levels did not change (Orset and Young 1999). A shift in the temperature from 28 to 20°C increased the 18:3 to 18:2 fatty acids considered as a common adaptive response to decreased temperature (Klyachko-Gurvich et al. 1997).

The optimum temperature for *D. tertiolecta* is 20°C (Wegmann 1968) was much lower than the 38°C suggested by Ukeles (1961). A sudden increase of temperature by 5-15°C above the preconditioning temperature, however, accelerated cell division temporarily (Jitts et al. 1964). Several species of *Dunaliella* have the unique capability of maintaining high levels of glycerol in their intracellular spaces which is released with heat treatment; the release was complete at 60°C and 50% at 50°C (Wegmann et al. 1980).

pH

Most *Dunaliella* species, except *D. acidophila*, grew at >6 pH. *D. tertiolecta* grew at 6.0 to 9.3 pH (Humphrey 1975) and the optimum is pH 6 (Gonzalez-Davila et al. 1995). Growth rate was inversely proportional to pH, with 9.2 pH as the upper boundary for maximal growth rate (Fábregas et al. 1994). The pH for maximum growth for *Dunaliella* isolated from evaporation salt ponds differed. For *D. viridis* it was between pH 7.2 and 7.6 and 8 to 9.1 pH for *D. salina* (Gibor 1956). Although *D. salina* tolerates a range of pH 5.5. to pH 10.0, higher and lower values result in eventual death (see Borowitzka and Borowitzka, 1988). At most salinities in the range 0.02 to 4.0 M Na Cl, growth rates of *D. tertiolecta* were affected by the pH; cells divided faster as the pH increased from pH 6.2 and to pH 8.2 at salinities from 0.5 to 1.5 M but at salinities > 2.5 M they grew better at pH 8.2 (Latorella and Vadas 1973). *Dunaliella* sp. isolated from Lake Tuz, a hypersaline lake in Turkey, grew steadily in Johnson's media with 1.71M, 3.42M and 5.13 M NaCl and attained maximum cell density at pH 9 (Çelekli and Dönmez, 2006).

There is considerable evidence to demonstrate that changes in pH of culture media, similar to other algal cultures, depend on the capacity of the cells to utilize dissolved inorganic carbon (DIC) and HCO_3 as carbon sources during photosynthesis (Latorella and Vadas 1973). The pH of a freshly inoculated batch culture of *D. tertiolecta* (10°C and 42.5μmol m^{-2} s^{-1}) was 7.89 and increased as the culture grew and utilized carbon dioxide during photosynthesis (Subba Rao unpublished). This caused a shift in the bicarbonate buffer system with an increase in the pH to 8.81 by day 3, and to pH 9.11 by day 7. Higher values were observed in senescent phase cultures. With a progressive increase in the inoculum i.e. 3, 6, 12, 24 and 48 ml per 300 ml medium, cultures grown only for 3 d under identical conditions registered pH values up to 7.96, 8.10, 8.35, 8.74 and 9.10, attributable to intense photosynthetic activity (Subba Rao unpublished).

In natural brine environments characterized by high salinity, pH of ~9.0 and high temperatures, the solubility of CO_2 is low and most of the inorganic carbon is in the form of HCO^{-3}. *D. salina* which lives under these conditions adapts to low CO_2 levels and accumulates inorganic C up to 20 times greater than in the medium (Zenvirth and Kaplan 1981). Such a high internal CO_2 concentration is facilitated by carbonic anhydrase activity (CA) on the cell surface as demonstrated in *D. tertiolecta* (Aizawa and Miyachi 1984). Ghoshal and Goyal (2001) showed that *D. tertiolecta* has a DIC pump that provides a mechanism for concentrating dissolved inorganic carbon. It is of interest to note that the cell surface of *D. tertiolecta* has high-affinity and low-affinity sites for Cu^{2+} absorption and factors such as pH, temperature and salinity affect this process (Gonzalez-Davila et al. 1995). *D. tertiolecta* cells adapt to salt stress by increasing their intracellular pH by 0.2 units, in both dark and light; the cells maintain a constant pH of about 7.5 over a range of 6.5 to 8.5 (Goyal and Gimmler 1989).

For growth of *D. salina, D. parva* and *D. pseudosalina* isolated from the salt marshes of Gave khoni of Isfahan-Iran the optimum pH was 7-8 but the first two species could grow at pH ranges from 6 to 9 while the third had an extended range from pH 5 to 10 (Sharati 2003).

Bubbling of CO_2 in air or addition of $NaHCO_3$ are some of the common practices used to stimulate growth of *Dunaliella* under cultivation. Addition of CO_2 lowers the pH thus making available CO_2 for photosynthesis (Van Auken and McNulty 1973). Ginzburg and Ginzburg (1981) reported some species of *Dunaliella* supplied with inorganic carbon are tolerant to high light and temperature.

For *D. tertiolecta*, aeration at a flow rate of 6.51 liters min^{-1} liter of $culture^{-1}$ sustained 12.46 x 10^6 cells per ml^{-1} compared to 7 x 10^6 cells ml^{-1} in cultures without aeration; the growth rate was inversely proportional to pH and attained a maximum of 0.94 day^{-1} at pH 9.2 (Fábregas et al. 1994).

Dunaliella acidophila usually lives at extremely low pH values (0.7 to 1.5), the lowest tolerated pH was 0.2; a period of adaptation was necessary in media with pH values lower than 1.0 (Fuggi et al. 1988).

For *D. marina* pH and EDTA appear to regulate its cell volume in hypotonic media containing 5 ppm copper; volume regulation is completely inhibited at pH 8.4 within 5 min , but not when the pH is changed to 7.4 and the cells are exposed for 30 min (Riisgard et al. 1980).

Salinity

In general members of the section *Tertiolecta* have optimum growth at a salinity of 2-4% whereas members of sections of *Dunaliella* and *Viridis* prefer higher salinities. In this study salinity S ‰ expressed as PSU (Provisional Salinity Units) is divided by 58.44 to obtain mols of NaCl). Salinity tolerance of *Dunaliella* is highly variable as revealed by their presence in habitats with wide ranges in salinity (Figs.2 and 3). For example massive floods from the Jordon River lowered the salinity in the Dead Sea and resulted in a bloom of *Dunaliella* (8.8 x 10^6 cells l^{-1}) in 1980 and 15 x 10^6 cells l^{-1} during April-May 1992 (Oren 1999). For example *D. salina* lives in a salinity range of 165-350 psu (2.823 to 5.98M NaCl) in the Salinas de Huacho (Montoya and Olivera, 1993) while several strains of *D. parva* (Utex 1983), CCMP 362, CCAP 19/9, *D. piercei* UTEX 2192, *D. tertiolecta* UTEX 999 and CCMP 1320 only attained their maximum division rates at about 0.86 M NaCl followed by a drastic decrease to a minimum at about 5.13M. Comparison of Figures 2 and 3 shows that between the strains of *D. tertiolecta* and *D. parva* division rated varied with variations in salinity. Marin et al (1998) showed their *D. salina* isolated from Venezeula adapts to large variations in salinity (1.54-3.39 M) and light at the expense of its growth rate. In *D. salina* optimum growth was at 3. 77M ; optimum carotenogenesis was at 5.99M NaCl, 30-40°C and pH 9.0 (Borowitzka 2005).Two Chilean strains of *D. salina* CONC-007 and CONC-006 responded differently to salinity changes (Cifuentes et al. 1996). In the former the highest division rates (0.76 and 0.65 d $^{-1}$) were at 0.87M NaCl while they were low (0.37 div d^{-1}) at 3.42M NaCl. The actual salinity for optimal growth in the laboratory experiments was much lower that that of the environments from which the strains have ben isolated (Oren 2005).

In the halotolerant *D. viridis* an increase in salinity from 1 M NaCl to 4 M NaCl slowed division rate by about 50% at 20, 25, 30 °C growth temperatures (Jiménez and Neill 1991). *Dunaliella* sp isolated from Oklahoma were progressively more heat tolerant with increasing salinity (Henley et al. 2002).

The optimum growth rate for *D. salina, D. parva* and *D. pseudosalina* isolated from salt marshes of Iran was at about 0.5 to 2.0 M NaCl (Shariati 2003).

Growth of *D. tertiolecta* was optimum at 0.2 to 0.5 M but was stressed at extreme salinities 0.05 M (hypo) to 3.0 M (hyper) NaCl (Jahnke and White, 2003). At the extreme high salinities the antioxidant enzyme and substrate activities were altered and the cells produced 260% more monodehydroascorbate reductase, doubled ascorbate peroxidase and the dark respiration increased by 3 fold. At the lowest salinities (0.05M) cell volumes doubled. *Dunaliella* sp isolated from an Antarctic hypersaline lake adjust their fatty acid levels; in the range 0.4M to 4.0M NaCl the proportion of total saturated and monostaurated fatty acids increased, while total polyunsaturated fatty acids decreased (Xu and Beardall, 1997). In *D. viridis* cell carbon and nitrogen increased with an increase in salinity but the cells accumulated glycerol, nitrate, structural proteins and free amino acids to adjust to high salinities (Jiménez and Niell 1991b).

Salinity has an effect on carotenogenesis; for example *D. bardawil* and *D. salina* tested for carotenogenesis in the range 1-3M NaCl, the highest cellular carotenoid level was at 2M (Ben-Amotz et al. 1982, Gómez-Pinchetti et al. 1992). Carotenogenesis increased slightly (Marin et al. 1998) at the highest salinity

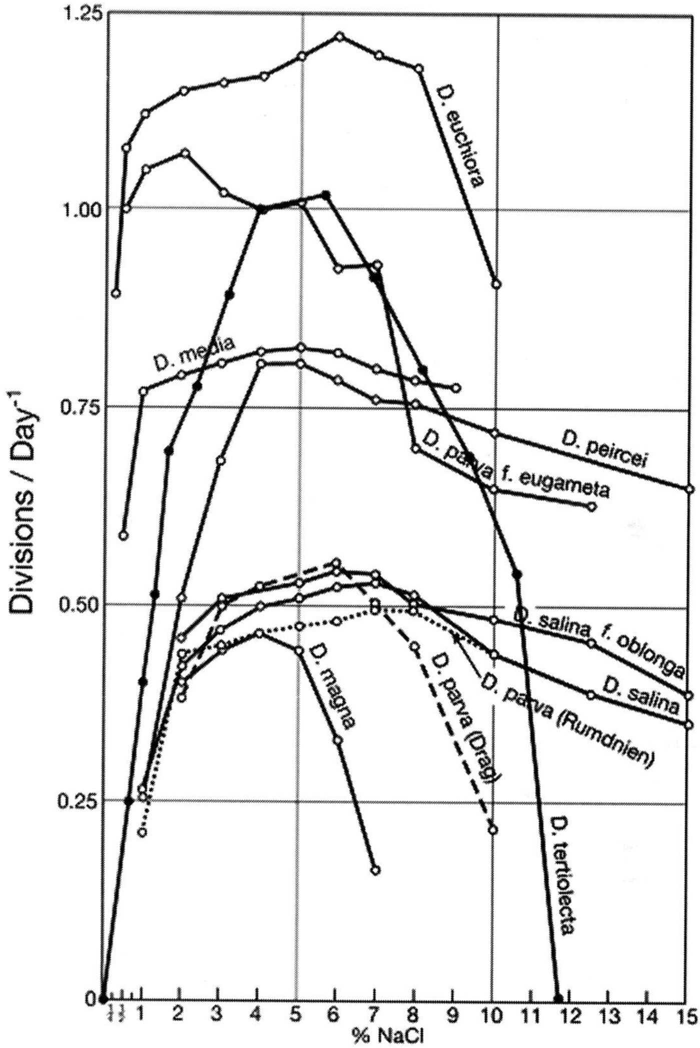

Figure 2: Division rates of *Dunaliella* species at various salt concentrations (based on Lerche 1937 and McLachlan 1960).

(3.59M) but not Chl *a* or total carotenoids. Changes of intensity from 50 μmol m^{-2} s^{-1} to 150 μmol m^{-2} s^{-1} caused an increase in carotenoid production; was in *D. salina* (strain CCAP) it was 2.4, 2.1 times while in *D. salina* (WT) it was and in *D. salina* (strain WT)1.4 times; carotenoids production peaked at salt concentration of 2M, while in *D. tertiolecta* it was at 0.7M (Fazeli et al. 2006).

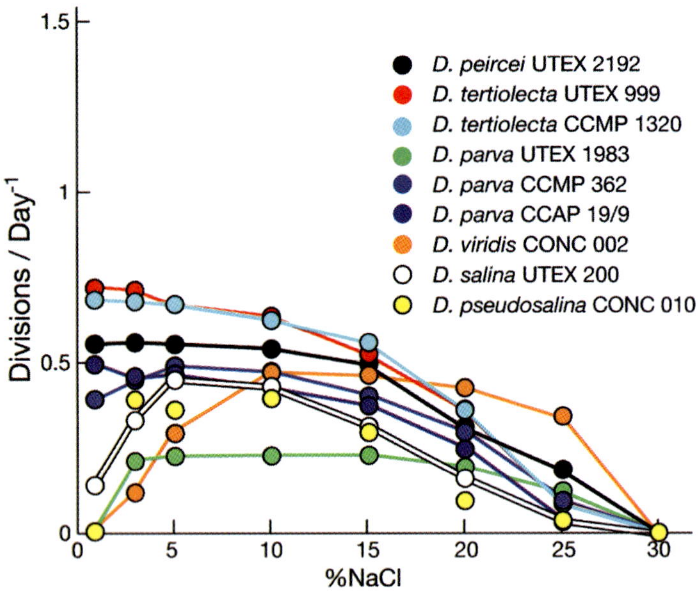

Figure 3: Division rates of *Dunaliella* species at various salt concentrations (based on Cifuentes et al. 2001).

D. tertiolecta previously adapted to 0.5M NaCl and then transferred to higher salinities (3.5M) needed an adaptation period (Latorella and Vadas 1973). Growth was restricted at salinities <0.02M while at salinities > 2.5M cells required continuous illumination for growth confirming the earlier observations (Ben-Amotz and Ginzburg, 1969). The carbonic anahydrase activity of cells adapted to higher salinities increased with salinity demonstrating HCO_3 incorporation occurs with Na+ extrusion.

As the salinities increased (0.5 to 3.5 M NaCl) along with high irradiance (~200 μmol m^{-2} s^{-1}) in the halotolerant strain of *D. parva* isolated from a Romanian Salt Lake glycerol production increased probably as a measure for osmoregulation in extreme environments; protein synthesis and cell division also continued (Ginzburg and Ginzburg 1993).

D. salina grows in 0.5 to 5M NaCl; with an increase in salinity there was an increase in 60-kDa protein - p60 (Fisher et al. 1994). The 60-kDa could also be a molecular chaperone (stress protein) known as Heat Stock Proteins. The location of p60 exclusively in the plasmamembrane, and the correspondence between its accumulation and growth at high salinities suggest that its role involves ionic homeostasis. Recent studies on this extremely halotolerant alga (Premkumar et al. 2003) showed the 60-kDa internally duplicated á-type carbonic anhydrase (Dca) that bestows halotolerance. The high salt tolerant species *D. viridis* and *D. salina* have higher activities of Dca than the less tolerant *D. tertiolecta* and *D. parva* (Goyal et al. 1992).

Phosphorus

Optimum growth of *D. salina* and *D. viridis* required 110-114 μM phosphate levels (See Borowitzka and Borowitzka 1998). *D. tertiolecta* accumulated more phosphorus than *Amphidinium, Cylindrotheca* or *Phaeodactylum* (Humphrey 1974) and the results of Graziano et al. (1996) also confirmed a luxury uptake of phosphate (Graziano et al. 1996). Cells grown under phosphorus limitation or starvation when transferred to phosphate rich medium phosphate uptake increased (Shelly et al. 2005). A ten fold increases in phosphorus increased photosynthesis: respiration ratios in *D. tertiolecta* (Humphrey 1974, 1975).

Due to the absence of a rigid cell wall in *Dunaliella*, the mechanism of assimilation of the major nutrients P, N is undoubtedly complex. Suzuki et al. (1997) showed that the intracellular P content in *D. parva* was highly dependent on the P concentration in the preculture medium. Phosphate uptake as well as sulphate were strictly and specifically dependent on Na+ ions; they are mediated by Na+/anion symporters and are driven by ΔμNa+ across the plasma membrane (Weiss et al. 2001). *D.viridis* carries a sodium-dependent phosphate transporter gene, DvSPT1, which is expressed upon phosphate depletion or NaCl hyperosmotic stress or (Li et al. 2006). Growth of *D. tertiolecta* cells in f/2 medium in phosphate remained static from d 0 till d 2 and then decreased rapidly (Roche et al.1993). Under phosphorus limitation *Dunaliella tertiolecta* produces an extracellular alkaline phosphatase (Wynne and Rhee 1988) and 200-kDa proteins accumulated and were membrane bound. Further studies by Graziano et al.(1996) revealed *D. tertiolecta* cells under all steady state phosphate-limited conditions produced high molecular weight (>200) proteins which increased with increasing P-limitation and decreasing growth rate. More recently Ghoshal et al. (2002) isolated and purified the osmoregulatory isoform of dihydroxyacetone phosphate (DHAP) reductase (Osm-DHAPR), unique to *Dunaliella* adapted to extreme high salinity environments. The Osm-DHAPR may play a significant role in the synthesis of free glycerol, an osmoregulatory compound used by *Dunaliella* species in extreme environments.

In *Dunaliella tertiolecta* phosphate starvation caused significant changes in the growth rates, the quantum efficiency of PSII electron transport, and P-uptake capacity (Shelly et al. 2005). In the presence of UV radiation and phosphate depletion the decline of growth rate and fluorescence were significantly faster. Phosphate depletion caused accumulation of ß-carotene but not glycerol in two isolates of *Dunaliella*, D1, obtained from GTCC and D2 an indigenous strain isolated from Sambhar salt lake, India (Phadwal and Singh 2003).

Dunaliella salina synthesizes from inorganic P(i) low molecular weight, probably cyclic, polyphosphate intermediates, which then condenses to high molecular weight polymers. These polymers are stored in the cell in a soluble form. Bental et al. (1990) hypothesize that due to differential volume changes of the cytoplasm and the chloroplast, the cytoplasmic orthophosphate concentration is changed. Results of Pick et al. (1990) suggest that the hydrolysis of large amounts of polyphosphates provides a pH-stat mechanism to counterbalance alkaline stress and hypoosmotic shock (Einsphar et al. 1988) in this alga.

Nitrogen

For the cultivation of *Dunaliella* species potassium nitrate, sodium nitrate, urea, ammonium nitrate, glycine, ammonium acetate, ammonium chloride, and histidine have been evaluated as nitrogen (N) sources. The physiological responses of *Dunaliella* varied with the species and with the N source (Giordano and Bowes 1997). In *D. tertiolecta* N assimilation is light dependent (Grant 1967) and nitrate is the most effective source of N (Kosmakova and Prozumenshchikova 1983, Grant 1968). Buffering capacity of the medium is also important in the effective assimilation of N (Grant 1968). The results of Goldman and Brewer (1980) showed that ammonium uptake results in acidification while nitrate and nitrite cause alkalinization of the medium.

For *D. salina* the results of Thakur and Kumar (1999) showed potassium nitrate supported better growth than did urea. Under limiting conditions of N steady state continuous cultures of *D. tertiolecta* utilized either nitrate, ammonia or urea for growth (Goldman and Peavey, 1979). A number of mutants of *D. tertiolecta* that could not utilize nitrate required ammonia (Latorella et al. 1981) and *D. tertiolecta* seems to preferentially utilize nitrous oxide (NO) rather than nitrate (Nagase et al. 2001). In a range of 0.25 to 16 mg atom.N/liter, under constant N: P, irrespective of the previous nitrogen source (nitrate, nitrite and urea), cellular protein and carbohydrates and gross energy per ml of *D. tertiolecta* culture increased proportionally with nitrogen concentration (Fabregas et al. 1989).

Cells seem to recover rapidly from N starvation. In *Dunaliella tertiolecta* (Butcher) in the early stages of N starvation, cell division was maintained despite reductions in cellular chlrophyll. Following resupply of NO_3, cells rapidly re-established photosynthetic function and cells divided with in 24h (Young and Beardall 2003). *D. viridis* also seems to adapt to changes in osmotic pressure or nitrate concentration approximately in 24 h (Jiménez and Niell 2003).

Nitrogen stress is reflected in physiological and biochemical responses, particularly in carotene synthesis. Growth of *D. salina* under nitrogen sufficient conditions was 2.5 times higher than when the cells are stressed for N. Although chlorophyll a per cell did not differ, the stressed cells contained 7.76 carotenoid pg 1^{-1} while cells with sufficient nitrogen had 5.33 pg 1^{-1} (Sa´nchez-Estudillo et al. 2006). The ß-carotene content of *D. salina* N starved cells increased from 1.65 pg/cell to 7.05 pg/cell (Pisal and Lele 2005).

D. salina cultures raised on high NH_4 levels were photosynthetically more effective than those grown in nitrate (Giordano et al. 1994). Changes in photosynthesis and C metabolism, cell size of *Dunaliella salina* cultured at air-equilibrium CO_2 concentration, at 1.5 M NaCl, were strongly impacted by either 10 mM NO_3/- or 10 mM NH_4/+ as the sole N source (Giordano 1997). While NH_4/+ caused a 20% increase in the cell volume, chlorophyll, ß-carotene, and protein content were higher by 49%, 75%, and 104%, respectively, than for those grown in media with NO_3. The algae cultured on NH_4/+ in vivo Rubisco activity was about three times higher and concentrations of Phosphoenolpyruvate (PEP) carboxylase, PEP carboxykinase, and carbonic anhydrase activities were greater than in cells incubated in the presence of nitrate alone.

Species of *Dunaliella* assimilated organic sources of nitrogen such as hypoxanthine, allantoate, urea and histidine (Hellio and Le Gal 1998, 1999). *D.salina* grew more rapidly when supplied with a mix of nitrate and glutamine (Kosmakova and Prozumenshchikova (1983). In axenic cultures of *D. tertiolecta* cell growth in medium with histidine as the sole source of nitrogen was comparable to that observed with the ammonium ion (Hellio and Le Gal, 1998). ß-carotene in *D. tertiolecta* differed by 145% when treated with different sources of nitrogen. Maximum ß-carotene and vitamin C values were found in cultures with urea as the nitrogen source but vitamin E was at a minimum (Herrero et al. 1991 *b*). Studies by Fabregas et al. (1995 a, b) showed *D. tertiolecta* has a great capacity for changing its biochemical composition in response to changes in the nitrogen source. For example under non-nitrogen-limited conditions the C: N ratio stabilized around 5.2-5.3 and the protein content of the organic fraction around 70%.

In *D. salina* the activities of four key enzymes are related to sulphur deficiency; while Adenosine 5' triphosphate sulphurylase activity increased 4-fold, nitrate reductase and phosphoenolpyruvate (PEP) carboxylase activities decreased 4- and 11-fold respectively, and carbonic anhydrase activity remained unchanged. As *D. salina* seems to differentially regulate these enzymes it is able to adapt to prolonged sulphur limitation (Giordano et al. 2000). Introduction of a functional nitrate reductase gene into a *D. viridis* mutant that lacked the gene, complemented the nitrate reductase activity (Sun et al. 2006). These studies have positive implications while using this alga as a bioreactor for recombinant proteins and for the management applications of *Dunaliella* in biotechnological processes.

Iron

As in most micro algae iron is necessary for growth of *Dunaliella* and in hypersaline habitats it may be a limiting factor. Maximal growth rate in *D.salina* occurred in samples enriched with boron-glass powder containing 1.7% Fe_2SO_3, 6.1% FeO and 52.5 % B_2O_3 (Yamaoka et al. 1997). Mil'ko (1962) reported 1.25 to 3.75 mg l^{-1} as optimal for *D. salina* and *D. viridis*; and its deprivation in *D. salina* resulted in a decrease in cell volume and chlorophyll (Varsano et al. 2003). More recent studies (Varsano et al. 2006) on iron deprivation in *D. salina* have demonstrated an enlargement of the diameter of its Photosystem I from 22nm to 31-37nm, changes in photosynthetic activities, and induction of Tidi – a major 45-KDa chloroplast protein. In *D. bardawil* growth and chlorophyll levels increased with the availability of iron ranging between 0.01iM and 5 iM; the cells utilize Fe^{3+} via a reduction mechanism similar to that of strategy-I in higher plants.(Keshtacher-Liebson et al. 1999). In *D. tertiolecta* iron kinetic parameters are dependent on temperature (Gonzalez-Devila 1995).

In the halotolerant *D. salina*, cell specific sites for iron assimilation are present on the plasma membrane. This is analogous to the high-affinity and low-affinity sites for Cu^{2+} absorption (Gonzalez-Devila 1995) and may play an important role in iron dynamics as a well as CO_2 assimilation (Gokhman et al. 1999). Pick (2004) with *D. salina* demonstrated that Antimycin A forms lipophylic Fe-AA complexes with Fe (III) ions that enhances bioavailability of Fe in hypersaline solutions leading to better growth and chlorophyll synthesis. In this species iron deficiency induces transferrin-like protein (Ttf), a 150 –kDa protein and iron deficiency-induced protein (idi-100 –kDa protein on the outer surface of the plasma membrane (Scwartz et al. 2003). Although these resemble transferrins their threshold levels are different ; TTf requires a higher Fe than that of idi-100 which demonstrates its Fe binding and uptake. In *D. salina* iron uptake appears to be transcriptionally regulated by a150-kDa protein p150 (another stress protein or molecular chaperon) found on the plasma membrane (Fisher et al. 1997) but absorption of transferrins could be phagocytotic in *Dunaliella* sp. Mori (1999).

Metals

Although the various media for cultivation of *Dunaliella* contain Zn, Co, Cu, Mo and V their absolute requirement for growth has not been established. For *D. tertiolecta* manganese in the range 0.1 to 0.5ppm. is optimal while at lower levels growth was inhibited (Noro 1981). At >10 ppm. manganese was toxic.

A few studies that utilized *Dunaliella* cultures in toxicological assessments show their tolerance for heavy metals. *D. salina* seems to tolerate 5.94 μM copper and 4.55 μM cadmium compared with another chlorophyte *Chalamydomonas bullosa* that has a tolerance of 0.78 μM copper and 0.025 μM cadmium (Visviki and Rachlin 1994). Pre-treatment with Cd enhanced its tolerance by 50%. When perturbed with 0.34 μM cadmium and 7.57 μM copper, cell volume and relative lipid content increased by 3.49% and 170% respectively but the relative volume of their pyrenoid decreased by 41.46%(Visviki and Rachlin 1992).

D. salina isolated from the salt ponds, Chennai, India seem to be effective in treatment of salt refinery effluent (Raja et al. 2004). Besides reducing potassium by 77.5% and sodium by 72.5%, *D. salina* cultures removed 56.5, 46.0, 32.4 and 4.8% of heavy metals Ba, Al, Ag and St respectively.

$Zinc^{2+}$ was a more effective inducer of phytochelation (PC) synthesis in *D. tertiolecta* than Cd^{2+} (Tsuji et al. 2003). Pre-treatment of *D. tertiolecta* with Zn enhanced tolerance for toxic heavy metals such as Cd, Hg, Cu, Pb and As and mitigated oxidative stress caused by the presence of paraquat – a widely used herbicide (Tsuji et al. 2002). In D.tertiolecta rate of uptake of dissolved selenium was ~ 4% of selenium in a 4.5nM solution of lysate selenium, similar to that obtained with 4.5 selenite (Bains et al. 2001).Uptake of selenite by Dunaliella sp. from the medium over a broad range of concentrations (10-10-10-5mol dm-3) was higher than that of selenate; arsenic accumulation did not affect uptake of selenium(Yamaoka et al. 2004). *D. tertiolecta* accumulated >60 times Co-cobalamine with a retention half-time of 4.4 d ; with $COCl_2$ the retention half-time was 0.6 d and (Nolan et al. 1992). High radiation doses of chromium resulted in a decrease in lower mitotic activity and cell division in *D. bioculata* (Peneda Saraiva 1976).Although chromium stimulated cell division at low concentrations complete inhibition was observed at 3.5 μM chromium.

D. tertiolecta raised in conditions containing 2.03 μm of mercury II, detoxified the medium to 84% within the cell possibly by precipation of highly insoluble mercury compounds (Davies 1976); growth continued, but was uncoupled with division resulting in giant cells.

Acknowledgements

For critical evaluation and constructive criticism of this manuscript I am indebted to Dr. James E. Stewart, Bedford Inst. Oceanography. I thank Professor. Ami Ben-Amotz, The National Institute of Oceanography, Haifa, Israel, Dr. Ravi V. Durvasula, Medical School, University of New Mexico and Prof. J. Polle, City University of New York, New York for several helpful comments on the manuscript. I am grateful to Professor Ami Ben-Amotz, Prof. J. Polle and Dr. S.W. Jeffrey for photographs of *Dunaliella* cultures used in Plate 1. My thanks are due to Prof. Ilya Blum, Department of Mathematics, Mount Saint Vincent University, Halifax, Nova Scotia, Canada for advice on calculation of *Dunaliella* cell densities based on division rates, and size of inoculum. I thank Bala T. Durvasula for her patience with formatting the text and to Mr. Francis Kelly, Technographics Division, Bedford Institute for drafting the illustrations.

References

Aizawa, K. and S. Miyachi. 1984. Carbonic anhydrase located on the cell surface increases the affinity for inorganic photosynthesis of *Dunaliella tertiolecta*. FEBS Letters 173: 41-44.

Andersen, R.A. (ed). 2005. Algal culturing techniques Academic Press, Elsevier Academic Press. Amsterdam. The Netherlands.

Andersen, R.A., J.A. Berges, P.J. Harrison and M.M. Watanabe. 2005. Appendix A-Recipes for freshwater and seawater media. pp 429-538 In Algal Culturing Techniques. R.A. Andersen (ed) Phycological Society of America, Elsevier Academic Press. Amsterdam. The Netherlands.

Baines S.B., N.S. Fisher, M.A. Doblin and G.A. Cutter. 2001. Uptake of dissolved organic selenides by marine phytoplankton Limnology and Oceanography, 46: 1936-1944.

Baroli, I. and A. Melis. 1996. A.Photoinhibition and repair in *Dunaliella salina* acclimated to different growth irradiances. Planta, 198: 640-646.

Behrens, P.W. 2005. Photobioreactors and fermentors: The light and dark sides of growing algae. pp 189-204. In Algal Culturing Techniques. R.A. Andersen (ed) Elsevier Academic Press, Amsterdam, The Netherlands.

Ben-Amotz, A., and B. Z Ginzburg 1969. Light-induced proton uptake in whole cells of Dunaliella parva. Biochim Biophys Acta.183:144-154.

Ben-Amotz, A., A. Katz and M. Avron, 1982. Accumulation of ß-carotene in halotolerant algae: purification and characterization of ß-carotene –rich globules from *Dunaliella bardawil* (Chlorophyceae). J. Phycol. 18: 529-537.

Ben-Amotz, A., J. Gressel and M. Avron, 1987. Massive accumulation of phytoene induced by norflurazon in *Dunaliella bardawil* (Chlorophyceae) prevents recovery from photoinhibition. J. Phycol. 23: 176-181.

Bental M., U. Pick, M.Avron and H. Degani.1991. Polyphosphate metabolism in the alga *Dunaliella salina* studied by 31P-NMR. Biochimica et Biophysica Acta - Molecular Cell Research 1092: 21-28.

Berges, J.A., D.J. Franklin, and P.J. Harrison. 2001.Evolution of an wrtificial seawater medium: Improvements in enriched seawater, artificial water over the past two decades. J. Phycol.37: 1138-1145.

Berner, T., Z. Dubinsky, K.Wyman and P.G. Falkowski 1989. Photoadaptation and the "package effect" in *Dunaliella tertiolecta* (Chlorophyceae). J. Phycol. 25: 70-78.

Introduction to the algae. Prentice-Hall, Princeton, New Jersey, USA.

Borowitzka, L.J. 1981. The microflora. Adaptation to life in extremely saline lakes. Hydrobiologia 81:33-46.

Borowitzka, M.A. 2005. Culturing microalgae in outdoor ponds. pp 205-218. In Algal culturing techniques.R.A. Andersen (ed) Elsevier Academic Press, Amsterdam, The Netherlands.

Borowitzka, L.J., M.A. Borowitzka and T.P. Moulton. 1984. The mass culture of *Dunaliella salina* for fine chemicals: from laboratory to pilot plant. Hydrobiologia 116/117:115-134.

Borowitzka, M.A. and L.J. Borowitzka, 1988. *Dunaliella*. pp 27-58 In Micro-algal biotechnology. M.A. Borowitzka and L.J. Borowitzka (eds). Cambridge University Press.Cambridge, UK.

Carballo-Cárdenas, E.C., P.M. Tuan, M. Hassen and R.H. Wijffels. 2003. Vitamin E (α-tocopherol) products by the marine microalgae *Dunaliella tertiolecta* and *Tetraselmis suecica* in batch cultures. Biomolecular Engineering 20: 139-147.

Çeleki, A. and G. Dönmez. 2006. Effect of pH, light intensity, salt and nitrogen concentrations on growth and B- carotene accumulation by a new isolate of *Dunaliella* sp. World Journal of Microbiology & Biotechnology 22: 183-189.

Chang, T., S.Ohta, N. Ikegami, H. Miyata, T. Kashimoto and M. Kondo. 1993. Antibiotic substances produced by a marine green alga, *Dunaliella primolecta*. Bioresource Technology 44: 149-153.

Cifuentes, A.S., M.A. González and O. Parra. 1996. The effect of salinity on the growth and carotenogenesis in two Chilean strains of *Dunaliella salina* Teodoresco. Biol. Res. 29: 227-236.

Cifuentes, A.S., M.González, O.Parra and M. Zúñiga. 1996. Cultivo de cepas de *Dunaliella salina* (Teodoresco1905) en diferentes medios bajo condiciones de laboratorio. Rev. Chil. Hist. Nat. 69, 105–112.

Cifuentes, A.S., M.A. González, I. Inostroza and A. Aguilera. 2001. Reappraisal of physiological attributes of nine strains of *Dunaliella* (Chlorophyceae): Growth and pigment content across a salinity gradient. J. Phycol. 37: 334-344.

Davies, A.G. 1976. An assessment of the basis of mercury tolerance in *Dunaliella tertiolecta*. J. Mar. Biol. Ass. U.K. 56: 39-57.

De Lonzeno, M.E., M. Leatherbury, J.A. Weiner, A.J. Lewitus and M.H. Fulton. 2004. Physiological factors contributing to the species-specific sensitivity of four estuarine microalgal species exposed to herbicide atrazine. Aquatic Ecosystems Health & Management 7: 137-146.

De Pauw, N., J. Morales and G. Persoone. 1984. Mass culture of microalage in aquaculture systems: Progress and constraints. Hydrobiologia,116/117, 121-134.

Einspahr K.J., T.C. Peeler and G.A. Thompson Jr. 1988. Rapid changes in polyphosphoinositide metabolism associated with the response of *Dunaliella salina* to hypoosmotic shock. Journal of Biological Chemistry, 263: 5775-5779.

Eppley, R.W. and P.R. Sloan. 1965. Carbon balance experiments with marine phytoplankton. J.Fish. Res. Bd. Canada 2, 1083-1097.

Eppley, R.W. and J.L. Coatsworth 1966. Culture of the marine phytoplankter *Dunaliella tertiolecta* with light-dark cycles. Arch. Mikrobiol. 55: 66-80.

Fábregas, J., J.Abalde, B.Cabezas and C. Herrero. 1989. Changes in protein, carbohydrates and gross energy in the marine microalga *Dunaliella tertiolecta* (Butcher) by nitrogen concentrations as nitrate, nitrite and urea. Aquacultural Engineering. 8: 223-239.

Fábregas, J, L. Ferron, Y. Gamallo, E. Vecino, A.Otero and C. Herrero. 1994. Improvement of growth rate and cell productivity by aeration rate in cultures of the marine microalga *Dunaliella tertiolecta*. Biosource Technology 48: 107-111.

Fábregas, J., L. Ferron, J. Abalde, B. Cabezas and A. Otero.1995a and b. Changes in the gross chemical composition of mass cultures of the marine microalga *Dunaliella tertiolecta* with different aeration rates. Bioresource Technology. 53: 185-188.

Fábregas, J., M. Patino, B.O. Arredondo-Vega, J.L. Tobar and A. Otero, A. 1995b. Renewal rate and nutrient concentration as tools to modify productivity and biochemical composition of cyclostat cultures of the marine microalga *Dunaliella tertiolecta*. Appl. Microbiol. Biotechnol. 44: 287-292.

Fazeli, M.R., H. Togighi, N. Samadi, H. Jamalifar and A. Fazeli. 2006. Carotenoids accumulation by *Dunaliella tertiolecta* (Lake Urmia isolate) and *Dunaliella salina* (CCAP 19/18 & WT) under stress conditions. Daru 14:146-150.

Fisher, M., U.Pick and A. Zamir, 1994. A salt-induced 60-Kilodalton Plasma Membrane Protein plays a potential role in the extreme halotolerance of the alga *Dunaliella*. Plant Physiology 106: 1359-1365.

Fisher, M., I. Gokhman, U. Pick and A. Zamir, 1997. A structurally novel tranmsferrin-like protein accumulates in the plasma membrane of the unicellular green alga *Dunaliella salina* grown in high salinities. Journal of Biological Chemistry 272: 1565-1570

Fuggi, A., G.Pinto, A. Pollio and R. Taddei, 1988. Effects of NaCl, Na_2SO_4, H_2SO_4 and glucose on growth, photosynthesis, and respiration in the acidophilic alga *Dunaliella acidophila* (Volvocales, Chlorophyta). Phycologia 27; 334-339.

Ghetti, F., H. Herrmann, D.-P. Hader and H.K. Seidlitz. 1999.Spectral dependence of the inhibition of photosynthesis under simulated global radiation in the unicellular green alga *Dunaliella salina*. Journal of Photochemistry and Photobiology B: Biology 48:166-173.

Ghosal, D., D. Mach, M. Agarwal, A. Goyal and A. Goyal. 2002. Osmoregulatory isoform of dihydroxyacetone phosphate reductase from *Dunaliella tertiolecta*: Purification and characterization. Protein Expression and Purification. 24: 404-411.

Ghoshal, D. and A. Goyal. 2001. Oxygen inhibition of dissolved inorganic carbon uptake in unicellular green algae. Phycological Research 49: 319-324.

Giardano, M. 1997. Adaptation of *Dunaliella salina* (Volvocales, Chlorophyceae) to growth on NH4+ as the sole nitrogen source. Phycologia 36: 345-350.

Giardano, M., V. Pezzoni and R. Hell, 2000. Strategies for the allocation of resources under sulfur limitation in the green alga *Dunaliella salina*. Plant Physiology 124: 857-864.

Gibor, A. 1956. The culture of brine algae. Biol. Bull. 111: 223-229.

Ginzburg, M. and B.Z. Ginzburg. 1981. Interrelationships of light, temperature, sodium chloride and carbon source in growth of halotolerant and halophilic strains of *Dunaliella*. British Phycological Journal. 16: 313-324.

Ginzburg, M. and B.Z. Ginzburg, 1993. Effect of light intensity on adaptation of Dunaliella to very high salt concentrations.1993. J.Exp. Bot.44: 1561-1567.

Giordano M. 1997. Adaption of *Dunaliella* salina (Volvocales, Chlorophyceae) to growth on NH4+ as the sole nitrogen source. Phycologia 36: 345-350.

Giordano M. and G. Bowes. 1997. Gas exchange and C allocation in *Dunaliella salina* cells in response to the N source and CO_2 concentration used for growth. Plant Physiology 115: 1049-1056

Giordano M., J.S.Davies and G. Bowes. 1994. Organic carbon release by *Dunaliella salina* (Chlorophyta) under different growth conditions of CO2, nitrogen, and salinity. Journal of Phycology 30: 249-257.

Gladue, R.M. and J.E. Maxey. 1994. Microalgal feeds for aquaculture. J. Appl. Phycol. 6: 131-141.

Gokhman, I., M. Fisher, U.Pick and A. Zamir. 1999. New insights into extreme salt tolerance of the unicellular green alga *Dunaliella*. pp 203-213. In Microbiology and Biogeochemistry of Hypersaline Environments. A.Oren (ed) CRC Press Boca Raton. USA.

Goldman, J.C. and D.G. and Peavey. 1979. Steady-state growth and chemical composition of the marine chlorophyte *Dunaliella tertiolecta* in nitrogen-limited continuous cultures. Applied and Environmental Microbiology 38: 894-901.

Goldman, J.C. and P.G. Brewer. 1980. Effect of nitrogen source and growth rate on phytoplankton-mediated changes in alkalinity. Limnol. Oceanogr. 25: 352-357.

Goldman, J.C., J.C. McCarthy, and D.G. Peavy, 1979. Growth rate influence on the chemical composition of phytoplankton in oceanic waters. Nature 279: 210-215.

Gómez, P.I., A. Barriga, A.S. Cifuentes and M.A. Gonzalez. 2003. Effect of salinity on the quantity and quality of carotenoids accumulated by *Dunaliella salina* (strain CONC-007) and *Dunaliella bardawil* (Strain ATCC 30861) chlorophyta. Biological Research 36: 185-192.

Gómez-Pinchetti, J.L., Z. Ramazanov, A. Fontes and G. Garcia-Reina. 1992. Photosynthetic characteristics of *Dunaliella salina* (Chlorophyceae, Dunaliellaes) in relation to B-carotene content. Jour. Applied Physiol. 4: 11-15.

Gonzalez-Davila, M. J. M. Santana-Casiano and J.P. Pena. 1995. Binding of Cu (II) to the surface and exudates of the alga *Dunaliella tertiolecta* in seawater. Environmental Science and Technology 29: 289-301.

Goyal, A. and H. Gimmler. 1989. Osmoregulation in *Dunaliella tertiolecta*: effects of salt stress, and the external pH on the internal pH. Arch Microbiol. 152: 138-142.

Goyal, A., Y. Shiraiwa, H. David Husic and N.E. Tolbert. 1992. External and internal carbonic anahydrases in *Dunaliella* species. Marine Biology 113: 349-355.

Grant, B.R. 1967. The action of light on nitrate and nitrite assimilation by the marine chlorophyte, *Dunaliella tertiolecta* (Butcher). Journal of General Microbiology 48: 379-389.

Grant, B.R. 1968. The effect of carbon dioxide concentration and buffer system on nitrate and nitrite assimilation by *Dunaliella tertiolecta.* Journal of General Microbiology 54: 327-336.

Graziano, L.M., J. La Roche and R.J. Geider. 1996. Physiological responses to phosphorus limitation in batch and steady-state cultures of *Dunaliella tertiolecta* (Chlorophyta): A unique stress protein as an indicator of phosphorus deficiency. J. Phycol. 32: 825-838.

Grobbelaar, J.A. 2003. Quality control and assurance: crucial for the sustainability of the applied phycology industry. Journal of Applied Phycology 15: 209-215.

Guillard, R.R.L. 1975. Culture of phytoplankton for feeding marine invertebrates. pp 29-60 In Culture of marine invertebrate animals W.L. Smith and M.H. Chanley (eds) Plenum Publishing Corporation, New York.USA

Hard, B.C. and D.J. Gilmour.1996. The uptake of organic compounds by *Dunaliella parva* CCAP 19/9. Eu.J. Phycol. 31: 217-224.

Harrison, P.J., R.E. Waters and F.J.R.Taylor. 1980. A broad spectrum artificial seawater medium for coastal and open ocean phytoplankton. J. Phycol. 16: 28-35.

Hellio, C. and Y. Le Gal. 1998.Histidine utilization by the unicellular alga *Dunaliella tertiolecta.* Comparative Biochemistry and Physiology - A Molecular and Integrative Physiology 119: 753-758

Hellio C. and Y. Le Gal. 1999. Histidase from the unicellular green alga *Dunaliella tertiolecta*: purification and partial characterization. Eur. J. Phycol. 34: 71-78.

Helliot, B. and A. Mortain-Bertrand. 1999. Accumulation of praline in *Dunaliella salina* (Chlorophyceae) in response to light transition and cold adaptation. Cryo-Letters 20: 287-296.

Henley, W.J., K.M. Major and J.L. Hironaka. 2002. Response to salinity and heat stress in two halotolerant chlorophyte algae. Journal of Phycology 38: 757-766.

Heraud, P., S. Roberts, K. Shelly and J. Beardall. 2005. Interactions between UV-B exposure and phosphorus nutrition. II. Effects on rates of damage and repair. J. Phycology 41: 12121-1218.

Herrero, C., A. Cid, J.Fábregas and J. Abalde. 1991a. Yields in biomass and chemical constituents of four commercially important marine microalgae with different culture media. Aquaculture Engineering 10: 99-110.

Herrero, C., J. Abalde and J. Fábregas.1991b. ß-carotene, vitamin C and vitamin E content of the marine microalga *Dunaliella tertiolecta* cultured with different nitrogen sources. Bioresource Technology, 38: 121-125.

Hermann, H; D-P.Haeder, M. Koefferlein, H.K.Seidlitz and F.Ghetti. 1996. Effects of UV radiation on photosynthesis of phytoplankton exposed to solar simulator light. J. Photochem. Photobiol.B: Biol. 34: 21-28.

Hezari, M.A., E. Holwerda and R.H. Wijffels. 2004. Milking microalga *Dunaliella salina* for ß-Carotene Production in Two-Phase Bioreactors. Biotechnology and Bioengineering, 85 (5): 475-481.

Hook, I.L., S. Ryan and H. Sheridan. 1999. Biotransformation of aromatic aldehydes by five species of marine microalgae. Phytochemistry. 51: 621-627.

Humphrey, G.F. 1974. Effects of carbon dioxide and phosphate supplied during growth, on phosphorus content and photosynthetic rates of some unicellular marine algae. J. Mar. Biol. Ass. India. 16: 358-366.

Humphrey, G.F. 1975. The photosynthesis: respiration ratio of some unicellular marine algae. J. Exp. Mar. Biol. Ecol. 18: 111-119.

Humphrey, G.F. 1979. Photosynthetic characteristics of algae grown under constant illumination and light-dark regimes. J. Exp.Mar. Biol. Ecol. 40: 63-70.

Jahnke, L.S. 1999.Massive carotenoid accumulation in *Dunaliella bardawil* induced by ultraviolet-A radiation. Journal of Photochemistry and Photobiology B: Biology 48: 68-74

Jahnke, L.S. and A.L. White. 2003. Long-term hyposaline and hypersaline stresses produce distinct antioxidant responses in the marine alga *Dunaliella tertiolecta*. Journal of Plant Physiology. 160: 1193-1202.

Janssen, M., L. De Bresser, T. Baijens, J. Tramper, L.R. Mur, J.F.H. Snel and R.H. Wijffels. 2000. Scale-up aspects of photobioreactors: Effects of mixing-induced light/dark cycles. Journal of Applied Phycology, 12: 225-237.

Janssen, M., P. Slenders, J.Tramper, L.R Mur and R.H. Wijffels. 2001.Photosynthetic efficiency of *Dunaliella tertiolecta* under short light/dark cycles. Enzyme and Microbial Technology 29: 298-305.

Jiménez, C. and F.X. Niell. 1991a. Influence of temperature and salinity on carbon and nitrogen content in *Dunaliella viridis* Teodoresco under nitrogen sufficiency. Bioresource Technology 38: 91-94.

Jiménez, C. and F.X. Niell. 1991b. Growth of *Dunaliella viridis* Teodoresco: effect of salinity, temperature and nitrogen concentration. J. Appl. Phycol. 3: 319-327.

Jiménez, C., and F.X. Niell. 2003.Influence of high salinity and nitrogen limitation on package effect and C/N ratio in *Dunaliella viridis*. Hydrobiologia 492: 201-206.

Jiménez, C., F.X. Niell and J.A. Fernandez. 1990. The photosynthesis of *Dunaliella parva* Lerche as a function of temperature, light and salinity. Hydrobiologia 197: 165-172.

Jitts, H.R., C.D. McAllister, K. Stephens and J.D.H. Strickland. 1964. The cell division rates of some marine phytoplankters as a functionof light and temperature. J. Fish. Res. Bd. Canada 21: 139-157.

Johnson, M.K., E.J. Johnson, R.D. McElroy, H.L. Speer and B.S. Bruff. 1968. Effects of salts on the halophilic alga *Dunaliella viridis*. Journal of Bacteriology 95: 1461-1468.

Jones T.W. and R.A. Galloway. 1979. Effect of light quality and intensity on glycerol content in *Dunaliella tertiolecta* (Chlorophyceae) and the relationship to cell growth/ osmoregulation. J. Phycol. 15: 101-106.

Keshtacher-Liebson, E., Y. Hadar and Y. Chen. 1999. Fe nutrition demand and utilization by the green alga *Dunaliella bardawil*. Plant and Soil 215: 175-182.

Klyachko-Gurvicha, G.L., L. N. Tsoglina, J.Douchab, J.Kopetskiib, I.B. Shebalina, and V.E. Semenenkoa. 1997. Desaturation of fatty acids as an adaptive response to shifts in light intensity Physiologia Plantarum 107: 240-249.

Kosmakova, V.E., and L.T. Prozumenshchikova. 1983. Growth and biochemical composition of the algae *Dunaliella salina* and *Platimonas viridis* fed on organic and inorganic forms of nitrogen. Soviet Journal of Marine Biology, 9 (1): 38-41.

Krol, M., D.P. Maxwell, and N.P.A. Huner. 1997. Exposure of *Dunaliella salina* to low temperature mimics the high light-induced accumulation of carotenoids and the carotenoid binding protein (Cbr). Plant and Cell Physiology 38: 213-216.

Kviderova, J. and W.J. Henley. 2005. The effect of ampicillin plus streptomycin on growth and photosynthesis of two halotolerant chlorophyte algae. J. Appl. Phycol.17: 301-307.

La Roche, J., R.J. Feider, L.M. Graziano, H. Murray and K. Lewis. 1993. Induction of specific proteins in eukaryotic algae grown under iron, phosphorus, or nitrogen-deficient conditions. J. Phycol. 29: 767-777.

Labbé, A. 1925. Contributions a l'etude de l'allelogenese 1.Memoire Les cycles biologiques des *Dunaliella*. Arch.d'Anat.Microsc 21: 314-399.

Lakeman, M.B. and R.A. Cattolico. 2007. Cryptic diversity in phytoplankton cultures is revealed using a simple plating technique. J. Phycol. 43: 662-674.

Latorella, A.H. and R.L. Vadas. 1973. Salinity adaptation by *Dunaliella tertiolecta*. I. Increases in carbonic anhydrase activity and evidence for light-dependent Na+/H+ exchange. Journal of Phycology 9: 273-277.

Latorella, A.H., S.K. Bromberg, K. Lieber and J. Robinson. 1981. Isolation and partial characterization of nitrate assimilation mutants of *Dunaliella tertiolecta* (Chlorophyceae). Journal of Phycology 17: 211-214.

Laws, E.A. and J. Caperon. 1976. Carbon and nitrogen metabolism by *Monochrysis lutheri*: Measurement of growth-rate-dependent respiration rates. Mar. Biol. 36: 85-97.

Laws, E.A. and D.C.L. Wong. 1978. Studies of carbon and nitrogen metabolism by three marine phytoplankton species in nitrate limited continuous culture. J. Phycol. 14: 406-416.

Lerche, W. 1937. Untersuchungen über Entwicklung und Fortpflanzung in Gattung *Dunaliella*. Arch.f. Protistenkd. 88: 236-268.

Li Q., X. Gao, Y. Sun, Q.Zhang, R.Song andZ. Xu Z. 2006. Isolation and characterization of a sodium-dependent phosphate transporter gene in *Dunaliella viridis*. Biochemical and Biophysical Research Communications 340: 95-104.

Lin S., C. J.Gobler and E.J. Carpenter. 2001.Cytological and biochemical responses of *Dunaliella tertiolecta* (Volvocales, Chlorophyta) to iron stress. Phycologia, 40(5) : 403-410.

Lin, S.L., J. Chang and E.J. Carpenter. 1995. Growth characteristics of phytoplankton determined by cell cycle proteins: PCNA immunostaining of *Dunaliella tertiolecta* (Chlorophyceae). J. Phycol. 31: 388-395.

Loeblich, A. 1975. A seawater medium for dinoflagellates and the nutrition of Cachonina niei. J. Phycol. 11: 80-86.

Madgwick, J.C. 1966. Chromatographic determination of chlorophylls in algal cultures and phytoplankton. Deep-Sea Research 13: 459-466.

Marín, N., F. Morales, C. Lodeiros and E. Tamigneaux. 1998. Effect of nitrate concentration on growth and pigment synthesis of *Dunaliella salina* cultivated under low illumination and preadapted to different salinities. Journal of Applied Phycology 10: 405-411.

Marques,A., Thi Dinh, C. Ioakeimidis, G. Huys, J. Swings, W.Verstraete, J. Dhont, P. Sorgeloos, and P. Bossier. 2005. Effects of Bacteria on *Artemia franciscana* Cultured in Different Gnotobiotic Environments. Applied and Environmental Microbiology 71: 4307-4317.

Massyuk, N.P. 1973. Morphology, Taxonomy, Ecology and Geographical distribution of the Genus *Dunaliella* Teod. and Prospects for its Potential Utilization. Akad.Sci. Ukraine. Institute of Botanics Naukova Dumka, Kiev.

McLachlan, J. 1960.The culture of *Dunaliella tertiolecta* Butcher – a Euryhaline organism, Can. J. Microbiol. 6: 367-379.

McLachlan, J. 1964. Some considerations of the growth of marine algae in artificial media. Can. J. Microbiol. 10: 769-782.

McLachlan, J. 1973. Growth media – marine. pp 25-51 In Handbook of Phycological Methods: Culture methods and Growth Measurements.J. Stein (ed) Cambridge University Press, Cambridge.UK

McLeroy-Theridge, S.L. and G.B. McManus. 1999. Food type and concentration affect chlorophyll and carotenoid destruction during copepod feeding. Limnology and Oceanography 44: 2005-2011.

Mil'ko, E.S. 1962. Study of the requirements of two *Dunaliella* spp. in mineral and organic components of the medium. Moscow University Vestnik, Biologiya 6: 21-23.

Montoya, T. and G.A. Olivera, 1993. *Dunaliella salina* from saline environments of the central coast of Peru. Hydrobiologia 267: 155-161.

Mori, S. 1999. Iron acquisition by plants. Current opinion in Plant Physiology 2: 250-253.

Muller-Feuga, A., J. Moal and R. Kaas. 2003. The microalgae of aquaculture. pp206-252 In Live Feeds in Marine Aquaculture StØttrup, J.G. and L.A. McEvoy (eds) Blackwell Science.

Muradyan, E.A., G. L. Klyachko-Gurvich, L.N. Tsonglin, T.V. Sergeyenko and N.A. Pronina. 2004. Changes in lipid metabolism during adaptation of the *Dunaliella salina* photosynthetic apparatus to high CO_2 concentration. Russian Journal of Plant Physiology 51: 53-62.

Murphy C.D., R.M. Moore and R.L. White .2000. An isotopic labeling method for determining production of volatile organohalogens by marine microalgae. Limnology and Oceanography 45: 1868-1871.

Nagase, H., K.I.Yoshihara, K. Eguchi, Y. Okamoto, S. Murasaki, R. Yamashita, K. Hirata and K. Miyamoto. 2001. Uptake pathway and continuous removal of nitric oxide from flue gas using microalgae. Biochemical Engineering Journal, 7 (3): 241-246.

Nelson, D. and L.E. Brand. 1979. Cell division periodicity in 13 species of marine phytoplankton on light-dark cycle. J. Phycol. 15: 67-75.

Nolan C.V., S.W. Fowler and J.L. Teyssie. 1992. Cobalt speciation and bioavailability in marine organisms . Marine Ecology Progress Series, 88 (2-3): 105-116.

Noro,T. 1981. Effect of Mn on the growth of a marine green alga, *Dunaliella tertiolecta.* Japanese Journal of Phycology 26: 69-72.

Ohta, S., F.Ono, Y. Shiomi, T. Nakao, O. Aozasa, T. Nagata, K. Kittamura, S. Yamaguchi, M. Nishi and H. Miyata. 1998. Anti-herpes simplex virus substances produced by the marine green alga, *Dunaliella primolecta.* Jour. Appl. Phycology 10: 349-355.

Okauchi, M. and K. Kawamura. 1997. Optimum medium for large-scale cultures of *Tetraselmis tetrathele.* Hydrobiologia 358: 217-222.

Oren, A. 2005. A hundred years of *Dunaliella* research: 1905-2005. Saline Systems 1: 2; doi:10.1186/1746-1448-1-2.

Oren, A. 1999. The rise and decline of a bloom of halophilic algae and Archaea in the Dead Sea:1992-1995. pp 129-138. In Microbiology and Biogeochemistry of Hypersaline Environments. A.Oren (ed) CRC Press Boca Raton.USA.

Orset, S and A. J. Young, 1999. Low temperature-induced synthesis of á-carotene in the microalga *Dunaliella salina* (Chlorophyta). Journal of Phycology 35: 520-527.

Osinga, R., E. P. Muela, J. Tramper and R.H. Wijffels. 1998. In Vitro cultivation of four marine sponge species, determination of the nutritional demands. pp 121-127. In Marine Microorganisms for industry. . Y.Le Gal and A.Muller-Feuga (eds). Proceedings of the meeting held in Brest-France on the 17-19th of September 1997, IFREMER.

Peneda Saraiva, M.C. 1976.Utilization of a nanoplanktonic alga as organism test in marine molysmology. Some reactions of *Dunaliella bioculata* to gamma irradiation and contamination with chromium and cadmium. Revue Internationale d'Oceanographie Medicale.43: 111-115.

Phadwal K. and P.K. Singh. 2003. Effect of nutrient depletion on ß-carotene and glycerol accumulation in two strains of *Dunaliella* sp. Bioresource Technology 90: 55-58.

Pick, U. 2004. The respiratory inhibitor antimycin A specifically binds Fe (III) ions and mediates utilization of iron by the halotolerant alga *Dunaliella salina* (Chlorophyta). Biometals 17: 79-86.

Pick, U., L. Karni and M. Avron.1986. Determination of ion content and ion fluxes in the halotolerant alga *Dunaliella salina*. Plant Physiol. 81: 92–96.

Pick, U., M. Bental, E. Chitlaru and M.Weiss.1990. Polyphosphate-hydrolysis - A protective mechanism against alkaline stress? FEBS Letters, 274: 15-18.

Pierson, M.D. and D.A. Corlett Jr. 1992. HACCP Principles and applications. Van Nostrand Reinhold, New York, USA.

Pisal, D.S.and S.S. Lele. 2005.Carotenoid production from microalga, *Dunaliella salina*. Indian Journal of Biotechnology 4: 476-483.

Polle, J.E.W., S. Kanakagiri, E.Jin, T.Masuda, and A.Melis. 2002. Truncated chlorophyll antenna size of the photosystems - A practical method to improve microalgal productivity and hydrogen production in mass culture International Journal of Hydrogen Energy 27:1257-1264.

Preisig, H.R. 1992. Morphology and taxonomy. pp. 1-15. In *Dunaliella*: Physiology, Biochemistry, and Biotechnology. M. Avron and A. Ben-Amotz(eds) CRC Press.Boca Raton, USA.

Premkumar, L., U.K. Bageshwar, I. Gokhman, A. Zamir and J.L. Sussman, 2003. An unusual halotolerant á -type carbonic anhydrase from the alga *Dunliella salina* functionally expressed in Escheria coli. Protein Expression & Purification 28: 151-157.

Puncocharova, M. 1994. A taxonomic study of two *Chlorella* strains Archiv fur Protistenkunde 144 (3): 237-247.

Raja R., C. Anbazhagan , V.Ganesan and R. Rengasamy. 2004.Efficacy of *Dunaliella salina* (volvocales, chlorophyta) in salt refinery effluent treatment Asian Journal of Chemistry 16: 1081-1088.

Riesing T.F. 2005. Cultivating algae for liquid fuel production. http://oakhavenpc.org/cultivating_algae.htm

Riisgård,H.U., K.NØrgård Nielsen and B. SØgaard-Jensen. 1980. Further studies on the volume regulation and effects of copper in relation to pH and EDTA in the naked marine flagellate *Dunaliella marina*. Marine Biology 56: 267-276.

Roche, J.L., R.J. Geider, L. M. Graziano and K. Lewis. 1993. Induction of specific proteins in eukaryotic algae grown under iron-phosphorus-, or nitrogen-deficient conditions. Journal of Phycology, 29: 767-777.

Ronquillo, J.D., J.R. Matias, T. Saisho and S. Yamasaki. 1997. Culture of *Tetrasemis tetratele* and its utilization in the hatchery production of different penaeid shrimps in Asia. Hydrobiologia 358: 237-244.

Sa´nchez-Estudillo, L., Y. Freile-Pelegrin, R. Rivera-Madrid, D. Robledo, and J.A. Narva´ez-Zapata. 2006. Regulation of two photosynthetic pigment-related genes during stress-induced pigment formation in the green alga, *Dunaliella salina.* Biotechnology Letters 28: 787-791.

Sanchez-Saavedra, M.P., C. Jiménez and F.L. Figueroa. 1996. Far-red light inhibits growth but promotes carotenoid accumulation in the green microalga *Dunaliella bardawil.* Physiologia Plantarum 98: 419-423.

Schwarz, M., N. Sal-Man, A. Zamir and U. Pick. 2003. A transferring-like protein that does not bind iron is induced by iron deficiency in the alga *Dunaliella salina.* Biocimica et Biophysica Acta- Proteins and Proteomics 1649: 190-200.

Sciandra A., J. Gostan, Y. Collos, C. Descolas-Gros, C. Leboulanger, V.Martin-Jézéquel, M. Denis, D. Lefèvre, C. Copin-Montégut and B. Avril. 1997. Growth-compensating phenomena in continuous cultures of *Dunaliella tertiolecta* limited simultaneously by light and nitrate. Limnol. Oceanogr. 42: 1325-1339.

Sharati, M. 2003. Characterization of three species of *Dunaliella salina, Dunaliella parva* and *Dunaliella psuedosalina* isolated from salt marsh of Gave khoni of Isfahan-Iran. Iranian Journal of Science and Technology, Transaction B: Technology 27: 185-190.

Shelly K., S. Roberts, P. Heraud and J. Beardall J. 2005. Interactions between UV-B exposure and phosphorus nutrition. I. Effects on growth, phosphate uptake, and chlorophyll fluorescence. Journal of Phycology, 41: 1204-1211.

Siron, R., G. Giutsi and B. Berland. 1989. Changes in the fatty acid composition of *Phaeodactylum tricornutum* and *Dunaliella tertiolecta* during growth and under phosphorus deficeiency. Mar. Ecol. Prog. Ser. 55: 95-100.

Sosik, H.M. and B.G. Mitchell, 1994. Effects of temperature on growth, light absorption, and quantum yield in *Dunaliella tertiolecta* (Chlorophyceae) J. Phycol. 30: 833-840.

Starr, R.C. and Zeikus, J.A. 1993. UTEX: the culture collection of algae at the University of Texas at Austin. J. Phycol. 29: (Supplment) 1-106.

StØttrup, J.G. and L.A. McEvoy, 2003. Live feed in Marine Aquaculture. Blackwell Science, Oxford, UK..

Stramski, D; G. Rosenberg and L. Legendre. 1993. Photosynthetic and optical properties of the marine chlorophyte *Dunaliella tertiolecta* grown under fluctuating light caused by surface-wave focusing. Mar. Biol. 115: 363-372.

Subba Rao, D.V. 1981. Growth response of marine phytoplankters to selected concentrations of trace metals. Botanica Marina 25: 369-379.

Subba Rao, D.V. 2002. (Ed) Pelagic Ecology Methodology. Swets & Zeitlinger and Balkema Publishers, Rotterdam, The Netherlands.

Subba Rao, D.V. 2006. (ed) Algal Cultures, Analogues of Blooms and Applications. Vols. 1 and 2. Science Publishers, New Hampshire, USA.

Sukenik, A., J. Bennett, A. Mortain-Bertrand and P.G. Falkowski. 1990. Adaptation of the photosynthetic apparatus to irradiance in *Dunaliella tertiolecta.* A kinetic study. Plant Physiology 92: 891-898.

Sun, Y., X.Gao, Q. Li, Q.Zhang and Z. Xu. 2006. Functional complementation of a nitrate reductase defective mutant of a green alga *Dunaliella viridis* by introducing the nitrate reductase gene. Gene 377: 140-149.

Suzuki T., M.Imai, and M. Ishida.1997. Influence of environmental factors on phosphorous content in the marine microalga *Dunaliella.*Toxicological and Environmental Chemistry 64: 173-181.

Teodoresco, E.C. 1906. Observations morphologiques et biologiquessur le genre *Dunaliella.* Rev. Gén. d. Bot. T18: 353-371.

Thakur, A., and H.D. Kumar. 1999. Effect of different nitrogen sources on growth and glycerol production by *Dunaliella salina.* Cytobios Vol. 97: 79-83.

Thompson, P.A., M. Guo and P.J. Harrison. 1992. Effects of variation in temperature. I. On the biochemical composition of eight species of marine phytoplankton. J. Phycol. 28: 481-488.

Torrentera, L., R.M. Uribe and R.E. Carrillo. 1994. Physical and biological characterization of a seawater ultraviolet radiation sterilizer. Radiation 43: 249-255.

Tsuji N., N.Hirayanagi, O. Iwabe, T. Namba, M.Tagawa, S.Miyamoto, H. Miyasaka, M.Takagi K. Hirata and K. Miyamoto. 2003. Regulation of phytochelatin synthesis by zinc and cadmium in marine green alga, *Dunaliella tertiolecta.* Phytochemistry 62: 453-459.

Tsuji N., N. Hirayanagi, M.Okada, H. Miyasaka, K.Hirata, M.H. Zenk M.H. and K. Miyamoto. 2002. Enhancement of tolerance to heavy metals and oxidative stress in *Dunaliella tertiolecta* by Zn-induced phytochelatin synthesis. Biochemical and Biophysical Research Communications 293: 653-659.

Ukeles, R. 1961. The effect of temperature on the growth and survival of several marine algal species. Biol. Bull. 120: 255-264.

Ukeles, R. 1965. A simple method for the mass culture of marine algae. Limnol. Oceanogr. 10: 492-495.

Uriarte, I., A. Farias, A.J.S. Hawkins and B.L. Bayne. 1993. Cell characteristics and biochemical composition of *Dunaliella primolecta* Butcher conditioned at different concentrations of dissolved nitrogen. Jour. Appl. Phycology 5: 447-453.

Van Auken, O.W. and I.B. McNulty. 1973. The effect of environmental factors on the growth of a halophilic species of algae. Biological Bulletin, Woods Hole 145: 210-222.

Varsano, T., D. Kaftan and U.Pick. 2003. Effects of iron deficiency on thalakoid membrane structure and composition in the alga *Dunaliella salina.* Journal of Plant Nutrition. 26: 2197-2210.

Varsano, T., S.G. Wolf and U.Pick. 2006. A chlorophyll a/b-binding protein homolog that is induced by iron deficiency is associated with enlarged photosystem I units in the eucaryotic alga *Dunaliella.* Journal of Biological Chemistry 281: 10305-10315.

Visviki I. and J.W. Rachlin.1992. Ultrastructural changes in *Dunaliella minuta* following acute and chronic exposure to copper and cadmium. Archives of Environmental Contamination and Toxicology 23 (4): 420-425.

Visviki I. and J.W.Rachlin.1994. Acute and chronic exposure of *Dunaliella salina* and *Chlamydomonas bullosa* to copper and cadmium: Effects on growth. Archives of Environmental Contamination and Toxicology 26: 149-153.

Vorst, P. 1995. Production of carotene with chemostat cultures of *Dunaliella.* Ph.D thesis, Research School BioCentrum Amsterdam, Institute for Molecular Cell Biology, section Plant Physiology, University of Amsterdam, The Netherlands.

Wada, M. A. Hirokawa and R.Fukumoto. 1996. Analysis of the expression of gene for P-type ATPase in the halotolerant alga *Dunaliella salina* (Chlorophyceae). J. Phycol. 32: 868-872.

Wegmann, K., A. Ben-Amotz and M.Avron. 1980. Effect of temperature on glycerol retention in the halotolerant algae *Dunaliella* and *Astromonas*. Plant Physiolgy 66: 1196-1197.

Weiss M., G. Haimovich and U. Pick. 2001. Phosphate and sulfate uptake in the halotolerant alga *Dunaliella* are driven by Na+-symport mechanism. Journal of Plant Physiology 158: 1519-1525.

Wynne, D. and G. -Y. Rhee. 1988. Changes in alkaline phosphatase activity and phosphate uptake in P-limited phytoplankton, induced by light intensity and spectral quality. Hydrobiologia 160:173-178.

Xu, X.-Q. and J. Beardall. 1997. Effect of salinity on fatty acid composition of a green microalga from an Antarctic hypersaline lake. Phytochemistry 45: 655-658.

Yamaoka,Y., O.Takimura, H. Fuse, and K. Kamimura. 2004. Accumulation of arsenic and sselenium by *Dunaliella sp.* Applied Organometallic Chemistry 4; 261-264.

Yamaoka, Y., O. Takimura, H. Fuse, K. Murakami, K. Kamimura, S. Ueda and M. Saeki. 1997. B- Carotene production by *Dunaliella salina* using boron-glass powder containing iron- Note. Seibutsu-kogaku Kaishi 75: 181-184.

Young, E.B. and J. Beardall. J.2003. Photosynthetic function in *Dunaliella tertiolecta* (Chlorophyta) during a nitrogen starvation and recovery cycle. Journal of Phycology 39: 897-905.

Zenvirth, D. and A. Kaplan, 1981. Uptake and efflux of inorganic carbon in *Dunaliella salina*. Planta. 152: 8-12.

Appendix

Appendix 1: Sources to obtain *Dunaliella strains*

1. http://www.ife.ac.uk/ccap/
   Culture Centre of Algae and Protozoa(CCAP), in Cambridge
2. http://www.gwdg.de/~epsag/phykologia/epsag.html
   Culture Collection of Algae at Goettingen University (SAG)
3. Woods Hole Culture Collection (WHOI), Woods Hole
   email dkulis@whoi.edu
4. http://ccmp.bigelow.org/index.html
   Provasoli - Guillard National Center for Culture of Marine Phytoplankton
5. http://www.botany.utoronto.ca/utcc
   University of Toronto Culture Collection of Algae and Cyanobacteria (UTCC)
6. http://www.bio.utexas.edu/research/utex/
   UTEX Culture Collection. A searchable source for the cultures at University of Texas
7. http://www.ocgy.ubc.ca/projects/nepcc/
   The North East Pacific Culture Collection (NEPCC) at the Department of Botany, University of British Columbia.

8. http://www.uc.pt/botanica/ACOI.htm
Coimbra Collection of Algae (ACOI)
9. http://www.atcc.org/
American Type Culture Collection
10. http://seaweed.ucg.ie/cost/MicroAlgalCultures.html
Microalgal culture collections (in Europe)
11. Culture Collection, School of Microbiology and Immunology
The University of New South Wales, NSW, 2052, Australia
12. Australian Collection of Microorganisms, Dept Microbiology
University of Queensland, Brisbane, Queensland, 4072, Australia
13. CSIRO Microalgae Research Centre
email ian.jameson@csiro.au
14. Harmful Algae Culture Collection (UTAS)
email Judi.Marshall@utas.edu.au
15. http://www.iam.u-tokyo.ac.jp/misyst/ColleBOX/IAMcollection.html
Culture Collection centre (IAM)
16. Cawthorne Culture Collection of Microalgae (CAW)
eamil Krystyna@cawthorn.org.nz
16. University of Rhode island Culture Collection (URI)
email pharg@gso.uri.edu
17. Plymouth Culture Centre (PLY)
email mgsj@mail.pml.ac.uk

4 Cultivation of *Dunaliella* for High-Value Compounds

René H. Wijffels[1], Maria Barbosa[1], Marcel Janssen[1]

Abstract

The need for mass cultivation of microalgae and the available technology for such cultivation are presented. In the design of photobioreactors used for mass cultivation of algae surface/volume ratio, light path, mixing frequency and light intensity are important and several designs of photo bioreactors are described. The possibility to adapt the light absorption capacities via genetic modification of microalgae is discussed. Milking the microalgae - a method to extract carotenoids while keeping cells alive making it possible to reuse stressed cells is recommended.

Introduction

The production of high-value compounds using light and seawater is a unique property of photosynthetic microorganisms. The basic idea of using sunlight to produce high-value compounds brings along several limitations such as light limitation and photoinhibition that are not found in heterotrophic fermentations. Consequently, specifically the light regime within any photobioreactor has to be considered in its design and scale-up, because the overall aim of the design of photobioreactors is to produce biomass and specific metabolites directly from solar irradiation at high photosynthetic efficiencies and high volumetric and surface productivities. In addition, one has to consider mass transfer, mixing rate and shear stress for control of metabolite production in order to produce high-value ingredients at high concentrations and with a high productivity.

[1] Wageningen University, Bioprocess Engineering Group, P.O. Box 8128, 6700 EV Wageningen, The Netherlands.
Author for correspondence, Email: Rene.wijffels@wur.nl

Research in the field of microalgae received much attention between 1950 and 1970 (Burlew 1953). It was anticipated that there would be an enormous shortage in food resources. Production of microalgae to be used as single cell proteins for direct human consumption or as feed was believed to be the solution. It was expected that cultivation systems could be developed in which 250 tons of algae (dry matter) per hectare per year could be produced (Richmond 2003). This is a yield many factors higher than ever could be obtained with agricultural crops (Radmer and Kok 1977).

After 1970 research in microalgae technology came into a recession because the costs of agricultural crops decreased (microalgae products could not compete with these products) and because the technology was not good enough to realize the promises of 250 tons of algae per hectare per year. Nevertheless, microalgae have an enormous potential. Comparisons with both development of microbial fermentations and agriculture can be made. Due to developments both in technology and in strains (in the case of fermentation) as well as in crops (in agriculture) the productivity of present systems is about 5000 times higher than the original natural production systems (e.g., Hersbach et al. 1984). In contrast, production of microalgae is still based on traditional technologies with wild type strains. It is a great challenge to realize breakthroughs in both photobioreactor technology and strain development. Species of the genus *Dunaliella* are currently still among the very few commercially used microalgae. For example, in various countries strains of the species *D. salina* are used for β-carotene production (see other chapters in this book). Also, strains of *D. tertiolecta* are grown for mariculture applications. With microalgae such as *Dunaliella* being photosynthetic, care has to be taken during cultivation to provide appropriate illumination which is always a balance between providing enough light for optimal cell growth versus providing too much light which can be damaging to cells.

Growth and Photosynthesis As a Function of Light Intensity

Growth can be defined as the orderly increase of all the cells constituents. The rate of increase of biomass in biomass concentration is generally expressed (Burlew 1953, Stein 1973) as the specific growth rate, μ that can be described by the following equation:

Equation 1 $$\mu = \frac{1}{C_x} \cdot \frac{dC_x}{dt}$$

where μ is the specific growth rate (h^{-1}), C_x is the biomass concentration ($g\ L^{-1}$), t is time (h), and dC_x/dt is the variation in biomass concentration with respect to time ($g\ L^{-1}\ h^{-1}$), that is the culture productivity per unit volume.

Integrating this equation from $C_x = C_0$ (initial biomass concentration) to $C_x = C_t$ (biomass concentration at time t) we obtain,

Equation 2 $$(\frac{C_t}{C_0}) = \mu \cdot t$$

When a microalga grows under suitable conditions in batch culture, its growth curve follows a pattern showing six principal phases (Figure 1). Each phase is characterized by a particular value of the specific growth rate and reflects a particular metabolic state of the cell population. Following inoculation growth does not necessarily start right away since most cells may be viable but they are not in condition to divide, especially when the parent culture was old. The interval necessary for the transferred cells to metabolically acclimate to the new situation and start to grow is called the lag phase (I). Following the lag phase, the culture enters into the acceleration phase, during which μ increases continuously (II). When a constant growth rate is reached, the culture is said to be in the exponential growth phase (III). During exponential growth, cells are not limited for light (if a saturating light, i.e. light intensity at which the growth rate is maximal, is provided at the culture surface) and growth proceeds according to Equation 2. The maximum μ is achieved in this phase, and in this respect microalgae do not behave differently from chemotrophs. However, the exponential growth phase of phototrophs in batch culture normally lasts only for a relatively short period because the cells start to shade each other as their number increases, thus growth becomes light limited and μ decreases (linear phase IV). Self-shading and the fact that the energy source is instantaneously provided at the culture surface and cannot be stored in the reactor are the two fundamental factors that differentiate photobioreactors from reactors for chemotrophs. During the linear growth phase μ is constant and the culture productivity remains stable, because of the continuous proportional increase in cell concentration (C_x). Following the linear growth phase, the cell population continues to increase, but μ decreases until it reaches zero, at which point the culture enters the stationary phase (V), during which the cell concentration remains constant at its maximum value. The stationary phase is followed by the death phase, in which the population of viable cells decreases (VI).

Photosynthesis

The rate of photosynthesis, i.e., microalgal growth, is not simply proportional to the rate of photon absorption, since photons may be captured by the pigments

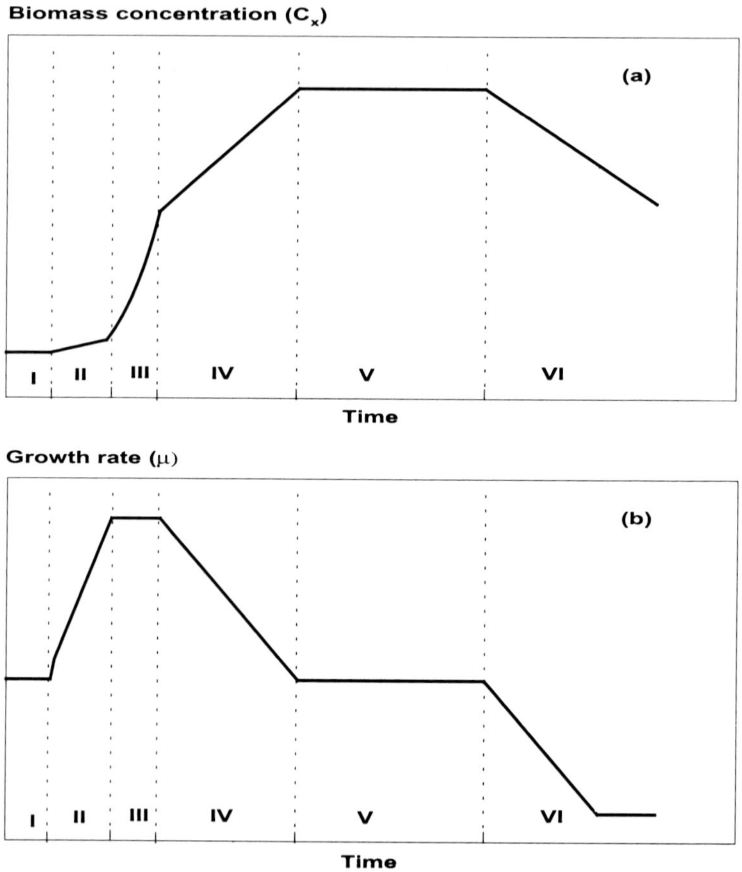

Figure 1: (a) Growth curve of a microalgae in batch culture, The following six phases are shown: I, Lag; II, acceleration; III, exponential; IV, linear; V, stationary; VI, death. (b) Changes in the growth rate (μ) during the six typical growth phases.

much faster than the photosynthetic apparatus can make use of them (Janssen et al. 2003). If photosynthesis is measured at different photon flux densities (PFD), the specific photosynthesis rate which can be measured as the specific oxygen production rate (PO_2), can be plotted as a function of PFD in the so called P-I or P-E curve (Figure 2).

In the dark, where there is no photosynthetic activity, O_2 is consumed and CO_2 is released due to cellular respiration. Under low PFDs oxygen production increases linearly with the PFD, illustrated with the dashed line in Figure 2. For microalgae, at PFDs higher than 100 μmol m^{-2} s^{-1} the increase of the oxygen production rate slows down and, eventually, the alga reaches its

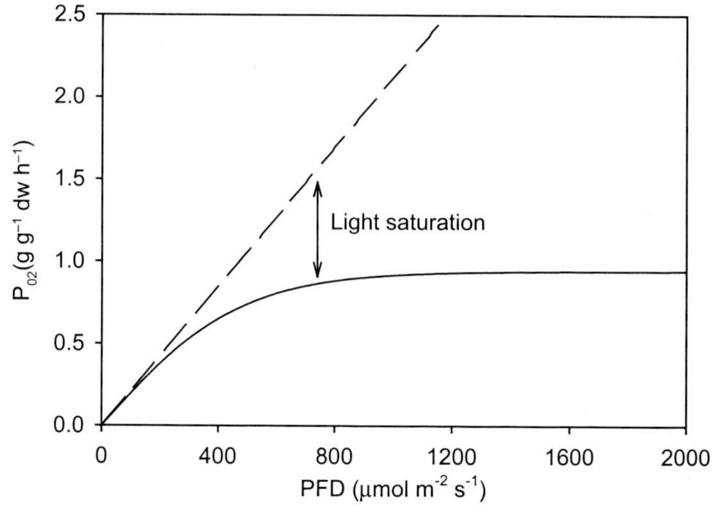

Figure 2: Specific oxygen production rate (P_{02} in g g^{-1} dw h^{-1}) as a function of the photon flux density (PFD in mmol m^{-2} s^{-1}). Data obtained for *Chlamydomonas reinhardtii* (green alga) cultivated under continuous light of 600-700 μmol m^{-2} s^{-1} (Janssen et al. 1999).

maximal oxygen production rate. This phenomenon is called light-saturation of photosynthesis. The discrepancy between the straight dashed line and the curved solid line is a measure of the PFD that cannot be utilized by the algae in the culture although this photon flux is absorbed by cells. In other words light energy is only utilized efficiently under low PFDs; lower than approximately 100 μmol m^{-2} s^{-1} in this case.

In contrast to low PFDs, at high PFDs the pigment antenna complexes of the photosystems absorb too much light. Alternatively, at high irradiance more photons are absorbed by the pigments of a photosystem per time than the reaction center can process during that time. A considerable part of the excitation energy cannot be utilized in the reaction centers because their capacity is limited. As a result, the surplus of excitation energy is predominantly dissipated as heat and fluorescence in the antenna. An overdose of excitation energy could damage the photosynthetic apparatus in a process called photoinhibition where the electron transport chain is shut down.

When phototrophic microorganisms such as *Dunaliella* are cultivated in a photobioreactor under a certain constant flux density (PFD) they will exhibit a corresponding specific growth rate. Also the size of the cellular light-absorbing surface, consisting of pigment molecules arranged in antenna complexes, will adapt to this PFD. The specific surface, i.e., the absorption coefficient, decreases if the PFD increases and this process is called photoacclimation. For example,

Figure 3 illustrates the decrease in the average absorption coefficient (a_c) of the microalgae *Dunaliella tertiolecta* when increasing the light intensity.

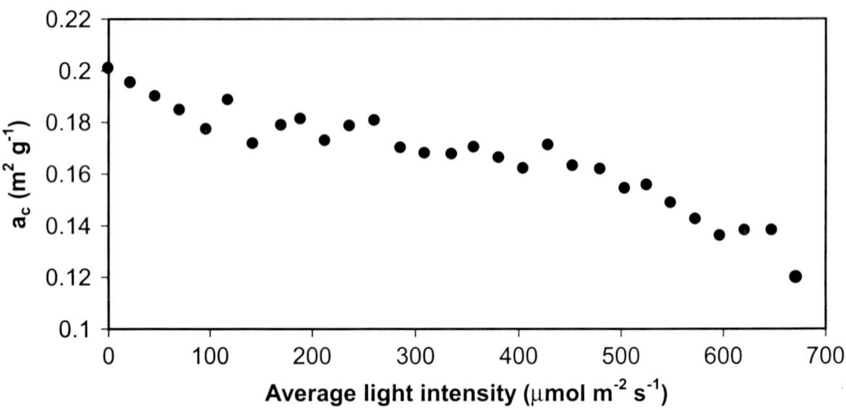

Figure 3: Photoacclimation to different light intensities by *Dunaliella tertiolecta* (Barbosa et al. 2004).

Cultivation of Phototrophs

As microalgae are autotrophic they can grow on simple and inexpensive media such as F/2 for *Dunaliella* species. The only problem is that algae are phototrophic and that light energy is the growth limiting 'substrate'. It is very difficult to expose a microalgal culture to a sufficient amount of light energy and have the algae cells utilize this energy efficiently for biomass production.

The 'biological' bottleneck, which algae utilize in low photon flux densities most efficiently, was explained earlier. However, besides the biological constraints we also have an engineering constraint due to light transfer problems inside the cultivation systems.

Light efficiency, volumetric and surface productivity

The two main parameters used to characterize the efficiency of microalgal cultivation systems are the volumetric productivity and the efficiency of light utilization.

The volumetric productivity is the product of the biomass density and the specific growth rate (Equation 3).

Equation 3 $$P_x = C_x \cdot \mu \qquad (\text{g L}^{-1}\ \text{h}^{-1})$$

A high volumetric productivity is beneficial because this means that one can suffice with a small cultivation system. Moreover, a high volumetric productivity usually is accompanied with a high biomass density, which is more attractive with respect to downstream processing. These considerations are very common in the field of bioprocess engineering. But working with photosynthetic microorganisms where high culture densities automatically result in strong self-shading, the efficiency of light utilization should be considered too. An increase in the efficiency of light utilization will lead to an increase of the volumetric productivity. But, even more important, light energy is the growth limiting substrate, and light falling on the cultivation system should be used as efficiently as possible. The efficiency of light utilization of photoautotrophic growth can be expressed in several ways. In algal biotechnology the term photosynthetic efficiency (PE) is used widely, and it is equal to the ratio between the caloric value of biomass produced and the light energy absorbed. The photosynthetic efficiency is usually expressed as a percentage. The maximum theoretical efficiency is about 20% which in practice can not be reached with sunlight because of several losses and limitations, which include the fact that PAR represents only 43% of the total solar radiation at the earth's surface, the loss of part of the incident radiation by reflection at the culture surface, and the fact that part of the absorbed energy is used for maintenance. Another major limitation is represented by the light saturating effect, i.e., the fact that the photosynthetic apparatus of most microalgae saturates at only 10-20% of full sunlight (about 2000 μmol m^{-2} s^{-1}).

On the other hand, the efficiency of light utilization can also be expressed as the biomass yield on light energy ($Y_{x,E}$) in grams of dry weight produced per amount of light energy absorbed (mol of photons absorbed). If light energy is expressed as mol quanta in the PAR range (400-700 nm), this yield can be determined easily using widely available PAR-based quantum sensors by using Equation 4.

Equation 4 $$Y_{x,E} = \frac{P_x \cdot V_{reactor}}{E_a \cdot 10^{-6} \cdot 3600} \quad (\text{g mol}^{-1})$$

where P_x is the biomass volumetric productivity (g L^{-1} h^{-1}), i.e., the amount of biomass produced per hour, $V_{reactor}$ is the reactor volume (L), E_a is the amount of light absorbed (μmol photons s^{-1}).

Another parameter to evaluate productivity is the surface yield ($P_{x,a}$), i.e., productivity per unit of occupied land area per unit time, expressed as g m^{-2} d^{-1} or tons ha^{-1} $year^{-1}$. This is often used in the case of algae cultivation in horizontal systems such as open ponds and tubular horizontal photobioreactors with no empty space between contiguous tubes. In these situations the surface productivity is proportional to the amount of light entering the system per unit illuminated area.

Light Distribution

Under optimal conditions of nutrients and temperature, the productivity of the microalgae cultures becomes a function of light availability to each cell in the culture. In dilute cultures, self-shading is minimal and all cells receive the same amount of light, independent of their positions. It is evident that, at these low densities, although the light intensity provided would permit the maximum growth rate, the culture will achieve low efficiency of light conversion and low productivity, since most of the impinging photon flux density will pass through the culture unabsorbed. For mass cultivation of microalgae, in order to obtain maximum productivity, cultures must be kept dense enough to absorb all the light falling on the culture surface. Under these conditions, self-shading, due to light absorption by cell pigments and scattering by the cells causes the photon flux density to fall off exponentially with the culture depth. In dense cultures, the light availability for each single cell is much reduced; this brings a significant decrease of the specific growth rate, but productivity is, in any case, higher than in diluted cultures because the decrease in growth rate is more than compensated by the increase in cell concentration. For achieving maximal productivity in mass cultures it is essential to operate at a proper combination of culture depth and cell concentration.

Open Systems

A huge variety of cultivation systems for microalgae have been developed, but the one used most frequently on a large scale and a commercial basis is the shallow open raceway-pond (Figure 4). These ponds are usually no more than 30 cm deep, and the water with nutrients and microalgae is circulated with a paddle wheel. Sunlight impinges on the surface and is absorbed inside the culture; the photon flux density will decrease with increasing depth. In a mixed mass culture with a depth of 10 cm or more, the largest part of the volume is exposed to very low *PFDs*. In this zone deeper within the culture, the light energy will be utilized efficiently. In contrast, close to the reactor surface the PFD is very high resulting in very low light energy conversion efficiency. In these pond systems, due to the long light path, biomass concentration is relatively low at about < 0.5 g l^{-1} and thus volumetric productivity is also low. The photosynthetic efficiency in these raceway-pond systems is maximal at about 3%

(Janssen et al. 2003). This is a very successful concept because of the simplicity and the relative low costs. However, only a few species can be grown in these systems open to the environment by means of a selective environment, high salinity for *Dunaliella salina* and high alkalinity in the case of *Spirulina platensis*. If non-selective media have to be applied, large amounts of inoculum produced under strictly controlled conditions need to be used and advantage can be taken of fast-growing species such as *Chlorella*. At present, commercial open pond production of microalgae is basically limited to these three strains mentioned.. Many other microalgae species hold promise for the production of a large variety of compounds, but to be able to grow monocultures for extended time periods, enclosed photobioreactors are preferred for reasons listed below.

Figure 4: Photograph of a raceway pond at the *Dunaliella* culture facility of NBT in Eilat, Israel (permission of Dr. Ben-Amotz).

Enclosed Photobioreactors

Photobioreactors are considered to have several major advantages over open ponds, because they can:

- Prevent or minimize contamination, allowing the cultivation of algal species that can not be grown in open ponds.

- Offer better control over cultivation conditions (pH, pCO_2, pO_2, temperature, etc).
- Prevent evaporation and reduce water use.
- Lower CO_2 losses due to outgassing.
- Permit higher cell concentration that may result in reduced operating costs.
- Attain higher productivity.

On the other hand, certain requirements of photobioreactors, including the need for cooling, the need for strict control of oxygen accumulation and biofouling, make these systems more expensive to build and to operate than ponds. Thus, despite their advantages, the use of photobioreactors are currently limited for production of inocula for open ponds and for the production of high value compounds from strains that can not be cultivated in open ponds.

Design criteria for photobioreactors should aim at achieving high efficiency of light conversion and at providing the necessary reliability and stability to the cultivation process by solving the main problems encountered in photobioreactor operation, such as overheating, oxygen build-up, biofouling and contamination. The fundamental design criteria for photobioreactors include reactor configuration in respect to light gradient and light/dark cycles, surface to volume ratio, mixing and degassing devices. In addition, one also has to consider if the photobioreactor will be located outside where sunlight will be the source of energy or if the reactor will be operated inside where artificial illumination would have to be used. Possible intermediates are photobioreactors within greenhouses where sunlight and artificial illumination may be used either simultaneously or in sequence.

Light Gradient and Light/Dark Cycles

An example of a photobioreactor is shown in Figure 5: a bubble column with transparent wall. Only the outer surface is exposed to sunlight or artificial irradiance. Light will be absorbed by cells inside the culture, and thus depending on the biomass concentration, a certain part of the interior of the bioreactor can be considered a 'dark zone'. Due to mixing, the algae will travel through both the dark zone and the 'photic' zone. In other words, within the photobioreactor the algae will experience a fluctuating light regime or light/dark cycles.

Although this example is based on a bubble column, the fluctuating light regime experienced by microalgae is a general characteristic of all types of photobioreactors. The interior of a photobioreactor will always be dark because it is not cost effective to operate photobioreactors at low biomass densities at which the light gradient is low. Based on the above example it is easy to imagine that each type of photobioreactor is characterized by a specific light regime. The light

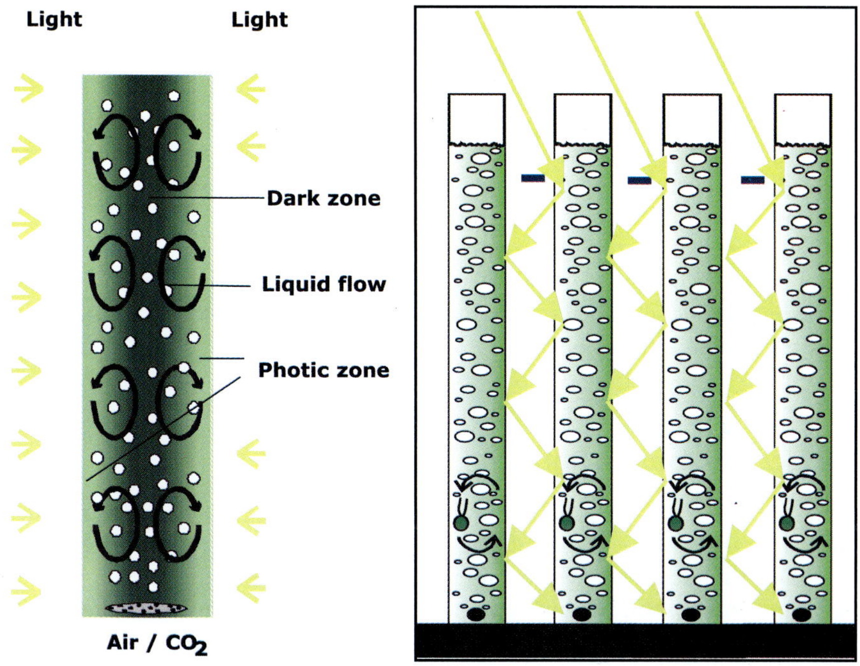

Figure 5: (a) Schematic presentation of light/dark zones and mixing in a bubble column. (b) Schematic showing several bubble columns arranged next to each other.

regime is characterized by the light gradient when going from the light-exposed surface to the interior of the reactor. This light gradient is dependent on reactor design and biomass density. A second characteristic is the frequency of the light/dark fluctuations, dependent again on reactor design and also the mixing intensity of the reactor liquid.

The efficiency of light utilization (or yield on light energy) is very important for an optimal photobioreactor design because light energy is the growth-limiting energy source. Inside photobioreactors algae are exposed to light/dark cycles with high light intensities close to the reactor surface and darkness in the interior of the reactor. But, as explained earlier, the photosynthetic efficiency is low under high light intensities. The fact that light energy cannot be stored and homogenized in a volume with photosynthetic microorganism is the one and only factor limiting the application of these microorganisms.

It was demonstrated, however, that very fast alternations between high light intensities and darkness could greatly enhance the photosynthetic efficiency under these high light intensities (Richmond et al. 2003, Hu et al. 1998b). This is called the flashing light effect and was observed under very short light/dark cycles from less than 40 μs up to 1 s (Kok 1953).

The short cycle-time flashing light effect is thought to result from the fast reduction of the e-acceptors, associated to PSII, followed by their oxidation in the dark period. This will result in a maximum 'photon-accepting capacity' of PSII during light flashes. The flashing light effect observed under relatively long cycle time, 100 ms to 1 s, can be explained by a similar mechanism: the alternating reduction and oxidation of the large pool membrane-soluble plastoquinone molecules during light and dark periods, respectively.

Cycle times of 10 to 100 s can be found in airlift reactors, 1 to 4 s can be found in bubble columns and 80 to 200 ms can be found in flat panels. These are approximate values obtained for commonly used microalgae cultivation conditions (flow rate, reactor dimensions).

Surface-to-Volume Ratio

The surface-to-volume ratio of the bioreactor, (i.e., the ratio between the illuminated surface of the reactor and its volume) determines the amount of light that enters the system per unit volume and the light regimen to which the cell population is exposed to, and is consequently one of the most important issues in photobioreactor design. The higher the S/V ratio, the higher the cell concentration at which the reactor is operated and the volumetric productivity of the culture (Janssen et al. 2003). The hydrodynamic behavior of the culture is also affected by this parameter. Higher S/V ratios can lead to shorter light/dark cycles. For these reasons, in recent years a general trend towards the reduction of the diameter of tubular reactors and the thickness of flat panels can be seen.

Mixing

The type of device used to mix and circulate the culture suspension is essential in the design of a successful photobioreactor. Both the productivity of a photobioreactor and the cost of its construction and operation are influenced to a great extent by the type of device used for mixing. Mixing is necessary to:

- Prevent cells from settling.
- Avoid pH and temperature gradients along the reactor.
- Distribute nutrients.
- Remove photosynthetically generated oxygen, which at a certain concentration inhibits photosynthesis.
- Supply CO_2.
- Ensure that all cells experience alternating periods of light and darkness.

Yet, it must be pointed out that excessive mixing can lead to cell damage and eventually cell death. For this reason the choice of mixing intensity and mixing system must be dictated by the characteristics of the organism to be cultivated (Barbosa et al. 2003, 2004). Specifically, cells of *Dunaliella* are more sensitive to shearing introduced by mixing and pumping than other microalgae, because *Dunaliella* cells do not contain the typical rigid cell wall that most other unicellular green algae possess. Instead, *Dunaliella* cells are surrounded by an elastic glycocalyx consisting of glyco-proteins. This glycocalyx provides cells flexibility regarding cell form and shape, but absence of a rigid cell wall also makes cells more fragile. Regardless of the type of photobioreactor used for cultivation of *Dunaliella*, this specific cell characteristic has to be kept in mind.

Photobioreactor Types

Photobioreactors can be classified on the basis of design. The main categories of reactors are:

1. Tubular photobioreactor
2. Tubular photobioreactor with light dilution
3. Flat panel photobioreactor

It is difficult to make comparisons of productivities between the different photobioreactors used as data are obtained under incomparable circumstances.

The Tubular Photobioreactor

Tubular photobioreactors consist of long transparent tubes with diameters ranging from 3 to 6 cm and lengths ranging from 10 to 200 m. The culture liquid is pumped through these tubes by means of mechanical or air-lift pumps. The tubes can be positioned in many different ways: in a horizontal plane as straight tubes with a small or large number of U-bends; vertical, coiled as a cylinder or a cone; in a vertical plane, positioned in a fence-like structure using U-bends or connected by manifolds; horizontal or inclined, parallel tubes connected by manifolds. In addition, horizontal tubes can be placed on different reflective surfaces with a certain distance between the tubes. Although tubular reactor design is very diverse, the predominant effect of the specific designs is on the light regime resulting in a diversity of photon flux densities incident on the reactor surfaces (PFD_{in}).

Regardless of the reactor design, the shape of the light gradient within the tubes is similar in most designs. Also with respect to liquid mixing, i.e., L/D cycling, the circumstances in most designs are similar.

The diameter of the tube and the biomass density in the culture liquid are the predominant factors determining the light gradient. Janssen et al. (2003) used the model of Evers (1991) to estimate the penetration depth of blue (440 nm), green (550 nm) and red light (678 nm), using Chl-*a* specific absorption coefficients determined for *D. tertiolecta*. Janssen et al. (2003) calculated that the biomass density in horizontal tubular systems theoretically can be higher by a factor of 5 in order to maintain the same light fraction (ε) as in larger diameter (0.2-0.4 m) vertical column reactors. Torzillo et al. (1993) used a culture density of 3.5 and 6.3 g dw L^{-1} for *Spirulina* in a horizontal tubular system of 2.6 cm diameter. Tredici and Zitelli (1998) used culture densities of 3.6 and 4.0 g L^{-1} in horizontal tubular systems with internal diameters of 3.4 and 3.0 cm. In addition, Molina Grima et al. (2000) compared horizontal tubular reactors of 3 cm and 6 cm diameters. The highest productivities of the microalga *Phaeodactylum tricornutum* were reached at biomass densities between 4 and 7 g L^{-1}. It is expected that *Dunaliella* can be grown in such systems at culture densities up to 5 g/l.

Tubular Photobioreactor with Light Dilution

The tubular photobioreactor has two disadvantages. The light/dark cycles are relatively long and there is build-up of oxygen. Due to the long light/dark cycles the photosynthetic efficiency is relatively low. The photosynthetic efficiency can be increased by reducing the intensity of light. Reduction of the light intensity falling on the tubes can be achieved by stacking the tubes and placing the stacked tubes in walls relatively close to each other. In this way the tubes shade each other and light is distributed across a larger surface area. Thus, light is diluted and as a consequence the photosynthetic efficiency in these tubular systems is increased.

Flat Panel Reactor

Flat panel reactors consist of a rectangular transparent box with a depth of only 1 to 5 cm. Height and width can be varied to some extent but in practice only panels with a height and width both smaller than 1 m have been studied. These photobioreactors are mixed with air introduced via a perforated tube at the bottom of the reactor. In order to create a high degree of turbulence 2.8 to 4.2 liters of

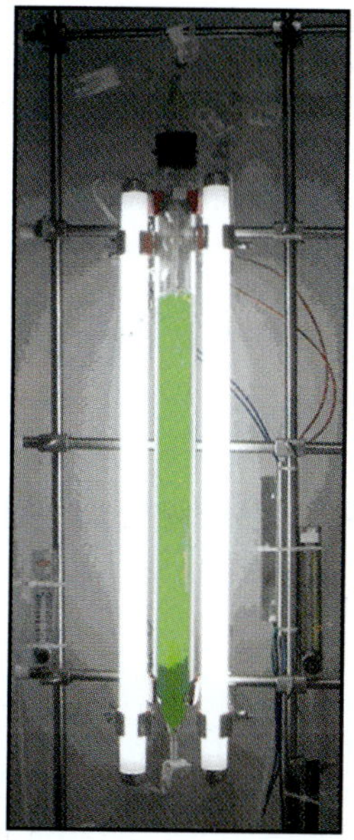

Figure 6: Bubble column photobioreactors for cultivation of *Dunaliella sp.* (permission of Dr. Jin).

Figure 7: Flat plate photobioreactor at Wageningen University.

air per liter of reactor volume per minute has to be provided corresponding to superficial gas velocities between 0.030 to 0.035 m s^{-1} (Hu et al. 1996, Hu and Richmond 1996, Hu et al. 1998a). Usually the panels are illuminated from one side by direct sunlight and the panels are placed vertically or inclined facing the sun.

In flat panel reactors biomass concentrations are in the range of 5 to 17.5 g L^{-1} at maximal productivity of the cyanobacterium *Spirulina platensis* (Hu and Richmond 1996, Hu et al. 1998a)., Gitelson et al. (1996) showed that the penetration depth for blue and red light is less than 1 mm at 100 g m^{-3} of Chl-*a* for *Spirulina platensis*.

Biological Methods to Improve Microalgae Productivity

As outline above, a major limitation to microalgae productivity is light utilization. To improve productivity through increased light energy conversion efficiency, it was proposed in the past to permanently truncate the chlorophyll antenna size of the photosystems of microalgae (Radmer and Kok 1977). The alga *Chlamydomonas reinhardtii* was used as one of the model systems to test the hypothesis that truncation of the chlorophyll antenna size will result in improved solar energy conversion efficiency (Melis et al. 1999, Melis 2002, Polle et al. 2003). Initial studies with the alga *D. salina* applying physiological methods to adjust the chlorophyll antenna size demonstrated that truncated antenna systems resulted in increased rates of maximal photosynthetic oxygen evolution (personal communication Jürgen Polle).

Milking of Microalgae

Natural carotenoids are fat-soluble stereoisomeric pigments. In the primary metabolism of microalgae, carotenoids play a role in the energy transfer to the chlorophyll in the photosystem (Tanada 1951). Carotenoids absorb most strongly light in the blue region (400-500 nm) and transfer the excitation energy to the chlorophylls, making photosynthesis efficient over a wider range of wavelengths. In the secondary metabolism, carotenoids are produced for protection against stress situations, such as high light intensities and high concentrations of salt, extreme pH, low temperature, and nutrient deficiencies. They protect the photosynthetic system in the situation that microalgae are exposed to high light intensities under limiting growth conditions. An overdose of excitation energy can lead to production of toxic components (e.g., oxygen radicals) and to photosystem damage. Carotenoids are capable of scavenging toxic photoproducts and/or prevent the formation of these products because an overdose of energy can be dissipated as heat (Demmig-Adams et al. 1995).

Green cells of *D. salina*, which are grown under normal growth conditions, have low carotenoid contents. The carotenoids are just like chlorophyll present in the chloroplasts of the cells in flattened vesicles, the thykaloid membranes. When stress is applied, much higher concentrations of carotenoids are obtained. In these red cells, carotenoids accumulate in fat globules outside the thylakoid membranes (Ben-Amotz 1995). Our data show that the globules containing carotenoids are located around thylakoid membranes and around the outer membrane of the cells (Hejazi and Wijffels 2004a).

A major bottleneck for the application of microalgae is the low productivity obtained with the processes used presently. The fundamental reason for this is that in those processes algae are grown and after that (usually after application of stress) cells are harvested and products are purified. The growth of cells is a slow process, mainly because much energy (usually from solar illumination) is needed for the fixation of CO_2 in biomass (photosynthesis). By 'milking' products from biomass, grown biomass is reused and does not have to be grown again.

In the past few years years we worked on extractive fermentation of carotenoids from microalgae (Hejazi and Wijffels 2004a). Carotenoids like β-carotene and astaxanthin are grown in the presence of organic solvents. Compounds extracted from the algae are produced again by the microalgae and as such the lipophilic compounds are 'milked' from the microalgae.

Effects of stress on the production of carotenoids have been studied extensively. Usually the maximum concentration of carotenoids in the cells reached is 10% of the dry weight. We observed in our experiments when solvents were used a concentration of carotenoids of 15%. Stress studies are usually done in a non-systematic way. Stressing is observed but poorly understood. It is known that due to stress stimuli, the expression levels of genes involved in carotenoid production are increased. Presently, we study stress in a more systematic way and look at stress effects both at the level of the bioreactor and also at the molecular level.

Summary

Microalgae are photosynthetic microorganisms with which many added value compounds can be produced for applications in food, feed, cosmetics and fine chemistry. Production systems for successful production of *Dunaliella salina* are discussed.

The growth kinetics of microalgae as a function of the intensity of light is described. The rate of photosynthesis, i.e., microalgal growth, is not simply proportional to the rate of photon absorption, since photons may be captured by the pigments much faster than the photosynthetic apparatus can make use of them and processes like light saturation and photoinhibition will take place at higher light intensities.

For cultivating microalgae at a larger scale, these phenomena need to be considered in order to obtain high productivities. For that, parameters like surface/volume ratio, light path, mixing frequency and light intensity are important in the design of photobioreactors.

A huge variety of cultivation systems for microalgae have been developed. The only one used on a large scale and a commercial basis is the shallow open raceway-pond. These ponds are usually no more than 30 cm deep and the water with nutrients and microalgae is circulated with a paddle wheel.

Tubular photobioreactors consist of long transparent tubes with diameters ranging from 3 to 6 cm and lengths ranging from 10 to 200 m. The culture liquid is pumped through these tubes by means of mechanical or air-lift pumps.

Flat panel reactors consist of a rectangular transparent box with a depth of only 1 to 5 cm. These photobioreactors are mixed with air introduced via a perforated tube at the bottom of the reactor.

It is also possible to adapt the light absorption capacities via genetic modification of microalgae by reducing the antenna size. In this way the light absorption capacity reduces allowing the productivity of cultivation systems to increase.

Green cells, which are grown under normal growth conditions, have low carotenoid contents. If stress is applied, much higher concentrations of carotenoids are obtained. In these red cells, carotenoids accumulate in fat globules outside the thylakoid membranes. A method is described to extract these carotenoids while keeping cells alive, making it possible to reuse stressed cells.

References

Barbosa, M.J., M. Albrecht and R.H.Wijffels. 2003. Hydrodynamic stress and lethal events in sparged microalgae cultures. Biotechnol. Bioeng. 83: 112-120.

Barbosa, M.J., Hadiyanto and R.H. Wijffels. 2004. Overcoming shear stress of microalgae cultures in sparged photobioreactors. Biotechnol. Bioeng. 85: 78-85.

Ben-Amotz, A. 1995. New mode of *Dunaliella* biotechnology: two-phase growth for β-carotene production. J. Appl. Phycol. 7: 65-68.

Bureow, J.S. 1953. Algal Culture from Laboratory to Pilot Plant, Carnegie Institute of Washington Publication 600 Washington D.C., USA.

Demmig-Adams, B., W.W. Adams, B.A. Logan and A.S. Verhoeven. 1995. Xanthophyll-cycle dependent energy dissipation and flexible photosystem II efficiency in plant acclimated to light stress. Aust. J. Plant. Physiol. 22: 249-260.

Evers, E.G. 1991. A model for light-limited continuous cultures: growth, shading and maintenance. Biotechnol. Bioeng. 38: 254-259.

Gitelson, A., H. Qiang and A. Richmond. 1996. Photic volume in photobioreactors supporting ultrahigh population densities of the photoautotroph *Spirulina platensis*. Appl. Environm. Microbiol. 62: 1570-1573.

Hejazi, M.A. and R.H. Wijffels. 2004. Milking of microalgae. Trends Biotechnol. 24: 189-194.

Hejazi, M.A., E. Holwerda and R.H. Wijffels. 2004. Milking microalga *Dunaliella salina* for b-carotene production in two-phase bioreactors. Biotechnol. Bioeng. 85: 475-481.

Hersbach, G.J.M., C.P. van der Beek and P.W.M. van Dijck. 1984. The penecillins: properties, biosynthesis and fermentation. pp 45-140: In Biotechnology of Industrial Antibiotics. E.J. van Damme (ed). Marcel Decker Inc. New York, USA.

Hu, Q., and A.Richmond. 1996. Productivity and photosynthetic efficiency of *Spirulina platensis* as affected by light intensity, algasl density and rate of mixing in a flat plate photobioreactor. J. Appl. Phycol. 8: 139-145.

Hu, Q., H. Guterman and A. Richmond. 1996. Physiological characteristics of *Spirulina platensis* (cyanobacteria) cultured at ultrahigh cell densities. J. Phycol. 32: 1066-1073.

Hu, Q, D. Faiman and A. Richmond 1998a. Optimal tilt angles of enclosed reactors for growing photoautotrophic microorganisms outdoors. J. Ferment. Bioeng.85: 230-236.

Hu, Q., Y. Zarmi and A. Richmond. 1998b. Combined effects of light intensity, light-path and culture density on output rate of *Spirulina platensis* (cyanobacteria). European Journal of Phycology 33(2): 165-171.

Janssen, M., T.C.Kuijpers, J. Tramper, L.R. Mur and R.H. Wijffels. 1999. Response of *Chlamydomonas reinhardtii* to intermittent illumination in loop reactors. J. Biotechnol. 70: 323-333.

Janssen, M., J. Tramper, R.L. Mur and R.H. Wijffels. 2003. Enclosed photobioreactors: light regime, photosynthetic efficiency, scale-up and future prospects. Biotechnology Bioengineering 81 : 193–210.

Kok, B. 1953. Experiments on photosynthesis by *Chlorella* in flashing light. pp. 63-75. In Algal Cultures. J.S. Burlow (ed). Carnegie Institutions.Washington DC, USA.

Melis, A. 2002. Green alga hydrogen production: progress, challenges and prospects. Intl. J. Hydrogen Energy 27: 1217-1228.

Melis, A., J. Neidhardt and J.R. Benemann. 1999. *Dunaliella salina* (Chlorophyta) with small chlorophyll antenna sizes exhibit higher photosynthetic productivities and photon use efficiencies than normally pigmented cells. J. Appl. Phycol. 10: 515-525.

Molina Grima, E.M., F.G.A. Fernandez, F.G. Camacho, F.C.Rubio and M.Y. Chisti. 2000. Scale-up of tubular photobioreactors. J. Appl. Phycol. 12: 355-368.

Polle, J.E.W., S. Kanakagiri and A. Melis. 2003. *tla1,* a DNA insertional transformant of the green alga *Chlamydomonas reinhardtii* with a truncated light-harvesting chlorophyll antenna size. Planta 217: 49-59.

Richmond, A. 2003. Handbook of microalgal cultures. Blackwell Publishers Boston, USA.

Richmond, A., Z. Cheng-Wu and Y. Zarmi. 2003. Efficient use of strong light for high photosynthetic productivity. Biomolecular Engineering 20:229-236.

Radmer, R. and B. Kok. 1977. Photosynthesis: Limited Yields, Unlimited Dreams BioScience 27 (9):599-605.

Stein, J.R. 1973. Handbook of Phycological Methods, Culture Methods and Growth Measurements, Cambridge University Press, Cambridge, UK.

Tanada, T. 1951. The photosynthetic efficiency of carotenoid pigments in *Navicula minima*. Am. J. Bot. 38: 276-283.

Torzillo, G., P. Carlozzi, B. Pushparaj, E. Montaini and R. Materassi. 1993. A two-plane tubular photobioreactor for outdoor culture of *Spirulin*. Biotechnol. Bioeng. 42: 891-898.

Tredici, M.R. and G.C. Zitelli. 1998. Efficiency of sunlight utilization. Comparison: tubular versus flar photobioreactors. Biotechnol. Bioeng. 57: 187-197.

5 Pigments of Green and Red Forms of *Dunaliella*, and Related Chlorophytes

Shirley W. Jeffrey[1] *and Einar Skarstad Egeland*[2]

Abstract

Particular strains of *Dunaliella salina, D. bardawil* and *Haematococcus pluvialis* (Chlorophyta) transform from green to red cells under extreme environmental conditions, such as high light intensity, high salinity, low nutrients and extremes of temperature. Massive quantities of either β,β-carotene (*Dunaliella*) or astaxanthin (*Haematococcus*) may be synthesized under these conditions, which are thought to be photoprotective responses to the photoinhibition induced. Commercial harvesting of these pigments (β,β-carotene and astaxanthin), which are both valuable food colouring agents in marine animal culture world-wide, mean that optimal conditions for successful production of these carotenoids in the farming situation are continually sought.

This brief review begins with new concepts on the origins of algal chloroplasts, then high-lights some key aspects of the biochemistry of pigments – chlorophylls and carotenoids – in the Chlorophyta. Particular attention is given to the role of stereoisomers in the functioning of the photosynthetic apparatus. Several methods of pigment analysis useful for monitoring the development of carotenogenesis in the farming situation are described, including methods for the HPLC analysis of *Dunaliella* and *Haematococcus* pigments. Pathways for chlorophyll and carotenoid biosynthesis in the Chlorophyta are discussed, noting that more information on gene regulation and enzyme transcription processes is needed. Finally, the biochemistry and ecology of the red-pigmented forms of *Dunaliella* and *Haematococcus* are compared.

[1] CSIRO Division of Marine and Atmospheric Research, GPO Box 1538, Hobart, Tasmania, 7001, Australia.

[2] Faculty of Biosciences and Aquaculture, Bodø University College, Main building, Mørkved, NO-8049 Bodø, Norway.

Introduction: the origin of algal chloroplasts

It is now generally accepted that the chloroplasts of eukaryotic algae were once endosymbiotic organelles, originally derived from a previously free-living ancestral cyanobacterium, which in early Precambrian time (Proterozoic epoch) became engulfed by a colourless non-photosynthetic host cell of unknown origin (Bhattacharya 1997, Delwiche 1999, McFadden 2001, Palmer 2003). Following evolutionary development of this ancient endosymbiosis, three permanent major monophyletic lineages eventually resulted: the Glaucocystophyta, the Chlorophyta and the Rhodophyta radiations (Delwiche 1999; see Figure 2.3 in Jeffrey and Wright 2006). Each lineage developed specific cellular ultrastructures, unique chloroplast pigment complements, genomic compositions and ecological preferences. Gene transfer, loss and substitution reduced the plastid genome, until the majority of plastid proteins were encoded by the nuclear genome of the host. Further secondary and tertiary endosymbioses also developed. The history of these events can be seen in present-day algal cells (Larkum and Vesk 2003) in the loss of cell compartments and organelles, the presence of vestigial nuclei, the number of residual membranes surrounding the chloroplast, and analysis of nuclear and plastid genomes. Evidence also suggests that in some groups (e.g. dinoflagellates) multiple plastid losses and replacements have occurred frequently, promoting further plastid diversity across algal groups.

The fossil record of the primitive oceans clearly shows the early dominance of the green algal super-family, containing chlorophylls *a* and *b* (Chls *a* + *b*), while in modern oceans the dominant phytoplankton are now the chromophyte algae from the red algal radiation (Chls *a* + *c*). What forces have promoted these massive changes are unknown, but are thoughtfully explored by Quigg et al. (2003) and Grzebyk et al. (2003).

The green algae occupying present-day ocean habitats include the three uni-cellular green algal phyla – the Prasinophyta, the Chlorophyta and the Euglenophyta – and two multi-cellular green algal phyla – the Charophyta and the Ulvophyta. These groups are considered the forerunners of the higher plants (the Viridiplantae; Cavalier-Smith 2000).

Melkonian (1990a, b, c) discusses the phylogeny and characteristics of the Chlorophyta to which the genus *Dunaliella* Teodoresco belongs. The Chlorophyta are classified into 11 orders, 350 genera and approximately 2,500 species with the taxonomy still in a state of flux. *Dunaliella* is a member of the order Dunaliellales, and approximately 30 species have been described (Preisig 1992, Melkonian 1990c). Many of these are available from the major Algal Culture Collections world-wide (Andersen 2005, Subba Rao 2006a, b). However, Borowitzka and Siva (2008) have recently updated the taxonomy of the genus *Dunaliella* based on morphological and biochemical characters, and evaluated their variability and usefulness. They also give a detailed examination of 36 *Dunaliella* strains held in the Murdoch University Algal Culture Collection, clarifying the origin of each strain and the now likely identifications. A key is also provided to the 22 *Dunaliella* species now recognized by these authors, which should assist future taxonomic recognition of strains.

Not all cultured species/strains of the Dunaliellales have been seriously studied, but *Dunaliella tertiolecta* and *D. salina* (which is thought equivalent to *D. bardawil*; Borowitzka and Siva 2008) are favoured species. Certain strains, isolated from highly saline environments (e.g. salt ponds), low nutrient and high solar radiation habitats, are able to synthesize massive amounts of β,β-carotene, converting from green motile cells to red cells in the process (see Section 5). This characteristic has made certain strains of *D. salina* and *D. bardawil* highly successful candidates for the commercial production of β,β-carotene, a carotenoid colouring agent used as a food additive in animal culture, and as an anti-oxidant for human medical applications. Successful farms for the production of β,β-carotene in which the conversion of "normal" green to red carotenoid-rich forms are promoted, are thriving in Israel, Australia, California, China and India (Borowitzka and Siva 2008).

Another group of chlorophyte microalgae, which include strains of *Haematococcus*, *Chlorococcum* and *Chlorella* species, produce red cells (aplanospores in *Haematococcus*) containing massive amounts of the commercially valuable carotenoid astaxanthin, under low nutrient, high light situations. The metabolism of these strains will be briefly compared with those of *Dunaliella* in Section 5.

2. Photosynthetic pigments of the Chlorophyta

The photosynthetic pigments of the Chlorophyta and terrestrial plants have been highly conserved throughout evolution, in contrast to those of the Chromophyte algae and the photosynthetic bacteria. The major chlorophylls (Figure 1) are the two green Chls *a* and *b* and the Chl *c*-like pigment Mg-3,8-divinylpheoporphyrin a_5 monomethyl ester (Mg DVP). The latter, while mostly found in traces in the Dunaliellales, occurs as a light-harvesting pigment in some members of the primitive green Prasinophyta (Ricketts 1966). Also present in the Chlorophyta are the major yellow, orange and red carotenoids: β,β-carotene, β,ε-carotene, lutein, violaxanthin, antheraxanthin, zeaxanthin and neoxanthin (Figure 2A), with additional chlorophyte carotenoids – echinenone, canthaxanthin, astaxanthin and loroxanthin, also found.

Some of the primitive uni-cellular Prasinophyta have identical pigment suites to those of the Chlorophyta (referred to as Prasinophyta 1 in Jeffrey and Wright 2006), while others contain the carotenoids siphonaxanthin and siphonaxanthin ester (= siphonein) (Prasinophyta 2), or prasinoxanthin, uriolide and micromonal (Prasinophyta 3); (see Egeland et al. 1997, Jeffrey and Wright 2006). The structures of these unique prasinophyte carotenoids are shown in Figure 2B.

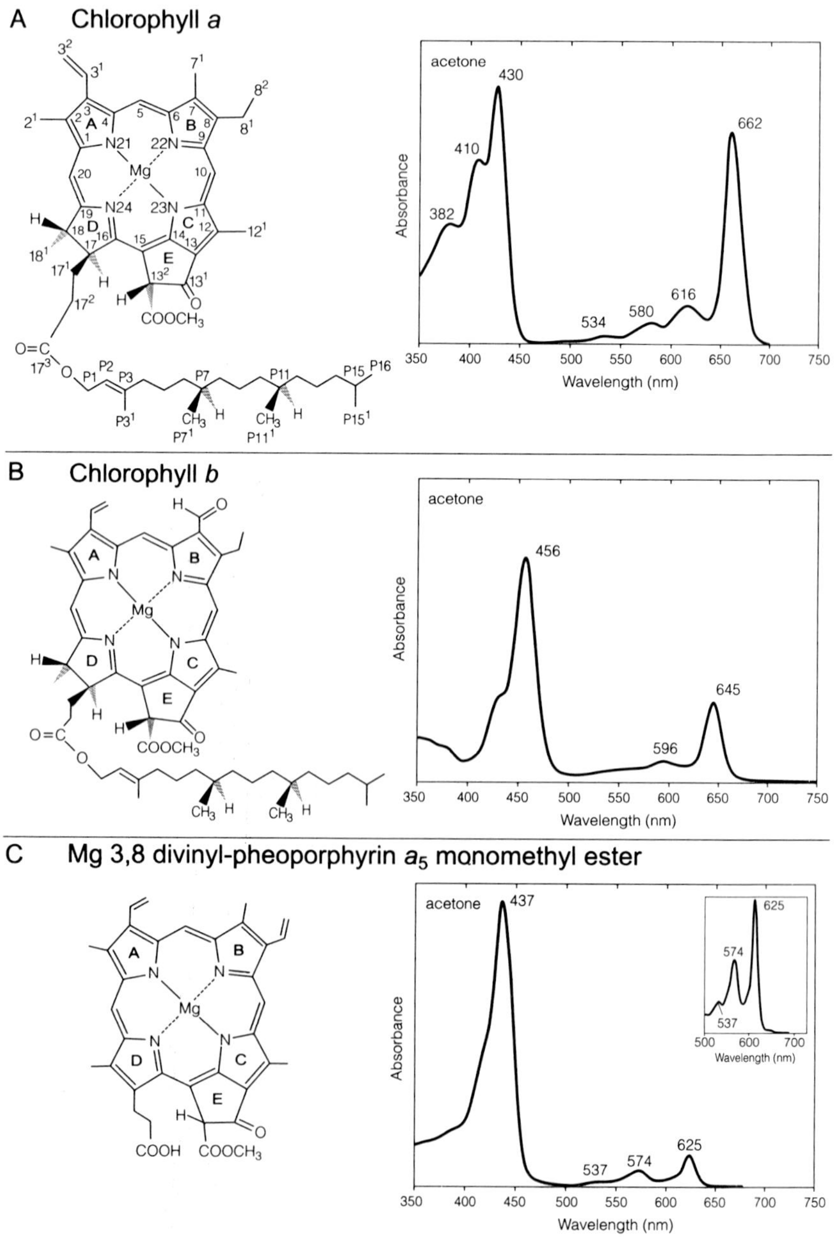

Figure 1: Structures of chlorophylls (Chls) *a*, *b* and Mg-3,8-divinylpheoporphyrin a_5 monomethyl ester (Mg DVP), and their visible absorption spectra in 100% acetone.

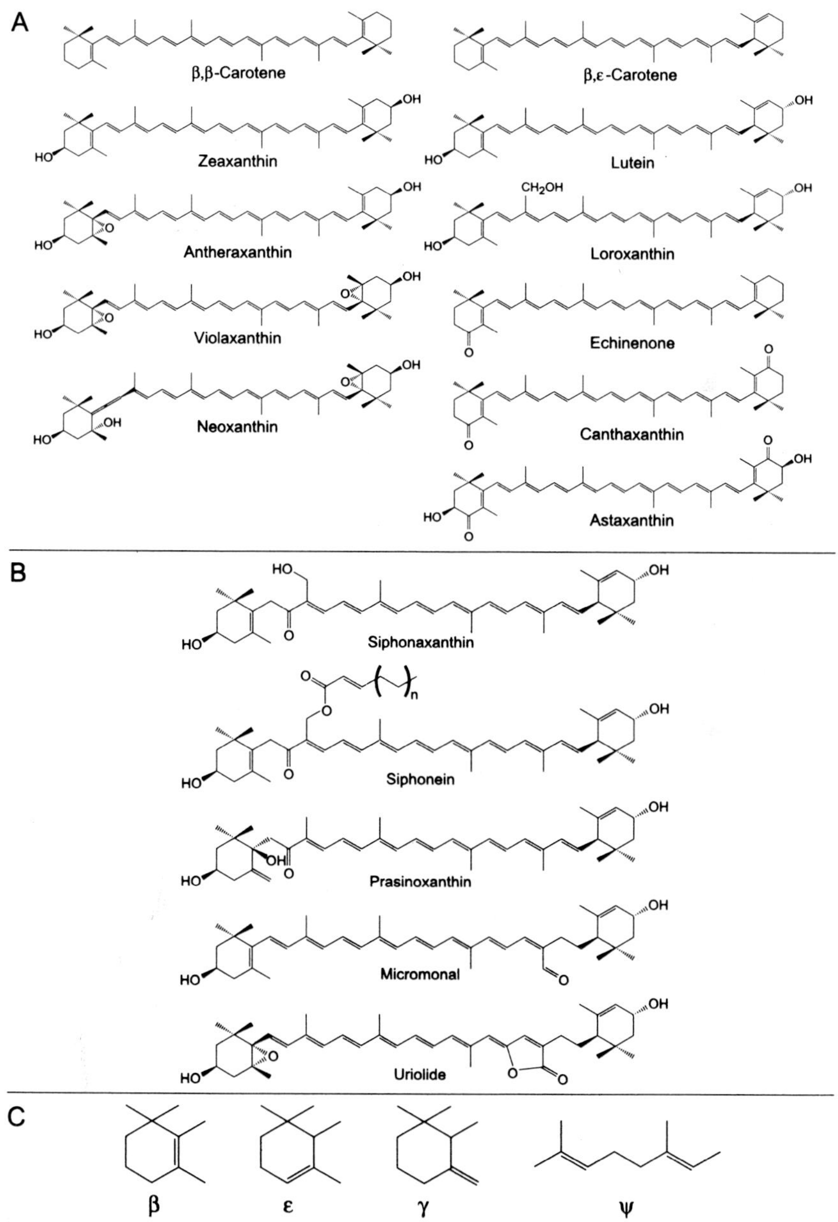

Figure 2: (A) Some common carotenoids of the Dunaliellales (Chlorophyta); (B) Five characteristic carotenoids of the Prasinophyta; (C) Basic end-groups of algal carotenoids.

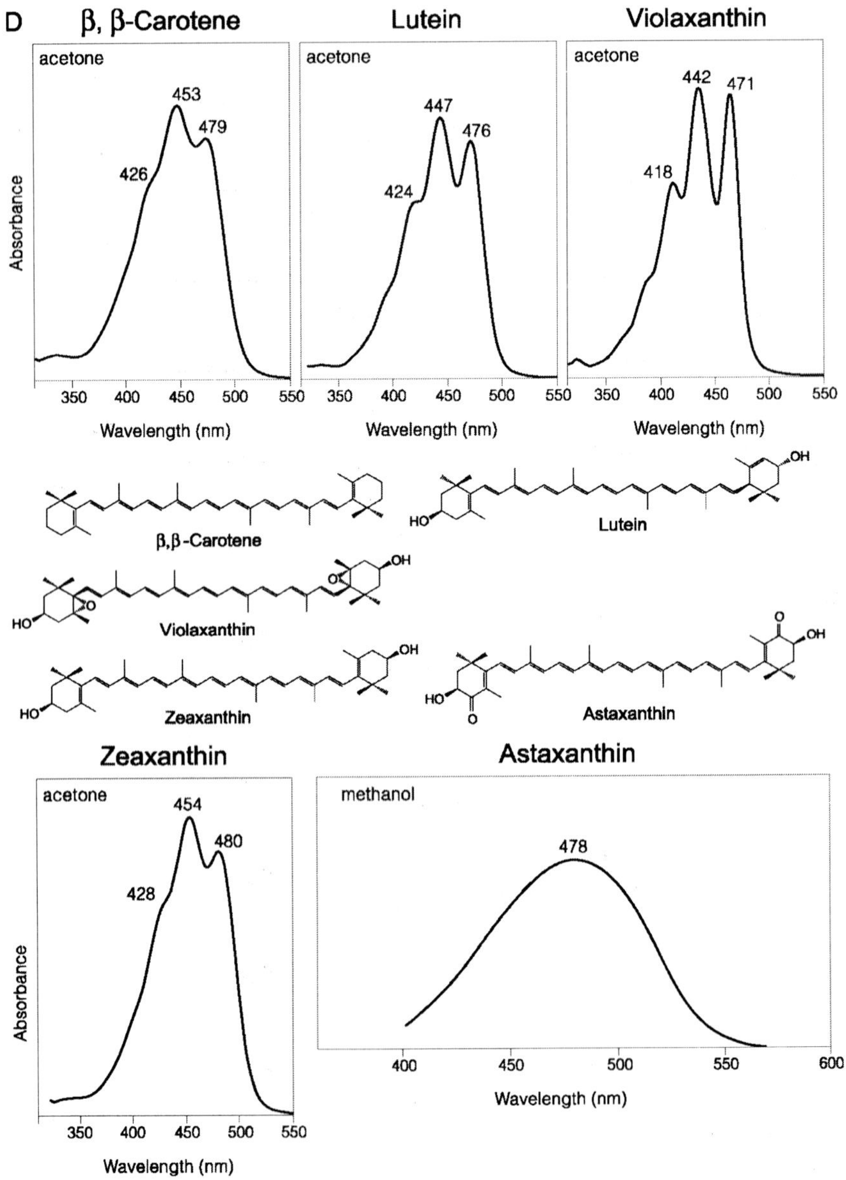

Figure 2: (D) Visible absorption spectra of β, β-carotene, lutein, violaxanthin and zeaxanthin in 100% acetone; and astaxanthin in 100% methanol.

Chlorophylls and carotenoids are essential components of the light-harvesting complexes (LHCs) of the plastids (chloroplasts), where they absorb excitation energy from sunlight and transfer it to the reaction centres (RC1 and RC2) of the two photosystems (PS1 and PS2). These are the sites of the initial chemical reactions of photosynthesis (Porra et al. 1997).

2.1 Chlorophyll Chemistry

The chemical structures of the photosynthetic pigments of terrestrial plants and the green chlorophyte algae have been extensively reviewed (Scheer 1991, 2003, 2006, Jeffrey et al. 1997b, Jeffrey and Wright 2006), and therefore only a very brief discussion will be given here.

Chls *a* and *b* are conjugated tetrapyrroles (Figure 1A, B) with ring D completely reduced at C-17 and C-18. Chls have a cyclopentanone ring (E) conjoint with ring C, and the propionic acid side chain at C-17 is esterified to a C_{20} alcohol, phytol. A central magnesium atom binds to the N atoms of the four pyrrole rings, and the macrocycle retains the capacity to bind electron donors on either side of the plane of the chlorin ring. Chls *a* and *b* can thus bind water, attach to proteins, and form self-aggregates by bonding between the Mg of one ring with the 13-keto group of another.

Chl *b* differs from Chl *a* by replacement of the methyl group of Chl *a* at position C-7 of ring B with an aldehyde group (CHO), which changes its spectral properties, polarity, and increases its stability to photo-oxidation.

Mg DVP, an intermediate in Chl *a* synthesis, is often found in trace quantities on HPLC chromatograms of *Dunaliella* extracts. It has a vinyl group at position C-3 on ring A and position C-8 on ring B, and a propionate at position C-17 on ring D (Figure 1C). Mg DVP also acts as a light-harvesting pigment in some Prasinophyta (Ricketts 1966, Jeffrey and Wright 2006).

Chl degradation products which also arise in trace quantities as artefacts of extraction or senescence (see HPLC chromatogram, Figure 4) may lose their Mg atom (pheophytins), the phytol chain (chlorophyllides, Chlides), both Mg and phytol (pheophorbides), modify ring E (pyropheophytins), form stereoisomers (epimers), or oxidation products (allomers). Examples of these structures are given in Jeffrey (1997a), Jeffrey et al. (1997a) and Porra et al. (1997), together with those of the cyanophyte divinyl Chls *a* and *b*, and the chromophyte Chl *c* family.

Chls *a* and *b* have intense absorption bands in the red (640–700 nm) and blue-green (400–470 nm) spectral regions (Figure 1) which allow them to act as efficient light-harvesting pigments. The positions of these bands (and those of the Chl *c* family, Zapata et al. 2006) are dictated by differences in the conjugation pathways of their respective macrocycles. For an up-to-date treatise on all aspects of the chlorophylls – biochemistry, biophysics, functions and applications – the reader is referred to Grimm et al. (2006).

2.2 Carotenoid Chemistry

Carotenoids are a diverse group of yellow, orange or red isoprenoid, polyene pigments (Figure 2A, B). They are synthesized in the chloroplast and are found in all photosynthetic tissues (DellaPenna 1999). While more than 700 carotenoids have been structurally characterized from the biosphere since the 1930s (Gray 1987; Britton et al. 2004), and about 60 are well-known from marine microalgae (Bjørnland, 1997a, b; Jeffrey et al. 1997a, Liaaen-Jensen 1998, Liaaen-Jensen and Egeland 1999), only a relatively small number are found in any one algal group. For example in the Dunaliellales the major carotenoids are lutein, β, β-carotene (previously known as α), violaxanthin, neoxanthin, antheraxanthin, zeaxanthin, β,α-carotene, each with specific functions. Minor amounts of biosynthetic intermediates may also accumulate under specific conditions (see Section 4.2).

Most carotenoids have a C_{40} skeleton with an alternating system of single and double bonds (Figure 2A, B), formed from condensation of eight C_5 compounds to form isoprenoid units (isopentenyl diphosphate; see Britton (1998) and Section 4.2). The two major groups of carotenoids are the carotenes, which are cyclized or uncyclized hydrocarbons, and the xanthophylls that are oxygenated derivatives of the carotenes. The base name for a carotene depends on the end groups, four types of which are usually encountered in the algae: β, ε, γ, ψ (Liaaen-Jensen and Egeland 1999 and Figure 2C). Carotenes may be modified to form xanthophylls by the addition of e.g. hydroxy, ketone or epoxide groups , variations in the 'in chain' unsaturation (acetylenic or allenic units), and esterification of hydroxyl groups with acyl or glycosidic groups. Loss of in-chain carbons may result in shortened skeletons.

Carotenoids have UV/Visible spectra in the blue-green spectral region (400–500nm), many having vibrational fine structures, usually showing three distinct peaks. Examples are given in Jeffrey et al. (1997a) and Britton et al. (2004). UV-Vis spectra of four common carotenoids from the Dunaliellales and astaxanthin from the Chlamydomonadales are given in Figure 2D.

Carotenoids have a variety of functions, acting as accessory light-harvesting pigments in photosynthetic processes, protecting the photosynthetic apparatus from photooxidation caused by high light stress, and maintaining the structural stability of pigment proteins in thylakoid membranes. Carotenoids may scavenge damaging singlet oxygen from the reaction centres or act to de-fuse reactive chlorophyll by various methods including thermal dissipation, and the operation of the photo-reversible violaxanthin epoxide cycle (Demmig-Adams et al. 1999, and Section 4.2).

2.3 Carotenoid Stereoisomers

Carotenoids are particularly labile molecules, spontaneously rearranging in solution. Most carotenoids found in microalgae have the carbon atoms arranged around the skeletal double bonds in the all-*trans*-form, producing linear molecules (Figure 2; Figure 3B). One exception is the universal occurrence of the 9′-*cis* form of neoxanthin in the chloroplasts of green algae and higher plants (Takaichi and Mimuro 1998), with its skeleton forming one in-chain 'bend' . In the *cis-trans* series of β,β-carotene (Figure 3A, 3B) major changes to the UV-visible absorption spectra occur in conjunction with the isomeric structural changes. In these cases, a so-called *cis*-peak of variable intensity, around 340 nm occurs (Britton 1995; Koyama and Fujii 1999). These steric changes can profoundly alter the function of the molecule (Ke 2001).

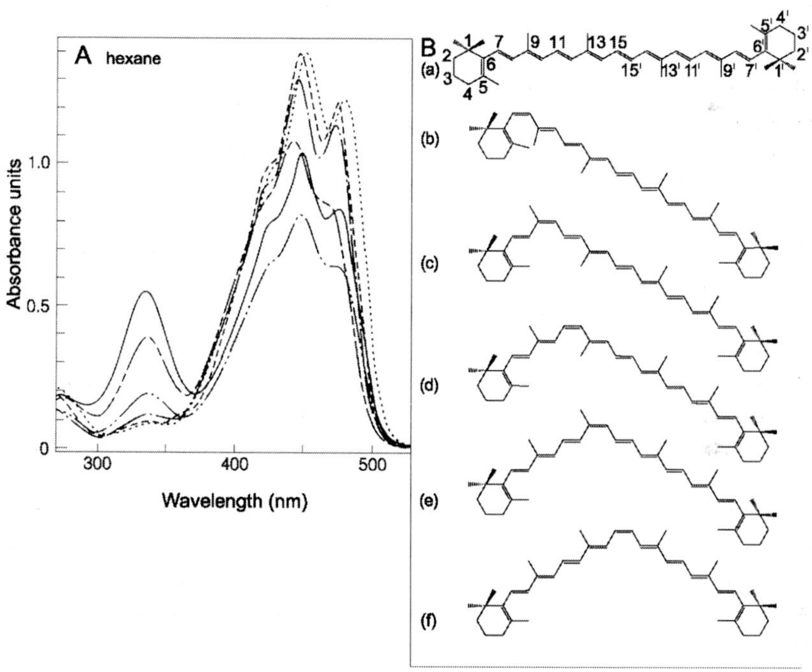

Figure 3: (A) Electronic absorption spectra of isomers of β,β-carotene in hexane and (B) their structural configurations (from Koyama and Fujii, 1999). (a) All-*trans* β,β-carotene (······), (b) 7-*cis* (– – – –), (c) 9-*cis*, (— · — ·), (d) 11-*cis* (— · · — · ·), (e) 13-*cis* (— – – — – –), and (f) 15-*cis* (——) isomers of β,β-carotene.

Due to the importance, precise location and function of carotenoid stereoisomers within the photosynthetic apparatus (Koyama et al. 1990, Kühlbrandt et al. 1994, Phillip et al. 2002, Ben-Shem et al. 2003) and the precise proportions of β,β-carotene stereoisomers formed in the carotenogenesis process in *Dunaliella* (Ben-Amotz et al. 1988, 1989, Ben-Amotz and Avron 1989, Shaish et al. 1990 and

Jiménez and Pick 1994), it is necessary to measure these isomers in any given situation. Suitable methods for HPLC analysis of stereoisomers are found in Jiménez and Pick (1994), Schierle et al. (1995) and Bononi et al. (2002).

3. Methods of Pigment Analysis

3.1 Spectrophotometry and Spectrofluorimetry

A number of spectrophotometric and fluorimetric methods of varying levels of accuracy have been used since the Arnon (1949) equations were published for the simultaneous determination of extracted Chls *a* and *b* (for reviews see Jeffrey and Welschmeyer 1997, Porra 2006, 2009). Their accuracy depends on

- the accuracy of the extinction coefficients used to derive the equations
- the spectral purity of these Chls
- the solvents used for 100% pigment extraction
- the relative proportions and concentrations of Chls in the extracts
- the presence of Chl degradation products with similar spectral and fluorescence properties to those of the parent compounds
- the precision of spectrophotometers and fluorometers used for the analysis

A one-step procedure in which Chls *a* and *b* are measured simultaneously in 80% acetone (Arnon 1949) has been, and still is, widely used. While rapid and simple, this method has proved grossly inadequate due to the inaccuracy of the Mackinney extinction coefficients used to construct the equations. Porra et al. (1989) have developed more accurate extinction coefficients for Chls *a* and *b*, in a variety of solvents, which are now highly recommended (Porra 2006, 2009). The use of these methods assumes that there are no other pigments in the extracts, such as carotenoids or other Chls, which could cause spectrophotometric or spectrofluorimetric interference. Porra (2006) lists a number of recommended 'post-Arnon' equations in various solvents – buffered 80% acetone, dimethyl formamide, dimethyl sulphoxide, chloroform, methanol etc. One set of such equations suitable for analyses of *Dunaliella* extracts are those in buffered 80% acetone (pH 7.8), with results expressed in [nmoles/ml].

$$[Chl\ a] = 13.71\ E^{663.6} - 2.85\ E^{646.6}$$
$$[Chl\ b] = 22.39\ E^{646.6} - 5.42\ E^{663.6}$$
$$[Chl\ a + b] = 19.54\ E^{646.6} + 8.29\ E^{663.6}$$

Porra (2006) also discusses the advantages and disadvantages of spectrofluorimetric assays, which, being non-discriminatory, are not recommended as highly as those for spectrophotometry. Detailed discussions of these methods can be found in Jeffrey and Welschmeyer (1997) and Porra (2006).

A simple method for measuring β,β-carotene spectrophotometrically in the presence of Chls *a* and *b* is regularly used for *Dunaliella* monitoring (Shaish et al. 1990, 1992). A small aliquot of *Dunaliella* culture is centrifuged and the pellet is extracted in a small volume (3 ml) of ethanol: hexane (2:1 v/v). After centrifuging, water and hexane (2:4 v/v) are added to the extract , which is well-mixed. Another centrifugation separates the hexane hyperphase containing the β,β-carotene from the aqueous hypophase. The β,β-carotene is measured spectrophotometrically at 450 nm in the hexane layer (see Shaish et al. 1990).

3.2 HPLC Analysis

If an accurate analysis of all pigments present in a plant extract is required, there is no alternative to using pigment separation methods, such as high performance liquid chromatography (HPLC). Great advances have been made in HPLC pigment technology since the 1980s, when new adsorbents became available and pigment standards were obtained from living log phase cultures of microalgae (Wright et al. 1991, Jeffrey and LeRoi 1997, Jeffrey and Wright 1997, Jeffrey et al. 1999, Garrido and Zapata 2006, Garrido et al. 2009). Separation of plant pigments by HPLC is particularly difficult because of the wide range of polarities of the pigments, from carotenoids diesterified with fatty acids to the polar Chlides and carotenoid diglycosides. Some closely related pigments that may need to be separated, such as the carotenoid families, differ only in the position of one double bond. The method of Zapata et al. (2000) can now separate most algal pigments (Chls, carotenoids and derivatives) in a single run from Chlide *b* (polar) to *cis*-β,β-carotene (non-polar; Wright and Jeffrey 2006).

The development of these more selective HPLC pigment methods results from the availability of new stationary phases (C_8–C_{30}), with reversed-phase monomeric octadecyl silica columns the most frequently used. Automated analysis with on-line identification using diode array detection is now regularly available. Gradient elution with buffering and ion-pairing reagents also improves separation selectivity. These topics are discussed in detail by Garrido et al. (2009) and Wright and Jeffrey (2006), and both list a range of HPLC methods for higher plant, chlorophyte and marine phytoplankton pigments. The resolution of a *Dunaliella tertiolecta* pigment extract is shown in Figure 4, using the monomeric C_8 method of Zapata et al. (2000), with a gradient elution profile from aqueous methanol/acetonitrile to methanol/acetonitrile/acetone, with pyridine acetate (pH 5.0) included as a modifier. Excellent separations are achieved with this and other

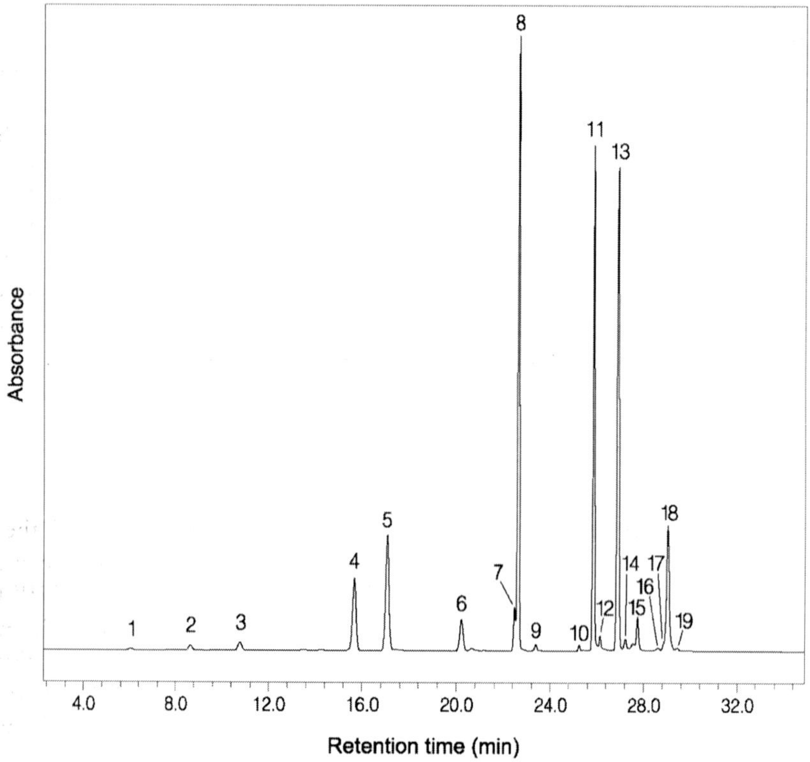

Figure 4: HPLC chromatogram of pigments of *Dunaliella tertiolecta* using the method of Zapata et al. (2000). Peak identities: 1, Mg DVP; 2, Chlide *a**; 3, methyl Chlide *a**; 4, *cis*-neoxanthin; 5, violaxanthin; 6, antheraxanthin; 7, zeaxanthin; 8, lutein; 9, *cis*-lutein*; 10, Chl *b* allomer*; 11, Chl *b*; 12, Chl *b* epimer*; 13, Chl *a*; 14, Chl *a* epimer*; 15, β, ψ-carotene; 16, pheophytin *a**; 17, β, ε-carotene; 18, β,β-carotene; 19, *cis*-β,β-carotene*. Peak names marked with an asterisk are probably artifacts of extraction. (Chromatogram courtesy Dr S.W. Wright).

methods (Garrido and Zapapta 2006) with traces of Chl degradation products and carotene isomers also being resolved. These may not always be seen on HPLC chromatograms using less precise methods (cf Wright and Jeffrey 1997).

For quantitative determination of pigments by HPLC, a list of pigment extinction coefficients is provided in Jeffrey (1997b), Britton et al. (2004) and Egeland (2009).

Suitable HPLC methods for separating stereoisomers of β,β-carotene are given in Section 2.3; and methods for separating astaxanthin esters are listed in Section 5.2.

4. Biosynthetic Pathways: Overview

The biosynthesis of chlorophylls and carotenoids have been thoroughly reviewed in recent times: those of the chlorophylls by Porra et al. (1997), Scheer (2003), Cahoon and Timko (2003), and Grimm et al. (2006), with detailed studies on various steps in the pathways by Rüdiger and Grimm (2006), Beale (2006), Jahn et al. (2006), Yaronskaya and Grimm (2006) and Rüdiger (2006).

Valuable reviews of carotenoids include those by Bjørnland and Liaaen-Jensen (1989), Britton (1998), Cunningham and Gantt (1998), Liaaen-Jensen (1998), DellaPenna (1999), Frank et al. (1999), Liaaen-Jensen and Egeland (1999), Britton et al. (2004) and Lohr (2009).

Enzyme isolations, enzyme inhibitors, gene activations and mutations, and theoretical considerations have assisted evaluating biosynthetic reaction sequences. The carotenoid sequences were more difficult to elucidate because many of the enzymes lose activity on isolation from their membrane locations. In the present treatment only a simple overview is given, hopefully sufficient to guide the reader into the wider literature cited above.

4.1 Chlorophylls

All tetrapyrroles are derived from a common biosynthetic precursor, δ-amino levulinic acid (ALA), formed by one of two known biosynthetic pathways (Figure 5). In animals and bacteria, ALA is formed from the condensation of glycine and succinyl-CoA (the ALA synthase pathway, Figure 5A), whereas higher plants and green algae use a four-step enzymic rearrangement of glutamate (the C_5 pathway, Figure 5B). The formation of Chl *a* from these precursors involves a series of some 25 reaction sequences (Porra et al. 1997), which are achieved very efficiently, although enzyme inhibitors and gene mutations can perturb the reaction pathways.

Following ALA formation, the biosynthetic process to tetrapyrroles proceeds by the condensation of two molecules of ALA to form the monopyrrole, porphobilinogen (PBG; Porra et al. 1997). A further sequence of reactions produces the tetrapyrrole macrocycle uroporphyrinogen III (Porra et al. 1997 and Figure 5). Further reactions yield protoporphyrin IX, which is the bifurcation point for iron insertion leading to haem, or magnesium insertion leading to Chl. The insertion of Mg into protoporphyrin IX gives the Mg-chelate, in an energetic process involving ATP and the enzyme Mg-chelatase. After four further enzymatic steps a fifth iso-cyclic ring E is formed (Mg-3,8-divinylpheoporphyrin a_5 monomethyl ester, Figure 1), a key intermediate in the Chl pathway. Ring D reduction yields Chlide *a* (Figure 5), which is then esterified to geranylgeranyl pyrophosphate, yielding, after three more reductase steps, a phytyl propionate ester at C-17, containing one

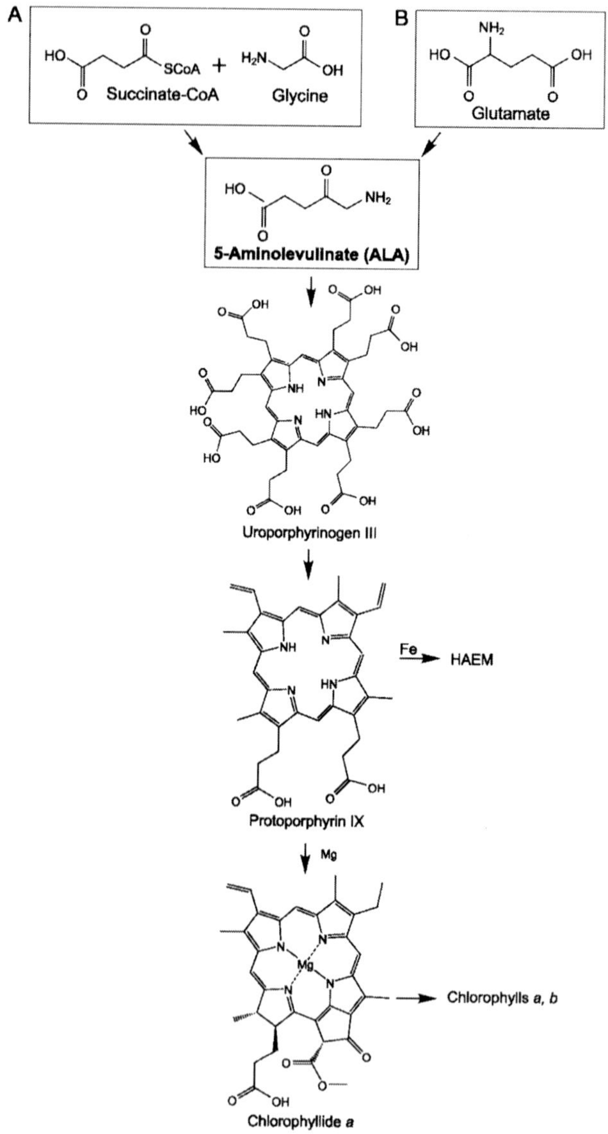

Figure 5: Scheme of tetrapyrrole biosynthesis from the formation of δ-aminolevulinate (ALA) by (A) the ALA synthase pathway (animals and bacteria) or (B) the C_5 glutamate pathway (algae, higher plants, some bacteria). Major stages to Chls include the branch points uroporphyrinogen III and protoporphyrin IX. For details of reaction sequences see Porra et al. (1997) and Rüdiger and Grimm (2006).

double bond between C-2 and C-3 of the phytol chain. The final step of Chl *a* synthesis is thus the esterification of a long chain C_{20} polyisoprene alcohol to the propionic side chain of ring D (Figure 1). Details of all these enzyme sequences are discussed by Cahoon and Timko (2003), Porra et al. (1997) and Porra et al. (2009).

Finally, Chl *b* (Figure 1, Figure 6) is formed from Chl *a* by oxidation of the methyl group at C-7 of ring B, yielding a formyl group by oxygenase activity (Porra et al. 1997, Porra et al. 2009). Chl *b* can then be reconverted to Chl *a* by hydratase and reductase activity, with Chlide *a, the central reaction product* following chlorophyllase activity (Oster et al. 2000 and Figure 6). These recently described reactions, the so-called Chlorophyll Cycle, (Ito et al. 1996, Rüdiger 2006, Porra et al. 2009) allow rapid changes to Chl *a:b* ratios, by a mechanism other than *de novo* synthesis. This cycle functions as a switching mechanism to balance biosynthesis and degradation pathways, allowing plants to quickly adjust their Chl *a:b* ratios to fit ambient conditions.

4.2 Carotenoids

De novo synthesis of carotenoids is confined to prokaryotes, fungi, algae and higher plants. Animals derive their essential carotenoids from ingested pigments, which are then often modified in the animal body (Schweigert 1998, Schiedt 1998).

In higher plants and the Chlorophyta, the earliest stages of isoprenoid biosynthesis are achieved by two independent pathways: (1) the classical acetate mevalonate pathway of the cytosol, and (2) the alternative recently discovered non-mevalonate 1-deoxy-D-xylose 5-phosphate (DOXP) pathway of the plastids (Lichtenthaler, 1999). Both pathways form the active C_5 intermediate, isopentenyl diphosphate (IPP, Figure 7), as the precursor from which all other isoprenoids, such as carotenoids, are formed by 'head-to-tail' addition. The DOXP pathway of IPP formation starts from D-glyceraldehyde 3-phosphate and pyruvate, with DOXP synthase as the starting enzyme. Using ^{13}C- and ^{2}H-labelling experiments, it is now known that the Chlorophyta and some other algal groups synthesize their IPP by the widely distributed chloroplast DOXP pathway (Lichtenthaler, 1999).

Since carotenoids are tetra-terpenoids, they are all made from isoprenoid units deriving from isopentenyl diphosphate. The enzymes responsible for these conversions are located in the chloroplast. Successive head to tail additions from IPP to form sequentially: geranyl diphosphate, farnesyl diphosphate, geranylgeranyl diphosphate, and the first C_{40} carotenoid, the colourless phytoene (Figure 8). Desaturation yields the red carotenoid lycopene, with 11 double bonds, and subsequent cyclization at both ends of the molecule brings about the two closely related hydrocarbons: β,β-carotene and β,ε-carotene (Figure 8). Carotenogenic enzymes constitute a delicate system, easily disturbed by enzyme fragility on isolation. The enzymology of these sequences is discussed in detail by Douce and Joyard (1996), Cunningham and Gantt (1998) and Hirschberg (1998).

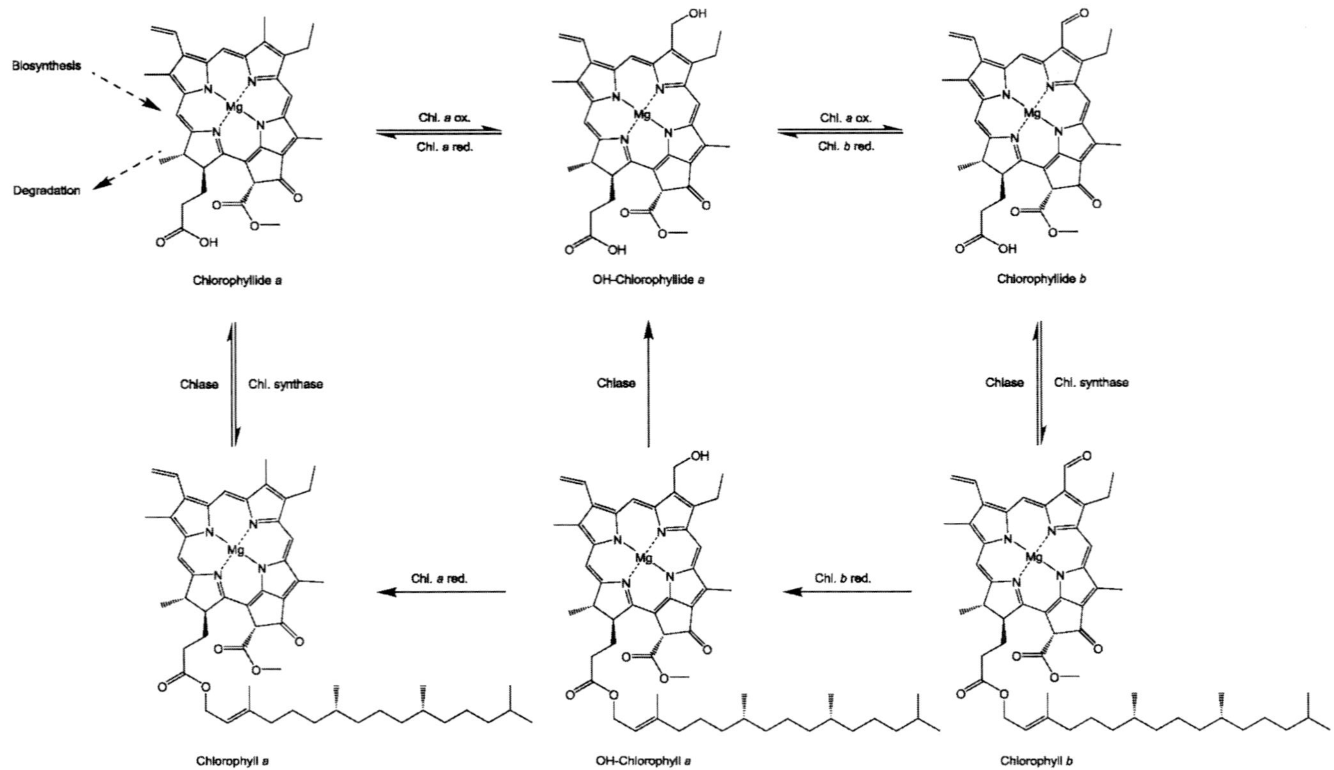

Figure 6: The Chlorophyll Cycle: the reversible transformation of Chl *a* into Chl *b* proceeds via the key intermediate Chlide *a*. Enzymes include: Chlase = chlorophyllase; Chl synthase = chlorophyll synthase; Chl *a* ox. = Chlorophyllide *a* oxygenase; Chl *b* red. = chlorophyll (ide) *b* reductase; Chl *a* red. = Hydroxy-chlorophyll(ide) *a* reductase (adapted from Rüdiger, 2006).

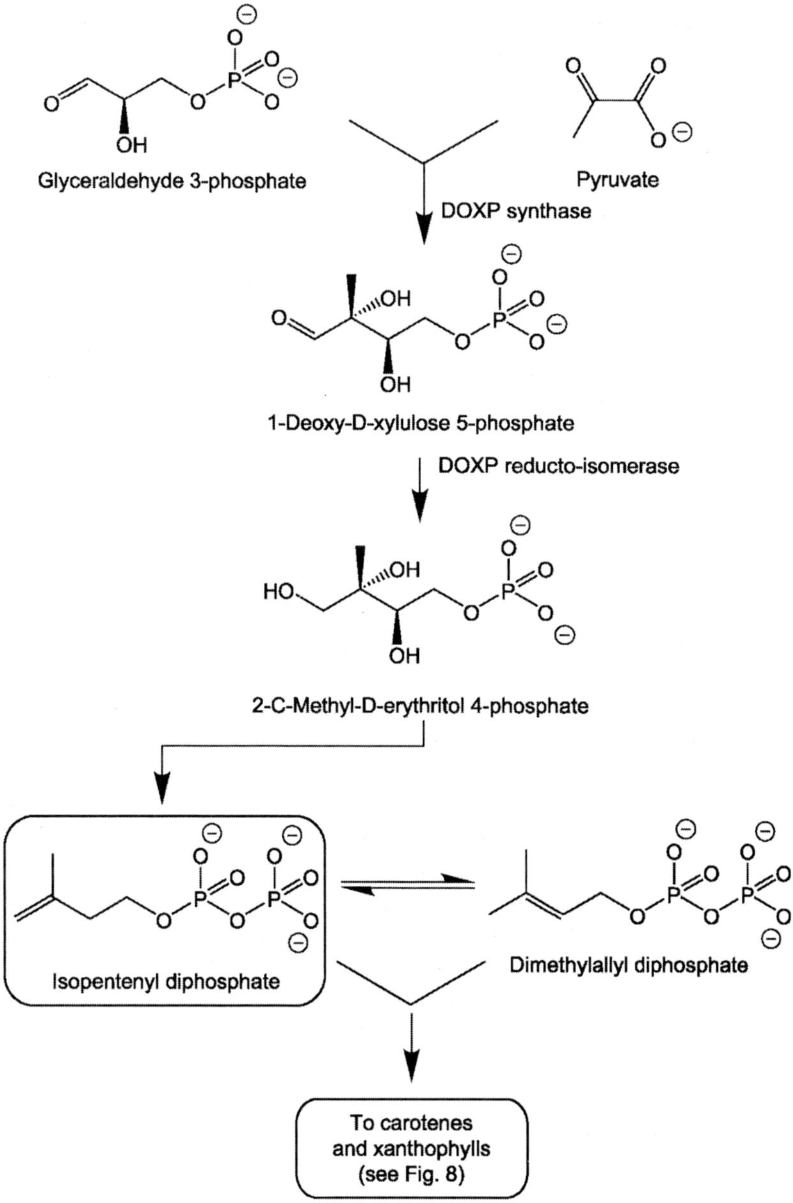

Figure 7: The biosynthesis of carotenoids in the Chlorophyta: early steps to isopentenyl diphosphate (IPP) formation, by the recently discovered non-mevalonate DOXP pathway. For details see Lichtenthaler (1999).

The common xanthophylls of the Chlorophyta are formed from the precursors of either β,ε-carotene (lutein and lutein epoxide), or β,β-carotene (zeaxanthin, antheraxanthin, violaxanthin, and neoxanthin, Figure 8). Antheraxanthin and violaxanthin are formed by successive dark epoxidation reactions, which can be recycled back to zeaxanthin by successive light-induced de-epoxidations. This reversible light-induced epoxide cycle effectively quenches excited Chl molecules produced by photosynthesis, and protects the photosynthetic apparatus from damaging singlet oxygen and excited triplet chlorophyll reactions (Demmig-Adams et al. 1999).

Once phytoene desaturation proceeds, the primary carotenoids of the green algal chloroplast are formed, although the regulation of this process is not well understood. Environmental aspects such as light intensity and the redox state of quinones etc. are of importance (Douce and Joyard 1996, DellaPenna 1999).

4.3 Pigment proteins

Carotenoids are non-covalently bound to protein complexes in the thylakoid membrane, some of which are structural components of the photosynthetic apparatus. In general, carotenoid levels are directly proportional to those of Chl in the photosynthetic tissues, with lutein and neoxanthin being correlated with Chl *b* levels and β,β-carotene with Chl *a*. Little is known about the sequence of reactions incorporating the Chls and carotenoids into the pigment-proteins of the light-harvesting complexes (LHC1 and LHC2). The pigments function in photosynthesis only when associated with specific *apo*-proteins in the thylakoid membranes. The Chls and carotenoids are tightly packed by their *apo*-protein binding sites and the Chl concentration may be as high as 0.3 M (Kühlbrandt et al. 1994), ensuring a high capacity for photon capture.

The photosystems (PS1 and PS2) are large pigment-protein complexes with a central reaction centre (RC1 and RC2) surrounded by antennae complexes that harvest and transfer the light energy to the reaction sites. These contain Chl *a* and β,β-carotene and also the specialist Chls: pheophytin *a* and Chl *a* epimer. The antennae LHCs are a large family of related proteins that bind the xanthophylls lutein, violaxanthin and neoxanthin and Chls *a* and *b* to the protein complexes. The assembly of these pigment proteins is discussed by DellaPenna (1999).

The carotenoids also have various Chl protective functions requiring precise juxtaposition to the Chls. A brilliant advance in structural analysis has been the solving of the atomic structure of plant LHC2 at 3.4 Å by electron crystallography (Kühlbrandt et al. 1994). Each LHC2 contained 12-13 Chl molecules plus two carotenoid (lutein) molecules in the *trans*-configuration, cross-braced in the centre of the molecule, giving structural stability. Lutein is present in the proportion of two molecules of carotenoid to one LHC monomer. The positioning of the Chls

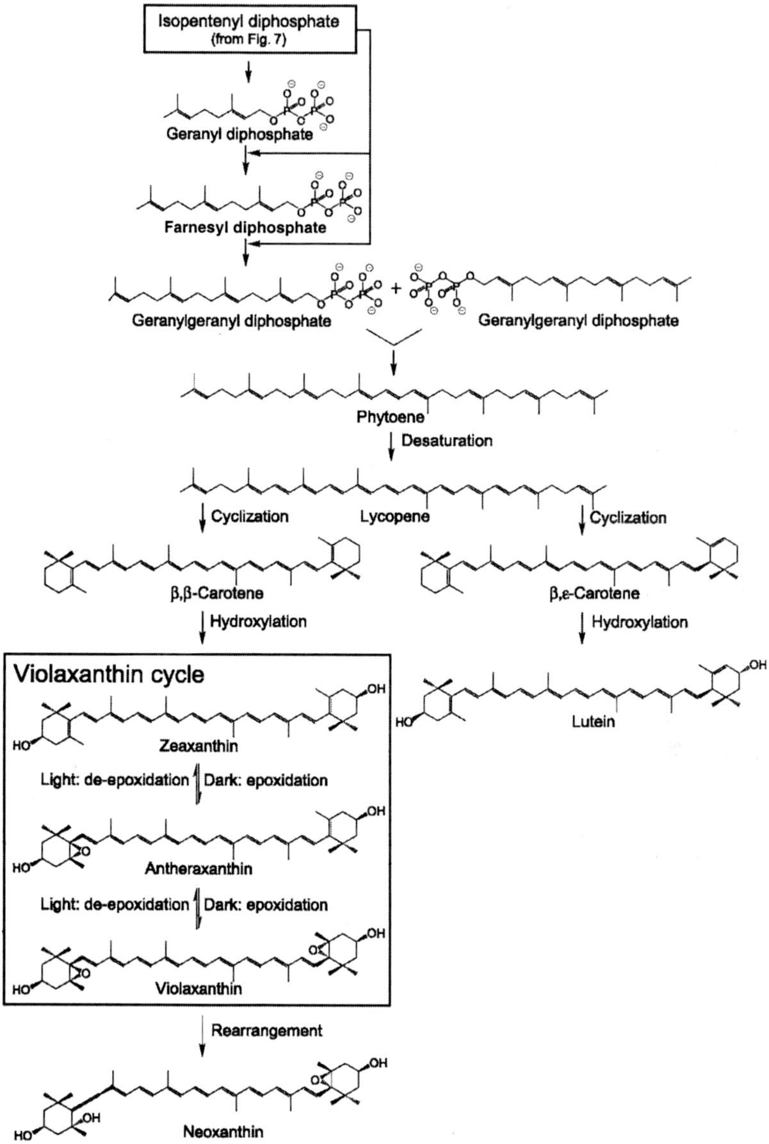

Figure 8: The biosynthetic pathway for the formation of β,β-and β,ε-carotene and some chlorophyte xanthophylls from the early intermediate isopentenyl diphosphate (IPP; see Figure 7). The boxed section shows the light-reversible violaxanthin epoxide cycle used by chlorophytes and higher plants to prevent photoinhibition (adapted from Porra et al. 1997).

and carotenoids in the complex suggest their importance in quenching excited Chl triplets during high light stress. The central position of lutein explains why this carotenoid is so essential for the light-harvesting efficiency of reconstitution experiments (Phillip et al. 2002), and why it is so conserved in the Chlorophyta and higher plants.

Similar studies of the crystal structure of plant PS1 at 4.4 Å resolution, showed an intricate structure of 12 core subunits, 4 different light-harvesting membrane proteins, 45 trans-membrane helices, 167 Chls, 3 Fe-S complexes, and 2 phylloquinones, with about 20 Chls strategically positioned between LHC1 and the core (Ben-Shem et al. 2003). These and other studies of xanthophyll binding to PS2 (Phillip et al. 2002) have revealed aspects of the precise molecular architecture of these complexes, providing a framework for understanding the evolutionary forces that must have shaped the photosynthetic apparatus of plants, following the divergence of photosynthetic pigment membranes more than one billion years ago (see Section 1, Introduction).

In summary, pigment biosynthesis in all plants and the Chlorophyta involves not only the biosynthetic pathways of building the individual Chl and carotenoid molecules themselves (Figures 5 to 8), but also the little understood regulation of achieving the precise pigment architectural arrangements within the protein complexes that make up the LHCs, the photosystems and the reaction centres (Grossman et al. 1995, Douce and Joyard 1996, DellaPenna 1999). Genetic and environmental controls, not considered here, also contribute to the regulation of these biosynthetic processes (Lers et al. 1990), and need to be further investigated.

5. Pigments of red forms of chlorophytes: environmental triggers, biology and function

Certain green algae (e.g. strains of *Dunaliella salina, D. bardawil, Haematococcus pluvialis* etc.) have the ability to convert to red forms under particular stressful environmental conditions. In these situations massive amounts of carotenoids are synthesized - mainly β,β-carotene in *Dunaliella* and the β,β-carotene derivative, 3,3'-dihydroxy-β,β-carotene-4,4'-dione, otherwise known as astaxanthin, in *Haematococcus.* These carotenoids may be 'over-produced' from a low of 0.3% dry weight in green forms to up to 10% dry weight in red cells, providing excellent biological sources for commercial development. Environmental triggers may be high light (up to 2,000 μEinstein/m^{2}s), low nutrients (nitrate, phosphate and/or sulphate), high salinity (e.g. salt ponds), excessively high or low temperatures, and combinations of these stressors. A very extensive literature exists, from which some highlights have been extracted (see below).

5.1 *Dunaliella*

Only certain strains of *Dunaliella* species are capable of high β,β-carotene biosynthesis. As described elsewhere in this volume, *Dunaliella* species are flagellated, halotolerant, motile, green unicells lacking a rigid cell wall, but having a glycocalyx-type cell envelope (Borowitzka and Borowitzka 1988). The cells have a single cup-shaped chloroplast (Figure 9A), containing a number of interconnected and appressed membranes (thylakoids), within which the highly pigmented photosynthetic apparatus is located. The chloroplast contains a large central pyrenoid surrounded by starch grains, and the large apical nucleus lies partly within the lobes of the chloroplast. The golgi apparatus lies beneath the apical flagella base. As there is no rigid cell wall, cell shape in *Dunaliella* can be very variable, often becoming spherical under unfavourable conditions. Under very extreme conditions (e.g. salt evaporation ponds, subject to high light intensities) carotenogenic strains of *Dunaliella* become red in colour, due to the massive accumulation of β,β-carotene-containing oil droplets within the chloroplast (Figure 9B). The cells do not lose their motility under these conditions.

The red forms of *D. salina* and *D. bardawil* have been studied in detail to learn more of the carotenogenic process, the nature of the transduction signals, and to optimize the harvest of the desired commercial product, β,β-carotene.

Selected data from the large literature include the following:

(a) Properties of red β,β-carotene globules from *D. salina* and *D. bardawil*

- β,β-Carotene globules from red cells are membrane-free droplets within the chloroplast but discrete from the thylakoids (Figure 9B, Shaish et al. 1991).
- The globules contain 65% β,β-carotene, 28% lipids, 3% protein and 0.5% Chl on a % dry weight basis (Shaish et al. 1991).
- The carotene isomer composition of the globules is 42% all-*trans*-β,β-carotene, 41% 9-*cis*-β,β-carotene and 10% 15-*cis*-β,β-carotene (Ben-Amotz et al. 1982).
- β,β-Carotene accumulation is dependent on gene activation; precise sites and mechanisms are unknown (Lers et al. 1990).
- Increased synthesis of oleic acid (18:1) and β,β-carotene are correlated at high light intensities (Mendoza et al. 1996, 1999).
- Triacylglycerol deposition stimulated β,β-carotene synthesis and created a localized sink in the chloroplast for carotenoid accumulation (Rabbani et al. 1998).
- Inhibition of fatty acid synthesis reduced β,β-carotene accumulation (Rabbani et al. 1998).
- A 38-kD protein was associated with β,β-carotene globules, stabilizing globule structure within the chloroplast. Trypsin treatment de-stabilized the globules (Katz et al. 1995).

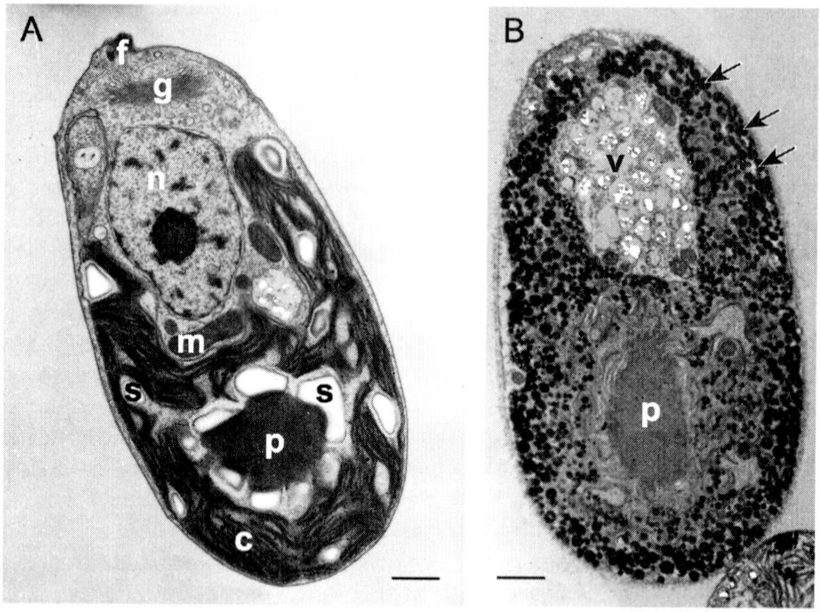

Figure 9: Electron micrograph (longitudinal section) of (A) a green cell of *Dunaliella tertiolecta* showing the cup-shaped chloroplast (c), pyrenoid (p), starch grains (s), mitochondrion (m), nucleus (n), golgi apparatus (g), and flagella base (f). (B) is a similar section of a red cell of *Dunaliella salina* showing a large vacuole (v) and many β,β-carotene containing osmiophilic globules within the chloroplast (arrows). Scale bars = 5 μm. (Micrographs courtesy of Dr Maret Vesk (A), and Professor Michael Borowitzka (B)).

(b) Environmental and physiological aspects of red forms of *Dunaliella*

- High salinity, high light intensity, low pH, nitrogen and phosphate deficiencies promoted growth of red *Dunaliella* cells (Loeblich 1982)
- High light intensity stimulated β,β-carotene biosynthesis (Ben-Amotz and Avron 1989).
- β,β-Carotene globules acted as a cellular blue-light filter (see Figure 2D) protecting the cell against blue light-induced photodamage. No protection was given to photodamage by red light, which was not absorbed by the β,β-carotene (Figure 2D and Ben-Amotz et al. 1989).
- Active photosynthetically-produced oxygen radicals triggered massive accumulation of β,β-carotene in *D. bardawil* (Shaish et al. 1993).
- Inhibitors of β,β-carotene synthesis showed that the isomerization reaction producing the 9-*cis*-isomers occurred at or before the production of phytoene in the biosynthetic pathway (Figure 8 and Shaish et al. 1990).

- Production of red cells, in the presence of multi-stressors, was the combined result of inhibition of cell division (high salinity, nutrient limitation etc.) and increased carotenogenesis (high light), and other factors.

Summary While the biochemical characteristics of the red β,β-carotene globules in *Dunaliella* are quite well understood, the nature of the transduction signals and the gene activation sites of the biosynthetic carotenogenic process are relatively unknown.

5.2 *Haematococcus*

Haematococcus pluvialis Flotow is a second chlorophyte that has been widely studied commercially for its ability to synthesize large amounts of the keto-carotenoid astaxanthin, a valuable food additive used to colour the flesh of salmon, trout and shrimp in the farming situation. It is also under development for its anti-oxidant medical applications (Boussiba 2000). Astaxanthin is a derivative of β,β-carotene (3,3'-dihydroxy-β,β-carotene-4,4'-dione, see Figure 2D), and, like β,β-carotene in *Dunaliella*, its accumulation is induced by a variety of environmental stresses, including high solar irradiance, nitrogen limitation, desiccation, high salinity and reactive oxygen species (ROS). Unlike *Dunaliella* however, carotenoid accumulation is accompanied by a remarkable encystment of the green motile vegetative cell to an inert red cyst (aplanospore). Boussiba (2000) and Kobayashi (2003) have provided schematic overviews of this process.

(a) Characteristics of aplanospore formation in *Haematococcus*

- Motile green vegetative cells of the freshwater chlorophyte *Haematococcus* enlarge and transform to red immotile cysts (aplanospores), during the stressful conditions that stimulate astaxanthin accumulation (Zlotnik et al. 1993, Boussiba 2000, Kobayashi 2003).
- Carotenoid biosynthesis to the β,β-carotene stage (Figure 8) initially takes place in the chloroplast, but secondary carotenoids (e.g. Figure 10) eventually accumulate as astaxanthin esters in the lipid vesicle compartment of the cytoplasm (Grünewald et al. 2001).
- Transport of carotenoid intermediates from the chloroplast to the cytoplasmic vesicles is proposed, supported by the locations of the oxygenase and hydroxylase enzymes (Grünewald et al. 2001), suggesting that two sites of carotenoid synthesis may operate (Boussiba 2000).
- Carotenoid distribution in vegetative cells and aplanospores calculated as a % of total carotenoid shows the clear change in carotenoid biosynthesis patterns from primary carotenoids in vegetative cells to secondary carotenoids in aplanospores (Harker and Young 1995 and Table 1).

Table 1: Pigment distribution between vegetative cells and aplanospores (from Harker and Young, 1995).

Pigment	Vegetative cells	Aplanospores
	(%)	(%)
Neoxanthin	13.0	ND
Violaxanthin	14.5	ND
Lutein-5,6-epoxide	5.2	ND
Lutein	53.8	1.0
β,β-Carotene	13.5	3.0
Canthaxanthin	ND	2.0
Adonirubin	ND	2.0
Echinenone	ND	3.0
Astaxanthin (unesterified)	ND	1.0
Astaxanthin (monoester)	ND	49.0
Astaxanthin (diester)	ND	39.0

ND = Not Detected

- Astaxanthin accumulates as mono- and diacylesters (70–80% and 20–30% respectively), up to 4% of the dry weight of the cell (Harker and Young 1995). Esterification may facilitate astaxanthin solubility (Bidigare et al. 1993).
- HPLC methods for separation of astaxanthin esters include those of Bidigare et al. (1993), Fraser et al. (1998), Hanagata and Dubinsky (1999), and Orosa et al. (2000).

(b) Astaxanthin biosynthesis and its regulation

- Studies of enzymes (Grünewald et al. 2001) and their inhibitors (Harker and Young 1995) indicate that the pathway to astaxanthin accumulation in *Haematococcus* starts at b,b-carotene and proceeds via the intermediates echinenone, canthaxanthin, adonirubin, astaxanthin and astaxanthin esters (Kobayashi 2003) i.e. stepwise addition of epoxides precedes that of the hydroxyls (Figure 10). An alternate pathway is also shown in Figure 10 although the evidence for this is less secure.
- Fatty acid biosynthesis facilitates increased astaxanthin deposition in the cytoplasmic vesicles.
- High growth temperatures retarded vegetative growth and promoted aplanospore formation (Krishna and Mohanty 1998).
- Inhibition of cell division caused massive accumulation of astaxanthin (Boussiba and Vonshak 1991).

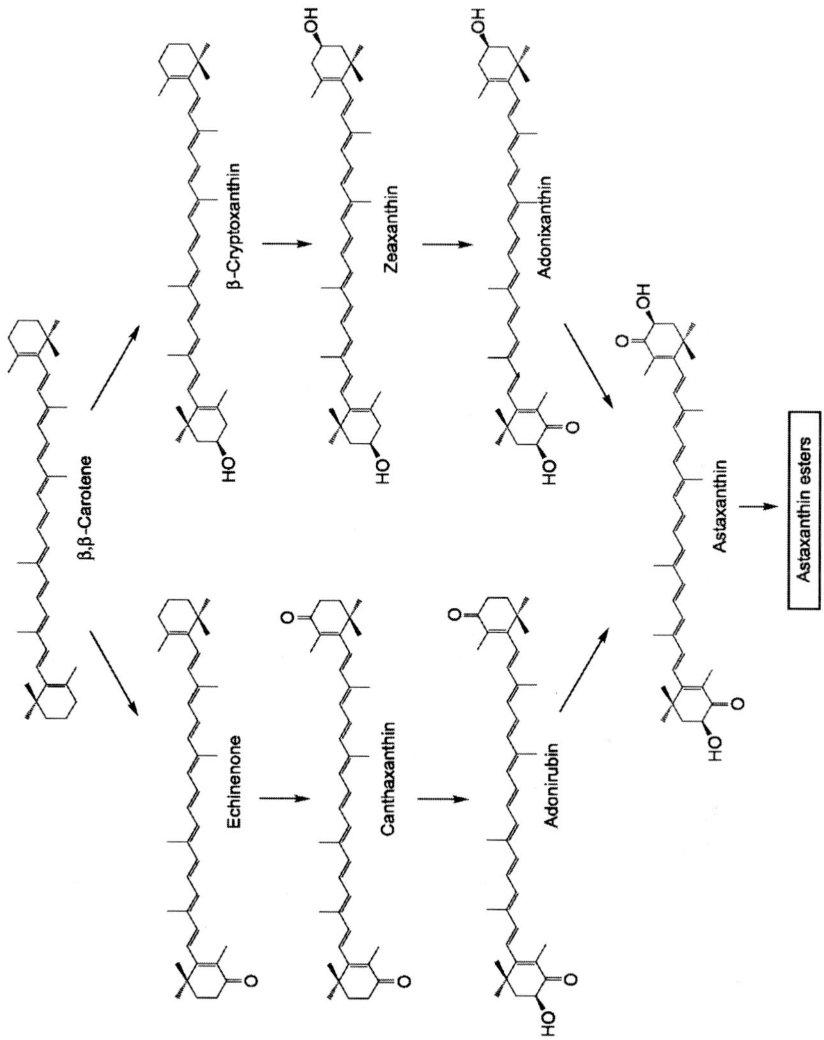

Figure 10: Possible biosynthetic pathways from β,β-carotene to astaxanthin in red aplanospores of *Haematococcus pluvialis* (from Kobayashi, 2003). The evidence favours the pathway via echinenone, canthaxanthin, adonirubin to astaxanthin and its esters.

(c) High light intensity and astaxanthin function

- Astaxanthin biosynthesis is stimulated by reactive oxygen species (ROS) generated by the photosynthetic process (Kobayashi et al. 1993).

- Singlet oxygen [$^1O_2$] is the dominant ROS involved in astaxanthin accumulation. It can cross biological membranes where it can leak into the cytoplasm and be quenched chemically by astaxanthin. However, not all investigators support this idea.
- Astaxanthin may act as a carotenoid screen protecting the whole cell from excessive blue light radiation that could damage the antennae Chls and nuclear processes. However, some researchers doubt whether the overlap between the absorption spectra of astaxanthin and Chl (Figs. 1 and 2D) is sufficient to afford protection.
- If the role of astaxanthin is not that of a sunshade, what is its function?
- For carotenoids to have a direct quenching role on triplet Chl [^{3}Chl] or singlet oxygen [$^1O_2$], there needs to be a close physical association of the carotenoids with the excited molecules. Some researchers think the location of the astaxanthin in the cytoplasmic vesicles is not close enough to the pigments of the LHCs, photosystems and reaction centres in the thylakoids (contrary to those of the thylakoid-located violaxanthin epoxide cycle, Figure 8), to afford protection.

Summary

In spite of the significant number of studies of *Haematococcus*, further biochemical and genetic investigations are needed to locate the precise sites of light activation of the biosynthetic pathway to astaxanthin. How these processes are linked to aplanospore formation also deserves attention.

5.3 Additional Chlorophytes

Additional chlorophytes that turn red in harsh environments have been surveyed as possible replacements for the slow growing *Haematococcus* as an astaxanthin source (Orosa et al. 2000). These include the green and red snow algae (e.g. *Chlamydomonas nivalis* (Bauer) Wille; Bidigare et al., 1993), a wall-inhabiting alga *Chlorococcum* sp. (Zhang et al. 1997, Masojidek et al. 2000, Liu and Lee 2000), the bark-inhabiting species *Scenedesmus komarekii* Hegewald (Hanagata and Dubinsky 1999), and other chlorophytes: *Chlorella zofingiensis, Chlorella emersonii* and *Ermosphera viridis* (Rise et al. 1994).

In all cases the cells converted from green to astaxanthin-containing brown, orange and red cells under high light intensity and nitrogen limitation. Other secondary carotenoids were detected in the non-green cells, accumulating in the lipoidal vesicles of the cytoplasm, which in some cases gradually filled the entire cell. This accumulation pattern, which differed from that of *Haematococcus,* was particularly studied in *Scenedesmus komarekii* and *Chlorella zofingiensis.*

Overall, the studies carried out on the above chlorophytes had some similar characteristics to *Haematococcus*:

- Secondary carotenoid accumulation was stimulated by nitrogen limitation and high growth irradiance.
- Esterified astaxanthins were the main carotenoids formed.
- No species tested formed a red aplanospore similar to that of *Haematococcus*, which returns to the green vegetative state when the light intensity is diminished and growth nutrients are restored.
- Most researchers suggested that esterification and lipid synthesis facilitated the solubility of the astaxanthin in the cytoplasmic vesicles.
- Very few studies of the biosynthetic pathway were carried out on these species, although most authors agreed that β,β-carotene was the starting point for astaxanthin synthesis.
- Most researchers suggested that secondary carotenoids functioned to protect the photosynthetic apparatus against photo-oxidative damage under high irradiance.
- No author recommended that the chlorophyte they studied was superior to *Haematococcus*, but many thought their species might be a useful adjunct to the field.

Summary

In order to maximize the biosynthesis of astaxanthin, more needs to be learnt about genetic and environmental activation of the biosynthetic enzymes involved, as well as the precise mechanism of ROS activation of this process.

No matter what the mechanism, there is no doubt that secondary carotenoids are a valuable survival strategy for life in a precarious environment.

Similar species and stressful environments may yet provide the biotechnology industry with further strains for exploitation.

6. Acknowledgements

The authors thank Ms Denise Schilling and Ms Leonie Wyld for typing support, Ms Louise Bell for design of the figures, the CSIRO librarians for instant and valuable assistance, and the Chief of the CSIRO Division of Marine and Atmospheric Research, Dr Greg Ayers, for continuing interest and support. Dr Shirley W. Jeffrey thanks her friend, Dr D.V. Subba Rao, a co-editor of this book, for his infinite patience.

Dr Simon Wright provided the HPLC chromatogram for Figure 4 and Dr Maret Vesk and Professor Michael Borowitzka provided the electron micrographs for Figs. 9A and 9B respectively.

References

Andersen, R.A. 2005. Algal culturing techniques. Elsevier Inc., Amsterdam.

Arnon, D.I. 1949. Copper enzymes in isolated chloroplasts: Polyphenoloxidase in *Beta vulgaris*. Plant Physiol. 24: 1-15.

Beale, S.I.2006. Biosynthesis of 5-aminolevulinic acid. pp. 147-158. In Chlorophylls and Bacteriochlorophylls: Biochemistry, Biophysics, Functions and Applications. B. Grimm, R.J. Porra, W. Rűdiger and H. Scheer (eds) Springer, Dordrecht, The Netherlands.

Ben-Amotz, A. and M. Avron. 1989. The wavelength dependence of massive β-carotene synthesis in *Dunaliella bardawil* (Chlorophyceae). J. Phycol. 25: 175-178.

Ben-Amotz, A., A. Katz and M. Avron. 1982. Accumulation of β-carotene in halotolerant algae: purification and characterization of β-carotene-rich globules from *Dunaliella bardawil* (Chlorophyceae). J. Phycol. 18: 529-537.

Ben-Amotz, A., A. Lers and M. Avron. 1988. Stereoisomers of β-carotene and phytoene in the alga *Dunaliella bardawil*. Plant Physiol. 86: 1286-1291.

Ben-Amotz, A., A. Shaish and M. Avron. 1989. Mode of action of the massively accumulated β-carotene of *Dunaliella bardawil* in protecting the alga against damage by excess irradiation. Plant Physiol. 91: 1040-1043.

Ben-Shem, A., F. Frolow and N. Nelson. 2003. Crystal structure of plant photosystem I. Nature. 426: 630-635.

Bhattacharya, D. 1997. Origins of algae and their plastids. Springer-Verlag, Vienna, Austria.

Bidigare, R.R., M.E. Ondrusek, M.C. Kennicutt II, R. Iturriaga, H.R. Harvey, R.W. Hoham and S.A. Macko. 1993. Evidence for a photoprotective function for secondary carotenoids of snow algae. J. Phycol. 29: 427-434.

Bjørnland, T. 1997a. Structural relationships between algal carotenoids. pp. 572-577. In Phytoplankton Pigments in Oceanography: Guidelines to Modern Methods. S.W. Jeffrey, R.F.C. Mantoura and S.W. Wright (eds) UNESCO Publishing, Paris, France.

Bjørnland, T. 1997b. UV-vis spectroscopy of carotenoids. pp. 578-594. In Phytoplankton Pigments in Oceanography: Guidelines to Modern Methods.S.W. Jeffrey, R.F.C. Mantoura and S.W. Wright (eds) UNESCO Publishing, Paris, France.

Bjørnland, T. and S. Liaaen-Jensen. 1989. Distribution of carotenoids in relation to chromophyte phylogeny and systematics. pp. 37-60. In The Chromophyte Algae: Problems and Perspectives. J.C. Green, B.S.C. Leadbeater and W.L. Diver (eds) Clarendon Press, Oxford, U.K.

Bononi, M. and I. Commissati, E. Lubian, A. Fossati and F. Tateo. 2002. A simplified method for the HPLC resolution of α-carotene and β-carotene (*trans* and *cis* isomers). Anal. Bioanal. Chem. 372: 401-403.

Borowitzka, M.A. and L.J. Borowitzka. 1988. *Dunaliella*. pp. 27-58. In Microalgal Biotechnology. M.A. Borowitzka and L.J. Borowitzka (eds)Cambridge University Press, Cambridge, U.K.

Borowitzka, M.A. and J.C. Siva. 2008. The taxonomy of the genus *Dunaliella* (Chlorophyta, Dunaliellales) with emphasis on the marine and halophilic species. J. Appl. Phycol. (in press).

Boussiba, S. 2000. Carotenogenesis in the green alga *Haematococcus pluvialis*: cellular physiology and stress response. Physiologia Plantarum 108: 111-117.

Boussiba, S. and A. Vonshak. 1991. Astaxanthin accumulation in the green alga *Haematococcus pluvialis*. Plant Cell Physiol. 32: 1077-1082.

Britton, G. 1995. UV/Visible spectroscopy. pp. 13-62. In Carotenoids, Vol. 1 B: Spectroscopy. G. Britton, S. Liaaen-Jensen and H. Pfander (eds) Birkhäuser, Basel, Switzerland.

Britton, G. 1998. Overview of carotenoid biosynthesis. pp. 13-147. In Carotenoids, Vol. 3: Biosynthesis and metabolism. G. Britton, S. Liaaen-Jensen and H. Pfander (eds) Birkhäuser, Basel, Switzerland.

Britton, G., S. Liaaen-Jensen, and H. Pfander (eds) 2004. Carotenoids: Handbook, compiled by A. Z. Mercadante and E. S. Egeland. Birkhäuser Verlag, Basel, Switzerland.

Cahoon, A.B. and M.P. Timko. 2003. pp. 95-131. Biochemistry and regulation of chlorophyll biosynthesis. In Photosynthesis in Algae.A.W Larkum, S.E Douglas and J.A. Raven (eds) Kluwer Academic Publishers, Dordrecht, The Netherlands.

Cavalier-Smith, T. 2000. Membrane heredity and early chloroplast evolution. Trends in Plant Sci. 5: 174-182.

Cunningham, F.X. and E. Gantt. 1998. Genes and exzymes of carotenoid biosynthesis in plants. Annu. Rev. Plant Physiol. 49: 557-583.

DellaPenna, D. 1999.Carotenoid synthesis and function in plants: insights from mutant studies in *Arabidopsis thaliana*. pp. 21-37. In The Photochemistry of Carotenoids. Advances in Photosynthesis, Vol. 8 H.A. Frank, A.J. Young, G. Britton and R.J. Cogdell (eds) Kluwer Acad. Publ., Dordrecht, The Netherlands.

Delwiche, C.F. 1999. Tracing the thread of plastid diversity through the tapestry of life. Amer. Nat. 154: S164-S177.

Demmig-Adams, B., W.W. Adams, V. Ebbert, V. and B.A. Logan.1999. Ecophysiology of the xanthophyll cycle. pp. 245-269. In The Photochemistry of the Carotenoids. H.A. Frank, A.J. Young, G. Britton and R.J.Cogdell (eds) Kluwer Academic Press. Dordrecht, The Netherlands.

Douce, R. and J. Joyard.1996. Biosynthesis of thylakoid membrane lipids. pp. 69-101. In Oxygenic photosynthesis: the light reactions.D.R. Ort and C.F. Yocum (eds) Kluwer Academic Publishers, Dordrecht, The Netherlands.

Egeland, E.S. 2009. Absorption coefficients. In Phytoplankton Pigments in Oceanography, Vol 2. S. Roy, E.S. Egeland, G. Johnsen and C. Llewellyn (eds) Publisher to be announced.

Egeland, E.S., R.R.L. Guillard and S. Liaaen-Jensen. 1997. Additional carotenoid prototype representatives and a general chemosystematic evaluation of carotenoids in Prasinophyceae (Chlorophyta). Phytochem. 44: 1087-1097.

Frank, H.A., A.J. Young, G. Britton and R.J. Cogdell. [eds.] 1999. The Photochemistry of Carotenoids. Kluwer Academic Publishers, Dordrecht, The Netherlands.

Fraser, P.D., H. Shimada and N. Misawa. 1998. Enzymic confirmation of reactions involved in routes to astaxanthin formation, elucidated using a direct substrate *in vitro* assay. Eur. J. Biochem. 252: 229-236.

Garrido, J.L. and M. Zapata. 2006. Chlorophyll analysis by new high performance liquid chromatography methods. pp. 109-121. In Chlorophylls and Bacteriochloropylls: Biochemistry, Biophysics and Biological Function. B. Grimm, R.J. Porra, W. Rűdiger and H. Scheer (eds) Springer, Dordrecht, The Netherlands.

Garrido, J.L., R.L. Airs, L. Van Heukelem, M. Zapata. 2009. New HPLC separation techniques. In Phytoplankton Pigments in Oceanography, Vol 2. S. Roy, E.S. Egeland, G. Johnsen, C. Llewellyn and S.W. Wright (eds) Submitted. Publisher to be announced.

Gray, J.C. 1987. Control of isoprenoid biosynthesis in higher plants. Adv. Bot. Res. 14: 25-91.β

Grimm, B., R.J. Porra, W. Rüdiger, and H. Scheer. (eds). 2006. Chlorophylls and Bacteriochlorophylls: Biochemistry, Biophysics, Functions and Applications. Springer, Dordrecht, The Netherlands.

Grossman, A.R., D. Bhaya, K.E. Apt and D. M. Kehoe. 1995. Light-harvesting complexes in oxygenic photosynthesis: diversity, control and evolution Annu. Rev. Genetics 29: 231-288.

Grünewald, K., J. Hirschberg and C. Hagen. 2001. Ketocarotenoid biosynthesis outside of plastids in the unicellular green alga *Haematococcus pluvialis*. J. Biol. Chem. 276: 6023-6029.

Grzebyk, D. and O. Schofield, V. Vetrani and P.G. Falkowski. 2003. The Mesozoic radiation of eukaryotic algae: the portable plastid hypothesis. J. Phycol. 39: 259-267.

Hanagata, N. and Z. Dubinsky. 1999. Secondary carotenoid accumulation in *Scenedesmus komarekii* (Chlorophyceae, Chlorophyta). J. Phycol. 35: 960-966.

Harker, M. and A.J. Young. 1995. Inhibition of astaxanthin synthesis in the green alga, *Haematococcus pluvialis*. Eur. J. Phycol. 30: 179-187.

Hirschberg, ˙J. 1998. Molecular biology of carotenoid biosynthesis. pp. 149-194. In Carotenoids, Vol. 3: , Birkhäuser,s and Metabolism. Birkhaűser, G. Britton, S. Liaaen-Jensen and H. Pfander (eds) Basel, Switzerland.

Ito, H., T. Ohtsuka and A. Tanaka. 1996. Conversion of chlorophyll *b* to chlorophyll *a* via α-hydroxy methyl chlorophyll. J. Biol. Chem. 271: 1475-1479.

Jahn, D., J. Moser, W. Schubert and D.W. Heinz. 2006. Transfer RNA-dependent aminolevulinic acid formation: structure and fu.ıction of glutamyl-tRNA synthetase, reductase and glutamate-1-semialdehyde-2,1-aminomutase. pp.159-171. In Chlorophylls and Bacteriochlorophylls: Biochemistry, Biophysics, Functions and Applications. B. Grimm, R.J. Porra, W. Rűdinger and H. Scheer (eds) Springer, Dordrecht, The Netherlands.

Jeffrey, S.W. 1997a. Structural relationships between algal chlorophylls. pp. 566-571. In Phytoplankton Pigments in Oceanography: Guidelines to Modern Methods. S.W. Jeffrey, R.F.C. Mantoura and S.W. Wright (eds) UNESCO Publishing, Paris. France.

Jeffrey, S.W.1997b. Chlorophyll and carotenoid extinction coefficients. pp. 595-596. In Phytoplankton Pigments in Oceanography: Guidelines to Modern Methods. S.W.Jeffrey, R.F.C. Mantoura and S.W. Wright (eds) UNESCO Publishing, Paris. France.

Jeffrey, S.W. and J-M. LeRoi. 1997. Simple procedures for growing SCOR reference microalgal cultures. pp. 181-205. In Phytoplankton Pigments in Oceanography:

Guidelines to Modern Methods. S.W. Jeffrey, R.F.C. Mantoura and S.W. Wright (eds) UNESCO Publishing, Paris, France.

Jeffrey, S.W. and N.A. Welschmeyer.1997. Spectrophotometric and fluorimetric equations in common use in oceanography. pp. 597-615. In Phytoplankton Pigments in Oceanography.S.W.Jeffrey, R.F.C. Mantoura and S.W. Wright (eds) UNESCO Publishing, Paris, France.

Jeffrey, S.W. and S.W. Wright. 1997. Qualitative and quantitative HPLC analysis of SCOR reference algal cultures. pp. 343-360. In Phytoplankton Pigments in Oceanography. S.W. Jeffrey, R.F.C. Mantoura and S.W. Wright (eds) UNESCO Publishing, Paris. France.

Jeffrey, S.W. and S.W. Wright. 2006. Photosynthetic pigments in marine microalgae: insights from cultures and the sea. pp. 33-90. In Algal Cultures, Analogues of Blooms and Applications. D.V. Subba Rao (ed). Science Publishers. Enfield (N.H.) USA.

Jeffrey, S.W., R.F.C. Mantoura and T. Bjørnland. 1997a. Data for the identification of 47 key phytoplankton pigments. pp. 449-559. In Phytoplankton Pigments in Oceanography. S.W. Jeffrey, R.F.C. Mantoura and S.W. Wright (eds) UNESCO Publishing, Paris, France.

Jeffrey, S.W., R.F.C. Mantoura and S.W. Wright (eds). 1997b. Phytoplankton Pigments in Oceanography: Guidelines to Modern Methods. UNESCO Publishing, Paris, France.

Jeffrey, S.W. and S.W. Wright and M. Zapata. 1999. Recent advances in HPLC pigment analysis of phytoplankton. Mar. Freshwater Res. 50: 879-896.

Jiménez, C. and U. Pick. 1994. Differential steroisomer composition of β,β-carotene in thylakoids and pigment globules in *Dunaliella.* J. Plant Physiol. 143: 257-263.

Katz, A., C. Jiménez and U. Pick. 1995. Isolation and chararcterization of a protein associated with carotene globules in the alga *Dunaliella bardawil.* Plant Physiol. 108: 1657-1664.

Ke, B. 2001. Photosynthesis: Photobiochemistry and Photobiophysics. Kluwer Academic Press, Dordrecht, The Netherlands.

Kobayashi, M. 2003. Astaxanthin biosynthesis enhanced by reactive oxygen species in the green alga *Haematococcus pluvialis.* Biotechnol. Bioprocess Engineering. 8: 322-330.

Kobayashi, M., T. Kakizono and S. Nagai. 1993. Enhanced carotenoid biosynthesis by oxidative stress in acetate-induced cyst cells of a green unicellular alga, *Haematococcus pluvialis.* Appl. Environ. Microbiol. 59: 867-873.

Koyama, Y. and R. Fujii.1999. *Cis-trans* carotenoids in photosynthesis: configurations, excited state properties and physiological functions. pp.162-188. In The Photochemistry of Carotenoids. H.A. Frank, A.J.Young, G.Britton and R.J Cogdill (eds). Kluwer Academic Publishers, Dordrecht, The Netherlands.

Koyama, Y., I. Takatsuka, M. Kanaji, K. Tomimoto, M. Kito, T. Shimamura, J. Yamashita, K. Saiki and K. Tsukida, K. 1990. Configurations of carotenoids in the reaction centre and the light- harvesting complex of *Rhodospirillum rubrum.* Natural selection of carotenoid configurations by pigment protein complexes. Photochem. and Photobiol. 51: 119-128.

Krishna, K.B. and P. Mohanty. 1998. Secondary carotenoid production in green algae. J. Sci. and Ind. Res. 57: 51-63.

Kühlbrandt, W., D.N. Wang and Y. Fugiyoshi. 1994. Atomic model of a plant light-harvesting complex by electron crystallography. Nature 367: 614-621.

Larkum, A.W.D. and M. Vesk.2003. Algal plastids: their fine structure and properties. pp. 11-28. In Photosynthesis in Algae. A.W.D. Larkum, S.E. Douglas and J.A. Raven (eds) Kluwer Academic Publishers, Dordrecht, The Netherlands.

Lers, A., Y. Biener and A. Zamir. 1990. Photoinduction of massive β-carotene accumulation by the alga *Dunaliella bardawil.* Plant Physiol. 93:389-395.

Liaaen-Jensen, S.1998. Carotenoids in chemosystematics. pp. 217-247. In Carotenoids, Vol. 3: Biosynthesis and Metabolism.G.Britton, S. Liaaen-Jensen and H. Pfander (eds) Birkhäuser, Basel, Switzerland.

Liaaen-Jensen, S. and E.S. Egeland. 1999. Microalgal carotenoids. pp. 145-172. In Chemicals from Microalgae. Z. Cohen (ed) Taylor & Francis, London, UK.

Lichtenthaler, H.K. 1999. The 1-Deoxy-D-Xylulose-5-phosphate pathway of isoprenoid biosynthesis in plants. Annu. Rev. Plant Physiol. Plant Mol. Biol. 50: 47-65.

Liu, B. and Y. Lee. 2000. Secondary carotenoid formation by the green alga *Chlorococcum* sp. J. Appl. Phycol. 12: 301-307.

Loeblich, L.A. 1982. Photosynthesis and pigments influenced by light intensity and salinity in the halophile *Dunaliella salina* (Chlorophyta). J. Mar. Biol. Ass. U.K. 62: 493-508.

Lohr, M.2009. New knowledge concerning carotenoid biosynthesis and metabolism. In Phytoplankton Pigments in Oceanography, Vol 2. S.Roy, E.S. Egeland, G.Johnsen and C. Llewellyn. (eds) Publisher to be announced.

Masojídek, J., G. Torzillo, J. Kopecký, M. Koblížek, L. Nidiaci, J. Komenda, A. Lukavská and A. Sacchi. 2000. Changes in chlorophyll fluorescence quenching and pigment composition in the green alga *Chlorococcum* sp. grown under nitrogen deficiency and salinity stress. J. Appl. Phycol. 12: 417-426.

McFadden, G.I. 2001. Primary and secondary endosymbioses and the origin of plastids. J. Phycol. 37: 951-959.

Melkonian, M.1990a. Phylum Chlorophyta: Introduction to the Chlorophyta. pp 597-599. In Handbook of Protoctista. L.Marguilis, J.Q. Corliss, M. Melkonian and D. Chapman (eds) Jones and Bartlett Publishers Inc. Boston, U.S.A.

Melkonian, M.1990b. Phylum Chlorophyta: class Prasinophyceae. pp. 600-607. In Handbook of Protoctista .L.Marguilis, J.Q. Corliss, M. Melkonian and D. Chapman (eds) Jones and Bartlett Publishers Inc. Boston, U.S.A.

Melkonian, M.1990c. Phylum Chlorophyta: Class Chlorophyceae. pp. 608-616. In Handbook of Protoctista.L.Marguilis, J.Q. Corliss, M.Melkonian and D. Chapman (eds) Jones and Bartlett Publishers Inc. Boston, U.S.A.

Mendoza, H., M. Jiménez Del Río, G. García Reina and Z. Ramazanov. 1996. Low-temperature-induced β-carotene and fatty acid synthesis, and ultrastructural reorganization of the chloroplast in *Dunaliella salina* (Chlorophyta). Eur. J. Phycol. 31: 329-331.

Mendoza, H., A. Martel, M. Jiménez Del Río and G. García Reina. 1999. Oleic acid is the main fatty acid related with carotenogenesis in *Dunaliella salina.* J.Appl. Phycol. 11: 15-19.

Orosa, M., E. Torres, P. Fidalgo and J. Abalde 2000. Production and analysis of secondary carotenoids in green algae. J. Appl. Phycol. 12: 553-556.

Oster, U., R. Tanaka, A. Tanaka and W. Rüdiger. 2000. Cloning and functional expression of the gene encoding the key enzyme for chlorophyll *b* biosynthesis (CAO) from *Arabidopsis thaliana.* Plant J. 21: 306-310.

Palmer, J.D. 2003. The symbiotic birth and spread of plastids: how many times and whodunit? J. Phycol. 39: 4-11.

Phillip, D., S. Hobe, H. Paulsen, P. Molnar, H. Hashimoto and A.J. Young. 2002. The binding of xanthophylls to the bulk light-harvesting complex of photosystem II of higher plants: a specific requirement for carotenoids with a 3-hydroxy-β-end group. J. Biol. Chem. 277: 25160-25169.

Porra, R.J. 2006. Spectrophotometric assays for plant, algal and bacterial chlorophylls. pp 95-107. In Chlorophylls and Bacteriochlorophylls: Biochemistry, Biophysics, Functions and Applications. B.Grimm, R.J. Porra, W. Rűdiger and H. Scheer (eds). Springer, Dordrecht, The Netherlands.

Porra, R.J. 2009. A proven simultaneous equation assay for chlorophylls *a* and *b* using aqueous acetone and similar assays for recalcitrant algae. In Phytoplankton Pigments in Oceanography, Vol 2.S.Roy, E.S. Egeland, G, Johnsen and C. Llewellyn. Submitted. Publisher to be announced.

Porra, R.J., W.A. Thompson and P.E. Kriedemann. 1989. Determination of accurate extinction coefficients and simultaneous equations for assaying Chls *a* and *b* extracted with four different solvents: Verification of the concentration of Chl standards by atomic absorption spectroscopy. Biochim. Biophys. Acta 975: 384-394.

Porra, R.J., E.P. Pfündel and N. Engel.1997. Metabolism and function of photosynthetic pigments. pp. 85-126. In Phytoplankton Pigments in Oceanography. S.W. Jeffrey, R.F.C. Mantoura and S.W. Wright (eds). UNESCO Publishing, Paris, France.

Porra, R.J., U. Oster and H. Scheer. 2009 Recent advances in chlorophyll and bacteriochlorophyll biosynthesis. In Phytoplankton Pigments in Oceanography, Vol 2. S. Roy, E.S. Egeland, G. Johnsen and C. Llewellyn. Submitted. Publisher to be announced.

Preisig, H.R. 1992. Morphology and taxonomy. pp. 2-15. In *Dunaliella*: Physiology, Biochemistry and Biotechnology. M.Avron and A. Ben-Amotz (eds) CRC Press.Boca Raton, USA.

Quigg, A. and Z.V. Finkel, A.J. Irwin, Y. Rosenthal, T-Y. Ho, J.R. Reinfelder, O. Schofield, F.M.M. Morel and P.G. Falkowski. 2003. The evolutionary inheritance of elemental stoichiometry in marine phytoplankton. Nature 425: 291-294.

Rabbani, S., P. Beyer, J.v. Lintig, P. Hugueney and H. Kleinig. 1998. Induced β-carotene synthesis driven by triacylglycerol deposition in the unicellular alga *Dunaliella bardawil.* Plant Physiol. 116:1239-1248.

Ricketts, T.R. 1966. Magnesium 2,4-divinylphaeoporphyrin a_5 monomethylester, a protochlorophyll-like pigment present in some unicellular flagellates. Phytochem. 5: 223-229.

Rise, M., E. Cohen, M. Vishkautsan, M. Cojocaru, H.E. Gottlieb and S. Arad. 1994. Accumulation of secondary carotenoids in *Chlorella zofingiensis.* J. Plant Physiol. 144: 287-292.

Rüdiger, W. 2006. Biosynthesis of Chlorophylls *a* and *b*: the last steps. pp. 189-200. In Chlorophylls and Bacteriochlorophylls: Biochemistry, Biophysics, Functions and Applications.B.Grimm, R.J. Porra, W. Rüdiger and H. Scheer (eds). Springer, Dordrecht, The Netherlands.

Rüdiger, W. and B. Grimm. 2006. Chlorophyll metabolism, an overview. pp. 131-146. In Chlorophylls and Bacteriochlorophylls: Biochemistry, Biophysics, Functions and Applications. B. Grimm, R.J. Porra, W. Rüdiger and H. Scheer (eds). Springer, Dordrecht, The Netherlands.

Scheer, H. 1991. Chemistry of Chlorophylls. pp. 3-30. In Chlorophylls.H.Scheer (ed). CRC Press, Boca Raton.USA.

Scheer, H. 2003.The Pigments. pp. 29-81. In Light-harvesting Antennas in Photosynthesis. Advances in Photosynthesis and Respiration, Vol. 13.B.R.Green and W.W. Parsons (eds). Kluwer Academic Publishers, Dordrecht, The Netherlands.

Scheer, H. 2006. An overview of chlorophylls and bacteriochlorophylls: biochemistry, biophysics, functions and applications. pp. 1-26. In Chlorophylls and Bacteriochlorophylls: Biochemistry, Biophysics, Functions and Applications. B. Grimm, R.J. Porra, W. Rüdiger and H. Scheer (eds). Springer, Dordrecht, The Netherlands.

Schiedt, K. 1998. Absorption and metabolism of carotenoids in birds, fish and crustaceans. pp. 285-358. In Carotenoids, Vol. 3: Biosynthesis and metabolism. G. Britton, S. Liaaen-Jensen and H. Pfander (eds). Birkhäuser, Basel, Switzerland.

Schierle, J., W. Härdi, N. Faccin, I. Bühler and W. Schüep.1995. Example 8: Geometrical isomers of β,β-carotene. pp. 265-272. In Carotenoids, Vol. 1 A: Isolation and analysis. G. Britton, S. Liaaen-Jensen and H. Pfander (eds) Birkhäuser, Basel, Switzerland.

Schweigert, F. J. 1998. Metabolism of carotenoids in mammals. pp. 249-284. In Carotenoids, Vol. 3: Biosynthesis and Metabolism.G. Britton, S. Liaaen-Jensen and H. Pfander (eds). Birkhäuser, Basel, Switzerland.

Shaish, A., M. Avron and A. Ben-Amotz. 1990. Effect of inhibitors on the formation of stereoisomers in the biosynthesis of β-carotene in *Dunaliella bardawil*. Plant Cell Physiol. 31: 689-696.

Shaish, A., A. Ben-Amotz and M. Avron. 1991. Production and selection of high β-carotene mutants of *Dunaliella bardawil* (Chorophyta). J. Phycol. 27: 652-665.

Shaish, A., A. Ben-Amotz and M. Avron. 1992. Biosynthesis of β-carotene in *Dunaliella*. Methods in Enzymol. 213: 439-444.

Shaish, A., M. Avron, U. Pick and A. Ben-Amotz. 1993. Are active oxygen species involved in induction of β-carotene in *Dunaliella bardawil*? Planta 190:363-368.

Subba Rao, D.V.2006a. Algal cultures. pp 917-930. In Algal Cultures, Analogues of Blooms and Applications D.V.Subba Rao (ed) Science Publishers, Enfield (NH), USA.

Subba Rao, D.V.2006b. Algal culture centres. pp 931-932. In Algal Cultures, Analogues of Blooms and Applications.D.V.Subba Rao (ed) Science Publishers, Enfield (NH), USA.

Takaichi, S. and M. Mimuro. 1998. Distribution and geometric isomerism of neoxanthin in oxygenic phototrophs: 9′- *Cis*, a sole molecular form. Plant Cell Physiol. 39: 968-997.

Wright, S.W. and S.W. Jeffrey. 1997.High resolution HPLC system for chlorophylls and carotenoids of marine phytoplankton. pp. 327-341. In Phytoplankton Pigments in Oceanography. S.W. Jeffrey, R.F.C. Mantoura and S.W.Wright (eds). UNESCO Publishing, Paris, France.

Wright, S.W. and S.W. Jeffrey.2006. Pigment markers for phytoplankton production. pp.71-104. In Handbook Envir. Chem. Vol. 2. J. Volkman (ed). Springer Verlag, Heidelburg, Germany.

Wright, S.W., S.W. Jeffrey, R.F.C. Mantoura, C.A. Llewellyn, T. Bjørnland, D. Repeta and N.A. Welschmeyer. 1991. Improved HPLC method for the analysis of cholorphylls and carotenoids from marine phytoplankton. Mar. Ecol. Prog. Ser. 77: 183-196.

Yaronskaya, E. and B. Grimm. 2006. The pathway from 5-aminolevulinic acid to protochorophyllide and protoheme. pp. 173-188. In Chlorophylls and Bacteriochlorophylls: Biochemistry, Biophysics, Functions and Applications. B. Grimm, R.J. Porra, W. Rüdiger and H. Scheer(eds). Springer, Dordrecht, The Netherlands.

Zapata, M., F. Rodríguez and J. L. Garrido. 2000. Separation of chlorophylls and carotenoids from marine phytoplankton: a new HPLC method using a reversed phase C_8 column and pyridine-containing mobile phases. Mar. Ecol. Prog. Ser.195: 29-45.

Zapata, M., J.L. Garrido and S.W. Jeffrey.2006. Chl *c* pigments: current status. pp 39-53. In Chlorophylls and Bacteriochlorophylls: Biochemistry, Biophysics, Functions and

Applications. B. Grimm, R.J. Porra, W. Rudiger and H. Scheer (eds). Springer, Dordrecht, The Netherlands.

Zhang, D.H., Y.K. Lee, M.L. Ng and S.M. Phang. 1997. Composition and accumulation of secondary carotenoids in *Chlorococcum* sp. J. App. Phycol. 9: 147-155.

Zlotnik, I., A. Sukenik and Z. Dubinsky. 1993. Physiological and photosynthetic changes during the formation of red alplanospores in the chlorophyte *Haematococcus pluvialis*. J. Phycol. 29: 463-469.

6 Carotenoid Biosynthesis in *Dunaliella* (Chlorophyta)

EonSeon Jin[1] and Jürgen E.W. Polle[2]*

Abstract

Carotenoids are mostly 40-carbon isoprenoids that are integral and essential components of photosynthetic pigment-protein complexes, but also membranes. In response to environmental stress, some species of *Dunaliella* accumulate high amounts of specific carotenoids, such as β-carotene and zeaxanthin. Specifically, in response to environmental stress, cells of the species *Dunaliella salina* TEODORESCO can over-accumulate β-carotene as a secondary carotenoid in lipid vesicles in the chloroplast stroma. The process of this over-accumulation of β-carotene is called carotenogenesis. Although the external environmental stimuli (high salt concentration, excess light, or nitrogen deficiency) leading to carotenogenesis have been identified, very little is known about the signal perception and signal transduction pathways that result in over-accumulation of carotenoids. This chapter reviews the current information about the carotenoid biosynthesis pathway and regulation of carotenogenesis in the alga *Dunaliella*.

Introduction

Carotenoids represent one of the most fascinating, abundant, and widely distributed classes of natural pigments. Photosynthetic organisms from anoxygenic photosynthetic bacteria to cyanobacteria, algae, and higher plants, as well as numerous non-photosynthetic bacteria and fungi, produce carotenoids (Goodwin 1980,

[1] Department of Life Science, College of Natural Science, Hanyang University, 133-791 Seoul, South Korea.

[2] Brooklyn College of the City University of New York, Biology Department, 2900, Bedford Ave, Brooklyn, NY 11210, USA.

* Author for correspondence, email: esjin@hanyang.ac.kr

Armstrong and Hearst 1996, Armstrong 1997). Most naturally occurring carotenoids are hydrophobic tetraterpenoids that contain a C_{40} methyl-branched hydrocarbon backbone. The polyene chains of carotenoids, consisting of conjugated double bonds, are responsible for the varying color of carotenoids and their ability to absorb photons in the blue-green region of the visible wavelengths. There are two main classes of naturally occurring carotenoids: carotenes, which are pure hydrocarbons that may be linear or contain rings at one or both ends of the molecule (such as β-carotene, α-carotene), and xanthophylls, which are oxygenated derivatives of carotenes. All xanthophylls produced by higher plants, for example violaxanthin, antheraxanthin, zeaxanthin, neoxanthin, and lutein, are also synthesized by green algae. However, in contrast to land plants, certain algae possess additional xanthophylls in their chloroplasts such as loroxanthin, prasinoxanthin, siphonaxanthin (Goodwin 1980). Other algae accumulate secondary xanthophylls such as astaxanthin, canthaxanthin, and echinenone in response to nutrient stress (Czygan 1968, Goodwin 1980, Kessler and Czygan 1967).

In photosynthesis, carotenoids have a variety of functions (Frank and Cogdell 1995, Fromme et al. 2006, Rabinowitch 1952, Telfer et al. 2002, Telfer 2005). For example, carotenoids function as integral structural components of pigment-protein complexes (Kühlbrandt et al. 1994) and as accessory light-harvesting pigments by absorbing photons and transferring them to chlorophyll (Chl) molecules. Carotenoids also participate in redox reactions (Tracewell et al. 2001, Frank and Brudvig 2004), the protection of organisms from photodamage by quenching singlet oxygen and triplet Chl species (Siefermann-Harms 1987, Frank and Cogdell 1993, Yamamoto and Bassi 1996), and the dissipation of excess absorbed light energy via interactions with singlet excited Chl molecules (Demmig-Adams 1990, Demmig-Adams and Adams 1992, Demmig-Adams et al. 1996, Yamamoto and Bassi 1996, Niyogi 1999, Baroli and Niyogi 2000). In general, carotenoids participating in photosynthesis are called primary carotenoids. All other carotenoids are referred to as secondary carotenoids.

As photosynthetic organisms, algae of the genus *Dunaliella* contain primary carotenoids that function in photosynthesis. In addition, some species of *Dunaliella* accumulate secondary carotenoids in response to environmental stress in vegetative cells or in zygotes. In particular, the process of β-carotene over-accumulation in *Dunaliella* cells upon exposure to environmental stress conditions is referred to as carotenogenesis. This chapter reviews the current knowledge of the carotenoid biosynthetic pathway and regulation of over-accumulation of β-carotene and xanthophylls in *Dunaliella*.

Carotenoid Biosynthesis in the Alga *Dunaliella*

Possible pathways of β-carotene biosynthesis in *Dunaliella* were postulated by Ben-Amotz and Shaish (1992). In the past 10 years, genes for enzymes of the carotenoid biosynthesis pathway were cloned for several green algae such as *Chlamydomonas*, *Haematococcus,* and *Dunaliella.* Moreover, considerable progress was made with respect to genomic analysis of *Chlamydomonas reinhardtii* (http://www.chlamy.org), and identification of the genes required for carotenoid biosynthesis was reported (Grossman et al. 2003, Im et al. 2003, Shrager et al. 2003, Zhang et al. 2004, Lohr et al. 2005, Yan et al. 2005, Zhu et al. 2005, Merchant et al. 2007). Identification of these genes allowed characterization of the enzymes of the carotenoid biosynthesis pathway and their regulation. In particular, knowledge gained from higher plants as well as from the algae *C. reinhardtii* and *Haematococcus pluvialis* was used to propose the carotenoid biosynthesis pathway for *Dunaliella,* which is shown in Figure 1. However, the pathway for biosynthesis of 9-cis-β-carotene in *Dunaliella* cells may already be branching off at the level of PSY or shortly after (see chapter of Ben-Amotz).

The key building blocks of carotenoids are isopentenyl pyrophosphate (IPP) and its allylic isomer dimethylallyl phyrophosphate (DMAPP). The enzyme Isopentenyl Pyrophosphate Isomerase (IPI) carries out this reversible isomerization reaction. In higher plants, isopentenyl pyrophosphate is produced by two different pathways, in the cytosol through the acetate/mevalonate pathway and in the chloroplast through the 2-C-Methyl-D-Erythritol-4-Phosphate (MEP) pathway (Disch et al. 1998, Rodriguez-Conception and Boronat 2002, Rohmer 1999). Genes for enzymes participating in the plastidal MEP biosynthetic pathway are localized in the nucleus, and gene products are imported into the chloroplast (Lichtenthaler 1999).

In contrast to higher plants, there is no evidence for activity of the cytosolic acetate/mevalonate biosynthetic pathway of isoprenoids in unicellular green algae belonging to the class of chlorophyceae such as *Dunaliella* (Lichtenthaler 1999, Schwender et al. 2001). In consequence, it was proposed that unicellular green algae including *Dunaliella* produce their isoprenoids only through the MEP pathway that operates in plastids (Disch et al. 1998, Schwender et al. 2001, Figure 1). It is believed that in green algae IPP and/or DMAPP are exported from the chloroplast into the cytosol (Lichtenthaler 2007). It is expected that the genes encoding for enzymes participating in the MEP pathway and possibly genes encoding for postulated transporters for export of the C_5-isoprenoid building block in *D. salina* will become available soon through the ongoing genome sequencing project.

In chloroplasts, the phytoene synthase (PSY) catalyzes the first committed step in carotenoid biosynthesis by head-to-head condensation (sometimes also referred to as tail-to-tail condensation) of two 20-carbon geranylgeranyl pyrophosphate molecules to form the 40-carbon molecule phytoene, which is the precursor molecule for all other carotenoids (Figure 1). Most organisms, particularly higher plants, algae, and fungi, synthesize 15,15'-cis-phytoene, although some microbes, including *Dunaliella bardawil* Ben-Amotz et Avron, produce mixtures of isomers that can include all-trans-phytoene and 9-cis-phytoene (Ben-Amotz et al. 1988, Ben-Amotz et al. 1989, Ben-Amotz and Shaish 1992, Jiménez and Pick 1994,

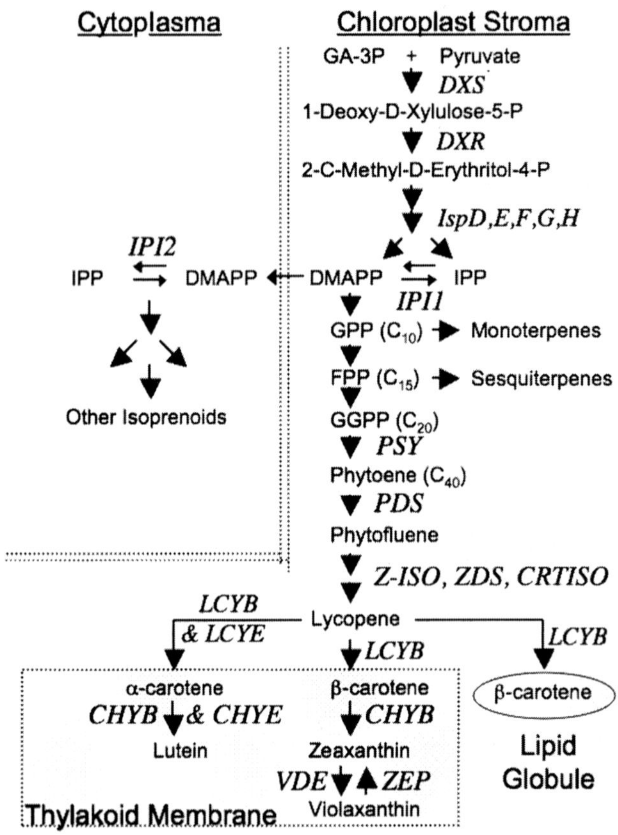

Figure 1: Schematic diagram of the proposed pathway of carotenoid biosynthesis in the microalgae *Dunaliella*. The proposed pathway is based on work previously published for higher plants (see text). Note, that the biosynthesis of 9-cis-β-phytoene may already occur at the level of PSY (see chapter by Ben-Amotz). Full enzyme names are as follows: DXS – 1-Deoxy-D-Xylulose Phosphate Synthase, DXR – 1-Deoxy-D-Xylulose Phosphate Reductase, IspD – 4-Diphosphocytidyl-2C-Methyl-D-Erythritol Synthase, IspE – 4-Diphosphocytidyl-2C-Methyl-Erythritol Kinase, IspF – 2C-Methyl-D-Erythritol-2,4-Cyclodiphosphate Synthase, IspG – 1-Hydroxyl-2-Methyl-2-(E)-Butenyl-4-Diphosphate Synthase, IspH – 1-Hydroxy-2-Methyl-2-(E)-Butenyl-4-Diphosphate Reductase, IPI1 & IPI2 – Isopentenyl Pyrophosphate Isomerase, PSY – Phytoene Synthase, PDS – Phytoene Desaturase, ZDS – ζ-Carotene Desaturase, Z-ISO – 15-cis-ζ-Isomerase, CRTISO – Carotenoid Isomerase, LCYE – Lycopene e-Cyclase, LCYB – Lycopene β-Cyclase, CHYE – Carotene ε-Hydroxylase, CHYB – Carotene β-Hydroxylase, ZEP – Zeaxanthin Epoxidase, VDE – Violaxanthin Deepoxidase.

Shaish et al. 1993). In some higher plants two PSY genes (PSY1 and PSY2) exist that are differentially regulated (Bartley and Scolnik 1993, Gallagher et al. 2004, Fraser et al. 1999, Wurtzel 2004). Also, for the alga *Haematococcus* the PSY gene was reported to be up-regulated in response to developmental and environmental cues (Steinbrenner and Linden 2001, Steinbrenner and Linden 2003). However, nothing is known about the number of PSY genes in *Haematococcus*. Further, the unicellular green alga *C. reinhardtii* contains only one PSY gene (McCarthy et al. 2004, Lohr et al. 2005).

It appears that for *D. salina* the sequence of at least one PSY gene is known (Yan et al. 2005). Unfortunately, it is not known if the strain used by Yan et al. (2005) can be classified as a strain of *D. salina*. As several strains of *D. salina* were misidentified in the past (see chapter by Gonzalez et al. in this book), there is a chance that the strain used by Yan et al. (2005) is some other species of *Dunaliella*. Another PSY gene sequence was published in NCBI under the accession number #U91900 for *D. bardawil*, and recently two more different PSY genes were partially cloned from *D. bardawil* (Tran and Polle, manuscript in preparation). Concomitantly, the *D. salina* genome sequencing project provided EST evidence for the existence of at least two different PSY genes (Tran and Polle, manuscript in preparation). If indeed the alga *D. salina* has more than one PSY gene, the situation may be similar to that in higher plants, which have several PSY genes (Gallagher et al. 2004, Wurtzel 2004).

In higher plants, PSY genes are differentially expressed depending on the developmental stage and the tissue (Hirschberg 2001, Cunningham 2002, Howitt and Pogson 2006, Taylor and Ramsay 2006). Figure 5 shows the results of a Reverse Transcription-PCR where general primers against PSY were used to amplify cDNA from non-stressed and nitrogen-stressed cells. In comparison to the actin control, the higher level of PCR product from nitrogen-stressed cells indicated that expression of PSY gene(s) is up-regulated at the mRNA level. As it appears that—similar to higher plants—carotenogenic species of *Dunaliella* have more than one PSY gene, it may be hypothesized that PSY genes in carotenogenic *D. salina* could be differentially expressed when cells are exposed to environmental stress. Possibly only one PSY gene is up-regulated and thus responsible for secondary carotenoid accumulation in carotenogenic *Dunaliella* species. As several genomes of unicellular green algae are available already or are being sequenced, in the future it should be possible to relate the number of PSY genes of various green algae to their ability of secondary carotenoid accumulation.

In general, successive introduction of conjugated double bonds during carotenoid biosynthesis lengthens the π-conjugated electron system of the chromophores, producing colored carotenoids. In higher plants, two structurally and functionally similar enzymes, Phytoene Desaturase (PDS) and ζ-Carotene Desaturase (ZDS), convert phytoene to lycopene via ζ-carotene. For each of the green algae *Chlamydomonas reinhardtii*, *Ostreococcus lucimarinus*, and *Ostreococcus taurii*, at least three different PDS genes exist in their genomes (http://genome.

jgi-psf.org/euk_home.html), but at this time the number of PDS genes for *Dunaliella* algae is not known. Recently, a cDNA for the PDS from *D. bardawil* was deposited in NCBI with the accession # Y14807. More recently, the full sequence of a PDS gene from *D. salina* was published (Zhu et al. 2005), but heterologous functional expression of this PDS in *E.coli* was not achieved (Zhu et al. 2007). Unfortunately, it is not known if the strain used by Zhu et al. (2005) and Zhu et al. (2007) actually is a 'true' carotenogenic *D. salina* strain or if this is a misidentified strain. Consequently, any speculation about the number of PDS genes within one species of *Dunaliella* cannot be addressed at this time.

In the alga *Dunaliella* carotenoids may exist in trans- and in different cis-configurations, indicating that at least one isomerization step must occur early on in the pathway (Ben-Amotz et al. 1989, Shaish et al. 1990, Shaish et al. 1993, also see the chapter by Ben-Amotz in this book). New data from higher plants show that in the dark, the reaction from phytoene to lycopene requires the participation of the two Carotenoid Isomerases CRTISO (Isaacson et al. 2002, Park et al. 2002, Isaacson et al. 2004) and Z-ISO (Li et al. 2007). In the light photo-isomerization of cis- to trans-carotenoids is possible without the specific activity of CRTISO, which allows plants to generate essential carotenoids in the light in the absence of CRTISO activity (Isaacson et al. 2002, Eckhardt 2002). However, it appears that also in light-exposed tissues of higher plants Z-ISO may be necessary for proper growth especially in plants under biotic stress (Li et al. 2007, Janick-Buckner et al. 2001). That photoconversion of carotenoids can occur in the light in the absence of CRTISO and possibly Z-ISO is congruent with the observation that cis- to all-trans conversion in *Dunaliella* does not appear to require distinct isomerase activity (Ben-Amotz and Shaish 1992,Orset and Young 2000, Armstrong 1997). However, Lohr et al. (2005) described a homolog of CRTISO for the alga *C. reinhardtii* and also the *D. salina* genome sequencing project very recently identified an Expressed Sequence Tag (EST) for a homolog of CRTISO (Polle, unpublished data). Consequently, further work is necessary to elucidate the mechanism of isomerization of carotenoids in *Dunaliella*.

Catalyzed by the enzyme Lycopene β-Cyclase (LCYB), lycopene undergoes cyclization at each end of the linear molecule to form β-carotene. It was shown previously that two different types of LCYB enzymes exist (Pecker et al. 1996, Ronen et al. 2000), and it remains to be discovered which type(s) of LCYB exists in *Dunaliella* algae. Following cyclization, the two β-ionone rings of β-carotene are subjected to identical hydroxylation reactions to yield zeaxanthin, which, in turn, undergoes epoxidation catalyzed by zeaxanthin epoxidase (ZEP), once to form antheraxanthin and a second time to form violaxanthin. An EST for a ZEP homolog was recently identified for *D. salina* through the genome sequencing project (Polle, unpublished data). Similar to some higher plants, through an additional molecular rearrangement, neoxanthin may then be derived from violaxanthin by Neoxanthin Synthase (NSY). However, it seems that the NSY is absent from the genomes of *C. reinhardtii* (Lohr et al. 2005) and *Volvox carteri* (http://genome.jgi-psf.org/euk_home.html). It may be that another type of gene unrelated to NSY has NSY activity. It is anticipated that once the complete *D. salina* genome sequence becomes available, the question of which enzyme produces neoxanthin in *Dunaliella* algae will be solved.

Additional carotenoids found in *Dunaliella* species are α-carotene and its derivatives (β,ε-carotenoids), which originate from lycopene through cyclization by two structurally related enzymes, LCYB and Lycopene ε-Cyclase (LCYE). Subsequent hydroxylation of the β- and ε-ionone rings of the α-carotene molecule by two related cytochrome P450 enzymes yields lutein (Pogson et al. 1996, Pogson et al. 1998, Jin et al. 2003b, Tian et al. 2004, Kim and DellaPenna 2006).

Accumulation of Xanthophylls in *Dunaliella* As a Response to Irradiance Stress

Long-term acclimation to irradiance stress of the green alga *Dunaliella* is accompanied by accumulation of zeaxanthin along with a lowering in the relative amount of other pigments, including chlorophylls and several carotenoids (Falkowski 1984, Krol et al. 1997, Jin et al. 2001, Jin et al. 2003a, Jin et al. 2003b, Yokthongwattana et al. 2005). A recent investigation using differential extraction of β-carotene and zeaxanthin indicated that in low light-grown *D. salina*, xanthophylls are structurally associated with and stabilized by the Chl-binding proteins in the thylakoid membrane (Yokthongwattana et al. 2005). In contrast, it appears that in cells exposed to high irradiance zeaxanthin is embedded in the lipid bilayer or in a domain of the chloroplast thylakoids that can easily be separated from the Chl-proteins upon mild surfactant treatment (Yokthongwattana et al. 2005).

Over-accumulation of β-carotene in *Dunaliella Salina* (= Carotenogenesis)

Vegetative cells of the green alga *Dunaliella* appear green under favorable growth conditions. Green cells contain only those pigments (chlorophylls and carotenoids) that are necessary for photosynthesis. In contrast, when vegetative green cells are exposed to environmental stress conditions, some species of *Dunaliella* produce and accumulate high amounts of β-carotene, with a concomitant change in coloration from green to orange (Ben-Amotz et al. 1982, Ben-Amotz et al. 1989, Lerche 1937, Orset and Young 1999, Teodoresco 1905). In general, the process of over-accumulation of β-carotene in *Dunaliella* is referred to as carotenogenesis. In addition to *D. salina*, *D. bardawil* (Ben-Amotz and Avron 1983) was reported to be carotenogenic (see also Figure 2). According to Borowitzka and Borowitzka (1988) the species of *D. bardawil* is only a strain of *D. salina*. However, currently this view is not generally accepted. Consequently, this chapter will still refer to *D. bardawil* as a separate species (see also earlier chapters in this book).

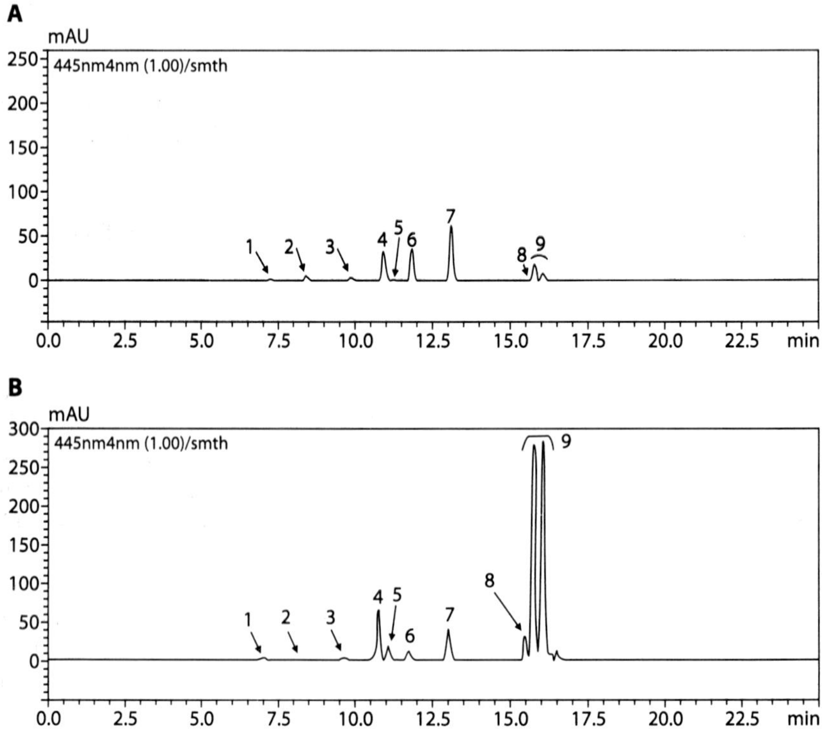

Figure 2: HPLC elution profile of total pigment extract from *D. bardawil.* (A) control condition in low light and (B) after three weeks high light (400 μE) exposure. Identification of numbered peaks is as follows: 1, neoxanthin; 2, violaxanthin; 3, antheraxanthin; 4, lutein; 5, zeaxanthin; 6, chlorophyll b; 7, chlorophyll a; 8, α-carotene; 9, β-carotene.

β-Carotene, which can account for up to about 10% of the dry cell weight of *D. salina*/*D. bardawil* can accumulate in oily globules located in the inter-thylakoid space of the chloroplast (Ben-Amotz et al. 1982, Ben-Amotz et al. 1989, Orset and Young 1999). The globules have a small diameter (~0.1-0.2 μm) and are exclusively composed of β-carotene, neutral lipids, and small amounts of protein (Katz et al. 1995).

Comparison of Strains

Numerous publications have reported on the over-accumulation of β-carotene in the alga *D. salina* (Lerche 1937, Lers et al. 1990, Rabbani et al., 1998, Orset and Young 1999, Gomez et al. 2003, Gomez and Gonzalez 2005, Ben-Amotz 1996, Bhosale 2004). It should be pointed out that the amount of β-carotene produced by different isolates of *D. salina* originating from various locations is not identical (Cifuentes et al. 1992, Markovits et al. 1993). An example for two strains and their abilities to over-accumulate β-carotene is shown in Figure 2 and 3. For this comparison, two strains were selected that are closely related when using the rDNA Internal Transcribed Spacer (ITS1 and ITS2) regions as molecular markers. Figure 3 shows that cells of the strain *D. bardawil* UTEX LB2835 originating from Israel turn orange specifically when grown in high salinity. This is due to over-accumulation of β-carotene which is visualized in the HPLC profile of pigments from control and light-stressed cells of *D. bardawil* which are shown in Figure 2.

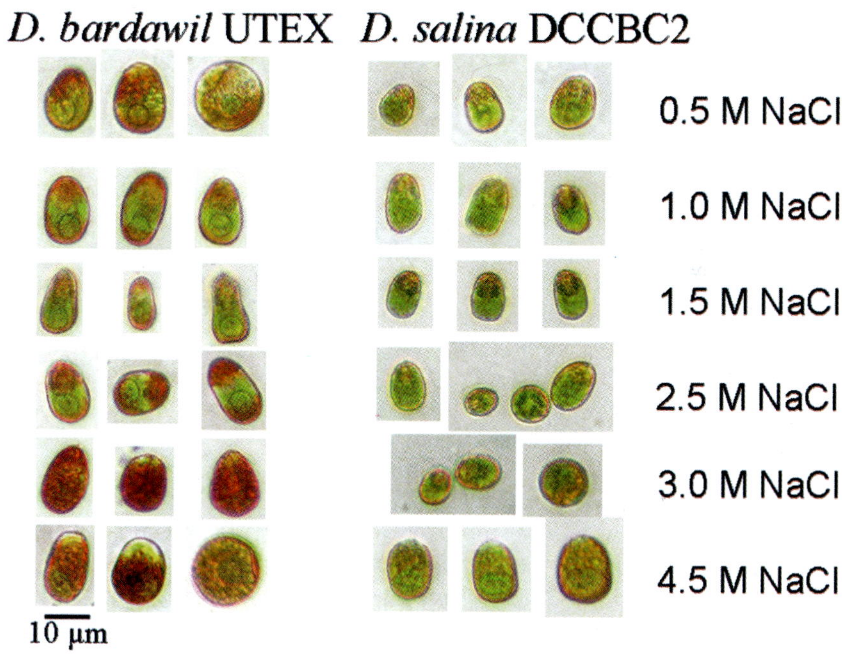

Figure 3: Shown are cells of *D. bardawil* and *D. salina* DCCBC2 grown under different salinities (Magnification 400x). The different salinities used for growth of cultures are indicated.

In contrast to D. *bardawil*, a new isolate from Korea (*D. salina* DCCBC2) showed significantly less content of β-carotene. Figure 3 demonstrates that both strains have the ability to over-accumulate β-carotene under high salinity. Both strains contain about 5-10 fold more β-carotene per cell when exposed to environmental stress, which is characteristic for strains of *D. salina* (Loeblich 1982). Nevertheless, comparison of these two strains showed that cells of *D. bardawil* already have a relatively high level of β-carotene even when grown under optimal conditions. As different strains of *D. salina* have dramatically diverse carotenoid accumulation capabilities, it is hypothesized that the extent to which β-carotene accumulates in various strains of carotenogenic *Dunaliella* depends on the capability of cells to store and sequester the carotenoid in lipid vesicles.

Regulation of Carotenogenesis

Several studies have focused on the mechanisms involved in cellular stress responses that result in β-carotene production. It is known that stressful conditions such as high salinity, extreme temperature, and deprivation of mineral nutrients, including nitrate, sulfate, and phosphate, enhance accumulation of β-carotene in *Dunaliella* (Ben-Amotz 1996, Ben-Amotz and Avron 1983, Borowitzka 1988, Ben-Amotz and Avron 1990, Bhosale 2004, Lers et al. 1990, Orset and Young 1999, Orset and Young 2000, Shaish et al. 1993). Currently, there is exists considerable debate concerning the regulation of the biosynthesis of β-carotene and its isomers in *D. salina*. There are conflicting data from separate studies (e.g., Ben-Amotz et al. 1988, Jiménez and Pick 1994) as to whether the synthesis of 9-cis-β-carotene is promoted at high or low irradiance. This may, however, be strain dependent, and Jiménez and Pick (1994) suggested major differences in the regulation of β-carotene synthesis between different strains of a single species. Currently, it is not known at which stage in the carotenoid biosynthesis the trans- and different cis-isoforms of carotenoids in *D. salina* are generated. A more detailed discussion about accumulation of the cis- and trans-stereoisomers of β-carotene can be found in this book in the chapter by Ben-Amotz.

In contrast to the fraction of β-carotene, which functions as a primary carotenoid in photosynthesis in the thylakoid membrane, during carotenogenesis in *Dunaliella*, additional β-carotene may be over-accumulated in lipid globules in the chloroplast of *Dunaliella* cells. In consequence, during carotenogenesis, cells not only up-regulate the level of cellular β-carotene, but they also change the location of β-carotene deposition. Formation of the globules in which β-carotene is deposited during carotenogenesis also requires lipids to be produced. Therefore, it is expected that the mechanisms of β-carotene accumulation and lipid production in cells are interdependent processes (Rabbani et al. 1998). Future cloning of genes for enzymes involved in fatty acid biosynthesis for carotenogenic *Dunaliella*, strains will allow comparative analysis of gene expression, enzyme levels, and enzyme activities between carotenoid and lipid biosynthesis.

Currently, it is largely unknown how various stress conditions are being sensed by *D. salina* and what mechanisms are involved in signal transduction that ultimately result in over-accumulation of β-carotene. At a first glance, sensing of the three stress factors salinity, irradiance, and nutritional deficiencies does not appear to have anything in common. Responses to salinity stress not only include production of β-carotene, but also other metabolic changes, for example, an increased cellular level of glycerol which is used as an osmoticum in cells of *D. salina* (Chitlaru and Pick 1991, Ben-Amotz and Avron 1990). As a basis for further discussion, Figure 4 shows a general working hypothesis, including sensing of environmental cues, intracellular signal transduction, and up-regulation of genes of the carotenoid biosynthesis pathway. For example, it was shown previously that various genes of the carotenoid biosynthesis pathway are up-regulated in response to environmental stress in *H. pluvialis* (Steinbrenner and Linden 2001, 2003, Linden 1999, Grünewald et al. 2000, Sun et al. 1998). It appears that some of the processes responsible for carotenogenesis in *H. pluvialis* are similar to those in *D. salina*, because based on unpublished data, Pick (1998) claimed that the genes of PSY and PDS were up-regulated at the transcriptional level in response to irradiance stress. In contrast to such transcriptional up-regulation, Rabbani et al. (1998) reported that the level of the PDS enzyme did not change in response

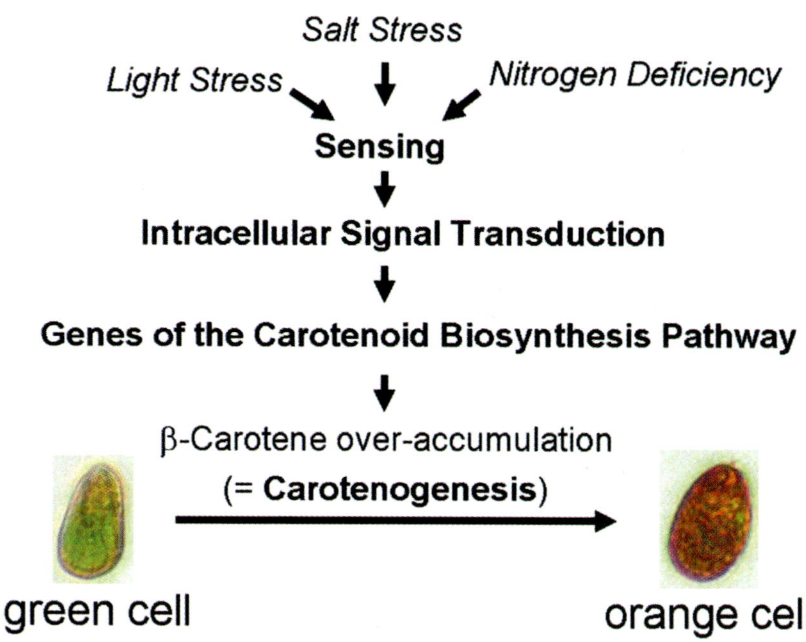

Figure 4: Summary of the postulated working hypothesis about carotenogenesis in the alga *D. salina*.

to light stress. Such differences between regulation at the transcriptional level and protein level of enzymes of the carotenoid biosynthesis pathway are not uncommon in carotenogenesis (Jin et al. 2006).

Currently, only genes for the PSY and the PDS are known for *D. salina*. Nevertheless, it is expected that the new *D. salina* genome sequencing project will provide the sequences for all genes of the carotenoid biosynthesis pathway soon. With more information about sequences for genes involved in carotenoid biosynthesis, detailed investigation of gene expression in response to environmental stress will be possible in the near future.

Carotenogenesis in response to changes in salinity

In past decades the effect of changes in salinity on carotenogenesis of *D. salina* was investigated extensively (Loeblich 1982, Ben-Amotz and Avron 1983, Borowitzka et al. 1990). Nevertheless, almost nothing is known about the processes as they relate directly to carotenogenesis (Figure 4). In contrast, progress was made in the dissection of the general process of osmo-regulation in response to changes in salinity. It was postulated, for instance, that sterols and a putative plasma membrane receptor in the plasma membrane are essential for the sensing of osmotic changes in *Dunaliella* (Zelazny et al. 1995). Signal transduction could then involve some factors located in the cytosol (Pick 1998). For example, one of the regulating factors involved in carotenogenesis in response to salt stress may be abscisic acid (Cowan and Rose 1991, Kobayashi et al. 1997).

Effect of irradiance on carotenogenesis

High levels of irradiance induce carotenogenesis in *D. salina* (Ben-Amotz and Avron 1983, Lers et al. 1990, Loeblich 1982), and Ben-Amotz et al. (1988) reported that the level of the 9-cis isomer of β-carotene accumulated in *D. salina* was proportional to the integral light intensity to which the alga was exposed during a division cycle. Also, carotenogenesis in response to high light in *Dunaliella* required de novo transcription and translation (Lers et al. 1990), suggesting a regulatory pathway for induction of carotenogenesis. This proposed regulatory pathway in response to irradiance requires sensory and transduction elements that ultimately up-regulate genes in the carotenoid and lipid biosynthesis pathways.

It was demonstrated that carotenogenesis in response to high irradiance is not dependent on any specific wavelength (Ben-Amotz and Avron 1989), suggesting that no specific photoreceptors are involved in carotenogenesis. Although carotenogenesis appears to be independent of light quality in *D. salina*, the photoreceptor phototrophin was recently shown to be involved in the regulation of gene expression for carotenoid biosynthesis in the alga *C. reinhardtii* (Im et al. 2006). However, at this time, it is still unknown if phototrophins exist in *D. salina*. In consequence, it is hypothesized that photosynthesis is responsible for sensing changes in irradiance levels resulting in carotenogenesis. As changes in irradiance have an impact on the redox state of plastoquinone, the mechanism for induction of carotenogenesis may be similar to the mechanism thought to be involved in chloroplast to nucleus redox signaling (Beck 2005, Fey et al. 2005a,b, Pfannschmidt 2003). Therefore, it is postulated that the redox state of the plastoquinone pool is the sensor for the initiation of carotenogenesis. As the redox state of the plastoquinone pool changes in response to increased irradiance, a retrograde signaling pathway could then be used for signal transduction from the plastid to the nucleus. Supporting evidence for this hypothesis comes from a recent study using the alga *Haematococcus,* because it was shown that use of an inhibitor of electron flow from photosystem II to plastoquinone prevented carotenogenesis by down-regulated expression of carotenogenic genes (Steinbrenner and Linden 2003). As similar results were obtained for the alga *D. bardawil* (Ben-Amotz, personal communication), retrograde signaling from chloroplast to nucleus which involves sensing of changes of the redox-state of plastoquinone is likely responsible for up-regulation of gene expression during carotenogenesis.

Effects of nutrient limitation on carotenogenesis

It was shown that nitrogen, phosphate, and sulfate limitation induce carotenogenesis in *D. salina* (Lerche 1937, Ben-Amotz and Avron 1983, Phadwal and Singh 2003). Also, iron deficiency may have an effect on carotenogenesis (Pick 1998).

Specifically when nitrogen is limiting, with cellular nitrogen becoming scarce, cells respond by increasing their content of carbohydrates and/or hydrocarbons. In *Dunaliella,* nitrogen deficiency leads to general enhancement of photosynthetic CO_2 assimilation (Liska et al. 2004). This response may be viewed as a valve to use excess energy from photosynthesis to produce reserve substances, and the processes involved in initiation of carotenogenesis may be similar to those postulated above as a response to high irradiance.

Recently, expression of genes coding for enzymes of the carotenoid biosynthesis pathway was investigated in cells that were exposed to nitrogen limitation. For example, one recent study (Sánchez-Estudillo et al. 2006) reported that expression of the 1-Deoxyxylulose-5-phosphate Synthase (DXS) was increased in response to nitrogen-limitation in a *Dunaliella salina* strain. In contrast, the expression of PSY was not significantly changed in these experiments (Sánchez-Estudillo et al. 2006). Unfortunately, it is not clear from the publication if the *D. salina* strain (CDBB500) investigated is carotenogenic. To clarify if PSY gene expression is stress-induced in the carotenogenic species *D. bardawil,* cultures of *D. bardawil* were grown under optimal conditions where all cells were green. Cells from green, non-induced cultures were collected by centrifugation and resuspended in medium containing only 1/10 nitrate. Resuspended cells were then exposed to a 10-fold increase in irradiance known to cause light stress for cells and induce carotenogenesis in *D. bardawil*. Before and after 16 hours of irradiance stress the same number of cells was harvested to isolate total RNA, used for reverse transcription followed by PCR amplification of Actin and PSY. Figure 5 shows preliminary Reverse-Transcriptase-(RT-)PCR results. Although the actin primers used gave multiple bands, the two main bands of expected size products had similar intensities regardless of the growth condition, indicating constant expression levels. At the same time, the band representing PSY was much more intense from stressed cells than from non-stressed cells, suggesting that PSY is in fact up-regulated when cells of *D. bardawil* are exposed to environmental stress.

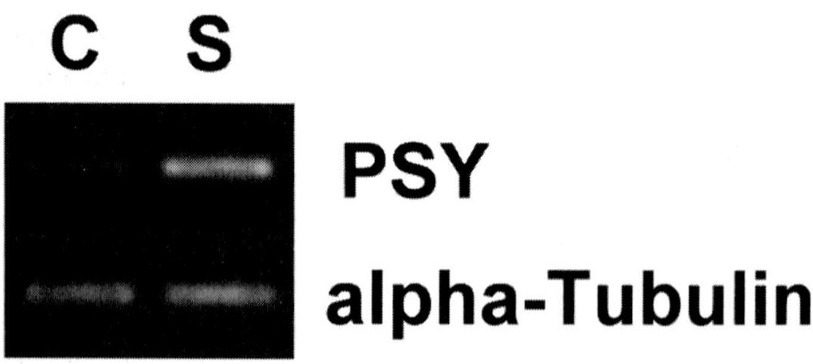

Figure 5: Shown are preliminary RT-PCR reaction results. C, unstressed cells as positive control; S, 16 hr combined high light & low nitrate stress. The alpha-tubulin gene was used as a control.

Carotenogenesis in maturing zygotes

According to Lerche (1937), a number of *Dunaliella* species form zygotes that may have an orange color when mature. One open question that still needs to be addressed is the process of carotenogenesis in zygotes. For example, it needs to be determined if carotenogenesis of vegetative cells and zygotes of the species *D. salina* is identical. It may be that during carotenogenesis in maturing zygotes, secondary carotenoids are accumulated that differ from β-carotene. Possibly, maturing zygotes accumulate canthaxanthin and/or astaxanthin in vesicles in the cytosol. Further studies of the pigment profile of mature zygotes, for example of *D. salina*, in combination with analysis of the genome will provide the answer to this question. Specifically, because other algae such as *Haematococcus* (Jin et al. 2006) and *C. reinhardtii* (Lohr et al. 2005) have genes for the hydroxylase and ketolase enzymes that are required to produce astaxanthin, it is expected that *Dunaliella* algae also contain genes for these enzymes and possibly produce astaxanthin in maturing zygotes.

Carotenoid Aberrant Mutants of *D. salina*

Despite the fact that pigment mutants of cyanobacteria and green algae have been used extensively to study the biogenesis and function of photosynthetic complexes (Chitnis et al. 1997, Hippler et al. 1998, Sun et al. 1998), so far, only a few mutants from *Dunaliella* have been reported. For example, mutants were selected for high content of β-carotene (Shaish et al. 1991, Ben-Amotz and Shaish 1992). These mutants were affected in the mechanism that regulates the activation of the carotene biosynthetic pathway, most probably at the metabolic steps which precede GGPP and which allow enhanced production of β-carotene, and possibly at a site after phytofluene (Ben-Amotz and Shaish 1992).

Recently, Jin et al. (2001, 2003a) employed mutagenesis of *Dunaliella* to manipulate the composition and quantity of carotenoids, especially that of zeaxanthin. In general, the level of irradiance regulates the zeaxanthin content in microalgae. Under photosynthetically active conditions, photosynthetic organisms, including *D. salina*, contain only trace amounts of zeaxanthin. Two different zeaxanthin-overproducing strains of *Dunaliella*, termed *dcd*1 and *zea*1, were generated upon mutagenesis. The *dcd*1 mutant was selected on the basis of its yellow-green coloration under moderate illumination. In the *dcd*1 mutant, zeaxanthin content was slightly greater than in the wild type (Jin et al. 2001). The *zea*1 mutant showed constitutive zeaxanthin accumulation under all growth conditions, and it lacked all zeaxanthin derivatives (Jin et al. 2003a) such as neoxanthin, violaxanthin, and antheraxanthin, strongly suggesting that *zea*1 is a zeaxanthin epoxidase (ZEP) mutant. Under normal growth conditions (low-light), the *zea*1 mutant had a zeaxanthin content that was 20 times larger than the wild type (Table 1). Such relative high levels of zeaxanthin found in *zea1* in low light conditions are most likely associated with light-harvesting complex (LHC) proteins, because zeaxanthin in antenna complexes from ZEP-deficient mutants appears to replace lutein and violaxanthin found in LHC-preparations from wild type cells (Polle et al. 2001). This is in contrast to zeaxanthin accumulation in photosynthetic membranes as a stress response.

Table 1: Total carotenoid (Car) and zeaxanthin (Z) content per cell dry weight in *D. salina* wild-type and *zea1* strains. Cells were grown under either 100 μmol photon $m^{-2}s^{-1}$ (LL) or exposed to 2000 μmol photon $m^{-2}s^{-1}$ (HL) for 24 h following a LL→HL shift. Values represent means ± SD (*n* = 3–5) (Jin et al. 2003a).

Pigment content	LL		LL→ HL	
	WT	*zea*1	WT	*zea*1
Car/cell mg/g DW	13.6 ± 2.8	12.8 ±2.4	10.7±0.1	11.0±0.4
Z/cell mg/g DW	0.23 ± 007	5.9± 0.5	2.62± 0.18	4.18 ±0.33

Metabolic Engineering of Carotenoids and Isoprenoids in Dunaliella

Overall, these above-mentioned successes have encouraged a growing, worldwide interest in the manipulation of the carotenoid biosynthetic pathways in plants and microalgae. Already carotenoid biosynthesis in higher plants (Gerjets and Sandmann 2006, Gerjets et al. 2007, Morris et al. 2006, Sandmann et al. 2006, Wurbs et al. 2007, Ye et al. 2000, Suzuki et al. 2007) and green algae (León et al. 2007) was altered by genetic engineering to produce either higher levels of carotenoids or novel carotenoids that are not endogenous. In addition to production of high-value carotenoids such as astaxanthin, plants and algae may also be considered as production platforms for other isoprenoids (Mahmoud and Croteau 2002). For example, monoterpenes (C10 carbon units) and sesquiterpenes (C10 carbon units) may be used for production of biodiesel. Carotenogenic species such as *D. salina*, which may accumulate up to 10% of their dry weight as β-carotene, should be excellent organisms for potential metabolic manipulation of secondary carotenoid production and/or generation of other isoprenoids. It may be possible to re-route the stress-induced isoprenoid metabolism in cells of *D. salina* towards generation of novel monoterpenes and sesquiterpenes. However, for metabolic engineering of isoprenoid biosynthesis in *Dunaliella* species it will be essential to understand the genetic basis of pathways in the context of their regulation within larger networks to be able to direct metabolic flux in a precise manner towards the desired product (Klein-Marcuschamer et al. 2007, Mahmoud and Croteau 2002, Stephanopoulos and Jensen 2005). This means that, for example, not only genes coding for enzymes involved in isoprenoid biosynthesis need to be known (Rodriguez-Concepcion and Boronat 2002, Rohdich et al. 2003, Sieiro et al. 2003), but also their regulation within the complex metabolic network with metabolic flux between different cellular compartments needs to be understood and used (Broun and Somerville 2001, Sandmann et al. 2006). Currently still little is known about mechanisms in algae that regulate metabolic flux though the MEP pathway. Further, as pathway intermediates or even isoprenoid end-products may be toxic to cells when present in large quantities and also because hydrophobic hydrocarbon molecules need sequestering structures, it is important to consider creation of metabolic sinks in cells (Li and van Eck 2007). Consequently, the mechanisms leading to formation of the β-carotene-sequestering globules which may become metabolic sinks for other engineered isoprenoid end-products in *Dunaliella* cells should be elucidated.

Equally important and essential to any attempt of metabolic engineering for use of cells as 'biorefineries' is that nuclear and/or plastidal transformation of *Dunaliella* cells can be easily accomplished. The current state of development of transformation systems for *Dunaliella* species is reviewed in this book in the chapter of Polle and Qin.

Summary

In photosynthetic organisms, carotenoids serve a variety of functions that are essential to the survival and ecological success of these organisms in their environments. Algae of the genus *Dunaliella* that over-accumulate β-carotene were used for a long time as model systems for studies of the mechanisms of cellular stress response. Unfortunately, at present, knowledge about carotenoid biosynthesis and its regulation in the alga *D. salina* is still rudimentary at best. However, it is expected that the genome sequencing project for *D. salina* will allow future breakthroughs in the knowledge about mechanisms that are involved in carotenogenesis.

Acknowledgements

This research was supported by a grant (B-2005-11) from the Marine Bioprocess Research Center of the Marine Bio 21 Center funded by the Ministry of Maritime Affairs & Fisheries, Republic of Korea. The authors thank Mr. D. Tran for his assistance and technical support.

References

Armstrong, G. A. 1997. Genetics of eubacterial carotenoid biosynthesis: a colorful tale. Ann. Rev. Microbiol. 51: 629-659.

Armstrong, G. A. and J. E. Hearst. 1996. Genetics and molecular biology of carotenoid pigment biosynthesis. FASEB J. 10: 228-237.

Baroli, I. and K.K. Niyogi. 2000. Molecular genetics of xanthophyll-dependent photoprotection in green algae and plants. Phil. Trans. R. Soc. Lond. *B*. 355: 1385-1394.

Bartley, G.E. and P.A. Scolnik. 1993. cDNA cloning, expression during development, and genome mapping of *PSY2*, a second tomato gene encoding phytoene synthase. J. Biol. Chem. 268: 25718-25721.

Beck, C.F. 2005. Signaling pathways from the chloroplast to the nucleus. Planta 222: 743-756.

Ben-Amotz, A. 1996. Effect of low temperature on the stereoisomer composition of β-carotene in the halotolerant alga *Dunaliella bardawil* (Chlorophyta). J. Phycol. 32: 272–275.

Ben-Amotz, A. and M. Avron.1983. On the Factors Which Determine Massive beta-Carotene Accumulation in the Halotolerant Alga *Dunaliella bardawil.* Plant Physiol. 72(3): 593-597.

Ben-Amotz, A. and M. Avron. 1989. The wavelength dependence massive carotene significance in *Dunaliella bardawil* (Chlorophyceae). J. Phycol. 25: 175-178.

Ben-Amotz, A. and M.Avron.1990. The biotechnology of cultivating the halotolerant alga *Dunaliella bardawil* (Chlorophyceae). J. Phycol. 25: 175-178.

Ben-Amotz, A. and A. Shaish. 1992. β-Carotene biosynthesis.pp.205-216. In: *Dunaliella*: Physiology, Biochemistry, and Biotechnology. M. Avron and A. Ben-Amotz (eds.). CRC Press, Boca Raton, FL, USA.

Ben-Amotz, A., A. Katz and M. Avron. 1982. Accumulation of beta-carotene in halotolerant algae: Purification and characterization of beta-carotene-rich globules from *Dunaliella bardawil* (Chlorophyceae). J. Phycol. 18: 529-537.

Ben-Amotz , A., A. Lers and M. Avron. 1988. Stereoisomers of β-carotene and phytoene in the alga *Dunaliella bardawil.* Plant Physiol. 86: 1286-1291.

Ben-Amotz, A., A. Shaish and M. Avron. 1989. Mode of action of the massively accumulated β-carotene of *Dunaliella bardawil* in protecting the alga against damage by excess irradiation. Plant Physiol. 91: 1040.1043.

Bhosale, P. 2004. Environmental and cultural stimulants in the production of carotenoids from microorganisms. Appl. Microbiol. Biotechnol. 63(4):351-361.

Borowitzka, M. A. 1988. Vitamins and fine chemicals from microalgae. pp. 153.196. In Micro-algal Biotechnology. M.A.Borowitzka and L.J. Borowitzka (eds.) Cambridge University Press, Cambridge, UK.

Borowitzka, M.A and L.J. Borowitzka. 1988. Micro-algal biotechnology. Cambridge University Press, Cambridge, UK.

Borowitzka, M.A., L. J. Borowitzka and D. Kessly. 1990. Effects of salinity increase on carotenoid accumulation in the green alga *Dunaliella salina.* J. Appl. Phycol. 2(2): 111-119.

Broun, P. and C. Somerville. 2001. Progress in plant metabolic engineering. Proc. Nat. Acad. Sci. USA 98 (16): 8925-8927.

Chitlaru, E. and U. Pick. 1991. Regulation of glycerol synthesis in response to osmotic changes in *Dunaliella.* Plant Physiol. 96: 50-60.

Chitnis, V. P., A. Ke and P. R. Chitnis. 1997. The PsaD subunit of photosystem I: Mutations in the basic domain reduce the level of PsaD in the membranes. Plant Physiol. 115: 1699-1705.

Cifuentes, A.S., M. González, M. Conejeros. V. Dellarossa and O. Parra. 1992. Growth and carotenogenesis in eight strains of *Dunaliella salina* Teodoresco from Chile. J. of Appl. Phycol. 4: 111-118.

Cowan, A.K. and P.D. Rose.1991. Abscisic Acid Metabolism in Salt-Stressed Cells of *Dunaliella salina.* Possible Interrelationship with β-Carotene Accumulation. Plant Physiol. 97: 798-803.

Cunningham, C.X. 2002. Regulation of carotenoid synthesis and accumulation in plants. Pure Applied Chemistry 74(8): 1409–1417.

Czygan, F.-C. 1968. Secondary carotenoids in green algae - I. Chemistry, occurrence, and factors controlling the biogenesis of these polyenes. [Sekundär-Carotinoide in Grünalgen - I. Chemie, Vorkommen und Faktoren, welche die Bildung dieser Polyene beeinflussen.] Arch. Mikrobiol. 61 (1): 81-102.

Demmig-Adams, B. 1990. Carotenoids and photoprotection in plants: a role for the xanthophyll zeaxanthin. Biochim. Biophys. Acta. 1020: 1-24.

Demmig-Adams, B. and W.W. Adams III. 1992. Photoprotection and other responses of plants to high light stress. Ann. Rev. Plant Physiol. Plant Mol. Biol. 43 (1): 599-626.

Demmig-Adams, B., A. M. Gilmore and W.W. Adams. 1996. Carotenoids 3: *In vivo* function of carotenoids in higher plants. FASEB. J. 10: 403-412.

Disch, A., J. Schwender, C. Mueller, H.K. Lichtenthaler and M. Rohmer. 1998. Distribution of the mevalonate and glyceraldehyde phosphate/pyruvate pathways for isoprenoid biosynthesis in unicellular algae and the cyanobacterium *Synechocystis* PCC 6714. Biochem. J. 333: 381-388.

Eckhardt N.A. 2002. Tangerine dreams: Cloning of carotenoid isomerase from *Arabidopsis* and Tomato. Plant Cell 14: 289-292.

Falkowski, P.G. 1984. Kinetics of adaptation to irradiance in *Dunaliella tertiolecta*. Photosynthetica (Praha) 18: 62-68.

Fey, V., R. Wagner, K. Brautigam and T. Pfannschmidt. 2005a. Photosynthetic redox control of nuclear gene expression. J. Exp. Bot. 56(416): 1491-1498.

Fey, V., R. Wagner, K. Brautigam, M. Wirtz, R. Hell, A. Dietzmann, D. Leister, R. Oelmueller and T. Pfannschmidt. 2005b. Retrograde plastid redox signals in the expression of nuclear genes for chloroplast proteins of *Arabidopsis thaliana.* J. Biol. Chem. 280(7): 5318-5328.

Frank, H.A. and R. J. Cogdell. 1993. The photochemistry and function of carotenoids in photosynthesis. pp. 253-326. *In* Carotenoids in Photosynthesis. A. Young and G. Britton (eds.) Chapman and Hall, London, UK.

Frank, H.A. and R.J. Cogdell. 1995. Carotenoids in photosynthesis. Photochem. Photobiol, 63: 257-264.

Frank, H.A. and G.W. Brudvig. 2004. Redox functions of carotenoids in photosynthesis. Biochem. 43: 8607-8615.

Fraser, P.D., J.W. Kiano, M.R. Truesdale, W. Schuch and P.M. Bramley. 1999. Phytoene synthase-2 enzyme activity in tomato does not contribute to carotenoid synthesis in ripening fruit. Plant Mol. Biol. 40: 687-698.

Fromme, P., H. Yu, Y.S. DeRuyter, C. Jolley, D.K. Chauhan, A. Melkozernov and I. Grotjohann. 2006. Structure of photosystems I and II. C. R. Chimie 9 (2): 188-200.

Gallagher, C. E., P.D. Mattews, F. Li and E. T. Wurtzel. 2004. Gene Duplication in the Carotenoid Biosynthetic Pathway Preceded Evolution of the Grasses. Plant Physiol. 135: 1776 -1783.

Gerjets, T., and G. Sandmann. 2006. Ketocarotenoid formation in transgenic potato. J. Exp. Bot. 57 (14): 3639-3645.

Gerjets, T., M. Sandmann, C. Zhu and G. Sandmann. 2007. Metabolic engineering of ketocarotenoid biosynthesis in leaves and flowers of tobacco species. Biotech. J. 2 (10): 1263-1269.

Gomez, P. I. and M. A. Gonzalez. 2005. The effect of temperature and irradiance on the growth and carotenogenic capacity of seven strains of *Dunaliella salina* (Chlorophyta) cultivated under laboratory conditions. Biol. Res. 38(2-3): 151-162.

Gomez, P.I., A. Gomez, A.S. Barriga, A.S. Cifuentes and M.A. Gonzalez. 2003. Effect of salinity on the quantity and quality of carotenoids accumulated by *Dunaliella salina* (strain CONC-007) and *Dunaliella bardawil* (strain ATCC 30861), *Chlorophyta.* Biol. Res. 36: 185-192.

Goodwin, T. W. 1980. The Biochemistry of the carotenoids. pp. 1-377. In Plants.Vol 1 Chapman and Hall, New York, USA.

Grossman, A.R., E. E. Harris, C. Hauser, P.A. Lefebvre, D. Martinez, D. Rokhsar, J. Shrager, C. D. Silflow, D. Stern and O. Vallon. 2003. *Chlamydomonas reinhardtii* at the crossroads of genomics. Eukaryot. Cell 2: 1137–1150.

Grünewald, K., M. Eckert, J. Hirschberg and C. Hagen. 2000. Phytoene desaturase is localized exclusively in the chloroplast and up-regulated at the mRNA level during accumulation of secondary carotenoids in *Haematococcus pluvialis* (Volvocales, Chlorophyceae). Plant Physiol. 122: 1261-1268.

Hippler, M., K. Redding and J.-D. Rochaix. 1998. *Chlamydomonas* genetics, a tool for the study of bioenergetic pathways. Biochim. Biophys. Acta. 1367: 1-62.

Hirschberg, J. 2001. Carotenoid biosynthesis in flowering plant. Curr. Opinion Plant Biol. 4: 210-218.

Howitt, C.A and B. J. Pogson. 2006. Carotenoid accumulation and function in seeds and non-green tissues. Plant Cell Env. 29: 435–445.

Im, C.S., Z. Zhang, J. Shrager, C.W. Chang and A. R. Grossman. 2003. Analysis of light and CO_2 regulation in *Chlamydomonas reinhardtii* using genome-wide approaches. Photosynth. Res. 75: 111–125.

Im, C.-S., S. Eberhard, K. Huang, C. F. Beck and A.R. Grossman. 2006. Phototropin involvement in the expression of genes encoding chlorophyll and carotenoid biosynthesis enzymes and LHC apoproteins in *Chlamydomonas reinhardtii.* Plant J. 48: 1–16.

Isaacson, T., G. Ronen, A. Zamir and J. Hirschberg. 2002. Cloning of *tangerine* from tomato reveals a carotenoid isomerase essential for the production of beta-carotene and xanthophylls in plants. Plant Cell 14: 333-342.

Isaacson, T., I. Ohad, P. Beyer and J. Hirschberg. 2004. Analysis *in vitro* of the enzyme CRTISO establishes a Poly-cis-Carotenoid biosynthesis pathway in plants. Plant Physiol. 136: 4246-4255.

Janick-Buckner, D., J.M. O'Neal, E.K. Joyce and B. Buckner. 2001. Genetic and biochemical analysis of the y9 gene of maize, a carotenoid biosynthetic gene. Maydica 46 (1): 41-46.

Jiménez, C. and U. Pick. 1994. Differential stereoisomer compositions of β,β-carotene in thylakoids and pigment globules in *Dunaliella.* J. Plant Physiol. 143: 257-263.

Jin, E., J.E.W. Polle and A. Melis. 2001. Involvement of zeaxanthin and of the Cbr protein in the repair of photosystem-II from photoinhibition in the green alga *Dunaliella salina.* Biochim. Biophys. Acta. 1506: 244-259.

Jin, E., B. Feth and A. Melis. 2003a. A mutant of the green alga *Dunaliella salina* constitutively accumulates zeaxanthin under all growth conditions. Biotech. Bioengin. 81:116-124.

Jin, E., J.E.W. Polle, H. K. Lee, S. M. Hyun and M. Chang. 2003 b. Xanthophylls in Microalgae: From Biosynthesis to Biotechnological Mass Production and Application. J. Microbiol. Biotechnol. 13: 165–175.

Jin, E., C.-K Lee and J.E.W. Polle. 2006. Secondary Carotenoid Accumulation in *Haematococcus* (Chlorophyceae): Biosynthesis, Regulation and Biotechnology. J. Microbiol. Biotechnol. 16: 1-11.

Katz, A., C. Jiménez and U. Pick. 1995. Isolation and Characterization of a Protein Associated with Carotene Globules in the Alga *Dunaliella bardawil.* Plant Physiol. 108: 1657-1664.

Kessler, E. and F.-C. Czygan. 1967. Physiologische und biochemische Beiträge zur taxonomie der Gattungen *Ankistrodesmus* und *Scenedesmus* - I. Hydrogenase, Sekundär-Carotinoide und Gelatine-Verflüssigung. Arch. Mikrobiol. 55 (4): 320-326.

Kim, J. and D. DellaPenna. 2006. Defining the primary route for lutein synthesis in plants: The role of *Arabidopsis* carotenoid β-ring hydroxylase CYP97A3. Proc. Nat. Acad. Sci. USA 103 (9): 3474-3479.

Klein-Marcuschamer, D., P.K. Ajikumar and G. Stephanopoulos. 2007. Engineering microbial cell factories for biosynthesis of isoprenoid molecules: beyond lycopene. Trends Biotechnol. 25 (9): 417-424.

Kobayashi, M., H. Nobuhiro, Y. Kurimura, H. Ohigashi and Y. Tsuji. 1997. Abscisic acid-dependent algal morphogenesis in the unicellular green alga *Haematococcus pluvialis*. Plant Growth Reg. 22(2): 79-85.

Krol, M., D. P. Maxwell and N. P. A. Huner. 1997. Exposure of *Dunaliella salina* to Low Temperature Mimics the High Light-Induced Accumulation of Carotenoids and the Carotenoid Binding Protein (Cbr). Plant Cell Physiol. 38: 213-216.

Kühlbrandt, W., D.N. Wang and Y. Fujiyoshi. 1994. Atomic model of plant light-harvesting complex by electron crystallography. Nature 367 (6464): 614-621.

León, R., I. Couso, and E. Fernández. 2007. Metabolic engineering of ketocarotenoids biosynthesis in the unicelullar microalga *Chlamydomonas reinhardtii*. J. Biotechnol. 130(2):143-152.

Lerche, W. 1937. Untersuchungen über Entwicklung und Fortpflanzung in der Gattung *Dunaliella*. Arch. Protistenkd. 88: 236-268.

Lers, A., Y. Biener and A. Zamir. 1990. Photoinduction of Massive beta-Carotene Accumulation by the Alga *Dunaliella bardawil*: Kinetics and Dependence on Gene Activation. Plant Physiol. 93(2): 389-395.

Li, L, and J. Van Eck. 2007. Metabolic engineering of carotenoid accumulation by creating a metabolic sink. Transgenic Res. 16 (5): 581-585.

Li, F., Murillo and E.T. Wurtzel. 2007. Maize Y9 encodes a product essential for 15-cis-ζ-carotene isomerization. Plant Physiol. 144 (2): 1181-1189.

Lichtenthaler, H. K. 1999. The 1-Deoxy-D-xylulose-5-phosphate pathway of isoprenoid biosynthesis in plants. Ann. Rev. Plant Physiol. Plant Mol. Biol. 50: 47-65.

Lichtenthaler, H.K. 2007. Biosynthesis, accumulation and emission of carotenoids, α-tocopherol, plastoquinone, and isoprene in leaves under high photosynthetic irradiance. Photosynth. Res. 92 (2): 163-179.

Linden, H. 1999. Carotenoid hydroxylase from *Haematococcus pluvialis*: cDNA sequence, regulation and functional complementation. Biochim. Biophys. Acta. 1999. 1446: 203–212.

Liska, A., J.A. Shevchenko, U. Pick and A. Katz. 2004. Enhanced photosynthesis and redox energy production contribute to salinity tolerance in *Dunaliella* as revealed by homology-based proteomics. Plant Physiol. 136: 2806 - 2817.

Loeblich, L.A.1982. Photosynthesis and pigments influenced by light intensity and salinity in the halophilic *Dunaliella salina* (Chlorophyta). J. Mar. Biol. Assoc. UK 62: 493-503.

Lohr, M. and C. Wilhelm. 1999. Algae displaying the diadinoxanthin cycle also possess the violaxanthin cycle. Proc. Natl. Acad. Sci. USA. 96: 8784-8789.

Lohr, M. and C. Wilhelm. 2001. Xanthophyll synthesis in diatoms: quantifications of putative intermediate and comparison of pigment conversion kinetics with rate constants derived from a model. Planta 212: 382-391.

Lohr, M., C. S. Im and A. R. Grossman. 2005. Genome-based examination of chlorophyll and carotenoid biosynthesis in *Chlamydomonas reinhardtii*. Plant Physiol. 138(1):490-515.

Ma, Y. Z., N. E. Holt, X. P. Li, K. K. Niyogi and G. R. Fleming. 2003. Evidence for direct carotenoid involvement in the regulation of photosynthetic light harvesting. Proc. Natl. Acad. Sci. USA 100: 4377-4382.

Mahmoud, S.S. and R.B. Croteau. 2002. Strategies for transgenic manipulation of monoterpene biosynthesis in plants. Trends Plant Sci. 7 (8): 366-373.

Markovits A., M.P. Gianelli, R. Conejeros and S. Erazo

M.P. Gianelli, P is Missing in text, please add P. after M.

Markovits, A., M. Glanelli, R. Conejers and S. Erazo. 1993. Strain selection for β-carotene production by *Dunaliella*. World J. of Microbiol. Biotech. 9: 534-537.

McCarthy, S. S., M. C. Kobayashi and K. K. Niyogi. 2004. White mutants of *Chlamydomonas reinhardtii* are defective in phytoene synthase. Genetics 168(3): 1249-1257.

Merchant, S.S., S.E. Prochnik, O. Vallon, E.H. Harris, S.J. Karpowicz, G.B. Witman, A. Terry, (...) and K. Zhou. 2007. The *Chlamydomonas* genome reveals the evolution of key animal and plant functions. Science 318 (5848): 245-251.

Morris, W.L., L.J.M. Ducreux, P.D. Fraser, S. Millam and M.A. Taylor. 2006. Engineering ketocarotenoid biosynthesis in potato tubers. Metabol. Eng. 8 (3): 253-263.

Niyogi, K.K. 1999. Photoprotection revisited: genetic and molecular approaches. Annu. Rev. Plant Physiol. Plant Mol. Biol. 50: 333–359.

Orset, S.C. and A.J. Young. 1999. Low-temperature induced synthesis of α-carotene in the microalgae *Dunaliella salina* (chlorophyta) J. Phycol. 35: 520-527.

Orset, S.C. and A.J. Young. 2000. Exposure to low irradiances favors the synthesis of 9-cis beta, beta-carotene in *Dunaliella salina* (Teod.). Plant Physiol.122: 609-618.

Park, H., S.S . Kreunen, A.J. Cuttriss, D. DellaPenna and B.J. Pogson. 2002. Identification of the carotenoid isomerase provides insight into carotenoid biosynthesis, prolamellar body formation, and photomorphogenesis. Plant Cell. 14: 321-332.

Pecker, I., R. Gabbay, F.X. Cunningham and J. Hirschberg. 1996. Cloning and characterization of the cDNA for lycopene beta-cyclase from tomato reveals decrease in its expression during fruit ripening. Plant Mol. Biol. 30 (4): 807-819.

Pfannschmidt, T. 2003. Chloroplast redox signals: How photosynthesis controls its own genes. Trends Plant Sci. 8(1): 33-41.

Phadwal, K. and P.K. Singh. 2003. Effect of nutrient depletion on beta-carotene and glycerol accumulation in two strains of *Dunaliella* sp. Bioresour. Technol. 90(1): 55-58.

Pick, U. 1998. *Dunaliella*- a model extremophilic alga. Israel J. Plant Sci. 46: 131-140.

Pogson, B.J. and H.M. Rissler. 2000. Genetic manipulation of carotenoid biosynthesis and photoprotection. Philos. Trans. R. Soc. Lond B. Biol. Sci. 355: 1395–1403.

Pogson, B. J., K.A. McDonald, M. Truong, G. Britton and D. DellaPenna. 1996. *Arabidopsis* carotenoid mutants demonstrate that lutein is not essential for photosynthesis in higher plants. Plant Cell 8: 1627-1639.

Pogson, B. J., K.K. Niyogi, O. Björkman and D. DellaPenna. 1998. Altered xanthophyll compositions adversely affect chlorophyll accumulation and non-photochemical quenching in *Arabidopsis* mutants. Proc. Natl. Acad. Sci. USA. 95: 13324-13329.

Polle, J.E.W., K. K. Niyogi and A. Melis. 2001. Absence of lutein, violaxanthin and neoxanthin affects the functional chlorophyll antenna size of photosystem-II but not that of photosystem-I in the green alga *Chlamydomonas reinhardtii*. Plant Cell Physiol. 42 (5): 482-491.

Rabbani, S., P. Beyer, J. V. Lintig, P. Hugueney and H. Kleinig. 1998. Induced β-Carotene Synthesis Driven by Triacylglycerol Deposition in the Unicellular Alga *Dunaliella bardawil*. Plant Physiol. 116: 1239-1248.

Rabinowitch, E. 1952. Photosynthesis. Ann. Rev. Plant Physiol. (3): 229-264.

Rodríguez-Concepción, M. and A. Boronat. 2002. Elucidation of the methylerythritol phosphate pathway for isoprenoid biosynthesis in bacteria and plastids. A metabolic milestone achieved through genomics. Plant Physiol. 130 (3): 1079-1089.

Rohdich, F., S. Hecht, A. Bacher and W. Eisenreich. 2003. Deoxyxylulose phosphate pathway of isoprenoid biosynthesis. Discovery and function of ispDEFGH genes and their cognate enzymes. Pure Appl. Chem. 75 (2-3): 393-405.

Rohmer, M. 1999. The discovery of a mevalonate-independent pathway for isoprenoid biosynthesis in bacteria, algae and higher plants. Nat. Prod. Rep. 16 (5): 565-574.

Ronen, G., L. Carmel-Goren, D. Zamir and J. Hirschberg. 2000. An alternative pathway to β-carotene formation in plant chromoplasts discovered by map-based cloning of Beta and old-gold color mutations in tomato. Proc. Nat. Acad. Sci. USA 97 (20): 11102-11107.

Sánchez-Estudillo, L., Y. Freile-Pelegrin, R. Rivera-Madrid, D. Robledo and J.A. Narváez-Zapata. 2006. Regulation of two photosynthetic pigment-related genes during stress-induced pigment formation in the green alga, *Dunaliella salina*. Biotechnol. Lett. 11: 787-791.

Sanchez Miron, A., M.C. Ceron, C. Garcia, M. G. Francisco and C. E. Yusufv. 2002. Growth and biochemical characterization of microalgal biomass produced in bubble column and airlift photobioreactors: studies in fed-batch culture. Enz. Microb. Technol. 3:11015–11023.

Sandmann, G., S. Römer and P.D. Fraser. 2006. Understanding carotenoid metabolism as a necessity for genetic engineering of crop plants. Metabol. Eng. 8 (4): 291-302.

Schwender, J., C. Gemuenden and H. K. Lichtenthaler. 2001. Chlorophyta exclusively use the 1-deoxyxylulose-5-phosphate/2-C-methylerythritol 4-phosphate pathway for the biosynthesis of isoprenoids. Planta 212: 416-423.

Shaish A., M. Avron and A. Ben-Amotz. 1990. Effect of Inhibitors on the formation of stereoisomers in the biosynthesis of β-carotene in *Dunaliella bardawil*. Plant Cell Physiol. 31(5): 689-696.

Shaish, A., A. Ben-Amotz and M. Avron. 1991. Production and section of high β-carotene mutants of *Dunaliella bardawil*. J. Phycol. 27: 652-656.

Shaish, A., M. Avron, U. Pick and A. Ben-Amotz. 1993. Are active oxygen species involved in induction of beta-carotene in *Dunaliella bardawil*? Planta 190:363.368.

Shewmaker, C.K., J. A. Sheehy, M. Daley, S. Colburn and D.Y. Ke. 1999. Seed specific overexpression of phytoene synthase: increase in carotenoids and other metabolic effects. Plant J. 20: 401-412.

Shrager, J., C. Hauser, C.W. Chang, E. H. Harris, J. Davies, J. McDermott, R. Tamse, Z. Zhang and A. R. Grossman. 2003. *Chlamydomonas reinhardtii* genome project. A guide to the generation and use of the cDNA information. Plant Physiol. 131: 401-408.

Siefermann-Harms, D. 1987. The light-harvesting and protective functions of carotenoids in photosynthetic membranes. Physiol. Plant 69: 561–568.

Sieiro, C., M. Poza, T. De Miguel and T.G. Villa. 2003. Genetic basis of microbial carotenogenesis. Internat. Microbiol. 6 (1): 11-16.

Steinbrenner, J. and H. Linden. 2001. Regulation of two carotenoid biosynthesis genes coding for phytoene synthase and carotenoid hydroxylase during stress-induced astaxanthin formation in the green alga *Haematococcus pluvialis*. Plant Physiol. 125: 810-817.

Steinbrenner, J. and H. Linden. 2003. Light induction of carotenoid biosynthesis genes in the green alga *Haematococcus pluvialis*: regulation by photosynthetic redox control. Plant Mol. Biol. 52:343-356.

Stephanopoulos, G. and K.L. Jensen. 2005. Metabolic engineering: Developing new products and processes by constructing functioning biosynthetic pathways in vivo. AIChE Journal 51 (12): 3091-3093.

Sun, J., A. Ke, P. Jin, V. P. Chitnis and P. R. Chitnis. 1998. Isolation and functional study of photosystem I subunits in the cyanobacterium *Synechocystis* sp. PCC 6803. Methods Enzymol. 297: 124-139.

Sun, Z., F.X. Cunningham Jr. and E. Gantt. 1998. Differential expression of two isopentenyl pyrophosphate isomerases and enhanced carotenoid accumulation in a unicellular chlorophyte. Proc. Natl. Acad. Sci. USA. 95: 11482-11488.

Suzuki, S., M. Nishihara, T. Nakatsuka, N. Misawa, I. Ogiwara and S. Yamamura. 2007. Flower color alteration in *Lotus japonicus* by modification of the carotenoid biosynthetic pathway. Plant Cell Rep. 26 (7): 951-959.

Taylor, M. and G. Ramsay. 2006. Carotenoid biosynthesis in plant storage organs: recent advances and prospects for improving plant food quality. Physiol. Plantarum 124: 143–151.

Telfer, A. 2005. Too much light? How β-carotene protects the photosystem II reaction centre. Photochem. Photobiol. Sci. 4 (12): 950-956.

Telfer, A., G.W. Brudvig, T.A. Moore, S. Styring, A.W. Rutherford, P. Fromme and E.-M. Aro. 2002. What is β-carotene doing in the photosystem II reaction centre? Phil. Trans. R. Soc. Lond. B 357 (1426): 1431-1440.

Teodoresco, E.C. 1905. Organization et developpement du *Dunaliella*, nouveau genre de Volvocacee-Polyblepharidee. Beih. Bot. Zentralblatt. Bd.18 Abt.1: 215-232.

Tian, L., V. Musetti, J. Kim, M. Magallanes-Lundback and D. DellaPenna. 2004. The Arabidopsis LUT1 locus encodes a member of the cytochrome P450 family that is required for carotenoid ε-ring hydroxylation activity. Proc. Nat. Acad. Sci. USA 101 (1): 402-407.

Tracewell, C.A., J. S. Vrettos, J. A. Bautista, H. A. Frank and G. W. Brudvig . 2001. Carotenoid photooxidation in photosystem II. Arch Biochem. Biophys. 385: 61-69.

Wurbs, D., S. Ruf and R. Bock. 2007. Contained metabolic engineering in tomatoes by expression of carotenoid biosynthesis genes from the plastid genome. Plant J. 49 (2): 276-288.

Wurtzel, E. 2004. Genomics, genetics and biochemistry of Maize carotenoid biosynthesis. pp. 86-110. In Recent Advances in Phytochemistry Vol 38, Secondary Metabolism in Model Systems. . Romeo (ed.) Elsevier, USA.

Yamamoto, H.Y. and R. Bassi.1996. Carotenoids: localization and function. pp 539–563. *In* Oxygenic Photosynthesis: The Light Reactions.D.R.Ort and C.F.Yocum (eds.) Kluwer Academic Publishers, Dordrecht, The Netherlands.

Yan, Y., Y.H. Zhu, J.G. Jiang and D.L. Song. 2005. Cloning and sequence analysis of the phytoene synthase gene from a unicellular chlorophyte, *Dunaliella salina*. J. Agric. Food Chem.53(5):1466-1479.

Ye, X. and Ye, X., S. Al-Babili, A. Klöti, J. P. Lucca, P. Beyer, and I. Potrykus. 2000 P. Beyer. 2000. Engineering the provitamin A (β-carotene) biosynthetic pathway into (carotenoid-free) rice endosperm. Science 287 (5451): 303-305.

Yokthongwattana, K., T. Savchenko, J.E.W. Polle and A. Melis. 2005. Isolation and characterization of a xanthophyll-rich fraction from the thylakoid membrane of *Dunaliella salina* (green algae). Photochem. Photobiol. Sci. 4(12): 1028-1034.

Zelazny, A.M., A. Shaish and U. Pick. 1995. Plasma membrane sterols are essential for sensing osmotic changes in the halotolerant alga *Dunaliella*. Plant Physiol. 109(4): 1395-1403.

Zhang, Z., J. Shrager, C.W. Chang, O. Vallon and A. R. Grossman. 2004. Genome based analysis of sulfur deprivation of wild-type cells and the *sac1* mutant of *Chlamydomonas*. Eukary. Cell 3: 1331-1348.

Zhu, Y.H., J.-G. Jiang, Y. Yan and X.-W. Chen. 2005. Isolation and characterization of phytoene desaturase cDNA involved in the beta-carotene biosynthetic pathway in *Dunaliella salina*. J. Agric. Food Chem. 53(14): 5593-5597.

Zhu, Y.-H., J.-G. Jiang and X.-W. Chen. 2007. cDNA for phytoene desaturase in *Dunaliella salina* and its expressed protein as indicators of phylogenetic position of the β-carotene biosynthetic pathway. J. Sci. Food Agricult. 87 (9): 1772-1777.

7 Acquisition and Metabolism of Inorganic Nutrients by *Dunaliella*

John Beardall[1] and Mario Giordano[2]

Abstract

Dunaliella, in common with other microalgae, possesses a range of systems to satisfy cellular demands for nutrients. In this chapter, we summarise what is known about the mechanisms involved in the acquistition of the major macronutrients N, P, S and C. We describe how uptake of these elements is affected by environmental conditions and the interactions that occur between C acquisition and the availability of the other commonly limiting macronutrients.

Introduction

Although chlorophyte algae of the genus *Dunaliella* are found in nature in many marine habitats, the genus never comprises a major component of algal populations in systems of normal seawater salinity (35 PSU). However, the ability of some *Dunaliella* species to tolerate high salt concentrations allows them to become the dominant microalgae in certain hypersaline situations such as the Dead Sea, Israel (Nissenbaum 1975), Pink Lake in Australia (Borowitzka 1981) and the Great Salt Lake in the USA (Post 1981). *D. parva* and *D. salina* have been reported as the dominant spring phytoplankton species in the athalassic lake of Fuente de Piedra in Spain (Jimenez et al. 1990). Due to their ability to grow and thrive at high salinities, *Dunalialla* species are also found as the predominant algae in artificially hypersaline environments such as saltworks. A full discussion of the ecology of *Dunaliella* may be found in the chapters by Polle et al. in this volume.

[1] School of Biological Sciences, Monash University, Clayton, Australia.
[2] Dipartimento di Scienze del Mare, Università Politecnica delle Marche, Ancona Italy.
Corresponding author: john.beardall@sci.monash.edu.au

In order to support adequate growth, *Dunaliella*, like all algae, requires a range of inorganic nutrients which are then incorporated into a plethora of organic compounds including the major macromolecular groups of proteins, lipids, carbohydrates and nucleic acids. Although the average values for elemental ratios in microalgae are close to the traditional Redfield ratio of 106:16:1, data for *Dunaliella tertiolecta* suggest somewhat higher ratios of 222:38:1 (Ho et al. 2003). S is present at a molar ratio of S:P of 0.28, again a slightly lower value than the average (1.3:1) for a range of microalgae examined (Ho et al. 2003).

Carbon

Dunaliella tertiolecta, in common with almost all other algae (but see Reinfelder et al. 2000, 2004) assimilates inorganic carbon via the enzyme ribulose bisphosphate carboxylase oxygenase (RUBISCO) and the C_3 carbon reduction cycle (Calvin cycle). All RUBISCOs have competitive carboxylase and oxygenase functions and have relatively low substrate-saturated carboxylase activities on a protein mass basis. The affinity of algal RUBISCOs is generally low, with half saturation constants for CO_2 ($K_{0.5}$ CO_2) for green algae in the order of 25-38 μM, although that for the lichen symbiont *Coccomyxa* (which has no ability to transport and accumulate inorganic carbon – see discussion of CO_2 concentrating mechanisms below) is as low as 12 μM (Palmqvist et al. 1995). The extent to which the two competitive reactions of RUBISCO occur depends on the O_2 and CO_2 concentrations at the RUBISCO active site and the molecular nature of the RUBISCO molecule involved. The selectivity factor defining the relative rates of carboxylase and oxygenase reactions is given by Eqn 1,

$$S_{rel} = \frac{K_{0.5}(O_2) \cdot k_{cat}(CO_2)}{K_{0.5}(CO_2) \cdot k_{cat}(O_2)} \qquad \text{(Eqn 1)}$$

where $K_{0.5}$ (CO_2) and $K_{0.5}$ (O_2) are the half saturation constants for the carboxylase and oxygenase functions, respectively, and K_{cat} (CO_2) and K_{cat} (O_2) are the corresponding substrate-saturated rates of catalysis. Although there are no reported direct measurements of the kinetic properties of isolated RUBISCO from *Dunaliella* species, all chlorophytes examined have Form 1B RUBISCOs, with S_{rel} values in the range 54 to 83, with the closest comparator being *Chlamydomonas* with a S_{rel} of 61 and a $K_{0.5}CO_2$ of 29 μM (Badger et al. 1998). As a consequence of these kinetic properties, inorganic carbon assimilation by cells relying on CO_2 diffusion and assimilation by RUBISCO would be seriously limited by CO_2 availability under present day atmospheric CO_2 concentrations (Giordano et al.2005a, b and c). However, in common with most microalgae,

Dunaliella species possess CO_2 concentrating mechanisms (CCMs) which use active transport of inorganic carbon species to improve the supply of CO_2 to the active site of RUBISCO and ameliorate the inherent inefficiencies of that enzyme. As a consequence, $K_{0.5}CO_2$ values for intact cells of *Dunaliella* species are < 1μM (Aizawa et al. 1985; Burns and Beardall 1987; Booth and Beardall 1991; Giordano and Bowes 1997; Giordano 1997, 2001; Giordano et al. 2000; Young et al. 2001).

The CCM of *Dunaliella tertiolecta* is reasonably well defined physiologically, although there is still some uncertainty over the membrane(s) involved in the active transport of dissolved inorganic carbon (DIC). The molecular characterization of the transporters in eukaryotic algae is still lagging somewhat behind that of cyanobacteria and those molecular studies that have been done have used *Chlamydomonas* rather than *Dunaliella* (see e.g. Wang et al. 2005; Mitra et al. 2005). *Dunaliella tertiolecta* is capable of active transport of both HCO_3^- and CO_2 (Amoroso et al. 1998) and internal pools of DIC of 1.2 to 5 mM have been reported (Burns and Beardall 1987; Beardall et al. 2002), corresponding to internal CO_2 pools of around 40 -167 μm. Utilization of bicarbonate is associated with an external carbonic anhydrase in *Dunaliella* (Burns and Beardall 1987; Booth and Beardall 1991; Colman et al. 2002) which, as in *Chlamydomonas*, is believed to supply CO_2 for transport (diffusive or via a CO_2 transporter) across the plasma membrane (Williams and Turpin 1987). Although active uptake of DIC at the plasma membrane undoubtedly occurs, DIC transport has also been shown by isolated chloroplasts of *Dunaliella* (Amoroso et al. 1998; Goyal and Tolbert 1989; Moroney and Chen 1998). A scheme summarizing possible mechanisms of entry of DIC in *Dunaliella* is presented in Figure 1.

In common with CCMs in other algae, CO_2 acquisition by *Dunaliella* is modulated by a range of environmental factors, notably the availability of CO_2. Thus, growth of cells at high CO_2 will down-regulate the CCM (see Beardall and Giordano 2002 and references therein). Lowering CO_2 levels, in contrast, leads to enhanced CCM activity. For instance, among the consequences of a high salinity environment are a decrease in CO_2 solubility and a shift in the equilibrium between CO_2 and HCO_3^-. *Dunaliella salina* responds to this decrease in CO_2 and increase in HCO_3^- at higher salinity by increasing the activity of a CO_2 concentrating mechanism (Zenvirth and Kaplan 1981; Booth and Beardall 1991). This is accompanied by elevated levels of an external carbonic anhydrase (CA_{ext}) (Booth and Beardall 1991). *D. salina* adapted to high salinity shows induction of an unusual external α-type CA which retains its activity over the range of salinities from 0–4 M NaCl (Fisher et al. 1996; Premkumar et al. 2003). This CA_{ext} consists of two internally duplicated tandem repeated sequences fused into a single polypeptide of 60 kDa, compared to the periplasmic CAs from *Chlamydomonas*, that consist of 75 kDa heterotetramers of 2 large and 2 small subunits and which are 90% inhibited by 0.6 M NaCl (Premkumar et al. 2003).

Other environmental factors such as low light (Young and Beardall 2005),

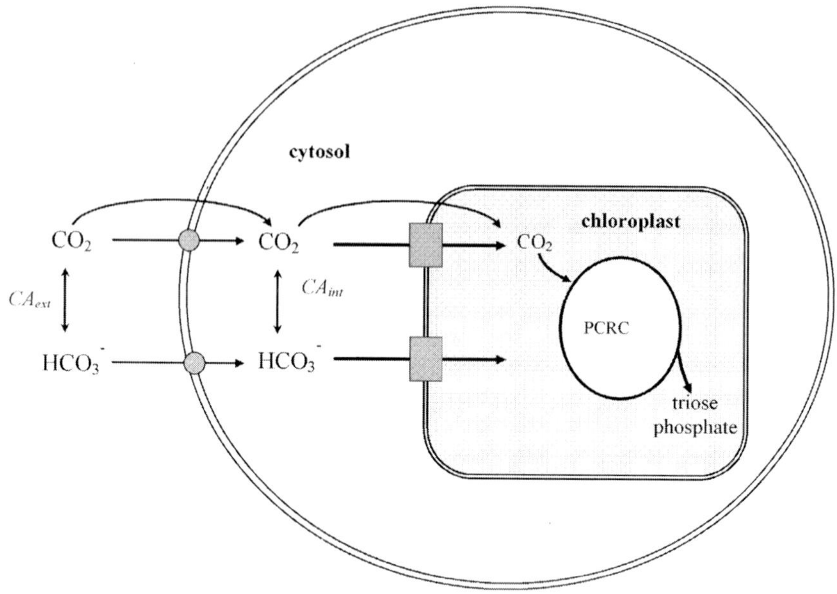

Figure 1: A model for inorganic carbon acquisition by microalgae. Transport of CO_2 can be achieved by diffusion across the plasmalemma and chloroplast envelope (curved arrows) or via transporters across either or both of these membranes (circles and squares respectively). Redrawn after Giordano et al. (2005a).

UVB exposure (Beardall et al. 2002) and nutrient availability can influence the acquisition of DIC via the CCM in *Dunaliella tertiolecta*. These are summarized by Beardall and Giordano (2002) and the interaction between DIC acquisition and the availability of other nutrients is specifically dealt with below.

Nitrogen

The range of environments that *Dunaliella* species inhabit is characterized by large variations in nutrient availability. For *D. salina* in saltworks, poor management of biological processes can lead to large variation in nutrient levels (Davis 2000; Dolapsakis et al. 2005) and this can lead to substantial physiological changes in *D. salina* cells (Giordano et al. 1994; Giordano and Bowes 1997).

Dunaliella species can take up both nitrate and ammonium (Giordano et al. 1994, 2002) though the latter is used preferentially if both are available. Indeed, the presence of ammonium stops transcription of nitrate reductase (NAR) genes and thus ammonium utilisation becomes the only way of inorganic entry of N under these circumstances (e.g. Fernandez et al. 1998). *Dunaliella* also has a capacity for using organic sources of nitrogen including hypoxanthine, allantoate and urea (Oliviera and Huynh 1989) and histidine (Hellio and Le Gal 1998, 1999). For a detailed review of organic N utilization by microalgae, see Antia et al. (1991) and Fernandez et al. (1998).

Ammonium is taken up via the AMT/Mep protein family of transporters (see e.g. Andrade et al. 2005; Fernandez et al. 1998; Hildebrand 2005; Khademi et al. 2004; Ludewig et al. 2003; Loqué et al. 2005; Soupene et al. 2002). The possible transport of NH_3 via homologues of aquaporins has also been suggested (Jahn et al. 2004).

Although some authors have reported that very high concentrations of ammonium (>2.5 M) were lethal for *Dunaliella* species (Grant 1968), ammonium concentrations in the range 0.5 to 10 mM had little effect on growth rates of *D. viridis* (Jiménez & Niell 1991), and *D. parva* (Giordano et al. 2002) and *D. salina* (Giordano et al. 1994) are both capable of adaptation to growth on N concentrations up to 5 and 10 mM, respectively, regardless of whether the N source is ammonium or nitrate. Cells acclimated to growth on high ammonium concentrations usually showed similar division rates, but had a larger cell size, than their counterparts grown at high nitrate levels (Giordano and Bowes 1997). Half saturation constants (Ks) for growth of *Dunaliella tertiolecta* on ammonia and nitrate have been reported as 18-30 μM and 19-71 μM respectively, depending on temperature (Reay et al. 1999). These are somewhat higher values than the reported Ks values for the high affinity nitrate uptake by *D. tertiolecta* of 11.1 μM, though *D. tertiolecta*, in common with other microalgae, appears also to possess a second, low affinity/high capacity, uptake system (Lomas and Glibert 2000). These values are, however, considerably higher than the 0.21 μM cited as the Ks value for *Dunaliella* NO_3^-, Cl^--activated ATPase activity (taken as an indirect measure of transport activity) by Falkowski (1975). Maximal nitrate uptake rates by the high affinity system have been reported as 18 fmol N h^{-1} $cell^{-1}$ (0.005 fmol N h^{-1} $fmol^{-1}$ C) (Lomas and Glibert 2000). Measurements of ammonium uptake kinetics in *Dunaliella* are rare – Syrett (1988) cites a value of ~ 1μM, almost an order of magnitude lower than the 7.5 μM cited for the high affinity ammonium carrier of the related freshwater chlorophyte *Chlamydomonas reinhardtii* (Franco et al. 1988).

Growth of *D. salina* on high concentrations of ammonium compared to nitrate results in nearly a doubling of the cellular chlorophyll content (Giordano and Bowes 1997). The effect of the N-source on cellular β-carotene content is much smaller (a 30-50% increase under ammonium), with a consequent increase in the chlorophyll to β-carotene ratio (Giordano and Bowes 1997). Algae cultured on ammonium rather than on nitrate increased not only the amount of chlorophyll per cell (~70%) but also the apparent chlorophyll concentration (~50%), despite an approximately 20% ammonium-related increase in cell size. The magnitude of the effect of the N-source on the cellular β-carotene content (~50%) and concentration (per unit cell volume) of β-carotene (~10%) was smaller, but this did not substantially affect the chlorophyll to β-carotene ratio (Giordano and Bowes, 1997). The amount of N available also plays an important role in the production of β-carotene. In combination with temperature, salinity and light, N-limitation is among the most potent inducer of carotenogenesis in the carotenogenic strains of *Dunaliella* (see e.g. Ben-Amotz and Shaish 1992). The production of β-carotene by *Dunaliella* is dealt with in detail elsewhere in this volume (see chapter by Ben-Amotz).

Nitrogen assimilation is carried out by *Dunaliella* using a conventional GS-GOGAT pathway (Figure 2). Cells using nitrate first reduce this to nitrite and then to ammonium. Reported K_m-NO_3^- values for nitrate reductase are 240 μM (Lomas and Glibert 2000). Nitrate reductases in *Dunaliella* species are unusual in that they can utilize either NADH, NADPH or FADH as electron donors (LeClaire and Grant 1972; Heimer 1975, 1976; see Berges 1997 for a review of algal nitrate reductases). Although nitrate assimilation is usually considered a process driven by photosynthetically generated energy, Del Rio et al. (1994) have, controversially, provided evidence that nitrate reduction in *D. salina* could also occur in the dark, thanks to the allegedly rather unusual (see Giordano et al. 2005b and references therein) properties of nitrate reductase from this microalga. The ability to take up and assimilate NO_3^- at night could make a big difference in terms of interpreting the strategy for N utilization by *D. salina* and further studies are needed to clarify this contentious aspect of NO_3^- reduction. The storage capacity for NO_3^- seems to be relatively small and is approximately 10 fold lower than in diatoms (Lomas and Glibert 2000). The ability to actively take up and store N as

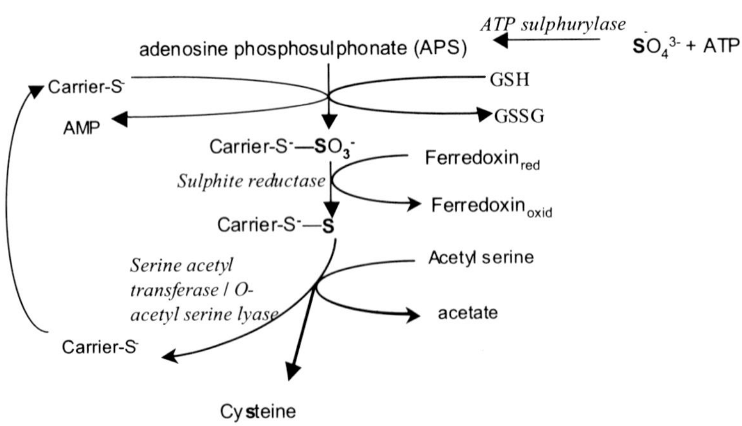

Figure 2: The pathway for assimilation of nitrate by microalgae and its incorporation into amino acids.

NH_4^+ is also still rather controversial (Pick et al. 1991; Lomas and Glibert 2000, Giordano and Bowes 1997; Giordano et al. 2000).

The enzyme glutamine synthetase plays a central role in the assimilation of ammonia by algae (Figure 2). It is present in two isoforms in *Dunaliella tertiolecta*, a cytosolic and a chloroplastic form, with the cytosolic form being present at higher levels than the chloroplastic one (Casselton et al. 1986; Robertson and Tartar 2006). While photophosphorylation will support ammonium assimilation (Turpin and Bruce 1990), there is also evidence from the diatom *Phaeodactylum tricornutum* that under some circumstances mitochondrial oxidative phosphorylation can provide the ATP to drive GS activity (Rees 2003). Whether this is also true of *Dunaliella* species remains to be seen.

Phosphate and Sulphur

Phosphate (P_i) and sulphate uptake by *Dunaliella salina* are both mediated by a Na^+-symport mechanism driven by the $\Delta\mu Na^+$ across the plasmamembrane. $K_s(P_i)$ has been reported as 1.4 μM and that for sulphate as ~ 13 μM (Weiss et al. 2001). Similar values for K_s(Pi) have been obtained for *D. tertiolecta* (Roberts 1998). Maximal uptake rates (V_{max}) for phosphate and sulphate by nutrient-replete *D. salina* were reported by Weiss et al. (2001) as 28 pmol 10^6 cells min^{-1} and 1.0 pmol 10^6 cells min^{-1} respectively. Maximal uptake rates for phosphate were stimulated considerably under nutrient starved conditions (V_{max} values increased to 730 pmol 10^6 cells min^{-1}) without a comcomittant change in K_s (P_i). In S-starved cells, increases in both V_{max} (3 fold) and K_s (50%) were recorded compared to nutrient replete controls. Increases in phosphate uptake rates under P-limitation or P-starvation have been noticed in a number of studies of *Dunaliella* (see e.g. Hirsch et al. 1993; Shelly et al. 2005). Under low phosphate supply conditions, *Dunaliella tertiolecta* will produce an extracellular alkaline phosphatase with activity up to ~ 9 x 10^{-6} nmol $cell^{-1}$ h^{-1} (Wynne and Rhee 1988).

Sulphur availability is usually not a factor in the ocean. The sulphur requirement for growth of *D. salina*, for instance, is well below the concentrations commonly found in the sea (Giordano et al. 2000). Sulphur, nevertheless, can vary appreciably in saltworks and in saline lakes and the sulphur requirement for growth may be directly related to the evolutionary trajectory of taxa and to their ecological success. Sulphur availability may also be related to the ability of algal cells to produce DMSP and its antigrazing derivative, acrylate, (Norici et al. 2005 and references therein).

Sulphur acquisition in *Dunaliella* is mediated through the pathway shown in Figure 3. APS sulphurylase is the key enzyme in this metabolic sequence and has, therefore, been the subject of considerable study in algae, though mostly in *Chlamydomonas* rather than in *Dunaliella* specifically (see Davies and Grossman 1998).

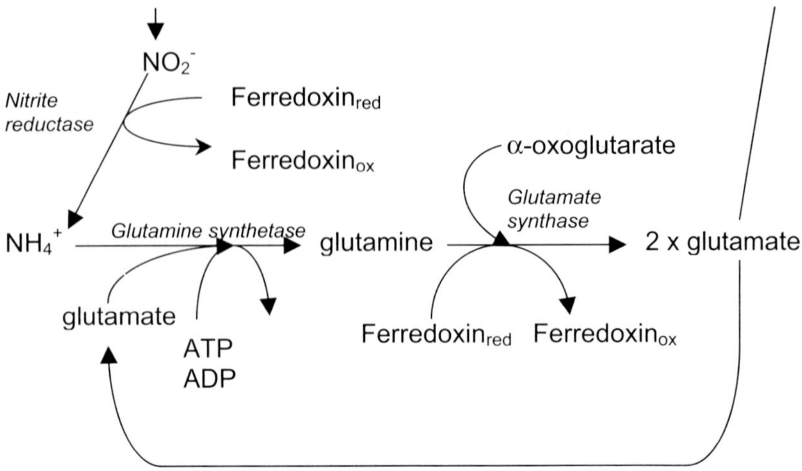

Figure 3: The pathway of sulphur assimilation in microalgae (GSH=reduced glutathione; GSSG=oxidized glutathione).

D. salina metabolism is strongly dependent on sulphur availability, and the maintenance of a sufficient intracellular sulphate concentration appears to play an important role in cell physiology: in response to a 8000-fold decrease of external sulphate, the apparent sulphate cell concentration (i.e. assuming an even distribution throughout the cell) only decreased by less than 2.5-fold (Giordano et al. 2000). Analogous to what is observed for *Chlamydomonas* (Davies and Grossman 1998; Yildiz et al. 1994), a specific sulphate transporter may be induced under S-limitation. This is indirectly confirmed by the fact that *D. salina* cells transferred from high to low sulphate medium in short term experiments showed an increase in sulphate uptake capacity (V. Pezzoni and M Giordano, unpublished). Modulation of sulphate transport and response to S-limitation is modulated, in *Chlamydomonas* at least, by a group of polypeptides known as Sac1, Sac2 and Sac3. Sac1 is involved in regulation of arylsulphatase activity (of which there are 2 periplasmic forms in *Chlamydononas* ; Davies and Grossman 1998), which cleaves sulphate from organic sulphur-containing compounds, allowing cells to make use of organic S in the environment. Sac1 may, additionally, also be involved in sensing the external level of sulphate (Davies et al. 1996; Davies and Grossman 1998). Sac2 plays a role in the regulation of genes encoding arylsuphatase and, possibly, of other genes induced during sulphur limitation. Sac3 is a protein kinase that appears to have both positive and negative effects on gene expression (Davies et al. 1999). Whether similar proteins modulate the responses to S-limitation in *Dunaliella* is not known.

When *D. salina* is exposed to low sulphate concentrations, growth rate, cell volume and photosynthetic activity decrease substantially, while the organism manages to maintain protein and chlorophyll levels similar to those of sulphate sufficient cells. This adaptation is mediated by an array of physiological responses that strongly suggest regulatory interactions between the assimilatory pathways of sulphur, nitrogen and carbon (Giordano et al. 2000).

In order to optimize the utilization of sulphur when S availability is reduced, S-rich pools are possibly mobilized, as suggested by a decline in RUBISCO protein. RUBISCO is in fact one of the main cellular reservoirs of reduced sulphur (up to 50 mM in the chloroplast, according to Ferreira and Teixeira, 1992); At the onset of sulphur limitation, RUBISCO may act as a source of fixed sulphur and carbon; upon acclimation, the decreased synthesis of RUBISCO would allow cells to shift resources to other, more needed proteins, as suggested by Gilbert et al. (1997) for acclimation of wheat leaves to low sulphate conditions. The strong reduction of photosynthesis under S-limitation may be correlated to the lower amount of RUBISCO. In contrast, no obvious changes were observed in the activity of periplasmic carbonic anhydrase, which was presumably not involved in the down regulation of photosynthesis.

A reduction of sulphate availability directly affects N utilization, as it is shown by the intracellular concentrations of NH_4^+ and NO_3^- and the changes occurring in N assimilation and anaplerotic pathways (Giordano et al. 2000). Moreover, when sulphur is limiting, a major reallocation of C occurs, with a diversion of C-skeletons towards 3-C compounds, possibly to support glycerol metabolism (Giordano et al. 2000).

Interactions Between the Acquisition of Carbon and Other Nutrients

In contrast to the effects of N source on acquisition of inorganic carbon, in *D. tertioleta* at least, N-limitation leads to diminished RUBISCO levels but enhanced CCM activity, leading to improved N-use efficiency of C assimilation under low N conditions (Beardall et al. 1991; Young and Beardall 2005). In iron-limited chemostat cultures of *Dunaliella tertiolecta*, Young (1999) demonstrated that affinity of cells for DIC was actually enhanced at lower growth rates. Enhanced affinity for DIC in iron-limited cells reflects an increased investment in an active DIC acquisition mechanism and $\delta^{13}C$ values supported the notion of a greater dependence upon active DIC acquisition during growth under iron-limitation.

Maintenance of high affinity DIC acquisition may confer improved energy and resource use efficiency in iron-limited cells when photosynthetic energy-harvesting capacity is impaired by iron deficiency. These findings suggest that maintenance of efficient DIC-uptake kinetics could be a general response to energy limitation imposed by nutrient-deficiency in this alga.

Although Fe-limitation of *D. tertiolecta* showed similar consequences to those of N-limitation (Young and Beardall 2005), P-limitation of *D. tertiolecta* leads to a decreased CCM activity, presumably due to a lower availability of ATP to drive active transport processes (Beardall et al. 2005). Sulphur limitation also results in down regulation of CCM activity in *D. salina* (Giordano et al. 2000).

The presence of either form of inorganic N, especially at high concentrations, plays a major role in the photosynthetic performances of *D. salina* and on the pattern of resource allocation (see below) (Giordano et al. 1994, 2002; Giordano and Bowes 1997; Giordano 2001; Norici et al. 2002).

Although growth on high concentrations of NH_4^+ rather than NO_3^- as the sole N source leads to a stimulation of both maximal rates of photosynthesis (P_{max}) and apparent quantum yield, in *D. salina*, this has no obvious repercussions for the photon flux density required to saturate photosynthesis or on the light compensation point, nor does it impact on the susceptibility of cells to photoinhibition (Giordano and Bowes, 1997).

D. salina cells adapted for several generations (Giordano et al. 1994) to growth on high NH_4^+ concentrations are more effective at utilizing light and CO_2, than those grown on NO_3^-, with NH_4^+ grown cells show a halving of their photosynthetic $K_{0.5}(CO_2)$. This is indirect evidence of the fact that the chemical source of N modulates the effectiveness of *D. salina* CO_2 concentrating mechanisms. The high P_{max} and low $K_{0.5}(CO_2)$ of photosynthesis in NH_4^+ grown cells are associated with an increase of both RUBISCO activity and RUBISCO protein abundance and concentration. The increase of RUBISCO protein under these growth conditions is not simply a result of a general increase of protein, since RUBISCO abundance increases even relative to total protein (Giordano and Bowes 1997). Periplasmic CA activity, which may also play an important role in facilitating CO_2 acquisition (see e.g. Giordano et al. 2005a), also increased when N was available as NH_4^+ rather than as NO_3^-. This increase in activity was substantially larger than the cell volume increase under the same conditions; however, its relevance with respect to the photosynthetic response remains somewhat obscure, since the actual rates of CA-catalyzed dehydration of HCO_3^- to CO_2 are far above the photosynthetic requirements, even for NO_3^--grown cells (Giordano and Bowes 1997).

It is apparent from the above discussion that *Dunaliella* species have evolved efficient systems for the acquisition of the macronutrients, C, N, P and S. Much of the work on the molecular biology of nutrient acquisition has, however, been done on the related chlorophyte and 'model organism', *Chlamydomonas*, and it has yet to be established that similar control mechanisms for nutrient uptake and assimilation also occur in *Dunaliella* species. It is also apparent that the responses to variation in availability of these macronutrients are complex and involve both acclimation and homeostatic (see Montechiaro et al. 2006) mechanisms. It is also evident that there are strong interactions between the acquisition of different resources, as is reflected particularly in the way the availability of N, P and S impinges on the ability of *Dunaliella* cells to acquire C.

Acknowledgement

John Beardall gratefully acknowledges the support of the Australian Research Council for his work on the ecophysiology of microalgae.

References

Aizawa, K., Y. Nakamura and S. Miyachi. 1985. Variation of phosphoenolpyruvate carboxylase activity in *Dunaliella* associated with changes in atmospheric CO_2 concentration. Plant Cell. Physiol. 26: 1199-1203.

Amoroso G., D.F. Sültemeyer, C. Thyssen and H.P. Fock.1998. Uptake of HCO_3^- and CO_2 in cells and chloroplasts from the microalgae *Chlamydomonas reinhardtii* and *Dunaliella tertiolecta*. Plant Physiol. 116: 193–201.

Andrade S.L.A., A. Dickmans, R. Ficner and O. Einsle. 2005. Crystal structure of the archaeal ammonium transporter Amt-1 from *Archaeoglobus fulgidus*. Proceedings of National Academy of Sciences (PNAS) USA 102: 14994-14999.

Antia N.J., P.J. Harrison and L. Oliveira. 1991. The role of dissolved organic nitrogen in phytoplankton nutrition, cell biology and ecology. Physiol. 30: 1-89.

Badger, M.R., T.J.Andrews, S.M. Whitney, M. Ludwig, D.C. Yellowlees, W. Leggat and G.D. Price. 1998. The diversity and coevolution of Rubisco, plastids, pyrenoids, and chloroplast-based CO_2-concentrating mechnisms in algae. Can. J. Bot. 76: 1052-1071.

Beardall J. and M. Giordano. 2002. Ecological implications of algal CO_2 concentrating mechanisms and their regulation. Funct. Plant Biol. 29: 333-347.

Beardall J., S. Roberts and J. Millhouse. 1991. Effects of nitrogen limitation on inorganic carbon uptake and specific activity of ribulose-1,5-P_2 carboxylase in green microalgae. Can. J. Bot. 69, 1146-1150.

Beardall J., P. Heraud, S. Roberts, K. Shelly and S. Stojkovic. 2002. Effects of UV-B radiation on inorganic carbon acquisition by the marine microalga *Dunaliella tertiolecta* (Chlorophyceae). Phycol. 41, 268-272.

Beardall, J., S. Roberts and J.A. Raven. 2005. Regulation of inorganic carbon acquisition by phosphorus limitation in the green alga *Chlorella emersonii*. Canadian. J. Bot. 83: 859-864.

Ben-Amotz A. and A.Shaish.1992. β-carotene biosynthesis. pp. 205-216. In *Dunaliella*: Physiology, Biochemistry and Biotechnology. M. Avron and A. Ben-Amotz (eds). CRC, Boca Raton, USA.

Berges J. 1997. Algal nitrate reductases. Eur. J. Phycol. 32: 3-8.

Booth W.A. and J. Beardall. 1991. Effects of salinity on inorganic carbon utilization and carbonic anhydrase activity in the halotolerant alga *Dunaliella salina* (Chlorophyta). Phycol. 30, 220-225.

Borowitzka L.J. 1981. The microflora. Adaptations to life in extremely saline lakes. Hydrobiol. 81: 33-46.

Burns D.B. and J. Beardall. 1987. Utilization of inorganic carbon by marine microalgae. J.Exp. Mar. Biol. Ecol. 107, 75-86.

Casselton P.J., G. Chandler, N. Shah, G.R. Stweart and N. Sumar. 1986. Glutamine synthetase isoforms in algae. New Phytol. 102: 261-270.

Colman B, I.E. Huertas, S. Bhatti and J.S. Dason. 2002. The diversity of inorganic carbon acquisition mechanisms in eukaryotic microalgae. Funct. Plant Biol. 29:261-270.

Davis J.S. 2000. Structure, function, and management of the biological system for seasonal solar saltworks. Global NEST: the Int. J. 2: 217-226.

Davies, J.P. and A.R. Grossman. 1998 . Responses to deficiencies in macronutrients. pp 613-633. In The Molecular Biology of Chloroplast and Mitochondria in *Chlamydomonas*. J.-D. Rochaix, M. Goldschmidt-Clermont and S.Merchant (eds) Kluwer Ac Publ., Amsterdam, The Netherlands.

Davies, J.P., F.H. Yildiz, and A. Grossman. 1996. Sac1, a putative regulator that is critical for survival of *Chlamydomonas reinhardtii* during sulfur deprivation. EMBO J. 15, 2150–2159.

Davies J.P., F. H. Yildiz and A.R. Grossman. 1999. Sac3, an Snf1-like serine/threonine kinase that positively and negatively regulates the responses of *Chlamydomonas* to sulphur limitation. Plant Cell 11: 1179–1190.

Del Rio M.J., Z.Ramazanov and G. Garcia-Reina. 1994. Dark induction of nitrate reductase in the halophilic alga *Dunliella salina*. Planta 192, 40-45.

Dolapsakis N.P., T. Tafas , T.J. Abatzopoulos, S. Ziller and A. Economou-Amilli. 2005. Abundance and growth response of microalgae at Megalon Embolon solar saltworks in northern Greece: An aquaculture prospect. J. Appl. Phycol. 17, 39-49.

Falkowski, P.G. Nitrate uptake in marine phytoplankton: comparison of half-saturation constants from seven species. Limnol. Oceanogr. 20: 412–417.

Fernandez E., A. Galvan and A. Quesada. 1998. Nitrogen assimilation and its regulation. pp 637-674. In The Molecular Biology of Chloroplast and Mitochondria in *Chlamydomonas*J.-D. Rochaix, M.Goldschmidt-Clermont and S. Merchant (eds). Kluwer Ac Publ., Amsterdam, The Netherlands.

Ferreira, R.M.B. and A.R.N. Teixeira. 1992. Sulfur starvation in *Lemna* leads to degradation of ribulose/bisphosphate carboxylase without plant death. J. Biol. Chem. 267: 6253-6257.

Fisher M., I. Gokhman, U. Pick and A. Zamir . 1996. A salt-resistant plasma membrane carbonic anhydrase is induced by salt in *Dunaliella salina* J. Biol. Chem. 271: 17718-17723.

Franco AR, J. Cárdenas, and E. Fernández . 1988. Two different carriers transport both ammonium and methylammonium in *Chlamydomonas reinhardtii*. J. Biol. Chem. 263: 14039-43.

Gilbert, S.M., D.T. Clarkson, M. Cambridge, H. Lambers and M.J. Hawkesford. 1997. SO_4^{2-} deprivation has an early effect on the content of ribulose-1,5-bisphosphate carboxylase/oxygenase and photosynthesis in young leaves of wheat. Plant Physiol. 115: 1231-1239.

Giordano M. 1997. Adaptation of *Dunaliella salina* (Volvocales, Chlorophyceae) to growth on NH_4^+ as the sole nitrogen source. Physiol. 36: 345-350.

Giordano M. 2001. Interactions between C and N metabolism in *Dunaliella salina* cells cultured at elevated CO_2 and high N concentrations. J. Plant Physiol. 158(5), 577-581.

Giordano M. and G. Bowes. 1997. Gas exchange and C allocation in *Dunaliella salina* cells in response to the N source and CO_2 concentration used for growth. Plant Physiol. 115: 1049-1056.

Giordano M., J.S. Davis and G.Bowes. 1994. Organic carbon release by *Dunaliella salina* (Chlorophyta) under different growth conditions of CO_2, nitrogen, and salinity. J. Physiol. 30: 249-257.

Giordano M, V.Pezzoni and R. Hell. 2000. Strategies for the allocation of resources under sulfur limitation in the green alga *Dunaliella salina*. Plant Physiol. 124: 857-864.

Giordano M., A. Norici and D.J. Gilmour. 2002. Influence of the nitrogen source and metabolites on phosphoenolpyruvate carboxylase activity in the unicellular green alga *Dunaliella parva* CCAP 19/9. Phycol. 41: 133-137.

Giordano M, J. Beardall and J.A. Raven. 2005a. CO_2 concentrating mechanisms in algae: mechanisms, environmental modulation and evolution. Ann. Rev. of Plant Biol. 56: 99-131.

Giordano M., Y.Chen, M. Koblizek and P. Falkowski. 2005b Regulation of nitrate reductase in *Chlamydomonas reinhardtii* by the redox state of the PQ pool. Eur. J. Phycol. 40 : 345-352.

Giordano M., A. Norici and R. Hell. 2005c. Sulfur and phytoplankton: acquisition, metabolism and impact on the environment. New Phytol. 166 : 371-382.

Goyal A. and N.E. Tolbert 1989. Uptake of inorganic carbon by isolated chlooroplasts from air-adapted *Dunaliella*. Plant Physiol. 89: 1264-1269.

Grant, B. R. 1968. Effect of carbon dioxide concentration and buffer system on nitrate and nitrite assimilation by *Dunaliella tertiolecta*. J. Gen. Microbiol. 54: 327-36.

Heimer Y.M. 1975. Nitrate reductase of *Dunaliella parva*: electron donor specificity and heat activation. Arch. Microbiol. 103: 181-183.

Heimer Y.M. 1976. Specificity for nicotinamide adenine dinucleotide and nicotinamide adenine dinucleotide phosphate of nitrate reductase from the salt-tolerant alga *Dunaliella parva*. Plant Physiol. 58: 57-59.

Hellio C. and Y. Le Gal. 1998. Histidine utilization by the unicellular alga *Dunaliella tertiolecta*. Comp. Biochem. Physiol. 119A: 753-758.

Hellio C. and Y. Le Gal. 1999. Histidase from the unicellular green alga *Dunaliella tertiolecta*: purification and partial characterization. Eur. J. Phycol. 34: 71-78.

Hildebrand M. 2005. Cloning and functional characterization of ammonium transporters from the marine diatom *Cylindrotheca fusiformis* (Bacillariophyceae). J. Phycol. 41: 105-113.

Hirsch R., M. deGuia, G. Falkner and H. Gimmler. 1993. Flexible coupling of phosphate uptake in *Dunaliella acidophila* at extremely low pH values. J. Exp. Bot. 44: 1321-1330.

Ho T.-Y., A. Quigg, Z.V. Finkel, A.J. Milligan, K. Wyman, P.G. Falkowski and F.M.M. Morel. 2003. The elemental composition of some marine phytoplankton. J. Phycol. 39: 1145-1159.

Jahn T.P., A.L.B. Møller, T. Zeuthen, L.M. Holm, D.A. Klærke, B. Mohsin, W. Kühlbrandt and J.K. Schjoerring. 2004. Aquaporin homologues in plants and mammals transport ammonia. FEBS Letters 574: 31-36.

Jiménez C. and F.X. Niell. 1991. Growth of *Dunaliella viridis* Teodoresco: effect of salinity, temperature and nitrogen concentration. J. Appl. Phycol. 3: 319-327.

Jiménez C., F.X. Niell and J.A. Fernández 1990. The photosynthesis of *Dunaliella parva* Lerche as a function of temperature, light and salinity. Hydrobiol. 197: 165-172.

Khademi S, J. O'Connell III, J. Remis, Y. Robles-Colmenares, L.J.W. Miercke and R.M. Stroud. 2004. Mechanism of ammonia transport by Amt/Mep/Rh: structure of AmtB at 1.35 Å. Science 305: 1587-1594.

LeClaire J.A. and B.R.Grant. 1972. Nitrate reductase from *Dunaliella tertiolecta*, purification and properties. Plant Cell Physiology. 13: 899-907.

Lomas M.W. and P.M. Glibert. 2000. Comparison of nitrate uptake, storage, and reduction in marine diatoms and flagellates. J. Phycol. 36, 903-913.

Loqué D., U. Ludewig, L. Yuan and N. von Wirén. 2005. Tonoplast intrinsic proteins AtTIP2;1 and AtTIP2;3 facilitate NH_3 transport into the vacuole. Plant Physiology. 137: 671-680.

Ludewig U., S. Wilken, B. Wu, W. Jost, E. Obrdlik, M. Bakkoury, A.-M. Marini, B. André, T. Hamacher, E. Boles, N. von Wirén and W.B. Frommer. 2003. Homo- and Hetero-oligomerization of Ammonium Transporter-1 NH_4^+ Uniporters. J. Biological. Chemistry. 278: 45603-45610.

Mitra, M., C.B. Mason, Y. Xiao, R.A. Ynalvev, S.m. Lato and J.V. Moroney. 2005. The carbonic anhydrase gene families of *Chlamydomonas reinhardtii*. Can. J. Bot. 83: 780-795.

Montechiaro F., C.J.Hirschmugl, J.A. Raven and M. Giordano. 2006. Homeostasis of cell composition during prolonged darkness. Plant, Cell Env. 29: 2198–2204.

Moroney J.V. and Z.Y.Chen. 1998. The role of the chloroplast in inorganic carbon uptake by eukaryotic algae. Can. J. Bot. 76:1025–34.

Nissenbaum A. 1975. The microbiology and biogeochemistry of the Dead Sea. Microbial Ecology. 2: 139-161.

Post F.J. 1981. Microbiology of the Great Salt Lake north arm. Hydrobiologia. 81: 59-69.

Norici A, A. Dalsass and M. Giordano. 2002. Role of phosphoenolpyruvate carboxylase in anaplerosis in the green microalga *Dunaliella salina* cultured under different nitrogen regimes. Physiologia. Plantarum. 116: 186-191.

Norici, A., R. Hell and M.Giordano. 2005. Sulfur and primary production in aquatic environments: an ecophysiological perspective. Photosynthesis. Research. 85: 409-417.

Oliviera L. and H. Huynh. 1989. Ultrastructure and cytochemistry of *Dunaliella tertiolecta* Butcher and *Pavlova lutheri* (Droop) Green grown on three different sources of organic nitrogen. New Phytol. 113: 481-490.

Palmqvist K., D. Sültemeyer, P. Baldet, T.J. Andrews and M.R. Badger. 1995. Characterisation of inorganic carbon fluxes, carbonic anhydrase(s) and ribulose-1,5-biphosphate carboxylase-oxygenase in the green unicellular alga *Coccomyxa*: comparisons with low-CO_2 cells of *Chlamydomonas reinhardtii*. Planta 197: 352–361.

Pick U., O. Zeelon and M. Weiss. 1991. Amine accumulation in acidic vacuoles protects the halotolerant alga *Dunaliella salina* against alkaline stress. Plant Physiology. 97:1226-1233.

Premkumar L., U.K. Bageshwar , I. Gokhman , A. Zamir and J.L. Sussman. 2003. An unusual halotolerant a-type carbonic anhydrase from the alga *Dunaliella salina* functionally expressed in *Escherichia coli*. Protein Expression. Purification. 28: 151-157.

Reay D.S., D.B. Nedwell, J. Priddle and J. Ellis-Evans 1999. Temperature dependence of inorganic nitrogen uptake: reduced affinity for nitrate at suboptimal temperatures in both algae and bacteria. Applied. and Env. Microbio. 65: 2577-2584.

Rees, T. A. V. 2003. Mitochondrial oxidative phosphorylation is required for ammonium assimilation in light in a marine diatom. Physiol. Plant. 117: 558-563.

Reinfelder J.R., A.M.L. Kraepiel and F.M.M. Morel. 2000. Unicellular C_4 photosynthesis in a marine diatom. Nature 407:996-99.

Reinfelder J.R., A.J. Milligan and F.M.M. Morel. 2004. The role of C_4 photosynthesis in carbon accumulation and fixation in a marine diatom. Plant Physiol. 135: 2106-11.

Roberts S. 1998. Physiological effects of phosphorus limitation on photosynthesis in two green algae. PhD Thesis, Monash University, Melbourne, Australia.

Robertson D.L. and A. Tartar. 2006. Evolution of glutamine synthetase in heterokonts: evidence for endosymbiotic gene transfer and the early evolution of photosynthesis. Mol. Biol. Evol. 23: 1048-1055.

Shelly K., S.Roberts, P.Heraud and J. Beardall. 2005. Interactions between UVB exposure and phosphorus nutrition: I. Effects on growth, phosphate-uptake and chlorophyll fluorescence. J. Phycol. 41, 1204-1211.

Soupene E., H.Leeand S. Kustu. 2002. Ammonium/methylammonium transport (Amt) proteins facilitate diffusion of NH_3 bidirectionally. Proceedings of National Academy of Sciences (PNAS) USA 99: 3926-3931.

Syrett P.J. 1988. Uptake and utilization of nitrogen compounds. pp 23-39. In Biochemistry of the algae and cyanobacteria. L.J. Rogers and J.R. Gallon (eds) Oxford University Press, Oxford, UK.

Turpin, D.H. and D. Bruce. 1990. Regulation of photosynthetic light harvesting by nitrogen assimilation in the green alga *Selenastrum minutum*. FEBS Letters 263: 99-103.

Wang Y., Z. Sun, K,M Horken, C-. Im, Y. Xiang, A.R. Grossman and D.P. Weeks. 2005. Analysis of CIA5, the master regulator of the carbon concentrating mechanisms in *Chlamydononas reinhardtii*, and its control of gene expression. Can. J. Bot. 83: 765-779.

Weiss M., G. Haimovich and U. Pick. 2001. Phosphate and sulphate uptake in the halotolerant alga *Dunaliella* are driven by Na^+-symport mechanism. J. Plant Physiol. 158: 1519-1525.

Williams T.G. and D.H. Turpin. 1987. The role of external carbonic anhydrase in inorganic carbon acquisition by *Chlamydomonas reinhardtii* at alkaline pH. Plant Physiol. 83: 92-96.

Wynne, D. and G. -Y. Rhee 1988. Changes in alkaline phosphatase activity and phosphate uptake in P-limited phytoplankton, induced by light intensity and spectral quality. Hydrobiol. 160: 173-178.

Yildiz, F.H., J.P. Davies and A.R. Grossman. 1994. Characterization of sulfate transport in *Chlamidomonas reinhardtii* during sulfur-limited and sulfur-sufficient growth. Plant Physiol. 104: 981-987.

Young E.B. and J. Beardall . 2005. Modulation of photosynthesis and inorganic carbon acquisition in a marine microalga by nitrogen, iron and light availability. Canadian. J. Bot. 83: 917-928.

Young E., J. Beardall and M. Giordano. 2001. Inorganic carbon acquisition by *Dunaliella tertiolecta* (Chlorophyta) involves external carbonic anhydrase and direct HCO_3^- utilization insensitive to the anion exchange inhibitor DIDS. Eur. J. Phycol. 36: 81-88.

Zenvirth D. and A. Kaplan. 1981. Uptake and efflux of inorganic carbon in *Dunaliella salina*. Planta 152: 8-12.

8

Bioactive Compounds: Glycerol Production, Carotenoid Production, Fatty Acids Production

*Ami Ben-Amotz**

Abstract

The biflagellate unicellular alga *Dunaliella* develops unique cellular adaptation to thrive in harsh environmental conditions. Consequently, the alga predominates other organisms in media with extreme concentrations of salt in desert lands exposed to high solar radiation. *Dunaliella* produces and accumulates large amounts of intracellular glycerol to counter the high extra-cellular osmotic concentration in the surrounding medium, a high content of a mixture of β–carotene stereoisomers to defend the cell from the damaging effect of excessive radiation, and membrane lipids to compensate for the lack of a cell wall with the presence of an elastic cell membrane. Both the continous search for biosolar energy conversion of carbon dioxide into photosynthetic products of commercial interest and current attempts to produce green bio-energy are focusing on *Dunaliella* as a key potential organism. The present chapter describes the algal production of glycerol, β–carotene, and lipids in *Dunaliella*.

Introduction

The taxonomic definition of algae is very broad, as algae display a tremendous diversity, ranging from microscopic prokaryotic cells to huge seaweeds. Within these extremes are unicellular forms from microns to millimeters in diameter, colonies of cells, filaments, sheets and complex thalli with differentiation into leaf-like blades and stem-like stipes. Algae are divided into a few major classes where the classical separation was generally based on visible color: Cyanophyceae (blue-green), Rhodophyceae (red), Phaeophyceae (brown), Chrysophyceae (gold, gold-

* Marine Biology, The National Institute of Oceanography, Israel Oceanographic & Limnological Research, POB 8030, Haifa 31080, Israel, Email: ami@ocean.org.il.

brown), Bacillariophyceae (diatoms), Chlorophyceae (green), and Xanthophyceae (yellow-green). Modern systematics maintain these major groups, but it relies on updated classification based on pigment analysis by liquid chromatography, as well as on other characters such as metabolic storage products, flagellation, structure, and cell wall components. In the last few years, modern molecular biology techniques have been used to confirm or modify the earlier classification of algae and have become a promising tool to exploit algal products. The early history of the varied uses of algae for commercial products other than food is limited to polysaccharides in macroalgae as a source for phycocolloids. Following the introduction of seaweed for human use in the early 17th century, the worldwide market expanded, offering many different seaweed products in a wide variety of applications. Microalgae utilization entered the market only later toward the end of the last century, when the appropriate technology was developed to cultivate and harvest the unicellular organisms. Four different microalga species—*Spirulina, Chlorella, Dunaliella, and Nannochloropsis*—reached the level of open large-scale controlled production, while a few more unicellular species were successfully scaled up in closed bioreactors. *Dunaliella* joins these species as a halotolerant green alga able to produce and accumulate three products of commercial interest: 1. glycerol, 2. β–carotene, and, to a lesser extent, 3. fatty acids. This chapter describes a few details on these bioactive constituents of the microalga *Dunaliella.*

The biflagellate unicellular alga *Dunaliella* develops unique cellular adaptation to thrive in harsh environmental conditions. As a consequence, in environments with extreme concentrations of salt, exposed to high solar radiation in desert lands, the alga predominates other organisms. *Dunaliella* produces and accumulates large amounts of intracellular glycerol to counter the high extra-cellular osmotic concentration in the surrounding medium and cellular lipids to compensate for the lack of a cell wall with the presence of an elastic cell membrane. Some *Dunaliella* species also contain a high content of a mixture of β–carotene stereoisomers to defend the cell from the damaging effect of excessive radiation.

Biology and Halotolerance

The biflagellated algal genus *Dunaliella* is classified under the *Chlorophyceae, Volvocales*, and the genus includes a variety of ill-defined unicellular species that thrive in hypersaline, marine, and freshwater (Avron and Ben-Amotz, 1992, Butcher, 1959, Post, 1977, Volcani, 1944, Oren, 2005, see also other chapters). *Dunaliella,* like the freshwater alga of the genus *Chlamydomonas,* is characterized by an ovoid cell volume usually in the shape of a pear, wider at the basal side and narrower at the anterior flagella top. The cellular organization of *Dunaliella* is no different than that of other members of the *Volvocales,* presenting one large chloroplast with a single-centered starch-surrounded pyrenoid, a few vacuoles, a nucleus, and a nucleolus. However, unlike other green algae, *Dunaliella* lacks a rigid polysaccharide cell wall and is enclosed by a thin elastic plasma membrane covered by a mucous surface coat. This lack of a rigid cell wall differentiates *Dunaliella* from *Chlamydomonas* and permits rapid flexibility to cell volume changes and cell membrane elasticity in response to extra cellular changes in osmotic pressure. Commonly, osmotically-treated *Dunaliella* vary from round enlarged volumes, shrinking to elongated ∞ shapes. Extreme hypotonic osmotic conditions will result in cell burst. Under high salt concentrations greater than 4M, *Dunaliella* cells lose their flagella and the surrounding mucous, as the cells round with a buildup of a thick surrounding wall to form a dehydration resistant 'aplanospore' (Loeblich, 1969) or rigid cyst. Detailed reports on the osmotic volume changes of *Dunaliella* were written earlier by Teodoresco (1905, 1906) and later by Lerche (1937) Massyuk (1973).

Dunaliella occurs in a wide range of marine habitats, such as oceans, brine lakes, salt marshes, and salt water ditches near the sea, predominantly in water bodies containing more than 2M salt and high levels of magnesium (Brock,1975). The effect of magnesium on the distribution of *Dunaliella* is not clear, but in many 'bittern' habitats of marine salt producers, *Dunaliella* usually flourishes as the only surviving eukaryotic organism. The phenomenon of orange-red algal bloom in such marine environments is usually related to combined sequential growth of *Dunaliella* first, and then halophilic bacteria, and occasionally protozoa, as commonly observed in concentrated saline lakes such as the Dead Sea in Israel, the Pink Lake in Western Australia, the Great Salt Lake in Utah, U.S.A., and in many other places around the globe. *Dunaliella* is one of the most halotolerant eukaryotic organisms known, showing a remarkable degree of adaptation to a variety of salt concentrations, from as low as 0.1M to salt saturation of > 5M (Borowitzka, 1991, Borowitzka, 1986).

Glycerol Production and Osmoregulation

The first information on the presence of high intracellular solute within *Dunaliella* was published by Marrè and Servettaz (1959) by measurements of the freezing point of the cytoplasm fluid. The results indicated an apparent 'salt' concentration that exceeded the 3.9 M salt in which the cells were grown. At that time it was postulated that salt is taken up into the cell under the conditions of hypertonic osmotic shock.

However, Johnson et al. (1968) realized that salt concentrations within *Dunaliella* cells were low in enzyme studies which demonstrated that a few key enzymes of the algal metabolism, such as pentose phosphate isomerase, ribulose bisphosphate carboxylase, glucose-6-phosphate dehydrogenase, and phosphohexose isomerase, were strongly inhibited by high salt concentrations. This first non-direct conclusion that the intracellular ionic concentration of *Dunaliella* is low and not different than that of other non-halotolerant organisms was verified by many studies which showed that the intracellular salt concentration is the same in both low-salt and high-salt-grown *Dunaliella* and is equivalent to other marine and freshwater organisms.

The first indication that glycerol is accumulated by *Dunaliella* cells and provides osmotic balance appeared in a 1964 publication by Craigie and McLachlan. As shown by autoradiography, *D. tertiolecta* transformed label carbon dioxide under light into glycerol. Glycerol amounted to most of the radioactivity of the neutral fraction extracted from cells incubated in high salt while the remainder consisted of soluble polysaccharides. In a subsequent study, Ben-Amotz and Avron (1973) confirmed the glycerol osmotic function in *D. parva* by using 14C-bicarbonate and light photosynthetic bio-fixation. The photosynthetic machinery of *Dunaliella* transforms the label carbon into glycerol in correlation to the extra-cellular osmotic concentration. When the salinity of the medium was increased, more radioactivity was found in label glycerol in negative equilibrium with starch. It was postulated that starch is the carbon storage and glycerol the osmotic agent in the halotolerant alga, and both can work in equilibrium in varying salty habitats. *Dunaliella* uses a special mechanism of osmoregulation to adapt to the various salt concentrations in the surrounding media by varying the intracellular concentration of glycerol in response to the extracellular osmotic pressure. The intracellular concentration of glycerol is directly proportional to the extracellular salt concentration and is sufficient to account for all the required cytoplasmic and most of the chloroplastic osmotic pressures. Glycerol biosynthesis and elimination occur in the light or dark by a novel glycerol cycle, which involves several specific enzymes working for the formation and degradation of glycerol in the cytoplasm (Avron, 1992). An osmoregulatory isoform of the dihydroxyacetone phosphate reductase was later identified in *Dunaliella*, an enzyme that is involved in the synthesis of free glycerol. The enzyme has different properties than the other chloroplast isoforms of dihydroxyacetone phosphate reductases from plants and algae that are responsible for glycerol phosphate formation and triglyceride synthesis (Goshal et al., 2002)

The *Dunaliella* osmotic cycle responds to high salinity by enhancement of photosynthesis through diversion of carbon and energy resources for synthesis of glycerol. The ability of *Dunaliella* to enhance photosynthetic activity at high salinity is remarkable, because in most plants and cyanobacteria, salt stress inhibits photosynthesis (Liska et al., 2004).

Dunaliella has various specific proteins responding to changes in the salinity of the medium. For example, two membrane proteins are strongly induced by salt up-shock, one with an apparent molecular mass of 60 kDa (Sadka et al., 1991) and one of 150 kDa (Fisher et al., 1994). These proteins have been purified and characterized. The 60 kDa protein is a carbonic anhydrase that apparently helps the cell to take up carbon dioxide in concentrated brines in which the solubility of gases is decreased. The 150 kDa protein is an unusual transferrin-like protein, involved in the transport of iron into the cell (Fisher et al., 1997). Liska et al. (2004) compared the protein profile of low-salt and high-salt-grown cells on two-dimensional gels and identified 76 salt-induced proteins. Among the proteins up-regulated following salinity stress were key enzymes in the Calvin cycle, enzymes involved in starch mobilization and in redox energy production, regulatory factors in protein biosynthesis and degradation, and a homolog of bacterial Na^+-redox transporters. The results indicate that *Dunaliella* responds on transfer to high salinity by enhancement of photosynthetic CO_2 assimilation and by diversion of carbon and energy resources for synthesis of glycerol.

The concept of 'compatible solutes' , a term used by Brown (1990) to indicate solutes that not only contribute to the osmotic status of the cell but also maintain enzyme activity under conditions of low water activity, was largely based on the study of the function of glycerol in *Dunaliella*. Intracellular glycerol concentrations in *Dunaliella* can be very high: cells grown above 4 M NaCl were reported to contain more than 6 M cellular glycerol, equivalent to more than 50% solution of glycerol in the cell water while the extracellular medium is totally ionic. Maintenance of such a high concentration gradient of glycerol and salt requires special properties of the cell membrane, specifically in view of the fact that most biological membranes are relatively permeable to glycerol. It has been confirmed that *Dunaliella* possesses a bi-phase membrane with an unusually low permeability for glycerol (Brown et al., 1982; Gimmler and Hartung, 1988). The structure and the construction of the *Dunaliella* membrane resulting in the low glycerol permeability are open for further research.

Recently, a direct molecular approach to osmoregulation in *Dunaliella* was published (He et al., 2007) showing a cDNA encoding a nicotinamide adenine dinucleotide (NAD+) -dependent glycerol 3-phosphate dehydrogenase that had been cloned by rapid amplification of cDNA ends from *Dunaliella salina*. The cDNA is 3032 base pairs long with an open reading frame encoding a polypeptide of 701 amino acids. The authors suggest that the phosphoserine phosphatase domain functions as glycerol 3-phosphatase and that, consequently, the NAD+ -dependent glycerol phosphate dehydrogenase from *D. salina* can catalyze the step from dihydroxyacetone phosphate to glycerol directly. Further reports on the molecular biology of osmoregulation in *Dunaliella* are summarized elsewhere in this book.

Attempts have been made to exploit the bio-solar machinery of *Dunaliella* to produce high concentrations of cellular glycerol for commercial production of this compound. Although technically laboratory-scale production of glycerol by *Dunaliella* was shown to be possible (Ben-Amotz and Avron, 1973, 1990, Chen, 1981), up-scaling, extraction, and purification at larger operations have not been reported. The technical and economic feasibility of bio-solar energy conversion by *Dunaliella* to glycerol and starch should attract more research in that line as an alternative photosynthetic energy source. The lack of a cellulose cell wall and other tough polysaccharides that usually surround algal and plant cells may attract modern research to study the use of *Dunaliella* glycerol-free biomass for alcohol fermentation.

β–Carotene Production

The pigment responsible for the bright red coloration displayed by certain species of *Dunaliella* and originally designated in the older literature as 'hematochrome' was already recognized very early as a carotenoid. It was identified as such by Blanchard (1891), Teodoresco (1906), and by Lerche (1937), while Ruinen (1938) confirmed this identification based on the solubility of the pigment in alcohol and in ether and by the blue color formed in the presence of vapor of concentrated sulfuric acid. Before modern electron microscopic observation showed β-carotene in *Dunaliella* in the form of globules, considerable differences of opinion existed regarding the intracellular location of this orange-red carotenoid pigment. Both Teodoresco (1905/6) and Labbé (1921) stated that the red pigment was distributed all over the cytoplasm. Hamburger (1905) could identify small droplets but could not clarify their location within the cell. The chloroplast location was first reported by Baas Becking (1928; 1931), and Lerche (1937) realized that the β-carotene masks the chlorophyll, so that the chloroplast color varies from orange-red to yellow-green, olive, and green (Massyuk,1968).

Among the many known strains of the genus *D. salina* Teod., only a few sub-species have been shown to produce and accumulate large amounts of β-carotene (Aasen et al.1969). In hypersaline lakes, which are generally low in available nitrogen and exposed to high solar radiation, these β-carotene-producing strains of *Dunaliella* predominate over all other organisms to a seasonal pigment bloom, in concentration of about 0.1 mg *â*-carotene per liter. Under such stressful environmental conditions, more than 12% of the algal dry weight is β-carotene, usually associated with a sharp decline in the thylakoids chlorophyll. The β-carotene in *Dunaliella* accumulates within distinctive oily globules in the interthylakoid spaces of the chloroplast periphery. Analysis of the globules showed that the β-carotene of *Dunaliella* is composed mainly of two stereoisomers: *all-trans* and 9-*cis*, with the rest comprised by a few other mono-*cis* and di-*cis* stereoisomers (Figure 1.). Both the amount of the accumulated β-carotene and the 9-*cis* to *all-trans* ratio depend on light intensity and on the algal division time, which is determined by the growth conditions. Thus, any growth stress, which will slow down the rate of cell division under light, will in turn increase β-carotene production in *Dunaliella*. In fact, high light and many environmental stress conditions, such as high salt, low temperature, pH extremes, nutrient deficiencies, and others, affect the content of β-carotene in *Dunaliella*. It was previously suggested that the equation of the amount of light absorbed by the cell during one division cycle integrates the effect of all growth variables on the content and isomeric ratio of β-carotene in *Dunaliella* (Ben-Amotz and Avron 1990). The exceptions to this integration are nitrogen deficiency and low growth temperatures, both of which induce extreme intracellular accumulation of β-carotene under any light intensity. Taking into account that nitrogen starvation inhibits chlorophyll production in algae as part of

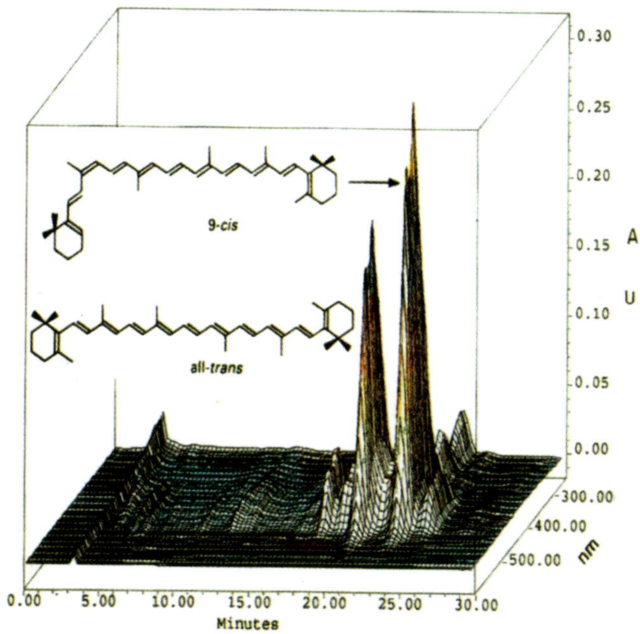

Figure 1: Three-dimensional HPLC profile of extracts from *Dunaliella bardawil* with resolution of cis- and trans-isomers of β-carotene.

its inhibitory effect on protein biosynthesis, the prolonged nitrogen independent biosynthesis of carotenoids will then protect the chlorophyll-reduced cells against the lethal damage of light. The effect of low temperatures on *Dunaliella* can be analyzed by measuring the specific stimulation effect of chilling on the biosynthesis of 9-*cis* β-carotene. The physicochemical properties of 9-*cis* β-carotene differ from those of *all-trans* β-carotene. *All-trans* β-carotene is practically insoluble in oil and is easily crystallized at low temperatures, while 9-*cis* β-carotene is much more soluble in hydrophobic-lipophilic solvents, very difficult to crystallize, and generally oily in its concentrated form. To avoid cellular crystallization of *all-trans* β-carotene and to survive at low temperatures, *Dunaliella* produces a higher ratio of 9-*cis* to *all-trans* β-carotene, where the 9-*cis* stereoisomer functions *in vivo* as an oily matrix for the *all-trans* form (Ben-Amotz, 1996).

Carotenogenesis and Isomerization

Although 272 geometric isomers of β-carotene can exist theoretically, 12 *cis* forms in total have been noted and recorded (Zechmeister, 1962). Physicochemical methods for stereomutation of β-carotene involve heat, light with no catalyst, and iodine catalysis under light yielding between one-third to one-half of the pigment in *cis* configuration. *Dunaliella* follows the same biosynthetic pathway of carotenoids with the same substrates and same intermediates as found in other eukaryotic organisms and plants. Pyruvate and Glyceraldehyde-3-Phosphate through isopentenyl diphosphate, geranyl diphosphate, and geranylgeranyl diphosphate form phytoene which undergoes a few desaturation steps and cyclization to β-carotene (Goodwin, 1988). Irrespective of many trials, the isomerization reaction, which produces 9-*cis* β-carotene in *Dunaliella*, is not identified as of yet. The observation of two stereoisomers of phytoene, 9-*cis* and *all-trans*, suggests that the biosynthesis of 9-*cis* β-carotene probably initiates early in the pathway of carotene biosynthesis, at or before the formation of phytoene. Thereafter, all the intermediates occur in two isomeric forms. The induction of β-carotene in *Dunaliella* as described above was successfully applied to accumulate any choice of stereoisomeric intermediates in the carotenogenesis pathway, yielding a large quantity of the twin stereoisomers, *all-trans* and 9-*cis*: phytoene, phytofluene, neurosporene, ξ-carotene, β-zeacarotene, γ-carotene and β-carotene, (Shaish et al., 1990, 1991). Subsequently, Ebenezer and Pattenden (1993) used ^{1}H and ^{13}C NMR on purified phytoene and phytofluene from *Dunaliella bardawil* extracts to verify the HPLC stereoisomeric analysis. Their NMR analysis showed 15-*cis* phytoene and 9-*cis* phytofluene as an indication that the phytoene is the branch point for the formation of 9-*cis* β-carotene. The contradiction between the HPLC and NMR analyses left the intriguing site of stereoisomerization open at any site between isopentenyl diphosphate to phytoene. Furthermore, the induced enzyme, enzymes, or enzyme complex responsible for the high accumulation and isomerization of β-carotene in *Dunaliella* has not been identified yet.

The function of the β-carotene globules in *Dunaliella* is assumed to be one of shading protection of the cell against injury by high intensity radiation under limiting growth conditions, since the β-carotene acts as a sunscreen, absorbing excess radiation. The harmful effect of the blue region of the spectrum is screened by the algal peripheral-located globules thus preventing cellular damage. Strains of *Dunaliella* and other algae unable to accumulate β-carotene die when exposed to high levels of radiation while at the same time the β-carotene-rich *Dunaliella* flourishes. Moreover, protection against photoinhibition by the massively accumulated β-carotene is observed only when the photoinhibitory light is composed of wavelengths absorbed by β-carotene, such as light in the blue region. No photoprotection is observed when red light, which is not absorbed by β-carotene, serves as the photoinhibitory agent. This is in agreement with the observation on the location of the β-carotene globules, distant from the thylakoid-located chlorophyll, and with the above-mentioned hypothesis that the mode of action of the massively accumulated carotene is mostly a screening effect. It was shown previously (Ben-Amotz et al., 1989b) that a series of sequential events lead to the photodestruction of chlorophyll in *Dunaliella bardawil*. 9-*cis* β-carotene is destroyed first, then the *all-trans* β-carotene, and later the chlorophylls, indicating a higher sensitivity of 9-*cis* β-carotene to direct elevated levels of light or to damaging chlorophyll-generated free radicals.

The induction of β-carotene biosynthesis in *Dunaliella* is light dependent. Photosynthetic Active Radiation (PAR) is absorbed by the algal chlorophylls triggering the massive synthesis of β-carotene in *Dunaliella*. The question regarding the involvement of chlorophyll-generated active oxygen species in the process of β-carotene accumulation was studied kinetically by adding promoters of oxygen radicals and azide, an inhibitor of catalase and superoxide dismutase, to *Dunaliella* during a selective induction period of β-carotene (Shaish et al., 1993). However, the actual action of active oxygen species in triggering the biosynthesis of β-carotene remains open for further confirmation.

Biotechnology of β-carotene and Phytoene Production by *Dunaliella*

Dunaliella is a most suitable organism for mass cultivation outdoors in open ponds (Figure 3). The ability to thrive in media with high sodium, magnesium, calcium, and the respective anions chloride and sulfate, in desert high solar-irradiated land with access to brackish water or seawater at extreme temperatures from around -5° to above 40°C, make *Dunaliella* most attractive for biotechnologists and venture capitalists. In fact, since 1980 several firms, government authorities, and industries have invested capital in the application of *Dunaliella* for the production of natural β-carotene. Large-scale *Dunaliella* production is based on autotrophic growth in media containing inorganic nutrients with carbon dioxide as exclusive carbon sources. Attempts to commercially develop heterotrophic strains or mutants of *Dunaliella* for growth on glucose or acetate, such as *Chlorella* or *Chlamydomonas*, respectively, were not successful. Due to the demand for high light intensity for maximal β-carotene production beyond that required for normal growth, production facilities are located in areas where solar output is maximal and cloudiness is minimal. Most of the present *Dunaliella* production plants are located close to available sources of salt water, e.g., sea-salt industries, usually in warm and sunny areas where the rate of water evaporation is high and non-agricultural land is abundant.

Two outdoor modes of cultivation, the extensive and intensive, were successfully used in large-scale production of *Dunaliella* over the last 20 years. The extensive cultivation method uses no mixing and minimal control of the environment. To decrease attacks by zooplanktonic predators, such as certain types of ciliates, amoebae, or brine shrimp, the growers employ very high salt concentrations. *Dunaliella* grows slowly in shallow lagoons in nearly saturated brine, and predators are largely eliminated. The naturally selected strain of *Dunaliella* is well adapted to salt conditions near saturation, partially loses its flagella, and produces an outer cell coat layer of amorphous mucilaginous nature and variable thickness upon transformation into a cyst form, 'Aplanospore' (Loeblich 1969). Extensive cultivation productivity is low, and the area needed for commercial production is very large, however, the low operating costs of such facilities have led to the development of two commercial plants in Australia. The second, termed intensive cultivation, uses high biotechnology to control all factors affecting cell growth and chemistry. The ponds are usually oblong lined and constructed raceways varying in size up to a production surface area of approximately 3,000m^2. The use of long-arm, slow-revolution paddle wheels is presently common in the large-scale facilities in Israel, USA, China, Chile, and Portugal. The current large-scale production of β-carotene under intensive and extensive cultivation varies from around 20 to 200 mg β-carotene/m^2/day, respectively, on a yearly average. Between the extensive and intensive modes, there are examples of a semi-intensive mode, where the ponds are enlarged ten times, to about 50,000m^2, each with partial control and no mixing. Highly intensive cultivation in closed photobioreactors has been experimented in the last decade to grow *Dunaliella* in different models of different designs, with attempts to build the best sunlight-harvesting unit for β-carotene optimization. The different designs include narrow, very long plastic tubes, plastic bags, trays, rotating trays, and more. However, to date, none of these photobioreactors have been successful beyond small to medium pilot plant volume, mainly due to contamination and economic limitations. The few industrial ventures of high-intensive closed photobioreactors of *Dunaliella* became insolvent and no longer exist.

Generally, large-scale optimization of β-carotene production is achieved in all modes of cultivation by high salt stress and by nitrogen deficiency in areas of high solar radiation. The first is applied in the extensive mode, while nitrogen starvation controls the intensive mode. Most species of *Dunaliella* grow optimally in a medium containing 1-2 M NaCl in accordance with the medium temperature, exhibiting closely similar growth rates at moderate temperatures of 25°C in 2M NaCl and at low temperatures of 15°C in 1M NaCl. The algal composition changes respectively by selective accumulation of glycerol and starch. This unique environmental adaptation of *Dunaliella* allows successful intensive outdoor growth in cold seasons and in cold areas. Most commercial *Dunaliella* ponds employ evaporated concentrated seawater or seawater augmented with dry salt to reach the desired concentration in the medium. Since a mixture of seawater and concentrated seawater is used, favored sites for *Dunaliella* cultivation are along the seashore or close to salt lagoons and salt-producing industries to meet the changing desired salt concentrations brought by seasons and temperatures. Recently, successful cultivation was noted in inland areas on underground salt and on lake salt. The use of recycled high-salt medium is common in a few plants, after harvesting of the algae by oxidative treatment of the organic load and filtration. The use of recycled medium enriches the medium with higher concentrations of magnesium, calcium, and sulfate. *Dunaliella* was found to grow well in seawater-based media containing around 1.5M NaCl, more than 0.4M $MgSO_4$ and 0.1M $CaCl_2$ under pH control.

Phytoene production by *Dunaliella* is a relatively more complicated target than β-carotene, as the colorless phytoene cannot protect the alga against high solar radiation and because the need exists to cultivate the alga in the presence of a bleaching herbicide (Ben-Amotz et al., 1988). Figure 2 shows an exemplary HPLC profile of extracts of *Dunaliella bardawil* with resolution of cis and trans isomers of the linear carotenoid phytoene. Current molecular approaches to improve phytoene production include genetic engineering of enzymes of the carotenoid biosynthesis pathway. Although, the gene of the phytoene synthase was cloned from various *Dunaliella* species (Yuan et al., 2005, for additional unpublished sequences see NCBI), genetic engineering cannot be applied yet for improving phytoene production with *Dunaliella*. At this time, a highly efficient transformation system and routine genetics with *Dunaliella* strains, which would be the basis for genetic engineering, are still in their infancy. Therefore, the various reports on large-scale production of phytoene-rich *Dunaliella* are still applying physiological and biochemical methodologies (Leon et al., 2005, see Oppen-Bezalel and Shaish in this book).

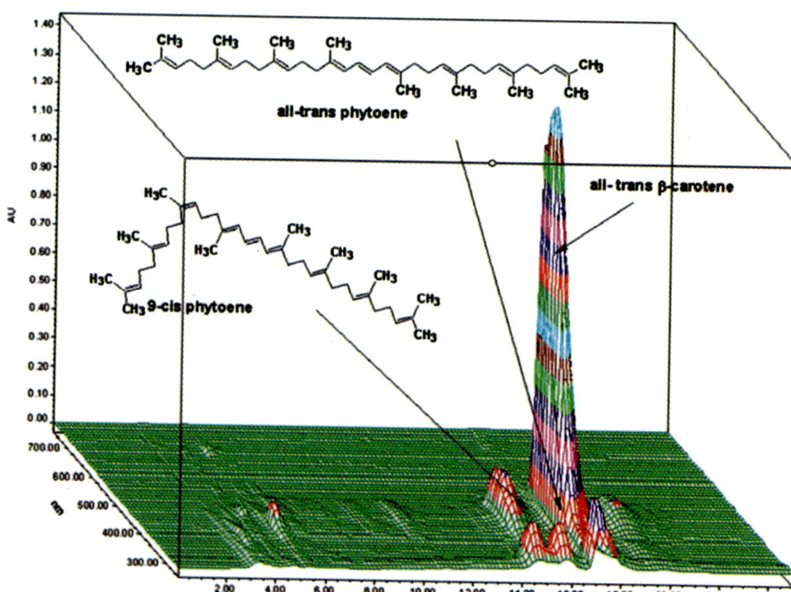

Figure 2: Three-dimensional HPLC profile of extracts from *Dunaliella bardawil* with resolution of the cis- and trans-isomers of the carotenoid phytoene.

Natural Versus Synthetic β-Carotene

β-Carotene has been used for many years as a food coloring agent, as pro-vitamin A (retinol) in food and animal feed, as an additive to cosmetics, multivitamin preparations, and in the last decade as a health food product under the 'antioxidant' claim. Many epidemiological and oncological studies suggest that humans who maintain higher than average levels of serum carotenoids and are fed a diet high in carotenoid-rich vegetables and fruits have a lower incidence of several types of cancer and cardiovascular disease (Krinsky, 1989). Two major studies, the Alpha-Tocopherol, β-Carotene (ATBC) Cancer Prevention Study Group and the β-Carotene and Retinol Efficacy Trial (CARET), clearly showed that not only does β-carotene fail to reduce the incidence of cancer and cardiovascular disease, but in fact it increases it in smokers and workers exposed to asbestos (Omenn et al., 1996; Hennekens et al., 1996). However, these USA National Cancer Institute-supported studies and numerous earlier trials dealing with the question of the protective role of β-carotene against chronic diseases used only synthetically formed *all-trans* β-carotene.

Despite the fact that the most convincing reports support a direct connection between high intake of fruits and vegetables and low incidence of cancer and cardiovascular disease, the literature lacks specific information on the possible medical contribution of natural carotenoids, isomers of carotenoids, and carotenoid fatty acid esters. Experimental nutrition and medical studies with natural carotenoids originating from different plants, fruits, vegetables, and algae have been very limited, and such research is in its infancy.

β-Carotene is present in most plants and algae in small amounts of ~0.2% of the dry weight with approximately 1/3 as 9-*cis* β-carotene. Observation of the high β-carotene content of *Dunaliella*, containing more than 50% 9-*cis* β-carotene, gave impetus to studies on the metabolism, storage, and function of 9-*cis* β-carotene in animals and humans, with emphasis on the possible role of the 9-*cis* stereoisomer in scavenging reactive oxygen species. The first dietary studies showed that low doses of the algal β-carotene are as potent as doses of synthetic *all-trans* β-carotene in providing retinol in rats and chicks (Ben-Amotz et al., 1989a). Later, preferential and selective uptake of either *all-trans* or 9-*cis* β-carotene was noted in different animals and humans. 9-*cis* β-Carotene is not detected in serum of chicks and rats nor in humans who are fed a diet rich in *Dunaliella* (Ben-Amotz and Levy, 1996) or the oil extract of *Dunaliella* (Stahl et al., 1993, Gaziano et al., 1995). The β-carotene detected in any of these sera was *all-trans* β-carotene. Rats and chicks that generally do not accumulate β-carotene in their tissues showed uptake of β-carotene into the tissues when the diet was supplemented with β-carotene-rich *Dunaliella*. The lack of 9-*cis* β-carotene in mammal serum and the different response to CNS toxicity (Bitterman and Ben-Amotz, 1994) and to *in vitro* oxidation (Levin and Mokady, 1994) led to

Figure 3: Various size race-way style ponds at the NBT facility in Eilat, Israel.

the hypothesis that the isomeric structure of 9-*cis* β-carotene acts as a quencher of singlet oxygen and other free radicals, thus serving as an effective *in vivo* antioxidant. Generally, carotenoids exert an antioxidative effect by a mechanism that results in the formation of new products that are more stable and more polar. Different carotenoids exhibit different anti-oxidative/anti-peroxidative capacities (Krinsky, 1989). Opening of the β-ionone ring, the addition of a chemical group on the ring, or replacement of the ring by various groups can modify the antioxidative activity. Therefore, structural variables besides the length of the polyene chain may direct the scavenging properties of the carotenoids. Insertion of any *cis* position along the conjugated double bond chain may lead to modified antioxidative properties. Assuming that the *cis* conformational change leads to a higher steric interference between the two parts of the carotene molecule, the *cis* polyenic chain will be less stable and more susceptible to low oxygen tension in *in vivo* oxidation. Earlier studies supported this hypothesis by measuring different antioxidative activity (Levy et. al., 1995, 1996) and much lower concentration of low-ultraviolet dienes in the serum of humans supplemented with β-carotene-rich *Dunaliella* as opposed to *all-trans* β-carotene (Ben-Amotz and Levy, 1996). However, the possibilities of low intestinal absorption, isomerization, or tissues uptake have to be considered as well (Johnson et al., 1996). The possibility of synergistic effects and the possible beneficial potency of the different plant nutrients and carotenoids are still obscure and warrant further research. Recent studies have shown that natural stereoisomer β-carotene may act by a different mechanism through the metabolic conversion of the carotene molecule to retinoic acid and retinoic acid isomers. A detailed summary about this action is included in this book under the chapter by Shaish et al.

Fatty Acids Production

Dunaliella, as other members of the *Chlorophyceae*, produces simple fatty acids with 16 and 18 carbons. However, in salt tolerance studies of *D. salina*, Azachi et al. (2002) isolated a cDNA for a salt-inducible mRNA encoding a protein homologous to plant beta-ketoacyl-coenzyme A synthases (Kcs). These microsomal enzymes catalyze the condensation of malonyl-CoA with acyl-CoA, the first and rate-limiting step in fatty acid elongation. Kcs activity, localized to a *D. salina* microsomal fraction, increased in cells transferred from 0.5 to 3.5 M NaCl, as did the level of the kcs mRNA. The function of the kcs gene product was directly demonstrated by the condensing activity exhibited by *Escherichia coli* cells expressing the *Kcs* cDNA. The effect of salinity on *Kcs* expression in *D. salina* suggested the possibility that salt adaptation entailed modifications in the fatty acid composition of algal membranes. Lipid analyses indicated that microsomes, but not plasma membranes or thylakoids, from cells grown in 3.5 M NaCl contained a considerably higher ratio of C18 (mostly unsaturated) to C16 (mostly saturated) fatty acids compared with cells grown in 0.5 M salt. Thus, the salt-inducible *Kcs*, jointly with fatty acid desaturases, may play a role in adapting intracellular membrane compartments to function in the high internal glycerol concentrations which balance the external osmotic pressure. Another equivalent study (Zelany, 1995) showed that the major plasma membrane sterol in *D. salina* and the high-light-induced sterol *in D. bardawil* co-migrate with ergosterol on thin-layer chromatography and on reversed-phase, high-performance liquid chromatography. These results suggest that the osmo-sensory mechanism in *Dunaliella* resides in the plasma membrane, and that sterols have an important role in sensing osmotic changes.

Productivity

The calculated productivity of *Dunaliella* biomass and its major constituents—glycerol, β–carotene, and lipids—are presented in Table 1. Our calculation is based on the accepted biosolar energy conversion efficiency and takes into consideration the limiting environmental factors of algal cultivation in open ponds. With the available photosynthetic efficiency of 3% the indicated maximum productivity of

Table 1: Environmental factors limiting long-term productivity and the calculated derived yield of algal biomass, glycerol, β-carotene, and lipids by *Dunaliella* in open algal ponds.

(%)	Reduction	Environmental Factor
100	———	Solar light
90	10%	Scattering and reflecting properties of surface
45	50%	Absorption spectrum (depth of culture)
11.3	75%	Photosynthetic efficiency (25%)
4.5	60%	Light saturation (7-95%)
4.3	5%	Respiration, photo respiration, excretion
3.8	10%	Photoinhibition
3.1	20%	Temperature
Productivity		
4,000 kcal/ m²/day		Mean daily solar intensity
120 kcal/ m²/day		Energy productivity at 3% efficiency
24 g/m²/day		**Algal biomass productivity (5 kcal/g)**
8.0 g/m²/day		*Dunaliella* glycerol productivity (30%)
1.92 g/m²/day		*Dunaliella* β–carotene productivity (8%)
2.4 g/m²/day		*Dunaliella* lipid productivity (10%)

algal biomass is 24 g/m²/day, of glycerol is 8 g/m²/day, and of β-carotene is 1.2 g/m²/day.

Summary

Currently, the alga *Dunaliella salina/bardawil* is one of the very few microalgae that is commercially exploited. Although it is now cultivated mainly for β-carotene production, *Dunaliella* grows well on CO_2, as well as on bicarbonate and carbonate. Therefore, global warming and the search for bio-fuels as green energy attract the possibility of using the CO_2 emitted from power plant flue gas to cultivate *Dunaliella* in arid desert lands. The possibility of cultivating *Dunaliella* on stack gas may be favorably considered, provided that the level of SO_2 in the gas is minimized to avoid competition with CO_2 in the process of photosynthesis.

References

Aasen A.J., K.E.Eimhjellen and S. Liaaen-Jensen. 1969. An extreme source of β-carotene. Acta Chem. Scand., 23: 2544-2545.

Avron, M. 1992. Osmoregulation pp.135-164. In *Dunaliella:* Physiol. Biochem, and Biotechno. M. Avron and A. Ben-Amotz (eds). CRC Press. Boca Raton, USA.

Avron, M. and A. Ben-Amotz. (eds.), 1992. *Dunaliella*: Physiology, Biochemistry, and Biotechnology.CRC Press. Boca Raton ,USA.

Azachi M., A.Sadka , M.Fisher , P.Goldshlag , I.Gokhman . and A. Zamir. 2002. Salt induction of fatty acid elongase and membrane lipid modifications in the extreme halotolerant alga *Dunaliella salina*. Plant Physiol. 129(3): 1320-1329.

Baas Becking, L.G.M.. 1928. On organisms living in concentrated brine. Tijdschr. Ned. Dierk Ver. Ser. III, 1: 6-9.

Baas-Becking, L.G.M.. 1931. Salt effects on swarmers of *Dunaliella viridis* Teod. J. Gen. Physiol., 14:765-779.

Ben-Amotz, A. 1996. Effect of low temperature on the stereoisomer composition of β-carotene in the halotolerant alga *Dunaliella bardawil* (Chlorophyta) J. Phycol., 32: 272-75.

Ben-Amotz, A. and M. Avron. 1973. The role of glycerol in the osmotic regulation of the halophilic alga *Dunaliella parva*. Plant Physiol., 51: 875-878.

Ben-Amotz, A. and M. Avron. 1990. The biotechnology of cultivating the halotolerant alga *Dunaliella*, Trends Biotechnol., 8: 121-26.

Ben-Amotz, A. and I. Levy. 1996. Bioavailability of a natural isomer mixture compared with synthetic *all-trans* β-carotene in human serum, Am. J. Clin. Nutr., 63, 729-34.

Ben-Amotz A., A.Lers and M. Avron. 1988. Stereoisomers β-carotene and phytoene in the alga *Dunaliella bardawil*. Plant Physiol., 86: 1286-1291

Ben Amotz, A., S. Mokady, S. Edelstein. and M. Avron. 1989a. Bioavailability of natural isomer mixture as compared with synthetic *all-trans* β-carotenein rats and chicks. J. Nutr., 119: 1013-19.

Ben-Amotz, A., A.Shaish and M. Avron. 1989b. Mode of action of the massively accumulated β-carotene of *Dunaliella bardawil* in protecting the alga against damage by excess radiation, Plant Physiol., 91: 1040-43.

Bitterman, N. and A. Ben-Amotz. 1994. β-Carotene and CNS toxicity in rats, *J.* Appl. Physiol., 76, 1073-76.

Blanchard, R. 1891. Résultats d'une excursion zoologique en Algérie. Mém. de la Soc. Zool. de France, IV.

Blanchard, R. 1888. Note préliminaire sur *Monas Dunalii*, flagellé, qui cause la rubéfaction des marais salants. Bull. de la Soc. Zool. de France, 13: 153.

Borowitzka, L.J. 1981. The microflora. Adaptations to life in extremely saline lakes. Hydrobiol., 81:33-46.

Borowitzka, M.A. 1986. Micro-algae as sources of fine chemicals. Microbiol. Sci., 3: 372-375.

Brock, T.D. 1975. Salinity and the ecology of *Dunaliella* from Great Salt Lake. J. Gen. Microbiol., 89: 285-292.

Brown, A.D. 1990. Microbial Water Stress Physiology. Principles and Perspectives. John Wiley & Sons. Chichester, UK.

Brown, F.F., I.Sussman , M.Avron and H. Degani. 1982. NMR studies of glycerol permeability in lipid vesicles, erythrocytes and the alga *Dunaliella*. Biochim. Biophys. Acta, 690: 165-173.

Butcher, R.W. 1959. An Introductory Account of the Smaller Algae of British Coastal Waters. Part I: Introduction and Chlorophyceae. Ministry for Agriculture, Fisheries and Food, Fishery Investigations, Series IV. Her Majesty's Stationery Office.London, UK

Chen, B.J. and C.H. Chi. 1981. Process development and evaluation for algal glycerol production. Biotechnol. Bioengin., 23: 1267-1287.

Craigie J.S. and J. McLachlan. 1964. Glycerol as a photosynthetic product in *Dunaliella tertiolecta* Butcher. Can. J. Bot., 42: 777-778.

Ebenezer, W. J. and G. Pattenden . 1993. *cis*-Stereoisomers of β-carotene and its congeners in the alga *Dunaliella bardawil*, and their biogenetic interrelationships, J.Chem. Soc. Perkin Trans., 1: 1869-1873.

Fisher M., U.Pick and A. Zamir. 1994. A salt-induced 60-kilodalton plasma membrane protein plays a potential role in the extreme halotolerance of the alga *Dunaliella*. Plant Physiol., 106: 1359-1365.

Fisher M., I.Gokhman ,U. Pick and A. Zamir. 1997. A structurally novel transferrin-like protein accumulates in the plasma membrane of the unicellular green alga *Dunaliella salina* grown in high salinities. J. Biol. Chem., 272: 1565-1570.

Gaziano, J.M., E.J.Johnson and E.M. Russel. 1995. Discrimination in absorbance or transport of β-carotene isomers after oral supplementation with either *all-trans*- or 9-*cis* β-carotene, Am. J. Clin. Nutr., 61, 1248-52.

Gimmler H. and W. Hartung. 1988. Low permeability of the plasma membrane of *Dunaliella parva* for solutes. J. Plant Physiol., 133: 165-172.

Goodwin, T.W. (ed.) 1988. Plant Pigments Academic Press. London, UK.

Goshal D., D. Mach, M.Agarwal . and A. Goyal. 2002. Osmoregulatory isoform of dihydroxyacetone phosphate reductase from *Dunaliella tertiolecta*: purification and characterization. Protein Expr. Purif. 24 (3): 404-11.

Hamburger, C. 1905. Zur Kenntnis der *Dunaliella salina* und einer Amöbe aus Salinenwasser von Cagliari. Arch. f. Protistenkd., 6: 111-131.

He Q., D.Qiao , Q.Zhang , W.Yang , Q.Li and Y. Cao. 2007. Cloning and characterization of a plastidic glycerol 3-phosphate dehydrogenase cDNA from *Dunaliella salina*. J Plant Physiol. 164(2): 214-20.

Hennekens, C.H., J.E. Buring and J.E. Mason, 1996. Lack of effect of long-term supplementation with beta-carotene on the incidence of malignant neoplasms and cardiovascular disease, N. Engl. J. Med., 334, 1145-49.

Johnson, J.E., N.I.Krinsky. and R.M. Russell. 1996. Serum response of *all-trans* and 9-*cis* isomers of β-carotene in humans, Circulation, 15: 620-4.

Johnson M.K., E.J.Johnson .,R.D. MacElroy , H.L.Speer and B.B. Bruff. 1968. Effects of salts on the halophilic alga *Dunaliella viridis*. J. Bacteriol., 95: 1461-1468.

Krinsky, N. I. 1989. Antioxidant functions of carotenoids, Free Radic. Bio. Med., 7 : 617-35.

Labbé, A. 1921. Sur les modifications adaptives de *Dunaliella salina* Dunal. C R Acad Sci, 172: 1074-1076.

León R., M.Vila , D.Hernánz and C. Vílchez. 2005. Production of phytoene by herbicide-treated microalgae *Dunaliella bardawil* in two-phase systems. Biotechnol. Bioeng. 92: 695-701.

Lerche, W. 1937. Untersuchungen über Entwicklung und Fortpflanzung in der Gattung *Dunaliella*. Arch. f. Protistenkd., 88: 236-268.

Levin, G. and S. Mokady. 1994. Antioxidant activity of 9-*cis* compared to *all-trans* β-carotene *in vitro,* Free Radic. Biol. Med., 17: 77-82.

Levy, Y., A.Ben-Amotz and M. Aviram. 1995. Effect of dietary supplementation of different β-carotene isomers on lipoprotein oxidative modification, J. Nutr. Environm. Med., 5: 13-22.

Levy, Y., M.Kaplan, , A.Ben-Amotz. and M. Aviram. 1996. Effect of dietary supplementation of β-carotene on human monocyte-macrophage-mediated oxidation of low density lipoprotein, Isr. J. Med. Sci. 32, 473-78.

Liska, A.J., A.Shevhenko , U.Pick and A. Katz. 2004. Enhanced photosynthesis and redox energy production contribute to salinity tolerance in *Dunaliella* as revealed by homology-based proteomics. Plant Physiol. 136: 2806-2817.

Loeblich, L.A. 1969. Aplanospores of *Dunaliella salina* (Chlorophyta). Protozool., (Suppl 16): 22-23.

Marrè, E. and A. Servettaz. 1959. Sul meccanismo di adattemento a condizioni osmotiche estreme in *Dunaliella salina*. II. Rapporto fra concentrazioni del mezzo esterno e composizione del succo cellulare. Atti Accad Naz Lincei Cl Sci Fis Mat Nat Rend Ser 8, 26: 272-278.

Massyuk, N.P. 1973. Morphology, Taxonomy, Ecology and Geographic Distribution of the Genus *Dunaliella* Teod. and Prospects for its Potential Utilization. Naukova Dumka. Kiev

G.E. Goodman, M.D. Thornquist, J. Balmes, M.R.Cullen, A.Glass, J.P. Keogh, F.L. Meyskens, B.Valaris, J.H. Williams, S. Barnhart and S. Hammer. 1996. Mass culture of the carotene bearing alga *Dunaliella salina*. Ukr. Bot. Zh. 1968, 23: 12-19.

Omenn, G. S., et al., 1996. Effects of combination of beta-carotene and vitamin A on lung cancer and cardiovascular disease, N. Engl. J. Med., 334: 1150-1155.

Oren, A. 2005. A hundred years of *Dunaliella* research: 1905–2005. Saline Systems, 1:2.

Post, F.J. 1977. The microbial ecology of the Great Salt Lake. Microb. Ecol., 3: 143-165.

Ruinen, J. 1938. Notizen über Salzflagellaten. II. Über die Verbreitung der Salzflagellaten. Arch. Protistenkd., 90: 210-258.

Sadka, A., S.Himmelhoch and A. Zamir. 1991. A 150 kilodalton cell surface protein is induced by salt in the halotolerant green alga *Dunaliella salina*. Plant Physiol., 95: 822-831.

Shaish, A., M.Avron, and A. Ben-Amotz. 1990. Effect of inhibitors on the formation of stereoisomers in the biosynthesis of β-carotene in *Dunaliella bardawil*, Plant Cell Physiol., 31: 689-96.

Shaish, A., A.Ben-Amotz and M. Avron. 1991. Production and selection of high β–carotene mutants of *Dunaliella bardawil* (Chlorophyta), J. Phycol., 27: 652-56.

Shaish, A., M.Avron, , U.Pick. and A. Ben-Amotz. 1993. Are active oxygen species involved in induction of β–carotene in *Dunaliella bardawil*? Planta, 190: 363-68.

Stahl, W., W.Schwartz and H. Sies. 1993. Human serum concentrations of *all-trans* and α-carotene but not 9-*cis* β-carotene increase upon ingestion of a natural mixture obtained from *Dunaliella salina* (Betatene), J. Nutr., 123: 847-51.

Teodoresco, E.C. 1905. Organisation et développement du *Dunaliella*, nouveau genre de Volvocacée-Polyblepharidée. Beih. z. Bot. Centralbl., Bd. XVIII:215-232.

Teodoresco, E.C. 1906. Observations morphologiques et biologiques sur le genre *Dunaliella*. Rev. Gén. Bot. 1906, 18: 353-371.

Volcani, B.E. 1944. The microorganisms of the Dead Sea. In Papers Collected to Commemorate the 70th Anniversary of Dr. Chaim Weizmann. Collective volume. Rehovoth: Daniel Sieff Research Institute, 71-85.

Yuan Y., Z.Yue-Hui ,J.Jian-Guo and S. Dong-Lin. 2005. Cloning and sequence analysis of the phytoene synthase gene from a unicellular Chlorophyte, *Dunaliella salina*. J. Agric. Food Chem., 53 (5): 1466 -1469.

Zechmeister, L. 1962. cis-trans Isomeric Carotenoids, Vitamin A, and Arylpolyenes, Springer-Verlag.Vienna, Austria.

Zelany, A.M., A.Shaish. and U. Pick. 1995. Plasma Membrane Sterols Are Essential for Sensing Osmotic Changes in the Halotolerant Alga *Dunaliella*. Plant Physiol. 109(4): 1395-1403.

9 Photosynthesis—Energy Relationships in *Dunaliella*

Pedro Duarte[1] *and D.V. Subba Rao*[2]

Abstract

Published data on photosynthesis (P) - energy (E) relationships of *Dunaliella* cultures by various authors, and our unpublished data were modeled. Comparison of the P-E parameters showed that cell densities are to be considered to avoid bias in comparing different experiments. Three to four day acclimation of algal cultures to low (~42 μmol m^{-2} s^{-1}) or high light energy (~144μmol m^{-2} s^{-1}) levels is necessary for stabilization of cell pigments and to prevent any negative effects of culturing. Photosynthetic rates were maximum at temperatures around 26°C, N: P 28, neutral pH and 500 (μmol m^{-2} s^{-1}) light energy. Over a 21 day period gross production decreased by about 7 fold while respiration decreased by 4 fold. Cultures grown at ~42 μmol m^{-2} s^{-1} or ~144 μmol m^{-2} s^{-1} showed a strong decrease in photosynthetic rate, rather than the respiration rate; as cells aged a marked decrease in photosynthesis: respiration resulted. The photosynthetic efficiency, carbon assimilation numbers and photosynthesis: respiration ratios of *Dunaliella* are of the same order of magnitude as those obtained on benign algae and toxigenic algae. In *D. salina,* acclimation to high light energy levels induced carotenoid production as protection from photoinhibition. For maximization of photosynthesis ammonia is the best nitrogen source. These findings have applications in mass cultivation of *Dunaliella* for commercial purposes.

[1] CIAGEB-University Fernando Pessoa, Prąca 9 de Abril, 349, 4249-004, Portugal. email:pduarte@ufp.edu.pt

[2] Ecosystem Research Division, Habitat Ecology Section, Bedford Institute of Oceanography, PO Box 1006, Dartmouth, Nova Scotia, Canada, B2Y 4A2.

Introduction

It was just 100 years ago that Teodoresco (1905) described a new genus *Dunaliella* and Blackman (1905) came up with the idea that light (*E*) is one of the factors that limits photosynthesis (*P*). Since then several equations were used to describe the *P-E* relationships in algae (for a revision see Duarte 2008). Those ranged from simple linear functions, with a level asymptote (Blackman 1919), through formulations of a Monod type (e.g. Jassby and Platt 1976) or response curves with a maximum (e.g. Steele 1962; Platt et al. 1980), to more complex ones with time dependent parameters (e.g. Pahl-Wostl and Imboden 1990; Franks and Marra 1994). An inhibition and a saturation curve are presented together with some usual *P-E* parameters (Figure 1); *P-E* functions including one light saturation model (Jassby and Platt 1976) and two photoinhibition models (Steele 1962; Eilers and Peeters 1988) are presented (Table 1). Most of these above mentioned formulations are empirical, capable of describing geometrically the observed results, not being based on physiological processes. However, models of Megard et al.(1984) and Rubio et al. (2003) are based on known sequences of metabolic transformations. The general aim of *P-E* equations is not a complete description of all the fine-controlled processes of photosynthesis, but rather a least sufficient representation of the main features of the *P-E* relationships. In most cases, these equations are merely used for predictive purposes and underlying physiologic processes are not considered explicitly.

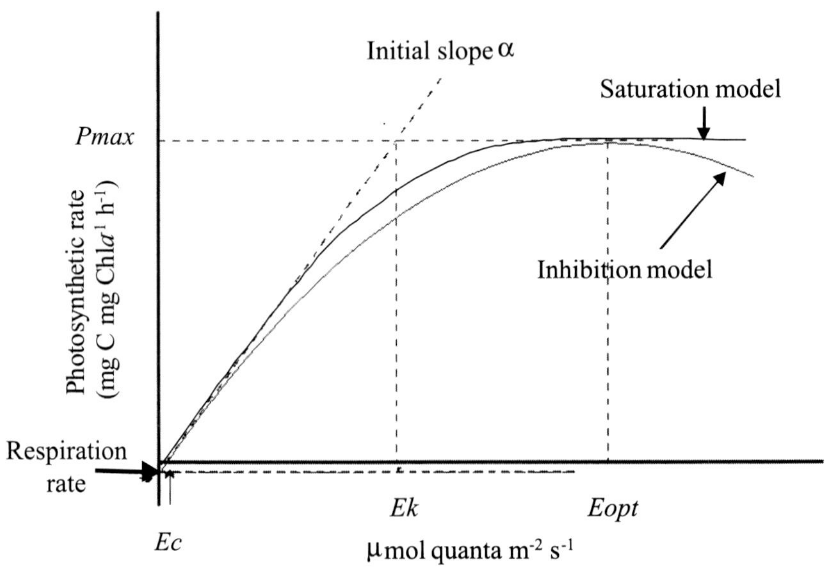

Figure 1: *P-E* inhibition and saturation curves and respective parameters (adapted from Parsons et al. (1984) and Duarte (2006)) (see text).

Some parameters are common to almost all models or can be derived from the models themselves, such as the initial slope or photosynthetic efficiency (α), optimal light intensity or the light level that maximizes photosynthesis under given nutrient and temperature conditions (*Eopt*), the light level at which the linear part of the *P-E* curve intercepts a plateau (light saturation index - *Ek*), the light level that saturates photosynthesis (in the absence of photoinhibition) (*Es*) and the maximal production rate or photosynthetic capacity (*Pmax*).

Dunaliella displays extreme tolerance to its environmental stresses such as hypersalinity, temperature and intense solar energy; studies were focused more on its biochemical adaptations and on how best to maximize its commercial production than on its photosynthetic adaptive mechanisms. Only a few photosynthesis-energy studies on *Dunaliella* are available compared to other micro algae living under normal environmental conditions. In this paper we present results of modelling photosynthesis-energy relationships in *Dunaliella* species to demonstrate their photosynthetic attributes.

Methodology

P-E curves obtained by different authors with *Dunaliella* species and some original data, obtained by one of the authors of this work (Rao, unpublished) are compared in this study. Rao's data include not only classical *P-E* results, but also data from experiments designed to test the effect of cell density and age of culture on photosynthesis measurements.

Our studies are based on batch cultures of *Dunaliella tertiolecta* grown at 10°C, over a bank of Grow-lux fluorescent lights. Growth irradiances were continuous and were 42, 144 and 340 μmol m^{-2} s^{-1}. Aged sea water was filtered and enriched to make Woods Hole medium F (Humphrey and Subba Rao 1967) and autoclaved at 125°C for 25 minutes. Cells centrifuged for 3 minutes at 450 *g* were harvested and suspended in calcium-free artificial sea water with 8.7 pH. Oxygen exchange was determined by Warburg manometry as detailed in Humphrey and Subba Rao (1967). Circular Warburg flasks (10 ml) with no side arms or central well were used with one ml slurry of cells. All manometers were calibrated for oxygen measurements. One of the flasks was covered with black tape and used for respiration measurements. Pre-cooled water (10 °C) was circulated in the Warburg tank. Flasks were shaken 90 cycles per minute in a horizontal arc of 7 cm. A bank of tungsten lams lamps mounted under the glass bottom tank provided light energies up to 3200μmol m^{-2} s^{-1}. The fasks were acclimatized for 10 minutes and the manometers were read every 30 minutes over 120 minutes; the readings were corrected for any thermobarometer changes and oxygen exchange was determined as μl O_2 per one ml of slurry. A photosynthesis quotient of unity was assumed.

Biomass was determined as cell numbers on suitably diluted slurries using a Haemocytometer and as chlorophyll *a*. Photosynthetic pigments were determined on one ml slurry ground at 500 rpm for 1 minute in a glass homogenizer containing about 0.1 g $MgCO_3$ and 3ml 100% acetone. The extract was centrifuged at 4200*g* for 10 minute and made up to 10 ml and the absorbancies determined in a spectrophotometer. The following two component equations (Jeffrey and Humphrey 1975) were used to calculate µg chl per sample.

Chl *a*	$13.31\ e_{663} - 0.27\ e_{630}$
Chl *b*	$-4.33e_{663} + 21.20\ e_{630}$

Standard carbon-14 method was used on 4 day log phase cultures acclimated to 340 µmol m^{-2} s^{-1} and at 9.5°C. The data are presented in Figure 6, bottom panel.

The above mentioned comparisons are based, not only on the general shape of obtained curves, but also on the corresponding *P-E* parameter values. Given the diversity of acclimation procedures, to which algal cultures were subjected prior to measuring their *P-E* responses, the variability on algal concentrations, light intensity regimes, water temperature and nutrient conditions, imposed during the experiments, as well the usage of different *Dunaliella* species and methods to measure primary production, a large variability on obtained results is anticipated. Therefore, one of the challenges is to try integrating available information into some general conceptual model of photosynthesis (*P*) and energy (*E*) relationships in *Dunaliella* sp.

Many authors described *P-E* curves with different degrees of detail. In some cases, only graphical information is provided, without a clear reference to *P-E* parameters. In these situations, for the present analysis available figures were digitized to convert experimental points into numerical data, using FindGraph© software. Figures similar to the original graphs were produced and a *P-E* model, among those presented in Table 1, was fitted to experimental data, using the non-linear regression Gauss method with Statistica© software. *P-E* models were selected, in each case, according to whether data exhibited a light saturation or a light inhibition behavior. After several trials with each data set, and several *P-E* models, those presented in Table 1 gave the best results. When the Jassby and Platt model was used (Table 1), *Ek* was calculated from *Pmax/α*. When a photoinhibition model was used, α was calculated from the limit of the derivative of production with respect to light, when this tends to zero. When *P-E* parameter data was available in the consulted papers, results were used directly, without any further attempt to fit a model. Considering that different authors presented *P-E* data on different units, we presented using the same units to keep figures as close as possible to the original. However, to facilitate comparison, *P-E* parameters are also presented in common units in a separate table. In converting photosynthesis oxygen data to carbon units a photosynthetic quotient of one was assumed.

Table 1: Formulations for the *P-E* relationship used in this work. *P* – photosynthetic rate (usually expressed as mg C mg Chl a^{-1} h^{-1}); *E* – light intensity (usually expressed as mmol quanta m^{-2} s^{-1}); α - initial slope or photosynthetic efficiency (for details on the parameters see text).

<table>
<tr><th>N°</th><th>Equation</th><th>Category</th><th>Source</th></tr>
<tr><td>1</td><td>$P = P_{max} \tanh\left(\alpha E / P_{max}\right)$</td><td>Saturation model</td><td>Jassby and Platt (1976)</td></tr>
<tr><td>2</td><td>$P = P\left[\frac{E}{E_{opt}\exp}\left(1-\frac{E}{E_{opt}}\right)\right]^n$
n=1 (n empirical integer)</td><td rowspan="2">Photoinhibition models</td><td>Steele (1962)</td></tr>
<tr><td>3</td><td>$P = \frac{E}{aE^2 + bE + c}$
where,
$\alpha = \frac{1}{\alpha Eopt^2}$
$b = \frac{1}{Pmax} - \frac{2}{\alpha Eopt}$
$c = \frac{1}{\alpha}$</td><td>Eilers and Peeters (1988)</td></tr>
</table>

Results and discussion

Comparative Analysis of Available Dunaliella sp. P-E Curves

A presentation of *P-E* curves and respective parameters obtained by different authors, will be presented in a chronological order and these are compared and discussed.

Table 2 presents a synthesis of *P-E* parameters obtained by several authors for *Dunaliella* species, between 1984 and present day. Original data were converted to the units depicted in the table for the sake of comparability. Figs 2 - 6, represent corresponding experimental data and model fits in the same order; exception to this are data from Giordano (1997) and Gordillo et al. (2001) who estimated and presented *P-E* parameters (cf. – Methodology).

The scarcity of *Dunaliella P-E* data in available literature is striking and the few data concern *D. salina* (Loeblich 1982, Borowitzka et al. 1984; Gómez

Pinchetti et al. 1992), *D. viridis* (Gordillo et al. 2001), *D. tertiolecta* (McAllister et al. 1964;Young and Beardall 2003), and *D. parva* (Jiménez et al. 1990). In fact, to our best knowledge, the unique work with *P-E* data on *D. parva* (Jiménez et al., 1990) includes an endnote suggesting that the species may be *D. viridis*. These authors focused their studies on the effects of pigment contents, nutrients, salinity and pH culture conditions and growth/acclimation photon flux density (PFD) on *P-E* curve parameters.

Pmax ranged between 0.86 and 26.25 mg C mg $Chl^{-1} h^{-1}$. (Table 2); the highest value was slightly above the theoretical maximum for photosynthesis of 25 mg C mg $Chl^{-1} h^{-1}$ (Falkowski 1981) was for *D. viridis* (Gordillo et al. 2001). The lowest value, obtained for nitrogen starved *D. tertiolecta* acclimated to high irradiance (Young and Beardall 2003). However, this apparent inconsistency may result from the conversion of original data from oxygen to carbon units (Gordillo et al. 2001). McAllister (1964) noticed that phosphorus deficiency may have a marked effect on reducing the rate of respiration as compared with photosynthesis. Wenquan et al. (2000) suggested that *Dunaliella* species photosynthetic rates are maximized under a temperature of 26°C and a cellular N/P ratio of 28.

Experiments combining acid and alkaline conditions (pH = ~6.5 and 8, respectively) with two different phosphate levels (65 and 645 μM P) did not lead to obvious differences in *Pmax* ranges (Table 2). However, photosynthesis: respiration (P: R) ratios were higher under alkaline conditions, especially under the higher phosphate level (Humphrey, 1974). In later studies on the P: R ratios of several marine algae as a function of pH, Humphrey (1975) showed photosynthesis and P: R ratio are maximized (P: R = 10.9) in *D. tertiolecta* at 7.1 and 8.1 pH respectively, suggesting rather a different effect of pH on photosynthesis than on respiration. These studies are relevant considering acidification of sea water caused by increasing atmospheric CO_2 (e.g. Santana-Casiano et al., 2007). It is not clear if the above pH dependence in *Dunaliella* is related to the availability of different inorganic carbon forms. *D. tertiolecta* cells from a stationary growth phase appeared to utilize the bicarbonate ion as a substrate for photosynthesis, whereas those from a mid-exponential phase appeared to utilize free carbon dioxide (Mukerji et al. 1978).

The highest photosynthetic efficiency α of c.a. 0.1670 mg C mg Chl a^{-1} h^{-1} μmol^{-1} m^2 s was obtained with *D. viridis* kept in the dark prior to the *P-E* measurements (Gordillo et al. 2001), and a minimum of 0.0008 mg C mg Chl^{-1} h^{-1} μmol^{-1} m^2 s with carotenoid rich *D. salina* (Loeblich 1982). Cao et al. (2001) reported with *D. salina* grown under two different salinity levels high *Pmax*(10.30) and high α (0.0115) in media with higher salinity (cf. – Table 2).

Saturation and inhibition *P-E* curves for *Dunaliella* obtained under a variety of conditions are given in Table 1 and Figures 2-6. Results of Gómez-Pinchetti et al. (1992) suggest a saturation curve for *D. salina* with high β-carotene (red form), whereas an inhibition curve is apparent for low β-carotene (green form) culture. High carotenoid content of the cells may be caused by nutrient limitation as well

Table 2: Synthesis of *P-E* parameters gathered from the literature and from Rao (unpublished) for *Dunaliella* species. With the exception of data from Giordano (1997) and Gordillo et al. (2001), *P-E* curves from which parameters were obtained are depicted in Figs 2-6 (cf. – Methodology). *Ek* and *Es* or *Eopt* presented according to the obtained *P-E* curve – saturation or inhibition, respectively. In the experiments by McAllister et al. (1964) 'High N' and 'High P' stand for initial concentrations of 500 μM N-NO_3 and 50 μM P, respectively. Low P and low N stand for initial concentrations similar to sea water levels used for the cultures of 22 μM N-NO_3 and 2 μM P, respectively.

Author	Species	Cultures			β-carotene (pg β-carotene $cell^{-1}$)	Ek/Es (μmol $m^{-2} s^{-1}$)	Eopt (μmol m^{-2} s)	α (mg C mg Chl^{-1} h^{-1} μmol $m^2 s^1$)	Pmax(mg C mg Chl^{-1} h^{-1})
McAllister et al. (1964)	*D. tertiolecta*	Low light preconditioned (321 – 481 μmol $m^{-2} s^{-1}$)	High P High N	1.9		3208(Es)			4.85
		High light preconditioned (962 – 1283 μmol $m^{-2} s^{-1}$)	High P High N	0.69		401(Es)			7.10
		Low P		1.9		962(Es)			3.30
		Low N		0.74		321(Es)			3.85

(Table 2 contd.)

(Table 2 contd.)

Humphrey (1974)	*D. tertiolecta*	Acid conditions	65 μM P						14.2-16.5*
			645 μM P						14.5-16*
		Alkaline conditions	65 μM P						12.5-17.1*
			645 μM P						13.2-14.4*
Loeblich (1982)	*D. salina*	Medium MH with salinity ranging from 0.86 till 4.8 M (see text)	1.6 – 3.4 (% Na Cl)	7 – 35	533-711 (Es)				2.62
Borowitza et al. (1984)	*D. salina*	Carotenoid-rich					2653.1	0.0008	6.25
		Carotenoid-poor					1214.5	0.0117	5.26
Jiménez et al. (1990)	*D. parva* or *D. viridis* (see text)	Salinity ranging from 1 till 3 M (see text)					600		3.56
Gómez-Pinchetti et al.(1992)	*D. salina*	Red forms		0.4	15	156/500		0.0276	4.31
		Green forms		4.0	0.5		311.4	0.0099	1.13
Giordano (1997)	*D. salina*	NH_4 grown cells		12.2		123/335		0.0207	2.16
		NO_3 grown cells		8.2		223/335		0.0059	1.76

(Table 2 contd.)

Giordano and Bowes (1997)	*D. salina*	NH_4 grown cells		11.5	4		226/500	0.0119	2.69
		NO_3 grown cells		6.8	2.7		197/500	0.0091	1.79
Gordillo et al. (2001)	*D. viridis*	Growth irradiance(μmol m^{-2} s^{-1})							
		0		0.12		>600 ES		0.1670	8.30
		30		0.375				0.0182	10.50
		250		0.25				0.0110	6.48
		700		0.12				0.0160	11.10
		1500		0.08				0.0135	26.25
Cao et al. (2001)	*D. salina*	1.3 M NaCl				892/3000		0.0115	10.30
		0.3 M NaCl						0.0045	5.41
Young and Beardall (2003)	*D. tertiolecta*	Nitrogen starved		0.36 – 2.42					0.86-2.77
Rao (un-published)	*D. tertiolecta*	Growth irradiance(μmol m^{-2} s^{-1})							
		Ex.I	42.5	0.81			627.39	0.0317	1.43
			144.4	0.63			742.38	0.0564	2.22
		Ex.II	340.4	0.56			1355.7	0.0106	3.94

* Results presented by Humphrey (1974) are in μL O_2 10^{-6} cell h^{-1}. It was not possible to convert these to the units indicated at the column heading due to the absence of chlorophyll data in the cited work. However Humphrey (1975) reported 7.43 mg C mg Chl $a^{-1}h^{-1}$.

as a response to high light stress (Lerche 1937, Zamir 1995). Carotenoids play an important role in protecting photosynthetic organisms from the photochemical generation of oxygen radicals and severe D1 photodamage that may be related with photoinhibition (Sandmann et al. 1993; Gordillo et al. 2001).

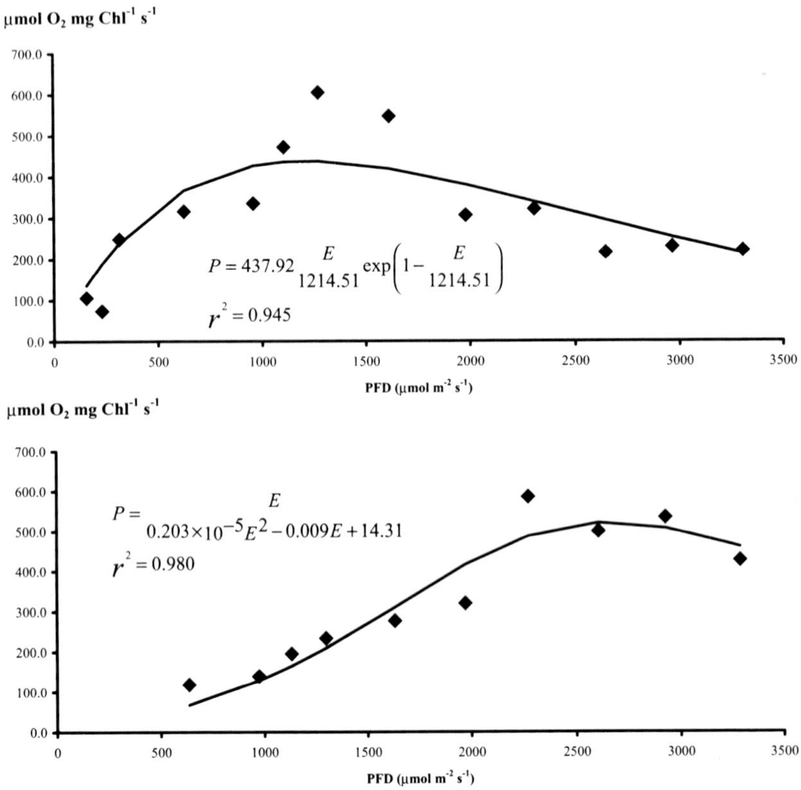

Figure 2: *P-E* curve based on *Dunaliella salina* data from Figure 5 of Borowitzka et al. (1984). Upper panel: results obtained with carotenoid poor cells and model fit with the Steele's photoinhibition model (Steele, 1962). Bottom panel: results obtained with carotenoid rich cells and model fit with Eilers and Peeters photoinhibition model (Eilers and Peeters, 1988) (cf. – Table 1 for equations and Table 2 for *P-E* parameters).

D. salina cultures grown with ammonium or nitrate as the nitrogen source, yielded higher *Pmax* and α (Table 2) in the former than in the latter case (Giordano 1997; Giordano and Bowes 1997). Algae raised in media with ammonium had higher cellular pigment levels, including β-carotene, suggesting better protection from

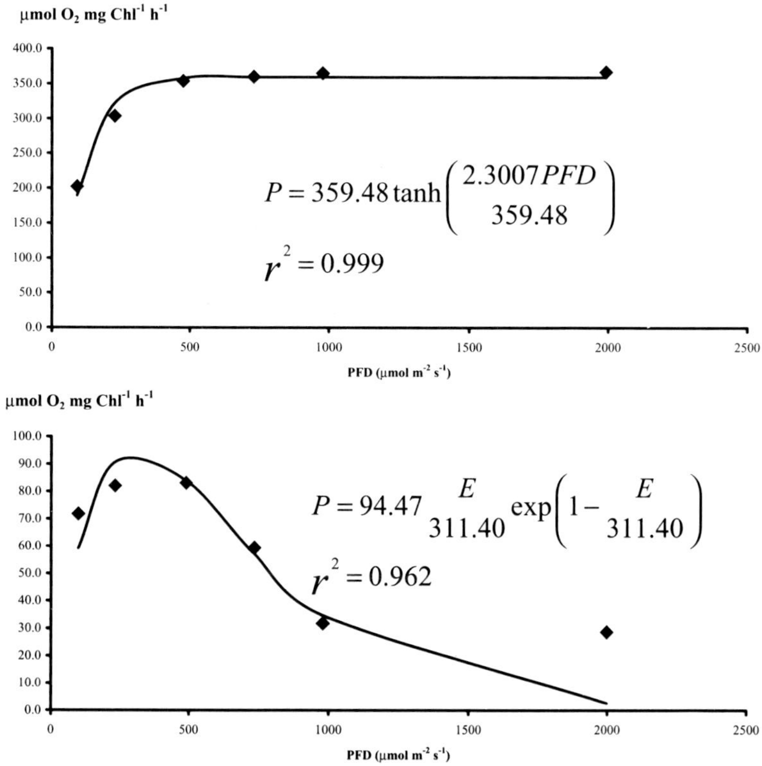

Figure 3: *P-E* curve based on *Dunaliella salina* data from Figure 2 of Gómez-Pinchetti et al. (1992). Upper panel: results obtained with high β-carotene (red form) cells and model fit with Jassby and Platt light saturation model (Jassby and Platt, 1976). Bottom panel: results obtained with low β -carotene (green form) cells and model fit with Steel's photoinhibition model (Steele, 1962) (cf. –Table 1 for equations and Table 2 for *P-E* parameters).

photodamage. In fact, ammonium is generally considered the preferred nitrogen source for algae (Parsons et al., 1984). In nitrogen starvation experiments, *D. tertiolecta* demonstrated a capacity to rapidly reestablish photosynthetic function and initiate cell division after enrichment with nitrogen (Young and Beardall 2003). Results from other experiments suggest a clear relationship between assimilation numbers and ammonium limited steady-state growth, with the former ranging between c.a. 1 and 2 mgC mg Chl a^{-1} h^{-1} (Turpin 1983). Turpin concluded that the addition of ammonium to ammonium limited cells induced a rapid temporary suppression of photosynthesis, followed by its enhancement.

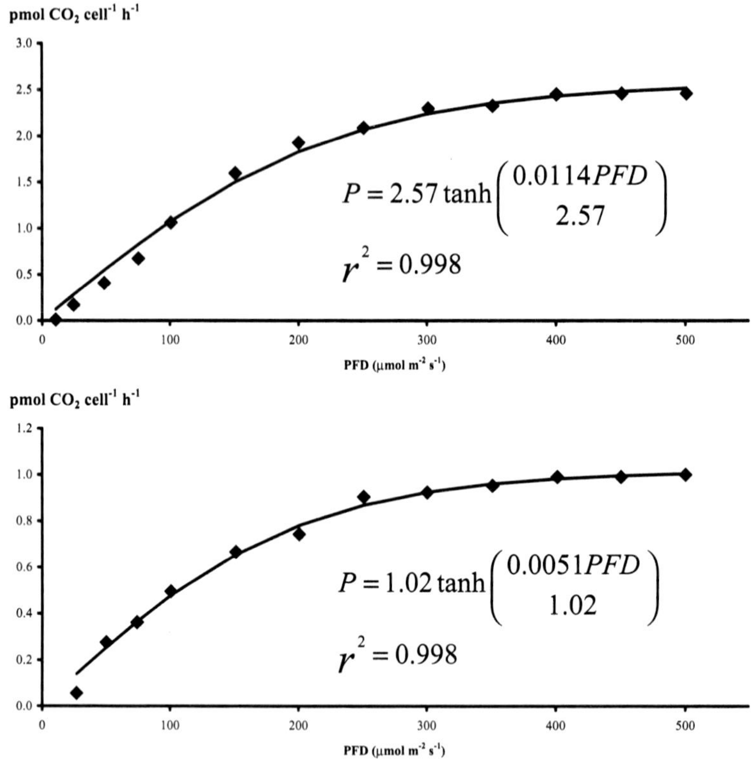

Figure 4: *P-E* curves based on *Dunaliella salina* data from Figure 3 of Giordano and Bowes (1997). Upper panel: results obtained with cells grown on 10µM NH_4^+. Bottom panel: results obtained with cells grown on 10µM NO_3^-. In both cases, the Jassby and Platt light saturation model was used (Jassby and Platt, 1976) (cf. – Table 1 for equations and Table 2 for *P-E* parameters).

Loeblich (1982) studied the effect of salinity (1.6 and 3.4 M NaCl) on the photosynthesis of *D. salina* and *D. parva* (or *D. viridis*, see above). Maximum photosynthetic rates (2.62 at mg C mg Chl $a^{-1}h^{-1}$) were at 1.7 M NaCl at light intensities above 500 µmol m^{-2} s^{-1} (Table 2). It is of interest that a linear relationship between light intensity and β-carotene accumulation existed, whereas chlorophyll *a* decreased with light, suggesting that β-carotene provides protection from phoinhibition. Furthermore, above 1.7 M NaCl both carotenoids and chlorophylls *a* and *b* increased with salinity. Jiménez et al. (1990) carried out a multifactorial experiment with *D. parva* (or *D. viridis*, see above) measuring photosynthetic rates as a function of light intensity, temperature and salinity. Photosynthesis was

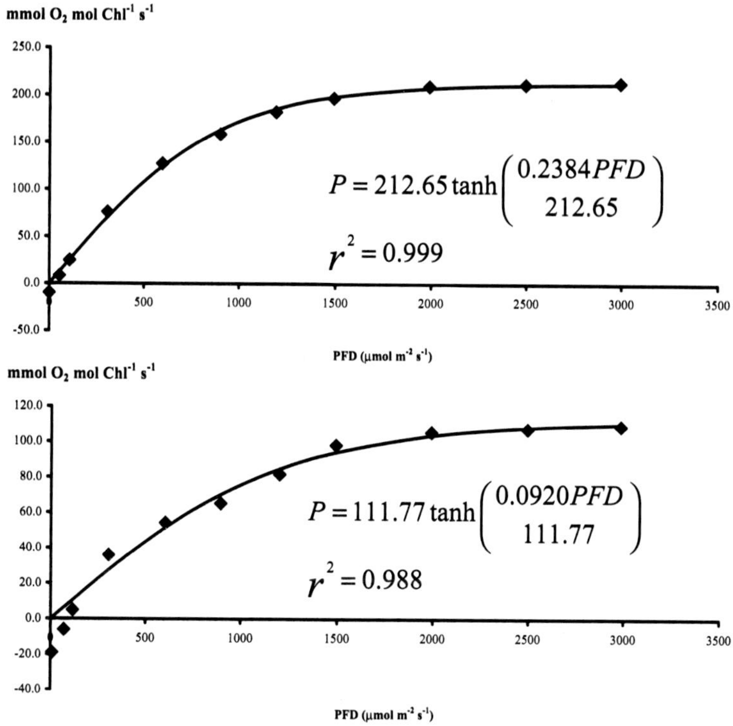

Figure 5: *P-E* curves based on *Dunaliella salina* data from Figure 2A of Cao et al. (2001). Upper panel: results obtained with cells grown on 1.5 M NaCl. Bottom panel: results obtained with cells grown on 0.3 M NaCl. In both cases, the Jassby and Platt light saturation model was used (Jassby and Platt, 1976) (cf. – Table 1 for equations and Table 2 for *P-E* parameters).

maximized at 2 M NaCl, under moderate light ~ 600 μmol m^{-2} s-1 - at 31°C. Obviously the fact that *D. parva* (or *D. viridis*) does not accumulate β-carotene explains the occurrence of photoinhibition unlike in *D. salina* (Loeblich 1982).

Gordillo et al. (2001) obtained *P-E* parameters for algae acclimated to various PFDs. However, their results must be considered with caution, because the range of PFDs used for the estimation of the mentioned parameters was not large enough for photosynthesis per cell to reach a plateau. Therefore, *Pmax* estimates are based on extrapolating the *P-E* curve behavior beyond the range of experimental light intensities (cf. – Table 2 and Figure 2B of Gordillo et al. (2001)). Results obtained in two experiments conducted by Rao (unpublished) (Table 2 and Figure 6), suggest that pigment cell concentration tends to decrease with acclimation

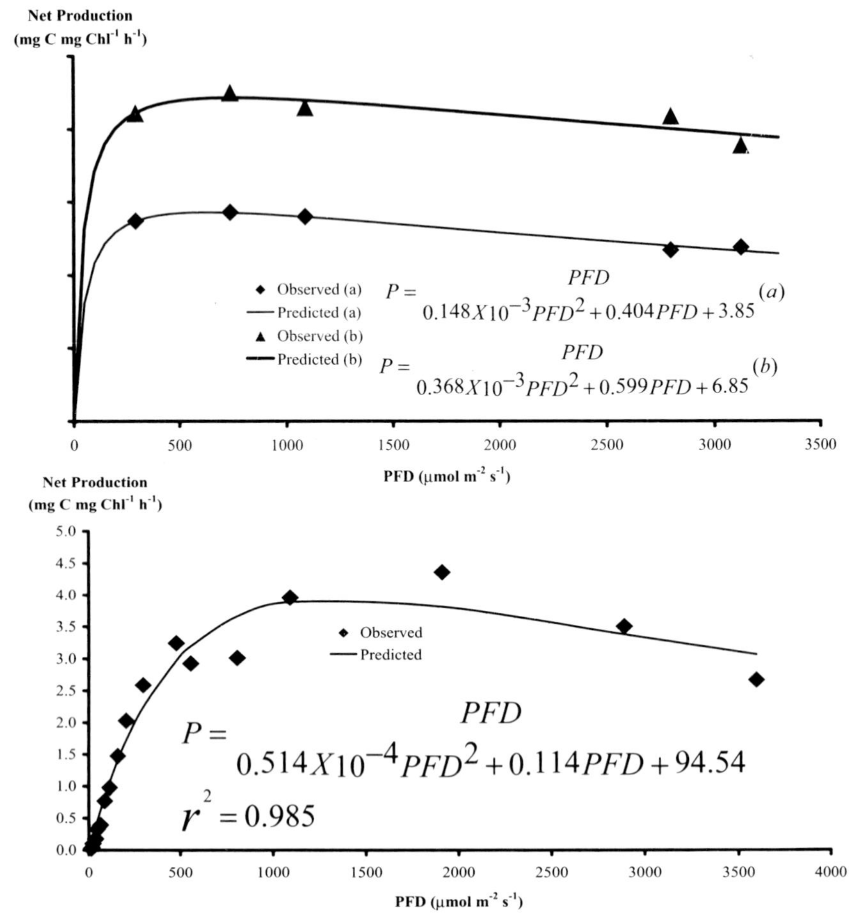

Figure 6: *P-E* curves based on data from Rao (unpublished). Upper panel: results obtained with *Dunaliella tertiolecta* acclimated to two different PFD levels – 42 (a) and 114 (b) μmol m^{-2} s^{-1} - and corresponding fit ($r^2 > 0.999$ for both fits) with the Eilers and Peeters photoinhibition model (Eilers and Peeters, 1988) (Experiment I). Bottom panel: results obtained with *D. tertiolecta* acclimated to 340.4 μmol m^{-2} s^{-1}, together with the Eilers and Peeters equation and respective fit (Experiment II). Equation parameters were used to calculate *Pmax* and *Eopt*, as described in Eilers and Peeters (1988) (cf. – Table 1 for equations and Table 2 for *P-E* parameters). Cultivation conditions were similar for both experiments, with the exception of cell density: c.a. $30X10^9$ and $7X10^6$ cell mL^{-1}, in the former and in the latter case, respectively. Incubation temperature was 10°C (see text).

PFD, with *Pmax* and *Eopt* following the opposite trend, similarly to the results of Gordillo et al. (2001). In both experiments conducted by Rao (unpublished) photoinhibition was apparent. These two experiments were conducted under quite different cell densities (cf. legend of Figure 6). This difference may, at least in part, explain the large difference in *Pmax* that is more than two fold higher in Experiment II, where cell density was lower by more than three orders of magnitude. Although chlorophyll levels and photosynthetic efficiency increase with decreasing acclimation light intensity (Falkwoski and Raven, 1997), results of Gordillo (2001) and Rao's (cf.- Table 2) on photosynthetic efficiency in particular do not support this.

Cell density and P-E curves

That cell density and culture age seem to be important sources of variability in *P-E* results is evident from Rao (unpublished) studies. In the case of cell density, *D. tertiolecta* cultures with various cell densities were acclimated to 42 and 144 μmol m^{-2} s^{-1} - for three days. Afterwards, several culture concentrations were incubated at the same PFDs to measure production and respiration (Figure 7). In the case of cultures from 42μmol m^{-2} s^{-1} gross production was reduced by about seven fold, whereas respiration was reduced by approximately 4 fold. Volume integrated net production does not exhibit such a clear reduction. In fact, there was an increase from the lowest cell density (3.6 X 10^6 cell mL^{-1}) to the second lowest one (5.3 X 10^6 cell mL^{-1}) – the reduction in gross production per unit of chlorophyll *a* was compensated by the larger number of production units. At higher cell densities volume integrated production decreased. In cultures at 144μmol m^{-2} s^{-1} the reductions in net production and respiration were not so large, probably because cell densities were higher (12 X 10^6 – 39 X 10^6 cell mL^{-1}) than those tested with algae acclimated to the lower PFD (3.6 X 10^6 – 12 X 10^6 cell mL^{-1}). Volume integrated net production was higher under the lowest cell concentrations (12 and 15 X 10^6 cell mL^{-1}). Analogous trends were obtained in continuous cultures, with larger cell division rates under lower cell densities (e.g. Sciandra et al., 2001). There are several possible explanations for the observed trends, such as a stronger nutrient and light limitation under higher cell densities. In spite of PFD levels being the same at the culture surface, higher cell densities will produce stronger light intensity decay within the incubation flasks, where production is measured. One common pattern to these results (Figure 7) is the decrease in the gross primary production:respiration (GPP:R) ratio and net production with cell density. This reduction with cell density was from 15 to 5, in the experiment under 42μmol m^{-2} s^{-1} PFD, and from 13 to 5 at the higher PFD (144 μmol m^{-2} s^{-1}), suggesting the larger sensitivity of photosynthesis than respiration to cell density.

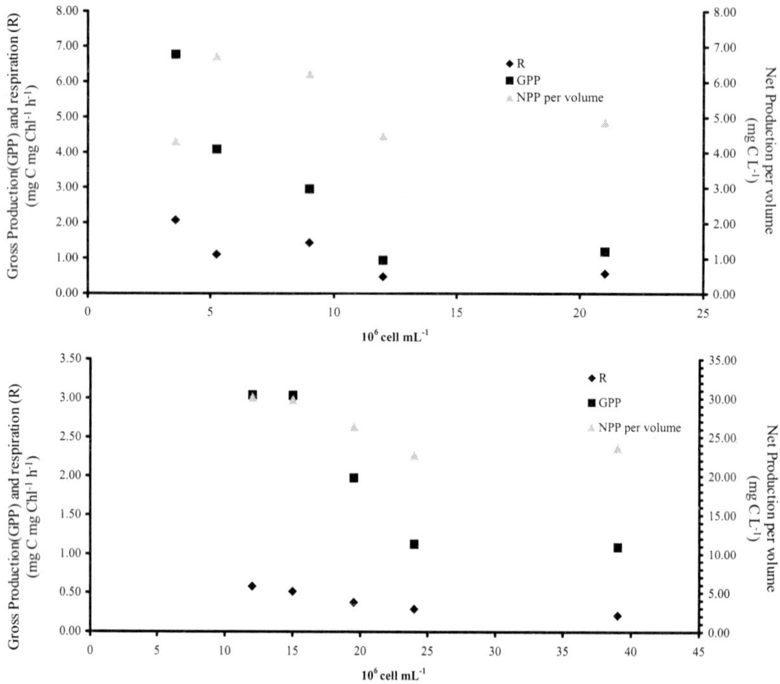

Figure 7: Production and respiration data obtained by Rao with *Dunaliella tertiolecta* (unpublished) as a function of cell concentration. Upper panel: results obtained with algae acclimated to a PFD of 42 mmol m^{-2} s^{-1} and incubated under the same PFD. Bottom panel: results obtained with algae acclimated to 144 µmol m^{-2} s^{-1} and incubated under the same PFD. Left axis refer to net production or respiration per unit of chlorophyll, whereas right axis refer to net production per unit of volume.

Growth and P-E Curves

In an experiment to investigate the effects of culture age, Rao (unpublished) used *D. tertiolecta* cultures acclimated to two PFDs 42 and 144 µmol m^{-2} s^{-1} and their photosynthesis and respiration measured at the acclimation PFDs after 3, 8, 14 and 21 days (Figure 8). Regardless of the acclimation PFD, there is a strong decline in gross production with culture age and almost no effect on respiration. Therefore, there was also a strong decline in the P:R ratio. Chlorophyll *a* per cell did not exhibit any clear pattern when the whole acclimation period is considered (21 days), initial and final values being very similar. However, when only the first eight days are considered, there is a clear increase in chlorophyll *a* content in cells acclimated to the lower light level, whereas the opposite is true for cells acclimated to the higher light level, in accordance with the findings of several authors. Results of Falkowski and Raven (1997) suggest a 13 hours acclimation to low light in *D. tertiolecta*. Berner et al. (1989) showed increasing cellular carotene levels and decreasing cellular chlorophyll levels as a result of adaptation to high light (700 μmol quanta m^{-2} s^{-1}); opposite was the case under low light

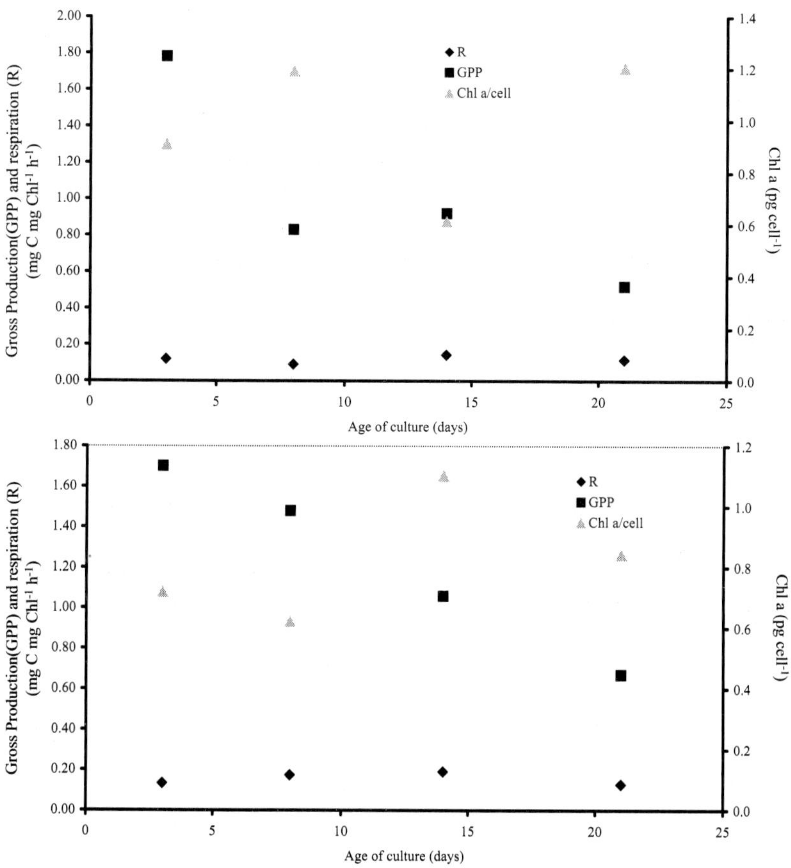

Figure 8: Production, respiration and chlorophyll data obtained by Rao with *Dunaliella tertiolecta* (unpublished) as a function of culture age. Upper panel: results obtained with algae acclimated to a PFD of 42 μmol m^{-2} s^{-1} and incubated under the same PFD. Bottom panel: results obtained with algae acclimated to 144 μmol m^{-2} s^{-1} and incubated under the same PFD. Left axis refer to net production or respiration per unit of chlorophyll, whereas right axis refer to chlorophyll concentration.

(70 μmol quanta m^{-2} s^{-1}). From Figure 2 presented in Berner and Dubinski (1989), it is apparent that chlorophyll contents per cell stabilized after c.a. 100 hours of exposure to the different light levels, whereas carotene contents stabilized faster, after c.a. 60 hours. The results of Sciandra et al. (1997) suggest that light acclimation occurs at the scale of up to four days in this species consistent with the results of Berner et al. (1989). Therefore, the variability in the photosynthetic response in aged cultures (Figure 8) is more likely a result of some limiting factor acting upon cells than resulting from a complete acclimation. If this is correct, perhaps it would be better to use always 'young' cultures (up to three or four days) or continuous steady state cultures, to prevent bias arising from limiting factors and to facilitate comparison of results obtained by different authors. It is noteworthy that acclimation periods were longer i.e. two weeks (e.g. Gordillo et al., 2001) or even two months (e.g. Gómez-Pinchetti et al., 1992) in batch cultures.

The large variability of experimental conditions under which *P-E* curves were obtained by different authors is an important constraint when attempts are made to arrive at some generalizations. The diversity of units used by various investigators for the *P-E* parameters (cf. – Figs 2 – 6) adds to this complexity. Ideally, production data should be expressed in terms of carbon per unit of phytoplankton biomass (expressed as carbon, or preferably as dry weight), to avoid bias arising from variability in chlorophyll contents, cell number and cell individual size. However, determination of phytoplankton biomass may present serious difficulties, especially in natural conditions. For example chlorophyll content and cell volume may change more than 100% over the light/dark cycle (Sciandra et al. 1997).

The ranges of photosynthetic efficiency α (0.8 -167) , *Pmax* (0.86 -11.1) and P:R of *Dunaliella* (Table 2) are comparable to those summarized for algae both benign and toxigenic (Subba Rao and Pan 1994).

Although *P-E* data are of utmost importance to determine conditions that may optimize cell growth, additional data on respiration and dissolved organic carbon (DOC) excretion are also necessary. That one of the main drawbacks of the carbon-14 method is the lack of measurements of respiration is well known. Direct oxygen exchange measurements as those in the manometric studies of Rao (unpublished) facilitate evaluation of factors such as temperature effecting gross photosynthesis and respiration. Furthermore, respiration is proportional to net photosynthesis (see Sciandra et al. 1997 and Figure 7). Gordillo et al. (2001) suggested that DOC release may act not only as a way to control C:N balance, but also as a secondary dissipation process against photodamage, when the xanthophyll cycle is saturated. This may explain why cell growth of their cultures decreased above a PFD of 250 μmol m^{-2} s^{-1} in spite of the higher *Pmax* of cells acclimated to these light intensities (cf. Table 2).

Conclusions

The main conclusions from this review of P-E relationships in *Dunaliella* are:

(i) Acclimation of algal cultures to low or high light levels, previous to *P-E* experiments, should last 3 – 4 days to ensure stabilization of cell pigment and to prevent any negative culture aging effects.

(ii) Acclimation to high light produces higher carotenoid levels in *D. salina* as protection from photoinhibition.

(iii) Photosynthetic rates appear to be maximized under temperatures around 26°C, a relatively high cellular N/P of c.a. 28, slightly alkaline pH and a light intensity above 500 μmol m^{-2} $s^{-1.}$

(iv) Ammonium is the best nitrogen source to maximize photosynthesis.

(v) Culture densities should be considered to avoid bias in comparing results from different *P-E* experiments.

References

Berner,T., Z. Dubinsky, K.Wyman and P.G. Falkowski. 1989. Photoadaptation and the "package" effect in *Dunaliella tertiolecta* (Chlorophyceae). J. Phycol. 25: 70-78.

Blackman.F.F.1905. Optima and limiting factors. Ann. Bot. 19: 281

Blackman, V.H. 1919. The compound interest law and plant growth. Ann. Bot. 33: 353-360.

Borowitzka, M., L.Borowitzka and T. Moulton. 1984. The mass culture of *Dunaliella salina* for fine chemicals: From laboratory to pilot plant. Hydrobiol. 116/117: 115-121.

Cao, H., L.Zhang and A. Melis. 2001. Bioenergetic and metabolic processes for the survival of sulfur-deprived *Dunaliella salina* (Chlorophyta). J. Appl. Phycol. 13: 25–34.

Duarte, P. 2006. Photosynthesis-Irradiance Relationships in Marine Microalgae. pp 639-670. In Algal Cultures Analogues of Blooms and Applications.Vol 2. D.V. Subba Rao(ed) Science Publishers Enfield (NH), USA, Plymouth,UK.

Eilers, P.H.C. and J.C.H. Peeters. 1988. A model for the relationship between light intensity and the rate of photosynthesis in phytoplankton. Ecological Modelling 42:199-215.

Falkowski, P.G. 1981. Light-shade adaptation and assimilation numbers. J. Plank. Res. 3: 203-216.

Falkowski, P.G. and J.A. Raven 1997. Aquatic photosynthesis. Malden, Massachusetts, pp. 375.

Franks, P. J. S. and J. Marra. 1994. A simple new formulation for phytoplankton photoresponse and an application in a wind-driven mixed-layer model. Mar. Ecol. Prog. Ser. 111: 143-153.

Giordano, M. 1997. Adaptation of *Dunaliella salina* (Volvocales, Chlorophyceae) to growth on NH_4^+ as the sole nitrogen source. Phycol. 36: 345-350.

Giordano, M. and G. Bowes. 1997. Gas exchange and C allocation in *Dunaliella salina* cells in response to the N source and CO_2 concentration used for growth. Plant Physiol. 115: 1049-1056.

Gordillo, F.J.L., C.Jiménez, J.Chavarría and F. X. Niell, 2001. Photosynthetic acclimation to photon irradiance and its relation to chlorophyll fluorescence and carbon assimilation in the halotolerant green alga *Dunaliella viridis*. Photosyn. Res. 68: 225–235.

Gómez-Pinchetti, J.L., Z.Ramazanov, A.Fontes and G. Garcia-Reina. 1992. Photosynthetic characteristics of *Dunaliella salina* (Chlorophyceae, Dunaliellales) in relation to β-carotene content. J. Appl. Phycol. 4: 11-15.

Humphrey, G.F. 1974. Effects of carbon dioxide and phosphate supplied during growth, on phosphorus content and photosynthetic rates of some unicellular marine algae. Journal of the Marine Biological Association of India 16: 358-366.

Humphrey, G.F. 1975. The photosynthesis: respiration ratio of some unicellular marine algae. J. Exp. Mar. Biol. and Ecol. 18: 111-119.

Humphrey, G.F. and D.V. Subba Rao 1967. Photosynthetic rate of the marine diatom *Cylindrotheca closterium*. Aust.J. mar. Freshwat. Res. 18: 123-127.

Jassby, A.D. and T. Platt. 1976. Mathematical formulation of the relationship between photosynthesis and light for phytoplankton. Limnol. Oceanogr. 21: 540-547.

Jeffrey, S.W. and G.F. Humphrey. 1975. New spectrophotometric equations for determining chlorophyll a,b,c_1 and c_2 in higher plants, algae and natural phytoplankton. Biochem. Physiol. Pflanzen 167: 191-194.

Jiménez, C., F.X. Niell and J.A. Fernández, 1990. The photosynthesis of *Dunaliella parva* Lerche as a function of temperature, light and salinity. Hydrobiologia 197: 165-172.

Lerche W. 1937. Untersuchungen über Entwicklung und Fortpflanzung in der Gattung *Dunaliella*. Arch. f. Protistenkd. 88: 236–268.

Loeblich, L.A., 1982. Photosynthesis and pigments influenced by light intensity and salinity in the halophyte *Dunaliella salina* (Chlorophyta). Journal of the Marine Biological Association of the United Kingdon 62: 493-508.

McAllister, C.D., N.Shah. and J.D.H. Strickland, 1964. Marine phytoplankton photosynthesis as a function of light intensity: A comparison of methods. J. Fisheries Research Board of Canada 21: 159-181.

Megard, R.O., D.W. Tonkyn and W.H. Senft. 1984. Kinetics of oxygenic photosynthesis in planktonic algae. J. Plankton Res. 6: 325-337.

Mukerji, D., H.E.Glover and I. Morris, 1978. Diversity in the mechanism of carbon dioxide fixation in *Dunaliella tertiolecta* (Chlorophyceae). J. Phycol. 14: 137-142.

Pahl-Wostl, C. and D. M. Imboden. 1990. DYPHORA - a dynamic model for the rate of photosynthesis of algae. J. Plankton Res. 12: 1207-1221.

Parsons, T.R., M. Takahashi and B. Hargrave. 1984. Biological oceanographic processes. Pergamon Press, Oxford, UK.

Platt, T., C.L. Gallegos and W.G. Harrison. 1980. Photoinhibition of photosynthesis in natural assemblages of marine phytoplankton. J. Mar. Res. 38: 687-701.

Rubio, F.C., F.G. Camacho, J.M.F. Sevilla, Y.Chisti and E. Molina. 2003. A mechanistic model of photosynthesis in microalgae. Biotechnol. Bioeng. 81: 459-473.

Sandmann, G., M.Kuhn and P. Böger. 1993. Carotenoids in photosynthesis: protection of D1 degradation in the light. Photosy. Res. 35: 185-190.

Santana-Casiano, J.M., M.González-Dávila, , M.-J. Rueda, O.Llinás, and E.-F. González-Dávila, 2007. The interannual variability of oceanic CO2 parameters in the northeast Atlantic subtropical gyre at the ESTOC site. Global Biogeochemical Cycles 21.

Sciandra, A., J.Gostan, Y.Collos, C.Descolas-Gros, C.Leboulanger, V.Martin-Jézéquel, M.Denis, D.Lefevre, C.Copin-Montégut and B. Avril, 1997. Growth-compensating phenomena in continuous cultures of *Dunaliella tertiolecta* limited simultaneously by light and nitrate. Limnol. Oceanogr. 42: 1325-1339.

Steele, J.H. 1962. Environmental control of photosynthesis in the sea. Limnol. Oceanogr. 7: 137-150.

Subba Rao, D.V. and Y. Pan. 1994. Apart from toxin production, are the toxigenic algae physiologically different? pp 675-680 In Proceedings of the 6th Conference on Toxic Phytoplankton. P.Lassus et al.(eds) Oct.18-22, 1993. Lavoisier Intercept Ltd.Nantes, France.

Teodoresco, E.C. 1905. Organization et development du Dunaliella, nouveau genre de Volvocacee-Polyblepharidee. Beih.Bot.Zentralblatt.Bd.18 Abt.1: 215-232.

Turpin, D.H. 1983. Ammonium induced photosynthetic suppression in ammonium limited Dunaliella tertiolecta (Chlorophyta). J. Phycology 19: 70-76.

Wenquan, L., H.Xianmang, C.Qinghua, Z.Airong, W.Xian and C. Agen. 2000. Study of effects of environmental factors on biochemical compositions of four species of marine single-celled algae. Acta Oceanologica Sinica 19: 69-77.

Young, E., B. and J. Beardall. 2003. Photosynthetic function in *Dunaliella tertiolecta* (Chlorophyta) during a nitrogen starvation and recovery cycle. J. Phycol. 39: 897-905.

Zamir, A. 1995. Plant defences against excessive light studied in the microalga *Dunaliella.* Endeavour 19: 152-156.

10

The Effects of Ultraviolet Radiation on *Dunaliella*: Growth, Development and Metabolism

Leland S. Jahnke[1], Andrea L. White[1] and Priya Sampath-Wiley[1]

Abstract

This chapter reviews the general literature on the physiological and developmental effects of solar UV and simulated solar UV radiation on *Dunaliella* and related microalgae. The depletion of stratospheric ozone allows increased UV-B (280-320nm) in the biosphere without affecting the quantity or quality of UV-A (320-400nm). These studies have provided impetus for new UV-B research, but risk diverting attention from the UV-A photobiology of *Dunaliella* and related species.

The physiological and metabolic UV effects on photosynthetic organisms are diverse. The growth, morphological, photomovement and metabolic responses of *Dunaliella* species to UV exposure are often, but not always, adverse. UV-A radiation exerts major developmental effects in algae acting through blue-light photoreceptors (cryptochromes and phototropins). Recent studies with algae closely related to *Dunaliella*, (i.e. *Chlamydomonas* spp.) have demonstrated UV-A acting through these photoreceptors will initiate biosynthesis of chlorophyll, carotenoids, light harvesting proteins, heat-shock proteins and chemotactic sensitivity. While UV-B appears to initiate important defensive signals for microalgae, its photoreceptor(s) remain unknown. UV-B radiation will directly damage DNA, and UV-A can activate DNA repair mechanisms. UV-A, but not UV-B causes some *Dunaliella* species to over-produce β-carotene. This increased β-carotene screens UV-A and this correlates with UV-B screening to reduce the damaging effects of those radiation bands.

The photosynthetic process is very sensitive to UV radiation. Some *Dunaliella* species appear to be more tolerant of UV-B than others. Photosystem II is a major site of UV-B induced damage in *Dunaliella* species as in other algae and higher plants. Nevertheless, some reports with *Dunaliella* have found minimal or no inhibition of PSII when UV-B similar to solar was used. The *rate* of repair of damage to PSII plays an important role in mitigating the loss of photosynthetic activity during and subsequent to UV-B exposure. The carbon fixing reactions of

[1] Department of Biological Sciences, Rudman Hall, University of New Hampshire, Durham, NH 03824.

photosynthesis (Rubisco) are also damaged by UV-B radiation. Studies with higher plants and microalgae suggest that the carbon-fixing reactions are inhibited *before* damage to the thylakoid reactions becomes apparent. One report indicates that the carbon concentrating mechanism in *Dunaliella* can be affected by UV-B. The relative UV-B sensitivities of the light and dark reactions of photosynthesis in *Dunaliella* species needs to be clarified. That UV-A radiation also damages photosynthesis in *Dunaliella* is well established. Both UV-A and UV-B radiation lead to oxidative stress in *Dunaliella* as measured in changes in the level of diverse antioxidant enzymes and substrates. Our understanding of the UV effects on the physiology, development and biochemistry of *Dunaliella* in the natural environment are inadequate, in part because experimental conditions that are often used differ greatly from sunlight both quantitatively and qualitatively. The ratios of UV-B to both UV-A and to visible radiation often differ drastically from solar, and UV-B lamps emit excess short wavelength radiation that makes it more difficult to extrapolate findings to ecological settings.

Introduction

Photosynthetic organisms have evolved in a fluctuating light environment that has included variable, but significant, amounts of ultraviolet (UV) radiation. At present, approximately 6 to 10% of ground-level photon flux from the sun is near UV (280-400nm), of which 95% are UV-A wavelengths (320-400nm) and the remainder are UV-B wavelengths (280-320nm; Häder and Tevini 1987). Solar UV-C radiation (<280nm) is negligible at ground level, and thus considered ecologically irrelevant. In contrast to UV-C, solar or near ultraviolet radiation (UV-A and UV-B combined) is an important component of the radiation in the biosphere. Even though the energy content of near UV is a small percentage of all solar radiation, the biological responses in photosynthetic organisms (including acclimation) are extensive and poorly characterized.

A great assortment of inhibitory effects have been attributed to near-UV and those impacting algal systems have been recently summarized in numerous reviews (Roy 2000, Vernet 2000, Vincent and Neale 2000, Day and Neale 2002, Häder 2001, Häder et al. 2003, Franklin et al. 2003, Beardall and Raven 2004). For earlier algal UV research, the reader is referred to other excellent sources (Halldal 1967, Halldal and Taube 1972). Generally, the effects of near UV radiation on algae can be divided into: I. damaging interactions, II. adaptive and acclimative (protective) responses, and III. aspects of normal photomorphogenesis. Damaging interactions involve numerous areas of algal metabolism including growth, DNA stability, photosynthesis, nitrogen metabolism and formation of toxic oxygen species. In contrast, adaptive and acclimative responses include formation of UV screening pigments, DNA repair, down-regulation of protein formation, protein replacement in photosystem II, increased activities of antioxidant enzymes, antioxidant substrates and signaling molecules. Various UV-A/blue-light effects influence normal developmental changes in microalgae including, motility, orientation, respiration, starch breakdown, photoperiodic control of enzyme levels and gametogenesis.

Although *Dunaliella* species cannot be considered major players in global photosynthetic balance relative to other autotrophs, they are ubiquitously found in marine and hypersaline ecosystems and thus, are critical contributors to those environments and their related carbon cycles. As green algae (phylum Chlorophyta) that exhibit a straightforward and easily accessible system of growth and analysis, they are widely used as a model system to broaden the scope of understanding UV effects on photosynthetic organisms, including land plants. Ultraviolet research on *Dunaliella* has covered a variety of experimental conditions, ranging from ecologically reasonable exposure to amplified intensities beyond those that would be found in the present environment. This review emphasizes near-UV investigations (i.e. UV-A and UV-B), and also includes limited, but pertinent, findings that examine the effects of UV-C exposure.

Molecular and Morphological Effects of UV

Although the majority of chromophores that interact directly with UV radiation are not clearly identified, the interaction with DNA and nucleic acid structure has been well established as a major target for damage by short-wavelength UV in algae (Halldal and Taube 1972). It has been shown that while DNA absorbs only weakly in the UV-A region, the molecule absorbs strongly within the UV-B, contributing to many of the lethal effects of UV-B exposure. Pyrimidine dimerizations are the most common type of lesion associated with UV-B exposure, leading to disruption of DNA replication and gene transcription if not repaired (Jeffrey and Mitchell 1997, Roy 2000). In contrast, UV-A exposure has been shown to have little direct effect on DNA, with the majority of indirect damage typically occurring through the formation of reactive oxygen species (ROS) and the subsequent production of modified bases (Peak and Peak 1986, Jeffrey and Mitchell 1997). Interestingly, UV-A has been shown to play a protective role in purging UV-B induced dimers through the process of photoreactivation, where UV-A and blue radiation interact with a photolyase enzyme, ameliorating damage (Britt 2004, Yi et al. 2006). Alternative DNA repair mechanisms include nucleotide excision and base excision (Taylor et al. 1997, Britt 2004).

Morphologically, exposure to UV-B radiation produced a doubling of cell volumes in *D. salina* Teod., a phenomenon that was attributed to DNA damage and subsequent cessation of cell cycle processes (Masi and Melis 1997). This effect has been observed in other UV-B exposed algae (Behrenfeld et al. 1992, Malanga and Puntarulo 1995) and has been attributed to the direct damage of microtubules resulting in a slowing of the G2 phase of the cell cycle (Zaremba et al. 1984, Staxén et al. 1993).

UV Radiation as a Developmental Signal

Many developmental processes in plants and algae are triggered by UV-B or UV-A/blue-light signals. The latter are better understood. The photoreceptor pigments for UV-A/blue-light are known as cryptochromes and phototropins and comprise families of flavoproteins related to the DNA-repairing photolyase enzymes (Cashmore et al. 1999, Briggs and Christie 2002). Low levels of UV-A or blue light were found to facilitate the breakdown of starch and decrease oxygen uptake in *D. tertiolecta* (Ruyters et al. 1984). Further studies showed that red light was able to reverse these effects, suggesting that a phytochrome-like pigment may be involved (Ruyters 1988). Most of the recent cryptochrome/phototropin investigations and UV-A/blue light signaling studies with algae have used *Chlamydomonas* as the model organism and investigate a wide range of developmental processes such as biosynthesis of chlorophyll, carotenoids, light harvesting proteins, heat-shock proteins and chemotactic sensitivity (Matters and Beale 1995, Kropat and Beck 1998, Bohne and Linden 2002, Ermilova et al. 2004, Teramoto et al. 2006). Thus, it is likely that many of the same processes are under UV-A/blue-light control in *Dunaliella* sp. Additionally, UV-A/blue light also appears to be responsible for triggering the expression of early light-inducible proteins (ELIPs) and UV screening pigments to be discussed later in this review.

UV-B appears to regulate signal transduction pathways in plants via reactive oxygen molecules, calcium and kinases. The photoreceptor pigments for UV-B are not known. The resultant signals affect multiple pathways of gene expression, but at present are poorly understood (A.-H.-Mackerness et al. 2001, Brosché and Strid 2003).

UV Radiation and Photomovement

Motile algae, including *Dunaliella* species, use solar radiation as a signal to direct their position and rate of movement (Vernet 2000). This response includes vertical migration in the water column to obtain optimal PAR/UV conditions for photosynthesis. While motility responses are typically beneficial, the UV-B portion of solar radiation has also been observed to harm light-directed swimming orientation and motility in many algae (Häder and Häder 1989). Ekelund found that non-motile diatoms were more tolerant of UV-B exposure than motile dinoflagellates (Ekelund 1990). Based on PAR flux density, *Dunaliella salina* demonstrated both positive and negative phototaxis, regulated by an exclusively blue-light absorbing chromophore (Wayne et al. 1991). In *D. bardawil* Ben-Amotz et Avron, growth in high-intensity photosynthetic photon flux density (PPFD) induces massive accumulation of β-carotene compared to low PPFD. It has been suggested that rapid accumulation of β-carotene could potentially act as a screen, altering phototactic response. Jimènez et al. (1996) investigated the role of radiation on swimming responses in *D. bardawil* and found no phototactic response in low β-carotene cells under low to moderate PPFD. In contrast, high-PPFD grown cells with increased β-carotene accumulation were negatively phototactic even at low PPFD. The group also found that the presence of UV-B or UV-A caused a loss of phototactic response in both high and low β-carotene cells, with a subsequent recovery in 24h following exposure. The typical negative gravitaxis observed in *D. bardawil* was unaffected by UV exposure, a finding that differs from other flagellated microalgae where UV-B removes the negative geotropic reaction (Häder and Liu 1990).

Intra- and Inter-Specific Sensitivities to UV Radiation

Within the genus *Dunaliella*, photosynthetic susceptibility to UV damage is observed (Figure 1). Six *Dunaliella* species were exposed to UV radiation under identical conditions and the UV-B flux density that produced a 50% loss of light-saturated photosynthesis was determined. A cell size correlation was noted (see discussion below). The large cells (*D. bardawil*, 300-470 μm volume) were found to be more resistant to UV damage to light-saturated photosynthesis than the small cells (*D. parva*, 40-120 μm volume).

A comparison of seven species of marine microalgae, each from a different phylum (Cyanophyta, Rhodophyta, Bacillariophyta, Cryptophyta, Primnesiophyta, Prasinophyta and Chlorophyta) showed the chlorophyte *D. salina* to be the most tolerant of UV-B exposure measured as change in effective quantum yield (Montero et al. 2002). Some studies have suggested that nutrient deficiencies may improve UV resistance in certain species of diatoms and decrease it in others (discussed later in this chapter). Nevertheless, a study of 67 species of microalgae from different phyla concluded that there were greater differences in short-term UV sensitivity between species within a phylum than between phyla (Xiong et al. 1996). Higher UV tolerance was found to be correlated with high altitiude and high flux density ecosystems (Xiong et al. 1996). Two structural factors that have been correlated with increased UV-B tolerance are relatively large cell volume and/or colonial growth forms (Karentz et al. 1991b, Xiong et al. 1996, Bracher and Wiencke 2000, Vernet 2000). The theoretical basis for this protection has been discussed (Garcia-Pichel 1994). Additionally, factors such as UV absorbing compounds (mycosporines and sporopollenins) appear to play an important protective role in some species (see later discussion in this chapter).

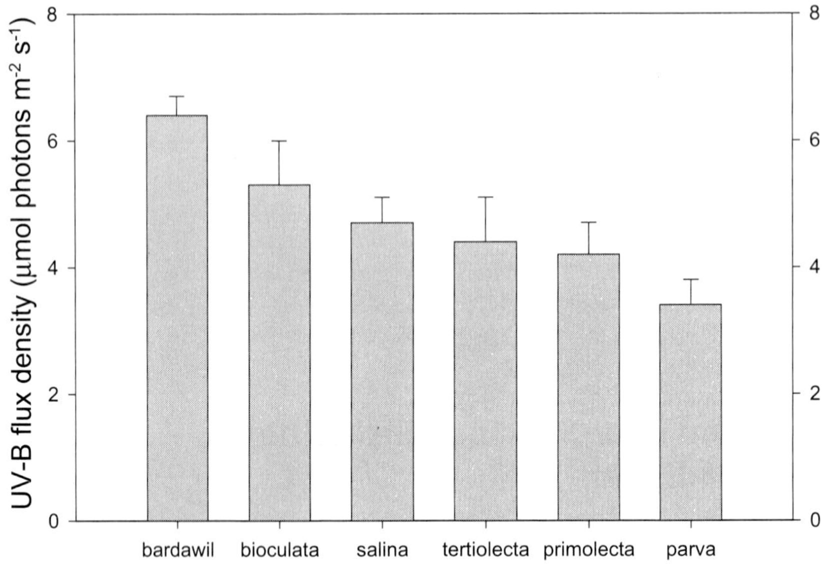

Figure 1: Flux density of UV-B radiation producing a 50% reduction in light saturated photosynthesis in six *Dunaliella* species. Two marine species (*primolecta* (UTEX 2355) and *tertiolecta* (UTEX 999) were grown and tested in artificial seawater (0.5 M NaCl) and hypersaline species (*bardawil* (UTEX 2538), *salina* (UTEX 200), *parva* (UTEX 1983) and *bioculata* (UTEX 199) were grown and tested in a 1 M NaCl medium. Note that this *D. salina* does not over-produce β-carotene as some varieties do. Cultures exposed for 24h with 110 µmol m^{-2} s^{-1} PAR with UV radiation filtered by pre-solarized cellulose acetate at 23°C. Other growth, UV exposure and photosynthetic measurement conditions as described in White and Jahnke (2004). Jahnke, unpublished data.

UV Effects on Photosynthesis

Photosynthetic Absorption in the Ultraviolet Region

The visible spectrum (400 to 700 nm) is defined as the region of Photosynthetically Active Radiation (PAR), giving the false impression that wavelengths shorter than 400 nm will not drive photosynthesis. Chlorophyll absorption extends well into the UV-C range and although UV radiation is generally detrimental to photosynthetic processes, under most circumstances, UV wavelengths are able to drive photosynthetic oxygen evolution and carbon fixation (Rabinowitch 1951, Halldal 1967). Previous research has demonstrated that UV-A radiation is capable of eliciting photosynthetic oxygen evolution in *Dunaliella* species during short-term exposure experiments (McLeod and Kanwisher 1962, Loeblich 1982). UV-A quantum efficiency was stable (0.065 O_2 per quantum) through 350 nm radiation in *D. tertiolecta*, but suffered a noticeable depression at 270 nm (0.025 O_2 per quantum, McLeod and Kanwisher 1962). Loeblich (1982) observed similar UV-driven photosynthetic oxygen evolution in *D. salina* down to 320 nm (Figure 2).

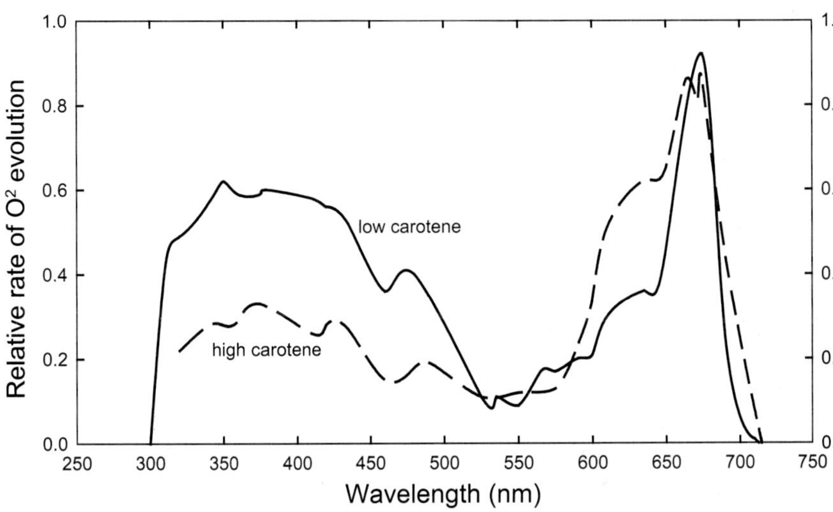

Figure 2: Action spectra for relative oxygen evolution rates of *D. salina*. Cells were grown in 10% NaCl (solid line) containing low carotene levels, or in 25% NaCl yielding cells with high carotene (dashed line). Replotted from L.A. Loeblich, 1982. J. Mar. Biol. Assoc. UK 62: 493-508.

PAR and UV-B induced Damage to Photosystem II

Photosynthetic organisms are prone to the inhibition of photosynthesis as a result of exposure to elevated PAR, as well as from exposure to near UV radiation (Neale et al. 1993). The correlation between PAR intensity and photoinhibition via damage to the D1 and D2 reaction center proteins during periods of excess excitation energy absorption has been well established (Baker et al. 1997, Ort 2001) and demonstrated in studies with *Dunaliella* (Kim et al. 1993, Gordillo et al. 2001). Congruous studies on UV-B inhibition in numerous *Dunaliella* species have confirmed the relationship between rapid photosynthetic damage and interruption of the electron transport chain emanating from photosystem II

(Herrmann et al. 1996, Masi and Melis 1997, Ghetti et al. 1998, 1999, Heraud and Beardall 2000, Montero et al. 2002, Shelly et al. 2003). The repair of UV-B damage to the thylakoid membrane involves the re-synthesis and replacement of the D1 protein which is slowed and/or eliminated during UV inhibition (Baker et al. 1997, Masi and Melis 1997, Chaturvedi and Shyam 2000, Heraud and Beardall 2000, Xiong 2001).

Both UV-B and PAR-induced damage to photosynthesis in *Dunaliella* species are typically seen as: (a) lowered rates of oxygen evolution in the light-limited region, (b) reduction of variable fluorescence as measured by Fv/Fm ratio and (c) reduced carbon fixation (Wolniakowski 1980, Herrmann et al. 1997, Ghetti et al. 1998, Heraud and Beardall 2000, Gordillo et al. 2001, White and Jahnke 2002, 2004). Despite the similarities between PAR and UV exposure effects on photosynthesis, research has suggested that different photosensitizers are implicated in different spectral zones (Jones and Kok 1966a, b, Greenberg et al. 1989, Melis et al. 1992, Neale et al. 1993, Jansen et al. 1996, Hideg et al. 2000). There are also conflicting reports in the literature regarding the primary site of UV-B damage in thylakoid membranes, with both donor and acceptor sides of PSII being implicated as the initial site of damage. Acceptor side advocates have shown that the absorption spectrum of reduced quinones on the acceptor side of PSII implies a direct involvement as a photosensitizer of UV damage (Jones and Kok 1966b, Greenberg et al. 1989, Melis et al. 1992). Likewise, comparisons of Hill Reactions in UV-B exposed *Chlamydomonas* Ehrenberg found that the acceptor side was the most sensitive, although this study did not rule out the contribution of donor side damage (Chaturvedi and Shyam 2000). Acceptor side damage has also been observed in *D. tertiolecta* following UV-B exposure (Beardall et al. 2002). Alternatively, evidence of donor side damage has suggested that the water oxidation complex is rapidly damaged by UV-B (Bornman et al. 1984, Renger et al. 1989, Sicora et al. 2003). The findings of these studies and others have led to suggestions that the mechanism of UV-B induced damage and repair in *D. salina* may be different from PAR produced damage (Masi and Melis 1997). While it is clear that PSII is rapidly inactivated by high PPFDs and by near-UV radiation, the initial sites of damage within the complex appear to be different (Vass 1997, Hideg et al. 2000, Sicora et al. 2003).

Logically, one might expect increases in respiratory oxygen uptake rates following UV radiation exposures. Indeed, this metabolic response has been observed following both PAR and UV-initiated photoinhibitory damage in many microalgae (Beardall et al. 1994, 1997, Singh et al. 1996). Whether this is merely a generalized response to UV damage remains unclear. To address this question, three studies looked for increased respiratory rates in *D. salina* and *D. tertiolecta* following UV exposures and found no elevation in rates of oxygen uptake over normal respiration (Herrmann et al. 1997, Ghetti et al. 1999, Heraud and Beardall 2002).

Reciprocity and Damage Assessment Following UV-B Exposure

In understanding photochemically driven reactions, a useful question is '*does reciprocity apply?*' Reciprocity assumes that the total cumulative dose (as opposed to the rate at which it is administered) is the dominant factor in determining the extent of the effect. The photoinhibition of photosynthesis becomes apparent when the rate of damage to PSII exceeds the rate of the repair processes. If it is assumed that inhibition is simply a function of total dose (independent of duration), then one can assume that the length of exposure becomes irrelevant. While research has demonstrated that isolated chloroplasts and phytoplankton show reciprocity following *short-term* exposure to UV radiation (Jones and Kok 1966a, Behrenfeld et al. 1993), other studies have demonstrated that longer UV-B exposure periods results in damage repair to counteract inhibition (Lesser et al. 1994). This observation reduces the observed effects of the total dose, indicating that reciprocity (i.e. total dose) does not accurately account for the observed rate of photosynthesis following longer term exposures, where flux density clearly plays a more critical role (Cullen and Lesser 1991). This shift from dose-dependent to flux density-dependent, UV-instigated photoinhibition has been confirmed in *D. tertiolecta* (Heraud and Beardall 2000). A recent study with four estuarine epipelic diatoms has demonstrated the complexity of algal responses to UV-B. Some of the damage in carbon metabolism was independent of damage to PSII as well as electron transport, and one species showed UV-B *dose*-dependent damage and three showed *rate*-dependent damage (Waring et al. 2006).

UV-A and Photosynthesis

Although UV-A comprises approximately 90% of the UV photons reaching ground level, the flux of this bandwidth is not altered by changes in stratospheric ozone, a fact that has likely minimized the research effort directed at investigating the effects of UV-A. In short-term solar exposures with *D. salina*, a UV-B photon was found to be about 50 times as damaging to photosystem II and oxygen evolution as a UV-A photon and 1000 times that of a PAR photon (Figure 3; Herrmann et al. 1997). The relative magnitude of photosynthetic inhibition caused by the UV-A and the UV-B bands of solar radiation in the natural environment remains unresolved, but research has implicated UV-A for more than 50% of the UV-induced inhibition of photosynthesis in surface phytoplankton populations in freshwater and marine environments (Bühlmann et al. 1987, Helbling et al. 1992, Holm-Hansen et al. 1993, Kim and Watanabe 1993). Concurrent reports have also been reported in studies conducted on macroalgal species. Dring et al. (1996) examined a broad range of red macroalgae exposed to various combinations of PAR

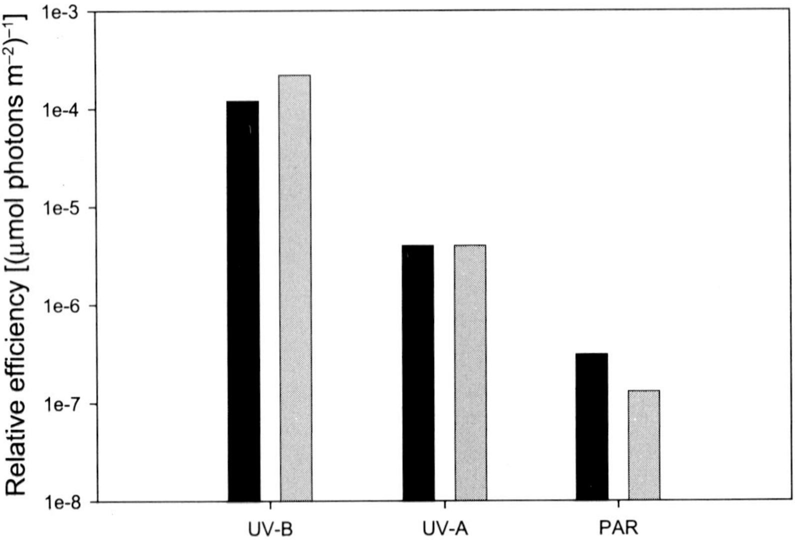

Figure 3: Relative efficiencies of UV-B, UV-A and PAR photons for inhibition of optimal quantum yield (black bars) and efficiency of oxygen production (gray bars) in *D. salina*. Replotted from H. Herrmann, D.-P. Häder and F. Ghetti. 1997. Plant Cell Envir. 20: 359-365.

and near UV and found that UV-A had a major effect in reducing Fv/Fm ratios, whereas the exclusion of the UV-B portion of the spectrum had no detectable effect. Likewise, Herrmann et al. (1995) reported no difference in the reduction of Fv/Fm for *Ulva laetevirens* when UV-B wavelengths were removed from natural irradiance. Further *in situ* studies with green algae have confirmed the correlation between UV-A exposure and decreases in both photosynthetic efficiency and Fv/Fm (Herrmann et al. 1995, 1997, West et al. 1999). West et al. (1999) observed significant effects of UV-A on PSII in natural freshwater lakes and the inclusion of UV-B with UV-A did not yield any further damage to the photosystem, as demonstrated by consistent Fv/Fm ratios. This research suggests that the majority of Fv/Fm damage is directly correlated to UV-A wavelengths rather than UV-B, as damage is still evident even in the absence of UV-B.

The possible synergistic effects of UV-A and UV-B on photoinhibition and damage to PSII are not always carefully scrutinized, particularly when solar radiation is not used. The filters normally applied to remove all UV-C and partial UV-B from UV lamps (i.e. cellulose acetate) do not remove any UV-A (Krizek and Koch 1979, Adamse and Britz 1992, Middleton and Teramura 1993). UV-A exposure (in the absence of UV-B) has been shown to have a significant detrimental effect on Fv/Fm ratios and photosynthetic capacity in *D. bardawil*, *D. parva* Lerche and *D. salina* (Jahnke 1995, Herrmann et al. 1997, White and

Jahnke 2002). The role of UV-A as a hindrance to photosynthetic light reactions has been observed as decreases in initial growth rates (decreased by 80% in 24 h) and Fv/Fm ratios (decreased by 50%) in *D. parva*, with recovery occurring after 48-60h at values somewhat below pre-UV-A exposure rates (Jahnke 1995). To remove residual UV-A, UV-C and blue light from standard UV lamps without decreasing the UV-B flux, a liquid filter using the chromate anion was developed (Darwin 1881, Hunt and Davis 1947). When UV-A and short wavelength UV-B was removed from UV lamps using this chromate filter, the remaining UV-B portion of radiation had little effect on lowering the Fv/Fm ratio in *D. salina* and *D. bardawil* until the UV-B flux density was significantly increased (Figure 5; White and Jahnke 2002, 2004). Other findings that support the role of UV-A as the primary inhibitor of electron transport include Hirosawa and Miyachi (1983) who demonstrated that inhibition of the Hill reaction in a cyanobacterium was mediated through a chromophore absorbing in the UV-A region and Turcsányi and Vass (2000) who concluded that UV-A blocks both the donor and acceptor sides of PSII by mechanisms that are different from visible light induced damage.

Herrmann et al. (1997) clearly demonstrated the major inhibitory effects of the UV-A portion of solar radiation on photosynthetic parameters in *D. salina*. By monitoring the inhibition of light-limited photosynthesis and Fv/Fm ratios they observed the same level of damage with or without the UV-B component of sunlight. The photon ratio of solar UV-A to UV-B is approximately 10:1 (Caldwell et al. 1994), however, most non-solar experiments use ratios of UV-A to UV-B that are unrealistically low. For example, Montero et al. (2002) did not find any change in the UV-B induced decrease in quantum yield in *D. salina* when additional UV-A was present during the exposure, but their ratio of UV-A to UV-B was well below that of solar radiation.

Nutrient Status and UV Effects on Photosynthesis

In natural ecosystems, organisms are generally subjected to multiple environmental stresses, simultaneously. The deficiency of fixed nitrogen is especially widespread in major oceans. Kolber et al. (1988) and Young et al. (2003) have shown that nitrogen deficiency produces a reduction in functional PSII reaction centers in *D. tertiolecta*. Furthermore, nitrogen or phosphate deficiencies interact with UV-B exposure in *D. tertiolecta* to intensify the damage to PSII (Shelly et al. 2002, 2005, Heraud et al. 2005). They found that a combination of either low phosphate or low nitrogen coupled with UV-B exposure accelerated the loss of photochemical efficiency and the repair of damage to PSII. These interactions appear to be quite unpredictable in other marine algae. Litchman et al. (2002) observed increased UV-B sensitivity in nitrogen-limited dinoflagellates. These damaging effects of nitrogen deficiency appeared to be multifaceted. Decreased rate of photosystem repair coupled with reduced mycosporine amino acid screens and smaller cell size all correlated with the increasing sensitivity to UV-B radiation. Lesser's study (Lesser et al. 1994) found nitrogen limited diatoms were more sensitive to UV-B. Contrary to these findings, diatoms subjected to various nutrient deficiencies (carbon, nitrogen, phosphorus, silicon or iron) were found to be *more* tolerant of UV stress than replete cultures in two other studies (Behrenfeld et al. 1994, van de Poll et al. 2005). This suggests that UV-B effects on diatom productivity may not be conspicuous in nutrient limiting oceans, but would likely be easily observed in nutrient replete oceans. These ostensible contradictions may indicate metabolic differences between phyla, such as dinoflagellates, diatoms and green algae, in the interaction between nutrient deficiencies and UV-B stress. The *duration* of the study may appreciably alter the observed effects. Evaluation of short-term (i.e. photosynthetic rates) versus long-term (growth) effects may have a major influence on the conclusions (see review by Vernet 2000). A recent mesocosm study found tropical algae were unaffected by increased UV-B irrespective of nutrient status whereas algae in temperate ecosystems did show significant UV-B damage related to nutrient limitations (Longhi et al. 2006).

UV Effects on Carbon Uptake and Fixation

Ribulose 1,5-bisphosphate carboxylase oxygenase (Rubisco) is the primary enzyme of CO_2 fixation and its activity is essential for carbon assimilation in photosynthetic organisms. Thus, any changes to the Rubisco enzyme with respect to total protein or activation can have a substantial impact on photosynthetic activity (Geider and MacIntyre 2002). The efficient operation of the Calvin-Benson cycle depends on linear (non-cyclic) electron transport in the thylakoids to supply reduced pyridine nucleotides and ATP. Damage to PSII will disrupt this flow of energy to the carbon reducing reactions. For this reason, direct and immediate effects of UV-B on the reactions associated with carbon-fixation have been examined extensively by a number of investigators.

Decreases in the activity of Rubisco have been observed in plants and algae following exposure to UV-B radiation (Vu et al. 1984, Strid et al. 1990, Jordan et al. 1992, Neale et al. 1992, Nogues and Baker 1995, Franklin and Forster 1997, Lesser 1996a,b, Bischof et al. 2002, Keiller and Holmes 2001, Keiller et al. 2003). These findings are consistent with the view that PSII is not the initial site of UV-B damage to photosynthesis. Nogues and Baker (1995) examined the effects of high intensity UV-B exposure (8-fold greater than the summer maximum) on photosynthetic parameters in pea plants and found that Fv/Fm ratios fell only after the photosynthetic capacity (carbon fixation) of the leaf had been substantially reduced. The work of Lesser (1996a, b, Lesser et al. 2002) further substantiated this observation in the dinoflagellates *Prorocentrum micans* Ehrenberg and *Symbiodinium bermudense* and the chlorophyte *Scenedesmus*. Lesser found that variable fluorescence remained unchanged; however, significant decreases in both Rubisco activity and photosynthetic capacity were observed following UV-B exposure. Similarly, in a recent study by Andreasson and Wängberg (2006), photosynthetic carbon fixation in *D. tertiolecta* exhibited a greater sensitivity to UV-B radiation compared to PSII, which are consistent with the findings of Sukenik et al. (1987) who reported that light saturated photosynthetic rates in *D. tertiolecta* were limited by carbon fixing reactions, not by electron transport rates. These data suggest that UV-B damage to carbon reduction reactions could have a greater impact on overall photosynthesis rates than the damage to PSII which slows non-cyclic electron transport.

It has been suggested that the decline in Rubisco activity may be a result of UV-B absorption by the holoenzyme (Gerhardt et al. 1999). Proteins rich in aromatic amino acids, such as Rubisco, are especially vulnerable to the direct effects of UV-B because aromatic amino acids absorb in this spectral region. Cross-linking between the large and small subunits of Rubisco as a direct consequence of UV-B radiation (via photolysis of tryptophan residues) may be a frequent protein modification (Gerhardt et al. 1999). Other studies have indicated that UV-B radiation disrupts the biosynthesis of Rubisco, decreasing the levels of Rubisco polypeptide subunits and their respective mRNA transcripts (Jordan et al. 1992, Keiller et al. 2003). Furthermore, UV-B radiation could interfere with the activation of Rubisco without directly altering the total quantity of the enzyme (Geider and MacIntyre 2002).

Another factor affecting the overall rate of photosynthesis is the availability and uptake of dissolved inorganic carbon (DIC). Most microalgae (including *Dunaliella* species) have carbon concentrating mechanisms (CCMs), which actively supply the site of Rubisco with adequate amounts of CO_2 (reviewed by Colman et al. 2002). Ultraviolet damage to CCMs could be expected to reduce carbon uptake and slow overall photosynthesis in most environments. Typically, algae growing in the presence of >0.2% CO_2 will lack CCMs but exposure to ambient (air) levels of CO_2 triggers a rapid induction of CCMs.

In *Chlamydomonas*, UV-A was shown to have no effect on blocking CCM formation, while low levels of UV-B inhibited CCM induction, with radiation at 290nm causing the greatest inhibition (Goyal and Tolbert 1991). More recent studies on *D. tertiolecta* have also reported that UV-B intensities facilitate a 50% loss in PSII activity without affecting DIC uptake or PSI activity (Beardall et al. 2002). It is assumed that unaffected PSI reaction centers supply the ATP necessary for the carbon pumps via cyclic electron transport while concomitantly sustaining a proton gradient (even though non-cyclic electron transport flow is inhibited). This is consistent with the findings of Naus and Melis (1991) who found that PSI:PSII ratios doubled in *D. salina* as cultures acclimated to higher external pHs and reduced access to dissolved carbon dioxide. However, very different responses to UV radiation were observed in the photosynthetic nanoplankton *Nannochloropsis gaditana* Lubián where the damages to the biocarbonate uptake mechanism appeared to be UV-A specific, with bands 330-370 nm generating CCM impairment (Sobrino et al. 2005).

Reactive Oxygen Species

The absorption of solar radiation by photosynthetic organisms is a necessary, yet potentially damaging, process. When photosynthetic cells are exposed to high intensity light or UV radiation, excess absorbed energy may drive a series of reactions that result in the formation of reactive oxygen species (ROS) which have the potential to cause lipid peroxidation, as well as protein and nucleic acid oxidation, adversely altering cellular structures and activities (Halliwell and Gutteridge 1999, Asada 2006).

UV-absorbing Chromophores and ROS Generation

ROS can be formed directly by pigments in excited states through the transfer of either a single electron or excitation energy to molecular oxygen (dioxygen). These transfer mechanisms are termed Type I and Type II, respectively. Type I pigment transfers produce superoxide, which is subsequently converted to hydrogen peroxide and other more stable ROS. In a Type II transfer, the excited pigment is deactivated by molecular oxygen, yielding the strong oxidant known as singlet oxygen (Niyogi 1999). Algae exhibit considerable *in vivo* absorption of near-UV wavelengths, as measured on an energy basis. Excited chlorophyll *a* at the reaction center of PSII can directly sensitize the formation of singlet excited oxygen in a Type II reaction driven by high PPFDs (Figure 4; Telfer et al. 1994). Recent research suggests that UV-B and UV-A damage to PSII is not caused via singlet oxygen formation, but by direct damage to the water oxidizing system (Hideg et al. 2000, Vass et al. 2002, Sicora et al. 2003).

In addition to photosynthetic pigments, other photosensitizers may be excited by near UV radiation and include porphyrins, flavins, pyridine nucleotides and quinones (Jagger 1985, Greenberg et al. 1989, Neale et al. 1993). Both NADH and NADPH are UV chromophores that will enhance DNA breakage when excited by 334 nm radiation (Peak et al. 1984). Cunningham et al. (1985) have demonstrated that monochromatic UV radiation (290 to 405 nm) can excite NADH and NADPH by photosensitization, resulting in the formation of superoxide. Whether this is a significant source of ROS in chloroplasts is not known. Nevertheless, apart from high PPFD induced photoinhibition, the major sources of ROS in photosynthetic organisms are not the result of either Type I or Type II reactions. The vast majority of ROS generated within cells emanates from redox reactions between dioxygen and reduced electron transfer agents including ferredoxin, iron-sulfur proteins in PSI and quinones (Figure 4; Niyogi 1999, Miyake and Asada 2003, Apel and Hirt 2004). The disruption of intended electron transfer recipients and the subsequent transfer to dioxygen (producing superoxide) is well established in all photosynthetic organisms (A.-H.-Mackerness et al. 1999; Langebartels et al. 2002, Apel and Hirt 2004, Asada 2006).

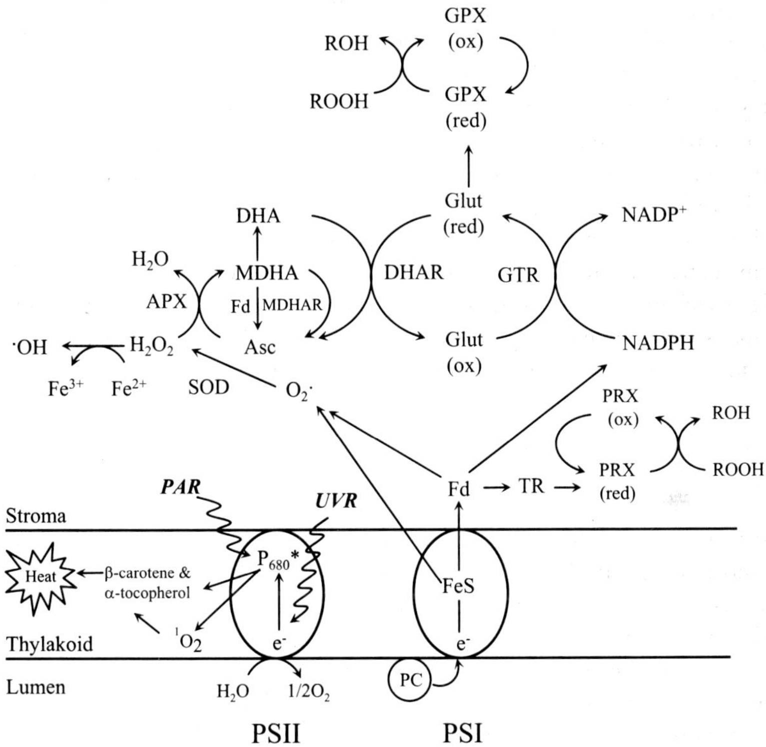

Figure 4: UV/visible radiation and ROS generation and scavenging within the chloroplast. Asc, Ascorbate; APX, ascorbate peroxidase; DHA, dehydroascorbate; DHAR, dehydroascorbate reductase; FeS, iron-sulfur center; Fd, ferredoxin; Glut, glutathione; GTR, glutathione reductase; GPX, glutathione peroxidase; MDHA, monodehydroascorbate; MDHAR, monodehydroascorbate reductase; $^1O_2$, singlet excited oxygen; $O_2^{\cdot -}$, superoxide; ·OH, hydroxyl radical; P_{680}, reaction center chlorophyll in PSII; PC, plastocyanin; PRX, peroxiredoxin; ROOH, organic peroxide; SOD, superoxide dismutase; UVR, UV radiation 280-400nm. Compiled from: Hideg et al. 2000, Dietz 2003, Miyake and Asada 2003, Sicora et al. 2003, Asada 2006, Fischer et al. 2006, Krieger-Liszkay and Trebst 2006.

UV Generation of ROS

Short wavelength UV (i.e. UV-C) damage to cells is not dependent on the presence of molecular oxygen and does not appear to contribute to the direct generation of ROS (Zetterberg 1964). In contrast with UV-C toxicity, molecular oxygen has long been known to strongly enhance the inhibitory effects of near UV radiation in both photosynthetic and non photosynthetic microorganisms (Webb and Brown 1979, Peak and Peak 1986, Tyrell 1991). Intuitively, it seems reasonable that photosynthetic organisms in general (and the chloroplast in particular) should be unusually prone to radiation-sensitized oxygen damage. Elevated oxygen tension generated as a consequence of water oxidation by photosystem II within plastids and the reduction of oxygen by photosystem I to generate superoxide (Mehler reaction, water-water cycle) facilitate the accumulation of various reactive oxygen intermediates within the chloroplast (Miyake and Asada 2003, Asada 2006). A.-H.-Mackerness et al. (2001) have argued that UV-B forms superoxide through multiple sources including peroxidases and NADPH oxidase, and the resulting ROS function in a number of signal transduction pathways.

Detection and Measurement of ROS

Most forms of ROS are difficult to detect directly, since they occur in micro to nanomolar concentrations and have short lifetimes (Halliwell and Gutteridge 1999). Under some conditions, hydrogen peroxide can accumulate and be measured in cells, but the presence of superoxide, singlet excited oxygen or hydroxyl radicals are inferred by the types of damage and the kinds of antioxidant enzymes and substrates that are formed in response to stresses. The hydroxyl radical (·OH), a very strong oxidant, can be formed from hydrogen peroxide by reduction. Several antioxidant systems have evolved to protect cells against their lethal effects (Figure 4). Enzymes including superoxide dismutase (SOD), catalase, ascorbate peroxidase (APX), monodehydroascorbate reductase (MDHAR), glutathione reductase (GTR), glutathione peroxidase (GPX), peroxiredoxin (PRX) and substrates ascorbate, glutathione, β-carotene and α-tocopherol are responsible for protecting cell membrane lipids, proteins and Calvin Cycle enzymes against excessive ROS (Figure 4; Halliwell and Gutteridge 1999, Vincent and Neale 2000, Miyake and Asada 2003, Asada 2006, Krieger-Liszkay and Trebst 2006). The most stable ROS are the hydroperoxides including hydrogen peroxide and alkyl hydroperoxides (ROOH). Most of the hydrogen peroxide results from the dismutation of superoxide formed at photosystem I (Miyake and Asada 2003). The alkyl hydroperoxides result from lipid and protein oxidation initiated either by singlet oxygen or by hydroxyl radicals (Halliwell and Gutteridge 1999). There are three families of enzymes that remove hydrogen peroxide and alkyl hydroperoxides in algal chloroplasts: ascorbate peroxidase, glutathione peroxidase and peroxiredoxin (Figure 4; Goyer et al. 2002, Miyake and Asada 2003, Fischer et al. 2006). Changes in concentration and types of these antioxidants are important sources of information relating to the occurrence of ROS.

UV-B and ROS Formation

Numerous reports have linked UV-B related damage to photosynthesis through the indirect formation of ROS (Lesser and Shick 1989, Malanga and Puntarulo 1995, Vass 1997, Langebartels et al. 2002). Lesser and Shick (1989) examined endosymbiotic algae, finding evidence to support UV-linked induction of superoxide dismutase and catalase, suggesting that oxygen radical formation is an important consequence of UV inhibition. Lesser (1996a,b) went on to observe the effect of both elevated temperature and UV exposure in symbiotic dinoflagellates and found that growth rates and chlorophyll *a* concentrations decreased as a result of UV exposure, while carbon to nitrogen ratios increased. In addition, concentrations of hydrogen peroxide (H_2O_2) and superoxide (O_2^-) increased with UV exposure, as did the activity of both superoxide dismutase (SOD) and ascorbate peroxidase (APX). Although there did not appear to be an effect on the quantum yield of photosynthesis from UV alone, photosynthetic effects were noted as decreased Fv/Fm ratios at elevated temperatures. Lesser (1996b) also observed decreases in Rubisco activity and maximum photosynthesis (P_{max}), effects that were minimized upon the addition of the antioxidants, ascorbate and catalase, an indication of ROS involvement.

Malanga and Puntarulo (1995) looked at the effects of UV-B on oxidative stress and antioxidant content in the green alga, *Chlorella vulgaris* Beirj and found dramatic increases in oxygen radical generation (284%) and lipid peroxidation (145%) following UV-B exposure. Significant increases were also observed for SOD and catalase activities (40% and 500%, respectively), as well as the antioxidants α-tocopherol and β-carotene. Although the photoreducing capacity of chloroplasts appeared to decrease following UV-B exposure, there was an increase in both the size and starch accumulation of cells. The researchers suggest that oxidative stress conditions related to low intensity UV-B exposure elicit a protective response through increased antioxidant activity.

Malanga et al. (1997) measured UV-B induced oxidative damage in sonicated cells of *Chlorella vulgaris.* High intensity UV-B appeared to increase radical production within the chloroplast membranes; however, SOD activity decreased following exposure accompanied by no change in ascorbate peroxidase activity and ascorbic acid levels. This research suggests that the oxidative damage from UV-B causes changes in chloroplast function and integrity and the antioxidant response resulted in a 4-fold increase in the concentration of the lipid antioxidant α-tocopherol.

Standard UV-B lamps emit roughly equal amounts of UV-A and UV-B radiation (Krizek and Koch 1979). To differentiate the effects caused by UV-A from those due to UV-B, White and Jahnke (2002) developed a chromate liquid filter. This filter allowed the removal of UV-A, UV-C and shorter wavelength

UV-B from standard UV-B lamp emissions. With this 'purified' UV-B radiation, they did not observe any changes in ascorbate peroxidase, ascorbate levels or photosynthetic parameters in two *Dunaliella* species. They did observe however, significant antioxidant effects due to UV-A (discussed below).

UV-A and ROS Formation

UV-A radiation, while producing less damage per photon than UV-B, appears to play a major role in solar inhibition of algal growth through formation of ROS. Jeffrey and Mitchell (1997) have reported that the majority of the aerobic lethal responses to UV-A result from photooxidative processes and the generation of reactive oxygen intermediates. Both the formation of ROS and antioxidants are considerably higher in the UV-A in the presence of oxygen compared with UV-B bands (Webb and Lorenz 1970, Tyrell 1991). Unfortunately, because UV-A intensity is not affected by changes in stratospheric ozone, less research has been focused towards UV-A specific effects. However, studies have provided strong support for the major role of UV-A in causing photosynthetic damage to natural algal populations (Helbling et al. 1992, Dring et al. 1996, Holm-Hansen 1997).

White and Jahnke (2002) found that exposing *D. bardawil* and *D. salina* to UV-A irradiance led to increased formation of reactive oxygen as exhibited by the increases (ca. 50%) in ascorbate and ascorbate peroxidase activity. Jahnke (1995) reported that long-term acclimated cells of *D. parva* exhibited a 30% increase in the antioxidant substrate, ascorbate and a doubling in activity of the ascorbate radical scavenger, monodehydroascorbate reductase. Correspondingly, Rijstenbil (2001) found considerable evidence of a link between exposure to low intensity UV-A radiation and oxidative stress in the planktonic diatom *Ditylum brightwellii* Grünow as confirmed by a greater than 2-fold increase in superoxide dismutase and oxidized glutathione levels and more subtle increases in malondialdehyde, an indicator of lipid peroxidation. These data suggest that ROS formation is an important component of UV-A inhibition.

UV-C and ROS

While UV-C exposure is typically not a consideration in UV exposure studies due to the lack of occurrence on the earth's surface, it is worth noting the similarities between UV-C toxicity in cells and the effects noted for near-UV, particularly with respect to ROS formation. Pulich (1974) explored UV-C tolerance in a unique strain of *Chlorella* that exhibits unusual tolerance for high oxygen environments. This work showed that the strain was not protected via photoreactivation, but rather, that antioxidant enzyme activity, particularly superoxide dismutase, increased dramatically.

UV Radiation and Pigments

The photosynthetic reaction center pigments and associated light harvesting complexes play a central role in the absorption and dissipation of light energy, particularly when absorption rates surpass oxygen evolution capabilities of the cell. The role of reaction center chlorophyll *a* as a single electron shuttle in an oxygen-rich environment creates a situation ripe for reactive oxygen formation. As a result, photosynthetic pigments, particularly those associated with the xanthophyll cycle and photoprotection, deserve particular attention when discussing ultraviolet effects and damage to the photosynthetic apparatus.

Hydrophillic and Cell Wall Screening Pigments

Many algae and higher plants acclimate to UV radiation by forming water soluble screening molecules. In some species, UV acclimation correlates with the production of these screens (Jeffrey et al. 1999, Roy 2000, Day and Neale 2002). Higher plants produce phenylpropanoid derivatives that accumulate in epidermal layers of leaves and act as effective screens of UV radiation (Caldwell 1981). Many algal phyla produce water soluble amino acid derivatives (i.e. mycosporine amino acids; MAAs) which screen UV-A and UV-B (Karentz et al. 1991a, Neale et al. 1998, Roy 2000, Carreto et al. 2001, 2005), most commonly in members of the Dinophyta, Phaeophyta and Rhodophyta. These specialized pigments are reputed to be induced by both UV-A and UV-B radiation (Jeffrey et al. 1999). MAAs appear to be uncommon in the Chlorophyta, as supported by the lack of any such compounds identified in extracts of various *Dunaliella* species exposed to UV-B and/or UV-A radiation (e.g. *parva*, *bardawil*, *salina*; Jahnke 1995, 1999, Jeffrey et al. 1999). Xiong et al. (1997) compared 8 tolerant and 8 susceptable species of freshwater algae and found a strong correlation between UV-B tolerance and sporopollenin in the cell wall. Sporopollenin is known to absorb short-wavelength UV radiation.

Carotenoids as Screening Pigments

There has been substantial investigation into the role of carotenoids as principle components of non-photochemical quenching (NPQ) processes, and far less examining the direct absorption of blue and UV light to serve as a protective filter for heme enzymes inactivated by these bands. Based on the absorption spectra of carotenoid compounds and their strong absorption in the blue to UV range, one would theorize that accumulation of these protective compounds would lend some positive component to the shielding of light sensitive enzymes. Various environmental stresses have been shown to trigger the massive accumulation of inter-thylakoid β-carotene in *D. bardawil* and *D. salina* including high intensity PAR, salinity and temperature extremes and deficiencies in either sulfate or nitrate (Mil'ko 1963, Loeblich 1982, Ben-Amotz and Avron 1983, Borowitzka and Borowitzka 1988, Lers et al. 1990). Interestingly, UV-B radiation did not increase chlorophyll and carotenoid levels in *D. tertiolecta* whereas UV-A exposure produced 30-40% increases in all chlorophylls and carotenoids except for α-carotene (Döhler et al. 1997). It is well established that the excess β-carotene accumulation is entirely within the chloroplast, but primarily in an inter-thylakoid location just inside the inner envelope (Ben-Amotz et al. 1989, Lers et al. 1990). Therefore, screening efficacy is limited to the thylakoids and stroma of the chloroplast, facilitated by the accumulation of β-carotene globules. It is significant that these supplementary carotenoids are *not* in physico-chemical contact with the antenna pigments, hence there cannot be a role in direct deactivation of excited light-harvesting pigments (Ben-Amotz et al. 1982, 1989, Shaish et al. 1993). Research has demonstrated that the globules of β-carotene accumulated by *D. bardawil* under stress conditions effectively screen blue light and offer significant protection against PAR induced photoinhibition (Ben-Amotz et al. 1989).

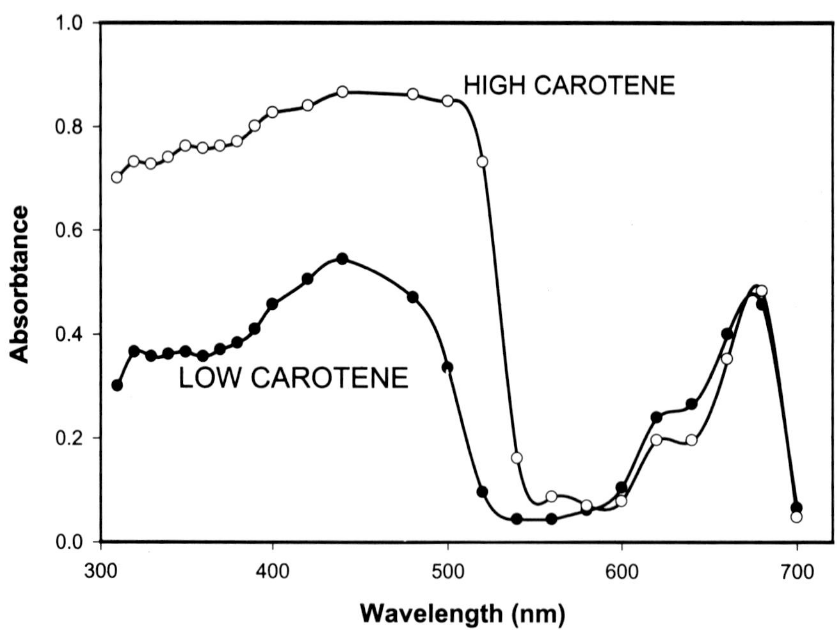

Figure 5: Absorptance spectra of intact cells of *D. bardawil* having high or low levels of carotenoids. Absorptance measured with integrating sphere and normalized at the chlorophyll a peak: 660 nm. Replotted from A.L. White and L.S. Jahnke. 2002. Plant Cell Physiol. 43: 877-884.

Based on the absorption spectrum of β-carotene as a chromophore, it becomes apparent that the molecule is responsible for significant absorption within the UV-A region, possibly preventing UV damage to photosynthesis through dissipation of excess energy. This mode of action (i.e. high energy blue light absorption) is consistent with the mechanism of photoinhibition observed in high PAR conditions. The major blue and near UV screening effect of β-carotene in *D. salina* was examined by Loeblich (1982) confirming that the wavelength screening effect of carotenoid pigments extends well into the UV region. This work compared the action spectra of photosynthesis (measured as oxygen evolution) in high and low β-carotene cells (Figure 2). The low β-carotene cells exhibited elevated photosynthetic excitation from 330–500 nm narrow band radiation wavelengths. In contrast, photosynthetic rates were reduced by 45-55% in cells with high concentrations of β-carotene following exposure to the same wavelength range. Furthermore, high carotenoid levels correlated with a two-fold higher absorptance measured in intact *D. bardawil* in both the UV-A and UV-B regions (Figure 5). This carotene screen correlated with reduced UV damage to photosynthesis in *D. bardawil* (White and Jahnke 2002). Further, UV-A exposure significantly decreased photosynthetic capacity as manifested by depressed Fv/Fm ratios (Figure 6) and light limited to light saturated photosynthetic ratios in low carotene cells of *D. bardawil.*

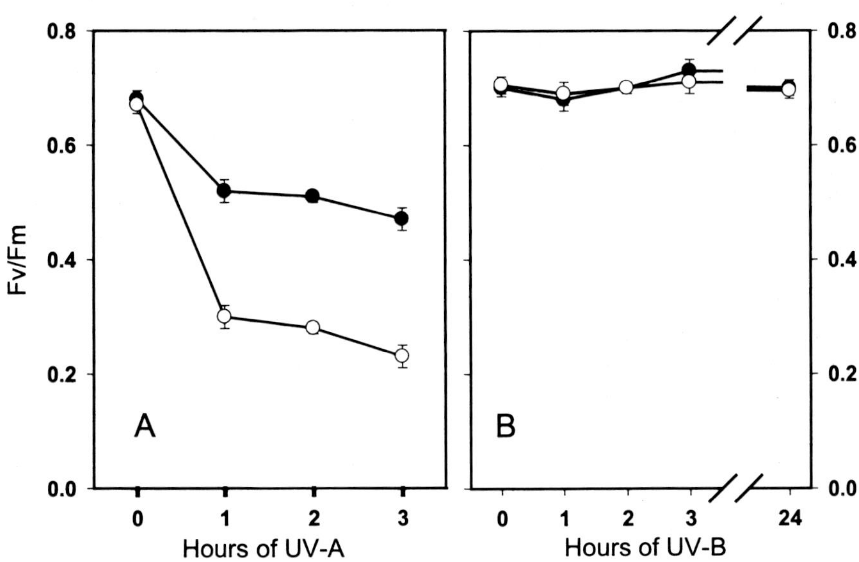

Figure 6: Ratio of variable to maximal fluorescence (Fv/Fm) ratios in *D. bardawil* with high (closed circles) or low levels (open circles) of carotene following exposure to 110 μmol m^{-2} s^{-1} UV-A (graph A) or 1.9 μmol m^{-2} s^{-1} UV-B (graph B). Redrawn from A.L. White and L.S. Jahnke. 2002. Plant Cell Physiol. 43: 877-884.

The effectiveness of typical *trans* forms of carotenoids in removing UV-B radiation is slight given their limited absorbance of UV-B wavelengths (Jimènez and Pick 1994). Nevertheless, *cis* forms of carotenes are known to accumulate during radiation stress in *Dunaliella bardawil* (Ben-Amotz and Avron 1989, Ben-Amotz et al. 1982, 1989, Jimènez and Pick 1994) and have significant absorption in the shorter wavelength UV region (Stahl and Sies 1994, Orset and Young 2000). Whether *cis* forms of carotenoids are significant in screening UV-B radiation in *Dunaliella* spp. is unresolved.

UV and Induction of *β*-Carotene Accumulation

Algae will form carotenoid-containing chloroplasts even when cultured in total darkness. Adamska et al. (1992) have shown that photoinhibitory levels of either visible light or UV radiation will induce the formation of early light-inducible proteins (ELIPs) in pea. ELIP transcription is induced selectively by either UV-A or strong blue light, but not by UV-B or by red light (Adamska et al. 1992). A gene resembling those identified for ELIP synthesis has been cloned from *D. bardawil* and a positive relationship has been established with massive carotenoid accumulation (Lers et al. 1991). This group found that the high PPFD induction of carotenes in *D. bardawil* required *de novo* transcription and translation following gene activation. ELIPs have been localized to thylakoid membranes in plants and in *Dunaliella* and function in the dissipation of excess absorbed excitation energy or binding of potentially damaging photoactive intermediates generated during pigment biosynthesis (Levy et al. 1992, Banet et al. 2000). A carotene biosynthesis related protein (Cbr) is formed only in response to UV-A/blue-light in *D. salina* and *D. bardawil* (Banet et al. 1999). Neither green nor red light would induce these proteins.

Jahnke (1999) has shown that *D. bardawil* accumulates massive quantities of β-carotene in response to UV-A exposure, but not in response to UV-B. Induced carotenoid levels were proportional to the intensity of UV-A when other environmental factors (salinity and PAR intensity) were held constant (Figure 7). Blue light (390-540nm, 200 $\mu mol\ m^{-2}\ s^{-1}$) was no more effective than white light in carotenoid induction suggesting that this is not a shared blue-light and near-UV effect. Shaish et al. (1993) suggested that photosynthetically produced oxygen radicals are involved in triggering massive β-carotene accumulation in *D. bardawil*. The work of Jahnke (1999) supports this idea as low levels of UV-A irradiance triggered the accumulation of β-carotene in *D. bardawil*, yet there was no noticeable decline in growth rates at any level of exposure. Highlighting the effectiveness of β-carotene as a metabolic shield, Salguero et al. (2005) demonstrated that *D. bardawil* growth rates actually increased following UV-A exposure.

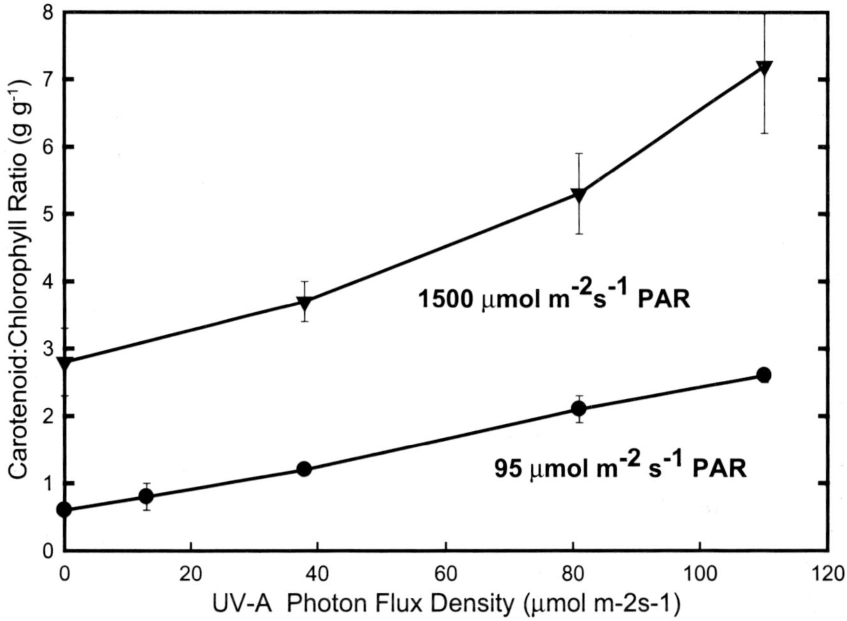

Figure 7: Carotenoid:chlorophyll ratios in cells of *D. bardawil* grown at either low or high flux densities of PAR (95 or 1500 µmol $m^{-2}s^{-1}$) at various UV-A flux densities (0-110µmol $m^{-2}s^{-1}$). Redrawn from L.S. Jahnke, 1999. J. Photochem. Photobiol. B. 48: 68-74.

The Xanthophyll Cycle

Carotenoids play a central role in antioxidant action through several critical mechanisms. First, they can decrease formation of singlet O_2 via absorption of excess excitation energy from chlorophyll by direct transfer. Second, carotenoids can quench singlet O_2 directly. Third, carotenoids are capable of dissipating excess excitation energy via the xanthophyll cycle (Demmig-Adams 1990, Young 1991, Niyogi 1999). In this light-dependent cycle, the de-epoxidated xanthophylls antheraxanthin and zeaxanthin are generated from violaxanthin when a large ΔpH builds up across the thylakoid membranes (Demmig-Adams et al. 1996, Niyogi 1999). Zeaxanthin and antheraxanthin facilitate the dissipation of excess absorbed light energy within LHCII either by accepting energy directly from singlet excited chlorophyll or by promoting aggregation quenching (Niyogi 1999).

Döhler found that xanthophyll cycle pigments in *D. tertiolecta* showed very different responses to UV-A and UV-B (Döhler et al. 1997). UV-A exposure induced large increases in zeaxanthin, violaxanthin and antheraxanthin. In contrast,

UV-B exposure resulted in a 30% decrease in violaxanthin without affecting levels of zeaxanthin and antheraxanthin. Banet et al. (1999) observed that the de-epoxidation of violaxanthin was triggered by UV-A in *D. salina* and *D. bardawil.* Pfündel et al. (1992) found that UV-B exerted an indirect inhibitory effect on the xanthophyll cycle in pea. They concluded that PSII damage lowered the redox level of thylakoid quinones, thus limiting the conversion of violaxanthin to zeaxanthin. The removal of UV-B from solar radiation produced a significant increase in the zeaxanthin portion of xanthophyll cycle pigments in the chlorophyte *Ulva lactuca* L. indicating a reduction in photoprotection when UV-B is present (Bischof et al. 2002). The chlorophyll c containing algae have a xanthophyll cycle consisting of diadinoxanthin and its de-epoxidzed form diatoxanthin (Franklin et al. 2003). In a study with the diatom *Phaeodactylum tricornutum* Bohlin, PAR (300 µmol m^{-2} s^{-1}) was supplemented with low levels of UV-B (3 µmol m^{-2} s^{-1}), triggering a significant increase in the diatoxanthin concentration (Goss et al. 1999). This observed change correlated with an increase in non-photochemical quenching of chlorophyll fluorescence, indicating that UV-B can induce increases in thermal energy dissipation. Mewes and Richter (2002) modified the experimental procedure of Goss slightly, yielding different results. By shortening the high-PPFD treatment, they observed that UV-B produced a decrease in diatoxanthin levels. A recent study with chlorophyll a containing *Nannochloropisis* observed increased de-epoxidation of violaxanthin induced by UV radiation (Sobrino et al. 2005). Their results indicated that short wavelength UV photons produced greater de-epoxidation than longer wavelength photons. Nevertheless, they concluded that PAR (given its much higher flux density in solar radiation) plays the major role in the induction of de-epoxidation. Overall, it appears that the inhibition of photosynthesis by UV radiation exposure is relatively greater than its induction of de-epoxidation in the xanthophyll cycle (Sobrino et al. 2005). For a more detailed discussion of algal xanthophyll cycles, the reader is referred to the review of Franklin et al. (2003).

Critical Concerns for Accurate Assessment and Characterization of UV Effects

The difference in sites of damage and the involvement of ROS in UV-B (as opposed to UV-A) stress needs to be clarified. Distinguishing between UV-B and UV-A effects has been made difficult in the past by problems in obtaining 'pure' UV-B exposure conditions (i.e. free of UV-A and UV-C). There are numerous methodological pit-falls in UV research. The difficulties become glaringly apparent when performing ecologically relevant research, particularly when predicting the effects of stratospheric ozone depletion on aquatic organisms (Vernet 2000). Natural waters allow variable penetration and scattering of PAR and UV radiation, frequently inducing a wide range of vertical movements for organisms that have specific light requirements. Due to the highly variable nature of these microenvironments, it is difficult, if not impossible, to truly mimic these variables in a laboratory setting. For a more detailed discussion of these complexities, there are a number of useful reviews (Kirk 1994, Santas et al. 1997, Tedetti and Sempéré 2005). Generally speaking, experimental UV photobiology generally uses two broad approaches to address the questions related to PAR and UV exposure, exclusion (attenuation) or enhancement, based on the experimental design (Day and Neale 2002).

UV Exclusion-based Analysis

The exclusion approach typically compares natural solar radiation with and without filters to remove UV radiation or specific regions within the UV range. Generally, mylar filters are used to remove UV-B from sunlight or alternative light sources, while various other types of plastic filters effectively block all UV from the exposure treatment (Manabe et al. 1986, Helbling et al. 1992, Holm-Hansen and Lubin 1994, Herrmann et al. 1996, 1997, Xiong et al. 1999). This procedure offers the advantage of relatively low cost and the ability to utilize ambient solar conditions to address the effects of near UV on metabolic processes. Unfortunately, a major limitation of this approach is that one cannot confidently extrapolate results to UV effects caused by an increase in UV flux density. In laboratory conditions lacking UV-specific supplemental lamps, photosynthetic organisms are typically grown under artificial lights that lack most, if not all, of the near-ultraviolet wavelengths (i.e. 280-400 nm), providing an inaccurate and incomplete picture of photosynthetic processes in nature.

UV Enhancement-based Analysis

In contrast to the exclusion-based approach to photosynthetic analysis, the enhancement-based method uses artificial lamps to provide either supplemental UV (in studies using natural solar conditions) or as the exclusive source of UV (in laboratory-confined studies). A major obstacle in achieving accuracy for ecologically-relevant UV research is designing experimental conditions that closely mimic the solar UV spectrum. A majority of published reports use fluorescent lamps as the primary source of UV-B, even though fluorescent UV-B lamps emit both UV-C and UV-A radiation in addition to UV-B (Krizek and Koch 1979, Caldwell et al. 1986, Middleton and Teramura 1993, Döhring et al. 1996, McLeod 1997). Determination of the unique physiological effects of UV-B radiation requires either (a) the removal of the UV-C and UV-A portions of the irradiance or (b) a method by which to account for the effects of extraneous exposure. Short-Wavelength Blocking (SWB) filters are the most widely used filters applied for specific UV removal but are only capable of removing limited wavelengths. Cellulose acetate (CA) is frequently used to exclude extraneous UV-C from UV-A and UV-B radiation sources, in an attempt to create a more ecologically relevant exposure treatment (Caldwell et al. 1986, Middleton and Teramura 1993). Since short wavelength UV-B is considered the most reactive part of the tropospheric solar spectrum, even small divergences between artificial and solar UV can produce enormous response differences. For this reason, evaluating the ecological relevance of UV using enhancement-based analysis is tentative.

Quantitative Discrepancies

The flux density ratio of UV-B to both PAR and UV-A has been well established as an important variable in assessing UV-B affects on physiological processes in algae and higher plants (Caldwell et al. 1994, Middleton and Teramura 1994, Deckmyn et al. 1994, Newsham et al. 1996, Ivanov et al. 2000, Krizek 2004, Waring et al. 2006). Solar radiation measured at ground level has roughly a 1:10:100 photon distribution of UV-B, UV-A, and PAR, respectively (Caldwell et al. 1994). Since UV-B comprises approximately 1% of the total PAR flux density from solar radiation, the ecological relevance of deviations from this ratio leaves much to question. Frequently, experiments have been conducted using unrealistically high ratios of UV-B: UV-A: PAR. Of additional concern is the inadequate understanding of interactions between UV-B and longer wavelength radiation (UV-A & PAR). Such interactions involve quantitative parameters such as inhibition of photosynthesis, DNA repair and alterations in enzyme action. Consequently, many studies report exaggerated effects because unrealistic levels of UV-B radiation are applied, and/or synergistic effects between UV-B and other wavelengths of light are ignored (Krizek 2004). Therefore, it is essential during studies involving organismal responses to increasing amounts of UV-B radiation, that an accurate and complete representation of solar flux is utilized (Cullen et al. 1992, Adamse and Britz 1992, Behrenfeld et al. 1994, Flint and Caldwell 1996, Krizek 2004). That is, the UV-B radiation level used should be comparable to 'natural' levels reaching the earth's surface and implemented along with other solar flux components (i.e. UV-A and PAR). Solar enhancement/exclusion studies generally have one clear advantage, over non-solar studies, as realistic ratios of PAR and UV-A are already present.

In addition to the consideration of accurate radiation quality (i.e. wavelength distribution) in the UV during experimental analyses, precise measurement of flux density is crucial, as well. The use of broad-band detectors to determine UV flux densities is fraught with dangers due to their non-linear response to wavelength and inconsistencies with extrapolation to both photon flux and energy densities.

For this reason, similar numerical readings from broad-band detectors representing UV flux density cannot be compared with any confidence unless the wavelength distribution of the incident radiation is known.

Biological Weighting Functions

Cellulose acetate improves the experimental quality of artificial UV-B radiation by removing UV-C, but compared to solar, a large residual fraction of shortwave radiation remains (Figure 8; Table 1). Physiological responses to UV are very dependent on both wavelength (qualitative assessment) and flux density (quantitative assessment). The responses to wavelength distribution in the UV-B range are particularly non-linear, with small changes in near-UV wavelength distribution generally eliciting large differences in responses. Mathematical functions termed Biological Weighting Functions (BWFs) are required to compensate these non-linear effects (Caldwell et al. 1986; Day and Neale 2002). These are determined with polychromatic spectra rather than with single wavelength responses (monochromatic *action spectra*). The latter approach is less useful in determining weighting functions since it does not take into account the significant interactions due to longer wavelength radiation that alters physiological responses (Caldwell et al. 1986, Caldwell and Flint 1997, Cullen and Neale 1997, Ghetti et al. 1998). It is well established that numerous variables affect the final BWF (Cullen and Neale 1997, Deckmyn and Impens 1997, Ivanov et al. 2000, Neale 2000, Banaszak and Neale 2001). Andreasson and Wängberg (2006, 2007) examined the effects of UV-B on photosynthetic processes in *D. tertiolecta* by determining the BWFs for three physiological responses (carbon fixation, PSII fluorescence and growth rates). They found that measuring carbon fixation (as ^{14}C incorporation) indicated greater damage at short wavelengths and in the UV-A, than was observed in fluorescence measurements of PSII. The BWF for growth rate showed more wavelength dependency than did the two photosynthetic parameters. They concluded that the fluorescence technique is slightly more wavelength dependent and slightly less sensitive than carbon fixation method. These findings illustrate some of the problems in creating BWFs that can be applied with confidence for a single process such as photosynthesis. Improvements in the generation of ecologically-accurate UV-B exposure conditions will improve the accuracy of BWFs, while also resulting in more consistent and realistic experimental results.

Qualitative Discrepancies - Mimicking Solar UV-B Radiation

One limitation in UV-B experimental design is caused by the spectral composition of radiation emitted by commonly used UV-B lamps. Supplemental UV-B

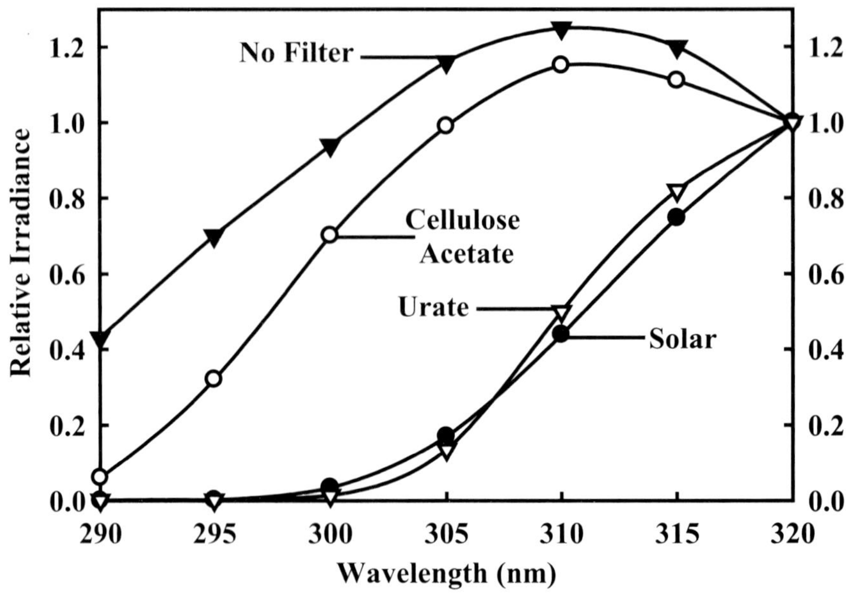

Figure 8: Irradiance spectra of UV-B solar flux (closed circles) compared with supplemental UV radiation from Q-Panel 313 lamps unfiltered (closed triangles) or filtered by solarized cellulose acetate (open circles) or urate solution (0.26 mM, 1.0 cm depth; open triangles). Azide (15 mM) is used in the UA filter to minimize the photooxidation of the urate. Solar flux data calculated from Gerstl et al. (1986). Lamp emission, CA and Urate values measured *in situ* using a spectroradiometer (International Light model 1700/760 D/783, Newburyport, MA) with a 2 nm bandpass and a fiber-optics probe calibrated by International Light. All values have been normalized to 320 nm.

radiation lamps give off shorter wavelengths that are not present in natural solar flux reaching ground level (i.e. UV-C) and emit excessive short wavelength UV-B compared to solar flux (Caldwell et al. 1986). The Q-Panel 313, the National Biological Corporation FS40/UV and the Phillips TL12 are the most widely used UV-B lamps. They have nearly identical emission spectra (peak at 313 nm) and significant emission into the UV-C. This emission is typically altered with either cellulose acetate or glass filters to remove UV-C and a portion of the excess short wavelength UV-B. In either case, the resultant emission in the UV-B region remains quite different from the solar spectrum (Table 1; Figure 8; Caldwell et al. 1986, Adams and Britz 1992, Döhring et al. 1996, McLeod 1997, Johanson and Zeuthen 1998, Holmes 2002).

There are available UV-B lamps (Q-Panel 340 lamps) that emit a spectral composition which more closely resembles that of natural flux. Unfortunately,

Table 1: Comparison of the relative amounts of UV-B radiation emitted from Q-Panel 313 lamps filtered by pre-solarized cellulose acetate (CA) and a liquid urate solution (0.26 mM, 1.0 cm depth; UA) against natural solar flux as measured by Gerstl et al. (1986). All values have been normalized at 320 nm. The percentage deviation from solar radiation is stated for each filter. Urate data from Figure 8.

Wavelength (nm)	Solar Flux	Q-Panel 313 Lamp			
		CA	% Solar	UA	% Solar
320	1.0	1.0	100	1.0	100
315	0.75	1.11	148	0.82	109
310	0.44	1.15	261	0.5	114
305	0.17	0.99	582	0.135	79
300	0.035	0.7	2,000	0.014	40
295	0.0023	0.32	13,900	0.0016	70
290	0.0001	0.06	60,000	0.0011	1,100

their output in the UV-B is not as substantial as the 313 lamps (representing only 12% of the total; < 6 μmol $m^{-2}s^{-1}$ equivalent to < 2.4 Wm^{-2}). is Thus, in experiments designed to simulate increasing UV-B (≥ 6 μmol $m^{-2}s^{-1}$, ≥ 2.4 Wm^{-2}), or to recreate UV-B conditions utilized in most published research, these lamps are inadequate.

The need to improve UV-B lamp filtration to more closely simulate the solar UV-B spectrum has been scrutinized (Caldwell et al. 1986, Adams and Britz, 1992, Middleton and Teramura 1993, Döhring et al. 1996, Kakani et al. 2003). Following a suggestion of Bowen (1946), a liquid urate (UA) filter was developed that has proven to be effective at removing UV-C and short-wave UV-B radiation from standard UV-B lamps (Sampath-Wiley and Jahnke 2006). Liquid filters with 1.0 cm depth containing an aqueous solution of sodium urate (pH 7.5) shows a sharp transmission cutoff in the middle of the UV-B range. A 0.26 mM urate filter has been shown to remove much of the excess short-wave UV-B from UV-B lamps rendering their spectrum nearly identical to that of sunlight (Figure 8; Table 1). This filter has been used in recent studies to compare the effects of traditional cellulose acetate filtration with UA filtration of standard UV-B lamps. To compare the consequences of CA versus UA filtration, *D. tertiolecta* cultures were exposed to 95 μmol $m^{-2}s^{-1}$ PAR with either zero or 6 μmol $m^{-2}s^{-1}$ UV-B for a 24h exposure. The UV-B radiation (280-320 nm) was filtered either by CA or UA and was adjusted to a total flux of 6 μmol $m^{-2}s^{-1}$ for both filters. The CA filtered radiation produced a total (100%) loss of light saturated photosynthesis, whereas the UA filtered treatment lost 19% compared to cells receiving no UV radiation (Jahnke and (Sampath-Wiley unpublished results). This implies that

BWFs may be unable to accurately compensate for excessive short-wavelength UV-B exposure in all circumstances.

Conclusions

Since the seasonal hole in the ozone layer was first identified by researchers, others have explored the implications of an increasing UV radiation environment on ecosystems of the world, particularly those organisms that depend upon photosynthesis for life. Research has indicated that there will be changes in growth, development and metabolism for many organisms, but the severity of these changes has yet to be accurately characterized and quantified based on inconsistencies in UV exposure treatments. While there is little doubt that near-UV radiation impact a range of metabolic and biochemical processes in photosynthetic organisms, many questions remain as to how these organisms will acclimate to their changed environment and the larger impact on the global nutrient status and ecological food web. Current methods offer an opportunity to gain a rough estimate of environmental impact of ozone depletion. It is crucial in studies involving organismal responses to near-UV that an accurate and complete representation of solar flux is utilized. To address the inconsistencies of previous studies, improved simulation of solar UV will be extremely important for evaluating the consequences of changing atmospheric ozone levels.

References

Adamse, P. and S.J. Britz. 1992. Spectral quality of two fluorescent UV sources during long-term use. Photochem. Photobiol. 56: 641-644.

Adamska, I., K. Kloppstech and I. Ohad. 1992. UV light stress induces the synthesis the synthesis of the early light inducible protein and prevents its degradation. J. Biol. Chem. 267: 24732-24737.

A.-H.-Mackerness, S., B.R. Jordan and B. Thomas. 1999. Reactive oxygen species in the regulation of photosynthetic genes by ultraviolet-B radiation (UV-B:280-320nm) in green and etiolated buds of pea (*Pisum sativum* L.). J. of Photochem. and Photobiol. B48: 180-188.

A.-H.-Mackerness, S., C.F. John, B. Jordan and B. Thomas. 2001. Early signaling components in ultraviolet-B responses: distinct roles for different reactive oxygen species and nitric oxide. FEBS Lett. 489: 237-242.

Andreasson, K.I.M. and S.-Å. Wängberg. 2006. Biological weighting functions as a tool for evaluating two ways to measure UVB radiation inhibition on photosynthesis. J. Photochem. Photobiol. B. 84: 111-118.

Andreasson, K.I.M. and S.-Å. Wängberg. 2007. Reduction in growth rate in *Phaeodactylum tricornutum* (Bacillariophyceae) and *Dunaliella tertiolecta* (Chlorophyceae) induced by UV-B radiation. J. Photochem. Photobiol. B. 86: 227-233.

Apel, K. and H.Hirt. 2004. Reactive oxygen species: metabolism, oxidative stress, and signal transduction. Ann. Rev. Plant Biol. 55: 272-399.

Asada, K. 2006. Production and scavenging of reactive oxygen species in chloroplasts and their functions. Plant Physiol. 141: 391-396.

Baker, N.R., S. Nogues and D. Allen.1997. Photosynthesis and photoinhibition. pp. 95-111. In Plants and UV-B. P.Lumsden (ed) Cambridge Univ. Press, New York USA.

Banaszak, A.T. and P.J. Neale. 2001. Ultraviolet radiation sensitivity of photosynthesis in phytoplankton from an estuarine environment. Limol. Oceangr. 46: 592-603.

Banet, G., U. Pick, S. Malkin and A. Zamir. 1999. Differential responses to different light spectral ranges of violaxanthin de-epoxidation and accumulation of Cbr, an algal homolog of plant early light-inducible proteins in two strains of *Dunaliella*. Plant Physiol. Biochem. 37: 875-879.

Banet, G., U. Pick and A. Zamir. 2000. Light-harvesting complex II pigments and proteins in association with Cbr, a homolog of higher-plant early light-inducible proteins in the unicellular green alga *Dunaliella*. Planta 210: 947-955.

Bracher, A.U. and C. Wiencke. 2000. Simulation of the effects of naturally enhanced UV radiation on photosynthesis of Antartic phytoplankton. Mar. Ecol. Progr. Ser. 196: 127-141.

Beardall, J. and J.A. Raven. 2004. The potential effects of global climate change on microalgal photosynthesis, growth and ecology. Phycologia 43: 26-40.

Beardall, J., T. Burger-Wiesma, M. Rijkeboer, A. Sukenik, J. Lemoalle, Z. Dubinsky and D. Fontvielle. 1994. Studies on enhanced post-illumination respiration in microalgae. J. Plankton Res. 16: 1401-1410.

Beardall, J., T. Berman, S. Markager, R. Martinez and V. Montecino. 1997. The effects of ultraviolet radiation on respiration and photosynthesis in two species of microalgae. Can. J. Fish. Aquat. Sci. 54: 687-696.

Beardall, J., P. Heraud, S. Roberts, K. Shelly and S. Stojkovic. 2002. Effects of UV-B radiation on inorganic carbon acquisition by the marine microalga *Dunaliella tertiolecta* (Chlorophyceae). Phycologia 41: 269-272.

Behrenfeld, M.J., J.T. Hardy and H. Lee. 1992. Chronic effects of ultraviolet-B radiation on growth and cell volume of *Phaeodactylum tricornutum* (Bacillariophyeae). J. Phycol. 28: 757-760.

Behrenfeld, M.J., J.W. Chapman, J.T. Hardy and H. Lee. 1993. Is there a common response to ultraviolet-B radiation by marine phytoplankton? Mar. Ecol. Progr. Series 102: 59-68.

Behrenfeld, M.J., H. Lee and L.F. Small. 1994. Interactions between nutritional status and long-term responses to ultraviolet-B radiation stress in a marine diatom. Mar.Biol. 118: 523-530.

Ben-Amotz, A. and M. Avron. 1983. On the factors which determine massive β-carotene accumulation in the halotolerant alga *Dunaliella bardawil*. Plant Physiol. 72: 593-597.

Ben-Amotz, A. and M. Avron, 1989. The wavelength dependence of massive carotene synthesis in *Dunaliella bardawil* (Chlorophyceae). J. Phycology 25: 175-178.

Ben-Amotz, A., A. Katz and M. Avron. 1982. Accumulation of β-carotene in halotolerant algae: purification and characterization of β-carotene-rich globules from *Dunaliella bardawil* (Chlorophyceae). J. Phycol. 18: 529-537.

Ben-Amotz, A., A. Shaish and M. Avron. 1989. Mode of action of the massively accumulated β-carotene of *Dunaliella bardawil* in protecting the alga against damage by excess irradiation. Plant Physiol. 91: 1040-1043.

Bischof, K., G. Kräbs, C. Wiencke and D. Hanelt. 2002. Solar ultraviolet radiation affects the activity of ribulose-1,5-bisphosphate carboxylase-oxygenase and the composition of photosynthetic and xanthophyll cycle pigments in the intertidal green alga *Ulva lactuca*. Planta 215: 502-509.

Bohne, F. and H. Linden. 2002. Regulation of carotenoid biosynthesis genes in response to light in *Chlamydomonas reinhardtii*. Biochim. Biophys. Acta 1579: 26-34.

Bornman, J.F., L.O. Bjorn and H.E. Akerlund. 1984. Action spectrum for inhibition by ultraviolet radiation of photosystem II activity in spinach thylakoids. Photobiochem. Photobiophys. 8: 305-313.

Borowitzka, M.A. and L.J. Borowitzka. 1988. Dunaliella. pp. 27-58. In Microalgal Biotechnology. M.A. Borowitzka and L.J. Borowitzka (eds) Cambridge University Press, New York. USA.

Bowen, E. J. 1946. The Chemical Aspects of Light. Clarendon Press, Oxford. UK

Briggs, W.R. and J.M. Christie. 2002. Phototropins 1 and 2: versatile plant blue-light receptors. Trends in Pl. Sci. 7: 204-210.

Britt, A.B. 2004. Repair of DNA damage induced by solar UV. Photosynth. Res. 81: 105-112.

Brosché, M. and Å. Strid. 2003. Molecular events following perception of ultraviolet-B radiation by plants. Physiol. Plantar. 117: 1-10.

Bühlmann, B., P. Bossard and U. Uehlinger. 1987. The influence of longwave ultraviolet radiation (u.v.-A) on the photosynthetic activity (14C-assimilation) of phytoplankton. J. Plankton Res. 9: 935-943.

Caldwell, M.M. 1981. Plant response to solar ultraviolet radiation. pp. 169-197. In Encyclopedia of Plant Physiology. Vol. 12A. O.L. Lange, P.S. Nobel, C.B. Osmond and H. Zieleger (eds) Springer Verlag, Berlin. Germany.

Caldwell, M.M. and S.D. Flint. 1997. Uses of biological spectral weighting functions and the need of scaling for the ozone reduction problem. Plant Ecology 128: 66-76.

Caldwell, M.M., L.B. Camp, C.W. Warner and S.D. Flint. 1986. Action spectra and their key role in assessing biological consequences of solar UV-B radiation change. pp. 87-111. In Stratospheric Ozone Reduction, Solar Ultraviolet Radiation and Plant Life. R. Worrest and M. Caldwell (eds). Springer-Verlag, New York, USA.

Caldwell, M.M., S.D. Flint and P.S. Searles. 1994. Spectral balance and UVB sensitivity of soybean: a field experiment. Plant Cell Envir. 17: 267-276.

Carreto, J.I, M.O. Carignan and N.G. Montoya. 2001. Comparative studies on mycosporine-like amino acids, paralytic shellfish toxins and pigment profiles of the toxic dinoflagellates *Alexandrium tamarense, A. catenella* and *A. minutum*. Mar. Ecol. Progr. Series 223: 49-60.

Carreto, J.I, M.O. Carignan and N.G. Montoya. 2005. A high-resolution reverse-phase liquid chromatography method for the analysis of mycosporine-like amino acids (MAAs) in marine organisms. Mar. Biol. 146: 237-252.

Cashmore, A.R., J.A. Jarillo, Y.-J. Wu and D. Liu. 1999. Cryptochromes: blue light receptors for plants and animals. Sci. 284: 760-765.

Chaturvedi, R. and R. Shyam. 2000. Degradation and de novo synthesis of D1 protein and psbA transcript in *Chlamydomonas reinhardtii* during UV-B inactivation of photosynthesis. J. Biosci. 25: 65-71.

Colman, B., I.E. Huertas, S. Bhatti and J.S. Dason. 2002. The diversity of inorganic carbon acquisition mechanisms in eukaryotic micro algae. Funct. Plant Biol. 29: 261-270.

Cullen, J.J. and M.P. Lesser. 1991. Inhibition of photosynthesis by ultraviolet radiation as a function of dose and dosage rate: results for a marine diatom. Mar. Biol. 111: 183-190.

Cullen, J.J. and P.J. Neale. 1997. Biological weighting functions for describing the effects of ultraviolet radiation on aquatic systems. pp. 97-118. In The Effects of Ozone Depletion on Aquatic Ecosystems. D.-P. Hader (ed) R.G. Landes Co., Austin, TX USA.

Cullen, J.J., P.J. Neale and M.P. Lesser. 1992. Biological weighting function for the inhibition of phytoplankton photosynthesis by ultraviolet radiation. Science 258: 646-650.

Cunningham, M.L., J.S. Johnson, S.M. Giovanazzi and M.J. Peak. 1985. Photosensitized production of superoxide anion by monochromatic (290-405 nm) ultraviolet irradiation of NADH and NADH coenzymes. Photochem. Photobiol. 42: 125-128.

Darwin, C. 1881. The Power of Movement in Plants. Appleton, New York, USA.

Day, T.A. and P.J. Neale. 2002. Effects of UV-B radiation on terrestrial and aquatic primary producers. Ann. Rev. Ecol. Syst. 33: 371-396.

Deckmyn, G., C. Martens and I. Impens. 1994. The importance of the ratio UV-B/photosynthetic active radiation (PAR) during leaf development as determining factor of plant sensitivity to increased UV-B irradiance: effects on growth, gas exchange and pigmentation of bean plants. Plant Cell Environ. 17: 295-301.

Deckmyn, G. and I. Impens. 1997. The ratio of UV-B/photosynthetic active radiation (PAR) determines the sensitivity of rye to increased UV-B radiation. Env. Exp. Botany 37: 3-12.

Deitz, K.-J. 2003. Plant peroxiredoxins. Ann. Rev. Pl. Biol. 54: 93-107.

Demmig-Adams, B. 1990. Carotenoids and photoprotection in plants: a role for the xanthophyll zeaxanthin. Biochim. Biophys. Acta 1020: 1-24.

Döhler, G., G. Drebes and M. Lohmann. 1997. Effect of UV-A and UV-B radiation on pigments, free amino acids and adenylate content of *Dunaliella tertiolecta* Butcher (Chlorophyta). J. Photochem. Photobiol. B 40: 126-131.

Döhring, T., M. Köfferlein, S. Thiel and J.K. Seidlitz. 1996. Spectral shaping of artificial UV-B irradiation for vegetation research. J. Plant Physiol. 148: 115-119.

Dring, M.J., A. Wagner, J. Boeskov and K. Luning. 1996. Sensitivity of intertidal and subtidal red algae to UVA and UVB radiation, as monitored by chlorophyll fluorescence measurements: influence of collection depth and season and length of irradiation. Eur. J. Phycol. 31: 293-302.

Ekelund, N.G.A. 1990. Effects of UV-B radiation on growth and motility of four phytoplankton species. Physiol. Plantar. 78: 590-594.

Ermilova, E.V., Z.M. Zalutskaya, K. Huang and C.F. Beck. 2004. Phototrophin plays a crucial role in controlling changes in chemotaxis during the initial phase of the sexual life cycle in *Chlamydomonas*. Planta 219: 420-427.

Flint, S.D. and M.M. Caldwell. 1996. Scaling plant ultraviolet spectral responses from laboratory action spectra to field spectral weighting factors. J. Plant Physiol. 148: 107-114.

Fischer, B.B., R.I.L. Eggen, A. Trebst and A. Krieger-Liszkay. 2006. The glutathione peroxidase homologous gene Gpxh in *Chlamydomonas reinhardtii* is upregulated by singlet oxygen produced in photosysytem II. Planta 223: 583-590.

Franklin, L.A. and R.M. Forster. 1997. The changing irradiance environment: consequences for marine macrophyte physiology, productivity and ecology. Eur. J. Phycol. 32: 207-232.

Franklin, L.A., C.B. Osmond and A.W.D. Larkum. 2003. Photoinhibition, UV-B and algal photosynthesis. pp. 351-384. In Photosynthesis in Algae. A. Larkum, S. Douglas and J.Raven (eds) Kluwer Acad. Publ. NY. USA.

Garcia-Pichel, F. 1994. A model for internal self-shading in planktonic organisms and its implications for the usefulness of ultraviolet screens. Limnol. Ocean. 39: 1704-1707.

Geider, R.J. and H.L. MacIntyre. 2002. Physiology and biochemistry of photosynthesis and algal carbon acquisition. pp. 44-77. In Phytoplankton Productivity. P.J. Williams, D.N. Thomas and C.S. Reynolds (eds) Blackwell Science, Malden MA, USA.

Gerhardt, K.E., M.I. Wilson and B.M. Greenberg. 1999. Tryptophan photolysis leads to a UV-B –induced 66 kDa photoproduct of ribulose-1,5-bisphosphate carboxylase/ oxygenase (Rubisco) in vitro and in vivo. Photochem. Photobiol. 70: 49-56.

Gerstl, S.A.W, A. Zardecki and H. L. Wiser. 1986. A new UV-B handbook, volume I. pp. 63-74.

Ghetti, F., S. Fuoco and G. Checcucci. 1998. UVB monochromatic action spectrum for the inhibition of photosynthetic oxygen production in the green alga *Dunaliella salina*. Photochem. Photobiol. 68: 276-280.

Ghetti, F., H. Herrmann, D.-P. Häder and H. K. Seidlitz. 1999. Spectral dependence of the inhibition of photosynthesis under simulated global radiation in the unicellular green alga *Dunaliella salina*. J. Photochem. Photobiol. 48: 166-173.

Gordillo, F.J.L, C. Jiménez, J. Chavarría and F.X. Niell. 2001. Photosynthetic acclimation to photon irradiance and its relation to chlorophyll fluorescence and carbon assimilation in the halotolerant green alga *Dunaliella viridis*. Photosynth. Res. 68: 225-235.

Goss, R., H. Mewes and C. Wilhelm. 1999. Stimulation of the diadinoxanthin cycle by UV-B radiation in the diatom *Phaeodactylum tricornutum*. Photosynth. Res. 59: 73-80.

Goyal, A. and N. Tolbert. 1991. U.V. inhibition of the dissolved inorganic carbon concentrating mechanism(s) and alternative respiration in unicellular green algae. pp. 95-107. In Impact of Global Climatic Changes on Photosynthesis and Plant Productivity. Y. Abrol (ed) IBB Publishing, New Delhi, India.

Goyer, A., C. Haslekås, M. Miginiac-Maslow, U. Klein, P. Le Marechal, J.-P. Jacquot and P. Decottignies. 2002. Isolation and characterization of a thioredoxin-dependent peroxidase from *Chlamydomonas reinhardtii*. Eur. J. Biochem. 269: 272-282.

Greenberg, B.M, V. Gaba, O. Cannani, S. Malkin, A.K. Matoo and M. Edelman. 1989. Separate photosensitizers mediate degradation of the 32 kDa photosystem II reaction center protein in the visible and UV spectral regions. Proc. Nat. Acad. Sci. USA 86: 6617-6620.

Häder, D.-P. 2001. Adaptation to UV stress. pp. 173-202. In Algal Adaptation to Environmental Stresses. L. Rai and J. Gaur (eds) Springer, New York, USA.

Häder, D.-P. and M. Tevini. 1987. General Photobiology. Pergamon Press, NY, USA.

Häder, D.-P. and M.A. Häder. 1989. Effects of solar UV-B irradiation on photomovement and motility in photosynthetic and colorless flagellates. Envir. Exp. Bot. 29: 273-282.

Häder, D.-P. and S.-M. Liu. 1990. Motility and gravitactic orientation of the flagellate *Euglena gracilis,* impaired by artificial and solar UV-B radiation. Curr. Microbiol. 21: 161-168.

Häder, D.P., H.D. Kumar, R.C. Smith and R.C. Worrest. 2003. Aquatic ecosystems: effects of solar ultraviolet radiation and interactions with other climatic change factors. Photochem. Photobiol. Sci. 2: 39-50.

Halldal, P. 1967. Ultraviolet action spectra in algology. A review. Photochem. Photobiol. 6: 445-460.

Halldal, P. and Ö. Taube 1972. Ultraviolet action and photoreactivation in algae. pp. 163-187. In Photophysiology. Vol. VII. A. Geise (ed). Academic Press, NY, USA.

Halliwell, B. and J.M.C. Gutteridge. 1999. Free Radicals in Biology and Medicine. Clarendon Press, Oxford, UK.

Helbling, E., V. Villafane, M. Ferrario and O. Holm-Hansen. 1992. Impact of natural ultraviolet radiation on rates of photosynthesis and on specific marine phytoplankton species. Mar. Ecol. Progr. Series 80: 89-100.

Heraud, P. and J. Beardall. 2000. Changes in chlorophyll fluorescence during exposure of *Dunaliella tertiolecta* to UV radiation indicate a dynamic interaction between damage and repair processes. Photosynth. Res. 63: 123-134.

Heraud, P. and J. Beardall. 2002. Ultraviolet radiation has no effect on respiratory oxygen consumption or enhanced post-illumination respiration in three species of microalgae. J. Photochem. Photobiol. B. 68: 109-116.

Heraud, P., S. Roberts, K. Shelly and J. Beardall. 2005. Interactions between UV-B exposure and phosphorus nutrition. II. Effects on rates of damage and repair. J. Phycol. 41: 1212-1218.

Herrmann, H., F. Ghetti, R. Scheuerlein and D.-P. Häder. 1995. Photosynthetic oxygen and fluorescence measurements in *Ulva laetevirens* affected by solar radiation. J. Plant Physiol. 145: 221-227.

Herrmann, H., D.-P. Häder, M. Köfferlein, H.K. Seidlitz and F. Ghetti. 1996. Effects of UV radiation on photosynthesis of phytoplankton exposed to solar simulator light. J. Photochem. Photobiol. B. 34: 21-28.

Herrmann, H., D.-P. Häder and F. Ghetti. 1997. Inhibition of photosynthesis by solar radiation in *Dunaliella salina*: relative efficiencies of UV-B, UV-A and PAR. Plant Cell Envir. 20: 359-365.

Hideg, É., T. Kálai, K. Hideg and I. Vass. 2000. Do oxidative stress conditions impairing photosynthesis in the light manifest as photoinhibition? Phil. Trans. Roy. Soc. Lond. B 355: 1511-1516.

Hirosawa, T. and S.A Miyachi. 1983. Effects of long-wavelength ultraviolet (UV-A) radiation on the growth of *Anacystis nidulans*. Plant Sci. Lett. 28: 291-298.

Holm-Hansen, O. 1997. Short- and long-term effects of UVA and UVB on marine phytoplankton productivity. Photchem. Photobiol. 65: 266-268.

Holm-Hansen, O. and D. Lubin.1994. Solar ultraviolet radiation: effect on rates of CO_2 fixation in marine phytoplankton. pp. 55-74. In Regulation of Atmospheric CO_2 and O_2 By Photosynthetic Carbon Metabolism. N.E. Tolbert and J. Preiss (eds) Oxford University Press, NY, USA.

Holm-Hansen, O., E.W. Helbling and D. Lubin. 1993. Ultraviolet radiation in Antarctica: inhibition of primary production. Photochem. Photobiol. 58: 567-570.

Holmes, M.G. 2002. An outdoor multiple wavelength system for the irradiation of biological samples: analysis of the long-term performance of various lamp and filter combinations. Photochem. Photobiol. 76: 158-163.

Hunt, R.E. and W. Davis. 1947. Filters for the isolation of the 3130 Å spectral group of the mercury arc. J. Amer. Chem. Soc. 69: 1415-1418.

Ivanov, A.G., E. Miskiewicz, A.K. Clarke, B.M. Greenberg and N.P.A. Huner. 2000. Protection of photosystem II against UV-A and UV-B radiation in the cyanbobacterium

Plectonema boryanum: the role of growth temperature and growth irradiance. Photochem. Photobiol. 72: 772-779.

Jagger, J. 1985. Solar UV Actions on Living Cells. Praeger Publ. NY, USA.

Jahnke, L.S. 1995. Acclimation to ultraviolet radiation in *Dunaliella parva*. pp. 279-282. In Photosynthesis From Light to Biosphere. Vol 4. P. Mathis (ed) Kluwer Acdemic Publishers, Dordrecht, The Netherlands.

Jahnke, L.S. 1999. Massive carotenoid accumulation in *Dunaliella bardawil* induced by ultraviolet-A radiation. J. Photochem. Photobiol. B. 48: 68-74.

Jansen, M.A.K, Greenberg, B.M., M. Edelman, A.K. Matoo and V. Gaba. 1996. Accelerated degradation of the D2 protein of photosystem II under ultraviolet radiation. Photochem. Photobiol. 63: 814-817.

Jeffrey, W.H. and D.L. Mitchell. 1997. Mechanisms of UV-induced DNA damage and response in marine microorganisms. Photochem. Photobiol. 65: 260-263.

Jeffrey, S.W., H.S. MacTavish, W.C. Dunlap, M. Vesk and K. Groenewoud. 1999. Occurrence of UVA- and UV β-absorbing compounds in 152 species (206) strains of marine microalgae. Mar. Ecol. Prog. Ser. 189: 35-51.

Jimènez, C. and U. Pick. 1994. Differential stereoisomer composition of β,β-carotene in thylakoids and in pigment globules in *Dunaliella*. J. Plant Physiol. 143: 257-263.

Jimènez, C., F.L. Figueroa, J. Aguilera, M. Lebert and D.-P. Häder. 1996. Phototaxis and gravitaxis in *D. bardawil*: influence of UV radiation. Acta Protozool. 35: 287-295.

Johanson, U. and J. Zeuthen. 1998. The ecological relevance of UV-B enhancement studies is dependent on tube type, tube age, temperature and voltage. J. Photochem. Photobiol. 44: 169-174.

Jones, L.W. and B. Kok. 1966a. Photoinhibition of chloroplast reactions. I. Kinetics and action spectra. Plant Physiol. 41: 1037-1043.

Jones, L.W. and B. Kok. 1966b. Photoinhibition of chloroplast reactions. II. Multiple effects. Plant Physiol. 41: 1044-1049.

Jordan, B.R., J. He, W.S. Chow and J.M. Anderson. 1992. Changes in mRNA levels and polypeptide subunits of ribulose 1, 5-bisphosphate carboxylase in response to supplementary ultraviolet-B radiation. Plant Cell Environ. 15: 91-98.

Kakani, V.G., K.R. Reddy, D. Zhao and K. Sailaja. 2003. Field crop responses to ultraviolet-B radiation: a review. Agri. For. Meteorol. 120: 191-218.

Karentz, D., F. McEuen, M. Land and W. Dunlap. 1991a. Survey of mycosporine-like amino acids compounds in antarctic marine organisms: potential protection from ultraviolet exposure. Mar. Biol. 108: 157-168.

Karentz, D., J.E Cleaver and D.J. Mitchell. 1991b. Cell survival characteristics and molecular responses of Antarctic phytoplankton to ultraviolet-B radiation. J. Phycol. 27: 326-341.

Keiller, D.R. and M.G. Holmes. 2001. Effects of long-term exposure to elevated UV-B radiation on the photosynthetic performance of five broad-leaved tree species. Photosynth. Res. 67: 229-240.

Keiller, D.R., S.A.-H. Mackerness and M.G. Holmes. 2003. The action of a range of supplementary ultraviolet (UV) wavelengths on photosynthesis in *Brassica napus* L. in the natural environment: effects on PS II, CO_2 assimilation and level of chloroplast proteins. Photosynth. Res. 75: 139-150.

Kim, D.-S. and Y. Watanabe. 1993. The effect of long wave ultraviolet radiation (UV-A) on the photosynthetic activity of natural population of freshwater phytoplankton. Ecol. Res. 8: 225-234.

Kim, J.H., J.A. Nemson and A. Melis. 1993. Photosystem II reaction center damage and repair in *Dunaliella salina* (Green algae). Plant Physiol. 103: 181-189.

Kirk, J.T.O. 1994. Optics of UV-B radiation in natural waters. Arch. Hydrobiol. 43: 1-16.

Kolber, Z., J. Zehr and P.G. Falkowski. 1988. Effects of growth irradiance and nitrogen limitation on photosynthetic energy conversion in photosystem II. Plant Physiol. 88: 923-929.

Krieger-Liszkay, A. and A. Trebst. 2006. Tocopherol is the scavenger of singlet oxygen produced by the triplet states of chlorophyll in the PSII reaction centre. J. Exper. Bot. 57: 1677-1684.

Krizek, D.T. 2004. Influence of PAR and UV-A in determining plant sensitivity and photomorphogenic responses to UV-B radiation. Photochem. Photobiol. 79: 307-315.

Krizek, D.T. and E.J. Koch. 1979. Use of regression analysis to estimate UV spectral irradiance from broad band radiometer readings under FS-40 fluorescent sunlamps filtered with cellulose acetate. Photochem. Photobiol. 30: 483-489.

Kropat, J. and C. F. Beck. 1998. Characterization of the photoreceptor and signaling pathway for light induction of the *Chlamydomonas* heat-shock gene HSP70A. Photochem. Photobiol. 68: 414-419.

Langebartels, C., M. Schraudner, W. Heller, D. Ernst and H. Sandermann Jr. 2002. Oxidative stress and defense reactions in plants exposed to air pollutants and UV-B radiation. pp. 105-135. In Oxidative Stress in Plants. D. Inze and M. Van Montegu (eds). Taylor and Francis, NY, USA.

Lers, A., Y. Biener and A. Zamir. 1990. Photoinduction of massive β-carotene accumulation by the alga *Dunaliella bardawil*. Plant Physiol. 93: 389-395.

Lers, A., H. Levy and A. Zamir. 1991. Co-regulation of a gene homologous to early light-induced genes in higher plants and β-carotene biosynthesis in the alga *Dunaliella bardawil*. J. Biol. Chem. 266: 13698-13705.

Lesser, M.P. 1996a. Acclimation of phytoplankton to UV-B radiation: oxidative stress and photoinhibition of photosynthesis are not prevented by UV-absorbing compounds in the dinoflagellate *Prorocentrum micans*. Mar. Ecol. Progr. Series 132: 287-297.

Lesser, M.P. 1996b. Elevated temperatures and ultraviolet radiation cause oxidative stress and inhibit photosynthesis in symbiotic dinoflagellates. Limnol. Oceanogr. 41: 271-283

Lesser, M.P. and J.M. Shick. 1989. Effects of irradiance and ultraviolet radiation on photoadaptation in the zooxanthellae of *Aiptasia pallida*: primary production, photoinhibition, and enzymic defenses against oxygen toxicity. Mar. Biol. 102: 243-255.

Lesser, M.P., J.J. Cullen and P.J. Neale. 1994. Carbon uptake in a marine diatom during acute exposure to ultraviolet B radiation: relative importance of damage and repair. J. Phycol. 30: 183-192.

Lesser, M.P., T.M. Barry and A.T. Banaszak. 2002. Effects of UV radiation on a chlorophyte algal (*Scenedsmus* sp.) isolated from the fumarole fields of Mt. Erebus, Antarctica. J. Phycol. 38: 473-481.

Levy, H., I. Gokhman and A. Zamir. 1992. Regulation and light-harvesting complex II association of a *Dunaliella* protein homologous to early light-induced proteins in higher plants. J. Biol. Chem. 267: 18831-18836.

Litchman, E., P.J. Neale and A.T. Banaszak. 2002. Increased sensitivity to ultraviolet radiation in nitrogen-limited dinoflagellates: Photoprotection and repair. Limnol Oceanogr. 47: 86-94.

Loeblich, L.A. 1982. Photosynthesis and pigments influenced by light intensity and salinity in the halophile *Dunaliella salina*. J. Mar. Biol. Assoc. UK 62: 493-508.

Longhi, M.L., G. Ferreyra, I. Schloss and S. Roy. 2006. Variable plankton response to enhanced UV-B and nitrate addition in mesocosm experiments at three latitudes (Canada, Brazil and Argentina). Mar. Ecol. Prog. Ser. 313: 57-72.

Malanga, G. and S. Puntarulo. 1995. Oxidative stress and antioxidant content in *Chlorella vulgaris* after exposure to ultraviolet B radiation. Physiol. Plantar. 94: 672-679.

Malanga, G., G. Calmanovici and S. Puntarulo. 1997. Oxidative damage to chloroplasts from *Chlorella vulgaris* exposed to ultraviolet B radiation. Physiol. Plantar. 101: 455-462.

Manabe, E., T. Hirosawa, M. Tsuzuki and S. Miyachi. 1986. Effect of solar near ultraviolet radiation on growth of *Chlorella* cells. Physiol. Plantar. 67: 598-603.

Masi, A. and A. Melis. 1997. Morphological and molecular changes in the unicellular green alga *Dunaliella salina* grown under supplemental UV-B radiation: cell characteristics and photosystem II damage and repair properties. Biochim. Biophys. Acta 1321: 183-193.

Matters, G.L. and S.I. Beale. 1995. Blue-light regulated expression of genes for two early steps of chlorophyll biosynthesis in *Chlamydomona reinhardtii*. Plant Physiol. 109: 471-479.

McLeod, A.R. 1997. Outdoor supplementation systems for studies of the effects of increased UV-B radiation. Plant Ecology 128: 78-92.

McLeod, G.C. and J. Kanwisher. 1962. The quantum efficiency of photosynthesis in ultraviolet light. Physiol. Plantar. 15: 581-586.

Melis, A., J.A. Nemson and M.A. Harrison. 1992. Damage to functional components and partial degradation of photosystem II reaction center proteins upon chloroplast exposure to ultraviolet-B radiation. Biochim. Biophys. Acta 1100: 312-320.

Mewes, H. and M. Richter. 2002. Supplementary ultraviolet-B radiation induces a rapid reversal of the diatinoxanthin cycle in the strong light-exposed diatom *Phaeodactylum tricornutum*. Plant Physiol. 130: 1527-1535.

Middleton, E.M. and A.H. Teramura. 1993. Potential errors in the use of cellulose acetate and Mylar filters in UV-B radiation studies. Photochem. Photobiol. 57: 744-751.

Middleton, E.M. and A.H. Teramura. 1994. Understanding photosynthesis, pigment and growth responses induced by UV-B and UV-A irradiances. Photochem. Photobiol. 60: 38-45.

Mil'ko, E.S. 1963. Effect of various environmental factors on pigment production in the alga *Dunaliella salina*. Mikrobiologiya 32: 299-307.

Miyake, D. and K. Asada 2003. The water-water cycle in algae. pp. 183-204. In Photosynthesis in Algae. A. Larkum, S. Douglas and J. Raven (eds) Kluwer Acad. Publ. NY, USA.

Montero, O.,M. Klisch, D.-P. Häder and L.M. Lubian. 2002. Comparative sensitivity of seven marine microalgae to cumulative exposure to ultraviolet-B radiation with daily increasing doses. Bot. Marina 45: 305-315.

Naus, J. and A. Melis. 1991. Changes of photosystem stoichiometry during cell growth in *Dunaliella salina* cultures. Pl. Cell Physiol. 32: 569-575.

Neale, P.J. 2000. Spectral weighting functions for quantifying effects of UV radiation in marine ecosystems. pp. 72-100. In The Effects of UV Radiation in the Marine Environment. S. de Mora, S. Demers and M. Vernet (eds) Cambridge Univ. Press, New York, USA.

Neale, P.J., M.P. Lesser, J.J. Cullen and J. Goldstone. 1992. Detecting UV-induced inhibition of photosynthesis in Antarctic phytoplankton. Antarct. J. U.S. 27: 122-124.

Neale, P.J., J.J. Cullen, M.P. Lesser and A. Melis. 1993. Physiological bases for detecting and predicting photoinhibition of aquatic photosynthesis by PAR and UV radiation. pp. 61-77. In Photosynthetic Responses to the Environment. H. Yamamoto and C. Smith (eds) Amer. Soc. Plant Physiol., Rockville, MD, USA.

Neale, P.J., A.T. Banasak and C.R. Jarriel. 1998. Ultraviolet sunscreens in *Gymnodinium sanguineum* (Dinophyceae): Mycosporine-like amino acids protect against inhibition of photosynthesis. J. Phycol. 34: 928-938.

Niyogi, K.K. 1999. Photoprotection revisited: Genetic and molecular approaches. Ann. Rev. Pl. Physiol. Pl. Molec. Biol. 50: 333-359.

Nogues, S. and N.R. Baker. 1995. Evaluation of the role of damage to photosystem II in the inhibition of CO2 assimilation in pea leaves on exposure to UV-B radiation. Plant Cell Envir. 18: 781-787.

Newsham, K.K., A.R. McLeod, P.D. Greenslade and B.A. Emmett. 1996. Appropriate controls in outdoor UV-B supplementation experiments. Global Change Biol. 2: 319-324.

Orset, S.C. and A.J. Young. 2000. Exposure to low irradiances favors the synthesis of 9-cis β,β-carotene in *Dunaliella salina* (Teod.). Plant Physiol. 122: 609-617.

Ort, D. 2001. When there is too much light. Plant Physiol. 125: 29-32.

Peak, M.J., J.G. Peak and M. MacCoss. 1984. DNA breakage caused by 334nm ultraviolet light is enhanced by naturally occurring nucleic acid components and nucleotide coenzymes. Photochem. Photobiol. 39: 713-716.

Peak, M.J. and J.G. Peak. 1986. Molecular photobiology of UVA. pp. 42-52. In Monograph on UVA. F. Urbach and W. Gange (eds) Plenum Press, New York, USA.

Pfündel, E.E., R.-S. Pan and R.A. Dilley. 1992. Inhibition of violaxanthin deepoxidation by ultraviolet-B radiation in isolated chloroplasts and intact leaves. Plant Physoiol. 98: 1372-1380.

Pulich, W.M. 1974. Resistance to high oxygen tension, streptonigrin, and ultraviolet irradiation in the green alga *Chlorella sorokiana* strain ORS J. Cell Biol. 62: 904 907

Rabinowitch, E.I. 1951. Photosynthesis and Related Processes. Vol. 2, part 1. Interscience Publ. New York, USA.

Renger, G., M. Volker, H.J. Eckert, R. Fromme, S. Hohm-Veit and P. Graber. 1989. On the mechanism of photosystem II deterioration by UV-B radiation, Photochem. Photobiol. 49: 97-105.

Rijstenbil, J.W. 2001. Effects of periodic, low UVA radiation on cell characteristics and oxidative stress in the marine planktonic diatom *Ditylum brightwellii.* Eur. J. Phycol. 36: 1-8.

Roy, S. 2000. Strategies for the minimisation of UV-induced damage. pp. 177-205. In The Effects of UV Radiation in the Marine Environment. S. de Mora, S. Demers and M.Vernet (eds) Cambridge Univ. Press, New York, USA.

Ruyters, G., T. Hirosawa and S. Miyachi. 1984. Blue light effects on carbon metabolism in *Dunaliella*. pp. 317-322. In Blue Light Effects in Biological Systems. H. Senger (ed) Springer-Verlag, Berlin, Germany.

Ruyters, G. 1988. Light-stimulated respiration in the green alga *Dunaliella tertiolecta*: involvement of the ultraviolet/blue-light photoreceptors and phytochrome? Planta 174: 422-425.

Salguero, A., R. León, A. Mariotti, B. de la Morena, J. M. Vega and C. Vílchez. 2005. UV-A mediated induction of carotenoid accumulation in *D. bardawil* with retention of cell viability. Appl. Microbiol. Biotechnol. 66: 506-511.

Sampath-Wiley, P.S. and L.S. Jahnke. 2006. Imitating solar UV-B radiation in the laboratory. Problems and solutions. Plant Physiol. Suppl. 164.

Santas, R., A. Koussoulaki and D.-P. Häder. 1997. In assessing biological UV-B effects, natural fluctuations of solar radiation should be taken into account. Plant Ecol. 128: 93-97.

Shaish, A., M. Avron, U. Pick and A. Ben-Amotz. 1993. Are active oxygen species involved in induction of β-carotene in *Dunaliella bardawil*? Planta 190: 363-368.

Shelly, K., P. Heraud and J. Beardall. 2002. Nitrogen limitation in *Dunaliella tertiolecta* (Chlorophyceae) leads to increased susceptibility to damage by UV-B radiation but also increased repair capacity. J. Phycol 38: 713-720.

Shelly, K., P. Heraud and J. Beardall. 2003. Interactive effects of PAR and UV-B radiation on PSII electron transport in the marine alga *Dunaliella tertiolecta* (Chlorophyceae). J. Phycol 39: 509-512.

Shelly, K., S. Roberts, P. Heraud and J. Beardall. 2005. Interactions between UV-B exposure and phosphorus nutrition. I. Effects on growth, phosphate uptake, and chlorophyll fluorescence. J. Phycol. 41: 1204-1211.

Sicora, C., Z. Máté and I. Vass. 2003. The interaction of visible and UV-B light during photodamage and repair of photosystem II. Photosynth. Res. 75: 127-137.

Singh, K.K., R. Shyam and P.V. Sane. 1996. Reactivation of photosynthesis in the photoinhibited green alga *Chlamydomonas reinhardtii:* role of dark respiration and of light. Photosynth. Res. 49: 11-20.

Sobrino, C., P.J. Neale and L.M. Lubián. 2005. Interaction of UV radiation and inorganic carbon supply in the inhibition of photosynthesis: Spectral and temporal responses of two marine picoplankters. Photochem. Photobiol. 81: 384-393.

Sobrino, C., P.J. Neale, O. Montero and L.M. Lubián. 2005. Bioloical weighting function for xanthophyll de-epoxidation induced by ultraviolet radiation. Physiol. Plantar. 125: 41-51.

Stahl, W. and H. Sies.1994. Separation of geometrical isomers of β-carotene and lycopene, pp. 388-400. In Methods in Enzymology. Vol. 234. L. Packer (ed) Acad. Press, NY USA.

Staxén, I., C. Bergounioux and J.F. Bornman. 1993. Effect of ultraviolet radiation on cell division and microtubule organisation in *Petunia hybrida* protoplasts. Protoplasma 173: 70-76.

Strid, A., W.S. Chow and J.M. Anderson. 1990. Effects of supplementary ultraviolet-B radiation on photosynthesis in *Pisum sativum*. Biochim. Biophys. Acta 1020: 260-268.

Sukenik, A.J. Bennett and P.G. Falkowski. 1987. Light saturated photosynthesis limitation by electron transport or carbon fixation? Biochim. Biophys. Acta 891: 205-215.

Taylor, R.M., A.K. Tobin and C.M. Bray. 1997. DNA damage and repair in plants. pp. 53-76. In Plants and UV-B. P. Lumsden (ed) Cambridge Univ. Press, New York, USA.

Tedetti, M. and R. Sempéré. 2005. Penetration of ultraviolet radiation in the marine environment. A review. Photochem. Photobiol. 82: 389-397.

Telfer, A., S.M. Bishop, D. Philips and J. Barber. 1994. Isolated photosynthetic reaction centre of photosystem II as a sensitizer for the formation of singlet oxygen: detection and quantum yield determination using a chemical trapping technique. J. Biol. Chem. 269: 13244-13253.

Teramoto, H., A. Ishii, Y. Kimura, K. Hasegawa, S. Nakazawa, T. Nakamura, S. Higashi, M. Watanabe and T. Ono. 2006. Action spectrum for expression of the high intensity light-inducible Lhc-like gene Lh14 in the green algal *Chlamydomonas reinhardtii*. Pl. Cell Physiol. 47: 419-425.

Turcsányi, E. and I. Vass. 2000. Inhibition of photosynthetic electron transport for UV-A radiation targets the photosystem II complex. Photochem. Photobiol. 72: 513-520.

Tyrrell, R. 1991. UVA (320-380nm) radiation as an oxidative stress. pp. 57-83. In Oxidative Stress: Oxidants and Antioxidants. H. Sies (ed) Acad. Press, NY, USA.

van de Poll, W.H., M.A. van Leeuwe, J. Roggeveld and A.G.J. Buma. 2005. Nutrient limitation and high irradiance acclimation reduce PAR and UV-induced viability loss in the Antarctic diatom *Chaetoceros brevis* (Bacillariophyceae). J. Phycol. 41: 840-850.

Vass, I. 1997. Adverse effects of UV-B light on the structure and function of the photosynthetic apparatus. pp. 931-950. In Handbook of Photosynthesis. M. Pessarakli (ed) Marcel Dekker, New York, USA.

Vass, I., E. Turcsányi, E. Touloupakis, D. Ghanotakis and V. Petrouleas. 2002. The mechanism of UV-A radiation-induced inhibition of photosystem II electron transport studied by EPR and chlorophyll fluorescene, Biochem. 41: 10200-10208.

Vernet, M. 2000. Effects of UV radiation on the physiology and ecology of marine phytoplankton. pp. 237-278. In The Effects of UV Radiation in the Marine Environment S. de Mora, S.Demers and M. Vernet (eds) Cambridge Univ. Press, New York, USA.

Vincent, W.F. and P.J. Neale. 2000. Mechanism of UV damage to aquatic organisms. pp. 149-176. In The Effects of UV Radiation in the Marine Environment. S. De Mora, S. Demers and M. Vernet (eds) Cambridge Univ. Press, New York, USA.

Vu, C.V., L.H. Allen Jr. and L.A. Garrard. 1984. Effects on enhanced UV B radiation (280 320 nm) on ribulose 1,5 bisphosphate carboxylase in pea and soybean. Environ. Exper. Bot. 24: 131-143.

Waring, J., G.J.C. Underwood and N.R. Baker. 2006. Impact of elevated UV-B radiation on photosynthetic electron transport, primary productivity and carbon allocation in estuarine epipelic diatoms. Plant Cell Envir. 29: 521-534.

Wayne, R., A. Kadota, M. Watanabe and M. Furuya. 1991. Photomovement in *Dunaliella salina*: fluence rate-response curves and action spectra. Planta 184: 515-524.

Webb, R.B. and J.R. Lorenz. 1970. Oxygen dependence and repair of lethal effects of near ultraviolet and visible light. Photochem. Photobiol. 12: 283-289.

Webb, R.B. and M.S. Brown. 1979. Action spectra for oxygen dependent and independent inaction of *Escherichia coli* WP2s from 254 to 460nm. Photochem. Photobiol. 29: 407-409.

West, L.J.A., B.M. Greenberg and R.E.H. Smith. 1999. Ultraviolet radiation effects on a microscopic green alga and the protective effects of natural dissolved organic matter. Photochem. Photobiol. 69: 536-544.

White, A.L. and L.S. Jahnke. 2002. Contrasting effects of UV-A and UV-B on photosynthesis and photoprotection of β-carotene in two *Dunaliella* spp. Plant Cell Physiol. 43: 877-884.

White, A.L. and L.S. Jahnke. 2004. Removing UV-A and UV-C radiation from UV-B fluorescent lamp emissions. Differences in the inhibition of photosynthesis in the marine alga *Dunaliella tertiolecta* using chromate versus cellulose acetate-polyester filters. Photochem. Photobiol. 80: 340-345.

Wolniakowski, K.U. 1980. The physiological response of a marine phytoplankton species *Dunaliella tertiolecta* to mid-wavelength ultraviolet radiation. M.S. Thesis, Oregon State University, Corvalis, Oregon. USA.

Xiong, F. 2001. Evidence that UV-B tolerance of the photosynthetic apparatus in microalgae is related to the D1-turnover mediated repair cycle in vivo. J. Plant Physiol. 158, 285-294.

Xiong, F., F. Lederer, J. Lukavsky and L. Nedbal. 1996. Screening of freshwater algae (Chlorophyta, Chromophyta) for ultraviolet-B sensitivity of the photosynthetic apparatus. J. Plant Physiol. 148: 42-48.

Xiong, F., L. Nedbal and A. Neori. 1999. Assessment of UV-B sensitivity of photosynthetic apparatus among microalgae: short-term laboratory screening versus long-term outdoor exposure. J. Plant Physiol. 155: 54-62.

Yi, Y., Y. Cao, L. Qian, L. Min, C. Long, B. Linhan, Y. Zhirong and Q. Dairong. 2006. Clonign and sequence analysis of the gene encoding (6-4)photolyase from Dunaliella salina. Biotech. Lett. 28: 309-314.

Young, A.J. 1991. The photoprotective role of carotenoids in higher plants. Physiol. Plantar. 83: 702-708.

Young, E.B. and J. Beardall. 2003. Photosynthetic function in *Dunaliella tertiolecta* (Chlorophyceae) during a nitrogen starvation and recovery cycle. J. Phycol. 39: 897-905.

Zaremba, T.G., T.R. LeBon, D.B. Millar, R.M. Smejkal and R.J. Hawley. 1984. Effects of ultraviolet light on the in vitro assembly of microtubules. Biochem. 23: 1073-1080.

Zetterberg, G. 1964 Mutagenic effects of ultraviolet and visible light. pp. 247-281. In Photophysiology, Vol. 3. A.C. Giese (ed) Academic Press, New York.USA.

11 Chloroplast Acclimation, Photodamage and Repair Reactions of Photosystem-II in the Model Green Alga *Dunaliella salina*

Kittisak Yokthongwattana[1], EonSeon Jin[2] and Anastasios Melis[3]*

Abstract

Photosynthesis is a light-driven process that sustains virtually all life on earth. However, light energy in excess of what is required for the saturation of photosynthesis causes photo-oxidative damage to the D1/32 kD (*psbA* gene product) reaction center protein of photosystem II. If not corrected, such photodamage could cause inhibition of photosynthesis and plant growth. Through evolution, organisms of oxygenic photosynthesis devised a unique repair mechanism by which to recover from this frequently occurring photo-oxidative damage. The repair process entails selective degradation and replacement of the photodamaged D1 reaction center protein in the thylakoid membrane of photosynthesis. In this chapter, up-to-date information about the photosystem II (PSII) damage and repair cycle in the model unicellular green alga *Dunaliella salina* Teod. [Eukaryota; Viridiplantae; Chlorophyta; Chlorophyceae; Chlamydomonadales; Dunaliellaceae; *Dunaliella*] is presented. The chapter first examines the physiology of cellular and chloroplast acclimation to irradiance. Examples of environmental and biotic factors that modulate the PSII photodamage are also provided. Lastly, the temporal sequence of events and current knowledge on the molecular mechanism of the PSII repair process are summarized in detail. Emphasis is placed on the role of the ELIP/Cbr protein and of the carotenoid zeaxanthin during D1 turnover and PSII repair. This chapter also examines the role of the chloroplast-localized heat-shock protein 70B (HSP70B) in facilitating the disassembly of photodamaged PSII and insertion of a *de novo* synthesized D1 protein in the PSII reaction center complex.

---

[1] Department of Biochemistry and Center for Excellence in Protein Structure and Function, Faculty of Science, Mahidol University, Rama 6 Rd. Bangkok 10400, Thailand.

[2] Department of Life Science, College of Natural Science, Hanyang University, 133-791 Seoul, South Korea.

[3] Department of Plant and Microbial Biology, 111 Koshland Hall, University of California, Berkeley, CA 94720-3102, USA.

* Author for correspondence, Email: tekyw@mahidol.ac.th

1. Introduction

Oxygenic photosynthetic organisms convert solar irradiance into chemical energy and ultimately store it in the form of carbohydrate, which subsequently serves to drive a number of biosynthetic reactions in the living cell. In the thylakoid membrane of algae and plants, light energy is absorbed by (Chl) *a* and *b* molecules contained within light-harvesting pigment-protein antenna complexes (LHCs), peripherally associated with the reaction centers of photosystem II (PSII, see also Wydrzynski and Satoh 2005) and photosystem I (PSI, see also Golbeck 2006). Photon energy absorbed by PSII helps generate a strong oxidant capable of oxidizing water molecules. Electrons from water, in turn, are passed through a series of electron transport intermediates and are delivered to $NADP^+$ to generate NADPH. The latter along with ATP, generated via photophosphorylation, is utilized in CO_2 assimilation, as well as inorganic nitrogen and sulfur reduction. In general, light absorption, rate of electron transport and carbon metabolism are synchronized to maximize the yield of photosynthesis. Levels of irradiance during plant growth, therefore, affect the structure and organization of the photosynthetic machinery, and the pigment composition in the chloroplast thylakoids (Anderson 1986, Melis 1991). Under limiting illumination, light energy must be captured and utilized with high efficiency. Whenever there is a change in the level of irradiance during growth, plants and algae respond with a highly-conserved acclimation mechanism (Melis 1998). Increasing light intensities will result in greater absorption of photons and enhanced rates of CO_2 assimilation. However, at the light intensity where the rate of CO_2 fixation becomes a limiting factor, photosynthesis is saturated (Stitt 1986). Absorption of sunlight in excess of that required for the saturation of photosynthesis can bring about photo-oxidative damage, a process that inhibits photosynthetic activity, specifically the activity of PSII (Powles 1984). This is a well-known phenomenon that entails irreversible chemical modification by singlet oxygen of the reaction center P680 molecule, thereby rendering the D1 protein inactive (Krause 1988, Aro et al. 1993, Melis 1999, see also Edelman and Mattoo 2006). Throughout evolution, oxygenic photosynthetic organisms have not been able to prevent this PSII photo-oxidative damage from occurring. Nevertheless, to moderate the adverse effect of the excess irradiance, all photosynthetic organisms have evolved photoprotective mechanisms, which range from the molecular to the whole plant level. Additionally, plants have also evolved a unique and specific repair mechanism by which they rectify the apparently unavoidable and irreversible PSII photodamage. Whenever the rate of photodamage exceeds the capacity of photoprotection and of the repair process, then PSII activity declines and photodamaged reaction centers accumulate in thylakoids. This condition is known as 'photoinhibition' of photosynthesis, manifested as an overall decline in the rate and quantum yield of photosynthesis (Kok 1956, Powles 1984, Greer et al. 1986, Long et al. 1994, see also

Adir et al. 2003, Osmond and Förster 2006). In addition to light, other environmental stress factors that lead to an imbalance in the light versus the carbon reactions of photosynthesis can also elicit photoinhibition, even under moderate light intensities. Such conditions include drought, chilling or freezing temperature and heat stress (Powles 1984, Havaux 1992, Król et al. 1997).

Unicellular green algae are commonly used as model organisms to study the mechanism of the PSII damage and repair process and the acclimation responses of the photosynthetic apparatus to different light regimes. In many cases, microalgae are preferred over higher plants, partly because they provide greater flexibility in terms of higher growth rates and partly because their responses to the environment are more pronounced than those of higher plants. This chapter summarizes the state of the art in green algal physiology, chloroplast acclimation to irradiance and the PSII damage and repair cycle in the model unicellular green alga *Dunaliella salina* Teod. [Eukaryota; Viridiplantae; Chlorophyta; Chlorophyceae; Chlamydomonadales; Dunaliellaceae; *Dunaliella.*]

2. Cell Physiology and Chloroplast Acclimation Strategies to Irradiance Stress

Changes in Pigment Composition and Light-harvesting Chl Antenna Size

Light-harvesting Chl antenna complexes (LHC) of photosynthesis are one of the popular subjects of recent and current research. One important aspect of research on the LHC antenna is its modulation in size and composition in response to different light intensities (Melis 1996). Such modulation occurs naturally in higher plants, green algae and photosynthetic bacteria in response to long-term changes in growth irradiance (Anderson 1986, Melis 1991, 1998). Since photosystem core proteins only bind Chl *a* while the LHC antenna contains both Chl *a* and Chl *b*, changes in Chl *a*/Chl *b* ratio can provide a first indication on the size of the Chl antenna associated with the photosystems. Modulation in size of the light-harvesting Chl antenna of both PSII and PSI has been extensively investigated in *D. salina* (Guenther et al. 1988, Smith et al. 1990, Webb and Melis 1995, Masuda et al. 2002). When *D. salina* cultures are grown under low light intensity (LL), cells possess a Chl *a*/*b* ratio of about 3 and the light-harvesting antenna of PSII (LHC-II) contains 230-300 Chl (*a* + *b*) molecules per PSII complex (Table 1). High-light (HL) grown *D. salina*, on the other hand, shows a Chl *a*/*b* ratio greater than 10 and the PSII antenna contains 40-60 Chl (*a* plus *b*) molecules per PSII complex (Smith et al. 1990). Chl antenna size of PSI follows the same trend with a size of about 220 and 110 molecules per PSI in LL and HL, respectively. In addition to the overall size of the Chl antenna, acclimation of *D. salina* to the level of irradiance also involves distinct changes in the composition of protein associated with the LHC (Webb and Melis 1995, Tanaka and Melis 1997). These results show that low levels of irradiance promote larger Chl antenna size for both PSII and PSI, whereas HL promotes smaller ones. Transition of *D. salina* from LL to HL resulted in decreasing Chl antenna size and increasing Chl *a*/*b* ratio (Kim et al. 1993, Webb and Melis 1995), while a HL→LL shift induced changes in the opposite direction (Neidhardt et al. 1998). The molecular mechanism that regulates this response of photosynthetic organisms to irradiance, e.g. compensatory adjustments of the Chl antenna size, is not yet known. However, downsizing of the Chl antenna does not appear to involve an active degradation of pre-existing LHCs. It was inferred, therefore, that the truncation of the Chl antenna in response to a HL condition may result from cell division and 'dilution' of the existing Chl-protein antenna complexes in the thylakoid membrane (Melis 1998).

Table 1: Parameters of photosynthetic apparatus organization in *D. salina* grown under LL vs. HL conditions. Cells were grown under either 100 ìmol photons m^{-2} s^{-1} (LL) or under 2,000 ìmol photons m^{-2} s^{-1} (HL). N is the number of Chl (*a* + *b*) molecules associated with the specified photosystem.

Parameters	LL-grown	HL-grown
Chl *a*/*b* ratio (mol: mol)	3-5	10-18
PSII per cell (10^{-18} mol $cell^{-1}$)	2.45	0.77
PSI per cell (10^{-18} mol $cell^{-1}$)	1.69	0.27
PSII/PSI (mol: mol)	1.4	2.8-3.2
N_{PSII} (molecules)	230-300	30-60
N_{PSI} (molecules)	220	110

Changes in the Stoichiometry between PSII and PSI Complexes

Efficiency of energy conversion in photosynthesis depends on the coordinated rate of linear electron transport through PSII and PSI in the thylakoid membrane. Due to differences in pigment composition, antenna configuration and protein structure, light absorption by PSII and PSI occurs in predominantly different regions of the solar spectrum. Imbalance in the rate of light absorption and utilization between PSII and PSI would lead to inefficient utilization of sunlight and suboptimal rates of linear electron flow. Such imbalance in light absorption/utilization by the photosystems is rectified upon adjustment and optimization in the stoichiometry between the two photosystems, which invariably deviates from the depicted 1:1 ratio. Stoichiometry of PSII/PSI in the thylakoid membrane of *D. salina* can range from 1.4/1 under LL growth condition to more than 3/1 under HL (Table 1) (Guenther et al. 1988, Guenther and Melis 1990, Smith et al. 1990, Vasilikiotis and Melis 1994, Baroli and Melis 1998). It has been shown that, under HL-conditions, as the majority of the PSII population is becoming photoinhibited, *D. salina* acclimates to the HL growth irradiance by lowering significantly the amount of assembled PSI. This apparent paradox can be explained upon consideration of the rate-limiting step in photosynthetic electron transport. It is widely accepted that this limitation occurs at the point of electron turnover through the cytochrome b_6f complex (Sukenik et al. 1987a, 1987b, Geider and Osborne 1992). Therefore, under photoinhibitory conditions, where PSII is frequently photodamaged, fewer PSI complexes are required to carry out electron transport from the Cyt b_6f complex to ferredoxin (Vasilikiotis and Melis 1994). Stoichiometry between PSII and PSI changes not only in response to different regimes of irradiance (light-intensity and light-quality), but also changes in response to mutations that differentially affect the Chl antenna size of PSII and PSI (e.g. Chl *b*-less mutations), and also during the various phases of cell growth (Naus and Melis 1991).

Upregulation of Gene Expression in Response to Irradiance Stress

Recently, a number of publications have reported on the HL-acclimation of microalgae at the molecular level using global gene expression analyses (Im et al. 2003, Ledford et al. 2004, Park et al. 2006). A large scale single-pass sequencing of cDNAs, generally known as 'expressed sequence tag' (EST), was employed to study the acclimation strategy of *D. salina* to irradiance stress (Park et al. 2006). In this study, 1112 randomly selected ESTs were analyzed. Approximately 1% of the analyzed ESTs were greatly up-regulated in response to irradiance stress. Specifically, after a 48 h of high-light exposure of the cells, an increase in the expression of antioxidant genes, such as Fe-superoxide dismutase and ascorbate peroxidase, was observed, as well as elevated levels of the ELIP/Cbr transcripts. In addition, the ATP-dependent Clp protease gene was also up-regulated in *D. salina* cells under irradiance stress. Consistent with the physiological and biochemical studies was a down regulation of the LHC-II gene expression under HL-conditions. A number of other genes, some with unknown function, showed moderately enhanced expression in HL stressed cells. Of those, about 18.5% are genes involved in general cellular metabolism, while 6.4% appeared to encode for components of the signal transduction pathway. These results suggested that the molecular response of the cells to irradiance stress involves coordinated expression of several genes, ranging from components of the signal transduction pathway to genes encoding structural and functional proteins of the photosynthetic apparatus.

3. Photosystem II Structure and Photo-oxidative Damage

The PSII reaction center from higher plants and green algae is a multicomponent protein complex consisting of more than 25 subunits. The individual protein components of PSII are named according to their encoding genes: PsbA-PsbZ and Lhcb1-Lhcb6, some of which are encoded by the chloroplast genome, while most are nuclear encoded (Hankamer et al. 1997, Swiatek et al. 2001). A simplified organizational schematic of the PSII holocomplex is shown in Figure 1. Among the PSII subunits, several participate in the light absorption and photochemical charge separation reaction. The most important component of the PSII holocomplex is the D1/D2 32/34 kD reaction center heterodimer proteins, which form the center of the PSII-core complex. The D1/D2 bind all the essential cofactors and reaction center molecules needed for the photochemical charge separation and electron transport reactions, including the Mn cluster, tyrosine Y_Z and Y_D, P680, Pheophytin, a non-heme iron, and the quinone electron acceptors Q_A and Q_B. Together with the core CP43 and CP47 chlorophyll-proteins, they form the PSII-core complex, which contains 37 Chl *a* molecules (Glick and Melis 1988) and about 8 β-carotenes.

Molecular mechanisms and the pathway leading to PSII photoinhibitory damage are fairly well understood (Melis 1999). Two possible pathways leading to photoinhibitory damage have been described in the literature: an acceptor-side-induced and a donor-side-induced process (Aro et al. 1993, Andersson and Aro 2002). The two mechanisms can be distinguished on the basis of differences in the site of electron transport malfunction. In the case of the acceptor-side photodamage, strong irradiance causes over-reduction of the PQ pool, which results in the Q_B site being devoid of oxidized plastoquinone. This situation leads to a longer lifetime of the reduced primary quinone acceptor (Q_A^-) (Vass et al. 1992), which in turn promotes charge recombination reactions between $P680^+$-$Pheo^-$, thus generating a long-lasting triplet excited state of P680 ($^3P680^*$). Under oxygenic photosynthesis conditions, this $^3P680^*$ can be quenched efficiently by triplet oxygen, resulting in the formation of singlet oxygen ($^1O_2$). $^1O_2$ is a highly reactive oxygen species that causes an apparently irreversible oxidative modification of P680 and/or of PSII reaction center components in the vicinity. Donor-side inhibition of the PSII reaction center, on the other hand, can occur when electron flow from the water oxidizing Mn complex cannot keep up with the rate of photochemical charge separation. This condition generates long-lived and strongly oxidizing radicals $P680^+$ and/or Y_Z^+, which can induce damage to the

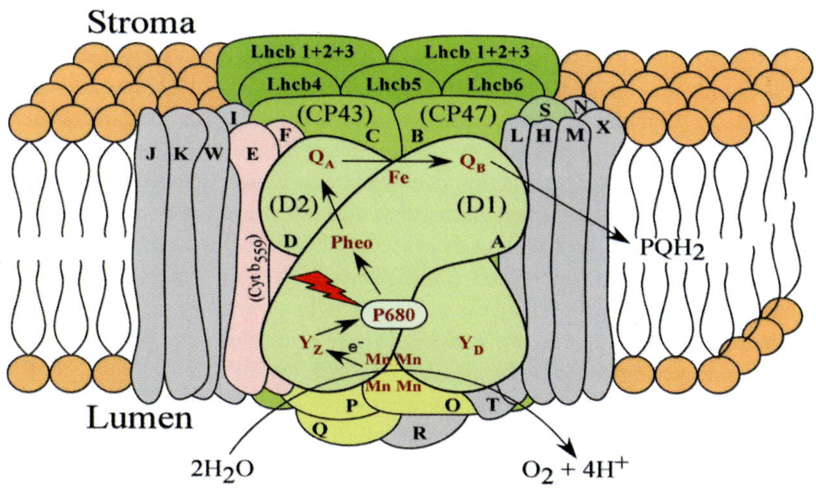

Figure 1: Schematic illustration of the structural arrangement of the various subunits in the PSII reaction center complex. Individual subunits were labeled by a single letter of the alphabet according to the respective gene product PsbA-PsbZ and Lhcb1-Lhcb6. The missing subunits (PsbG, U, V, Y and Z) are proteins that either do not exist in higher plant PSII or their exact localization within the complex are presently unknown. (Adapted from Yokthongwattana and Melis 2006.)

reaction center proteins (Barber and Andersson 1992). Contrary to the acceptor-side photoinhibition, which manifests itself under saturating irradiance, the donor-side-induced mechanism can occur at any light intensity.

4. Modulation of PSII Photodamage by Environmental and Biotic Factors

Linear Relationship between Increasing Light Intensity and PSII Photodamage

Studies have shown that the rate constant of PSII photodamage depends on light intensity. Specifically, the rate constant for photodamage was found to be a linear function of growth irradiance in the physiological range (Figure 2), both in higher plants (Park et al. 1995, Tyystjärvi and Aro 1996) and in the unicellular green alga *Dunaliella salina* (Baroli and Melis 1996). The linear dependence of the rate constant of photodamage on light intensity may suggest that PSII has a tendency to undergo photodamage every time excitation energy is absorbed, regardless of the rate of photosynthesis. According to this model, the rate of light absorption and excitation energy reaching a reaction center increases linearly with light intensity, causing a correspondingly linear increase in the rate constant for photodamage. This interpretation is consistent with evidence showing that the same degree of PSII inactivation can be obtained when plants are exposed to equal dosage of irradiance (reciprocity of short time at higher light intensities or longer time at lower irradiances) (Park et al. 1995, 1996a). These results lead to a 'photon counter' hypothesis, implying that there is a fixed probability of photodamage to PSII occurring after a certain amount of photons (approximately 10^6 to 10^7 photons) have been absorbed, irrespective of the photochemical or electron transport status of the electron transport intermediates (Anderson et al. 1997, Melis 1999). Nevertheless, this hypothesis cannot fully explain the fact that the rate constant of PSII photodamage can be modulated by physiological and metabolic parameters, e.g. carbon dioxide availability, drought and temperature stress.

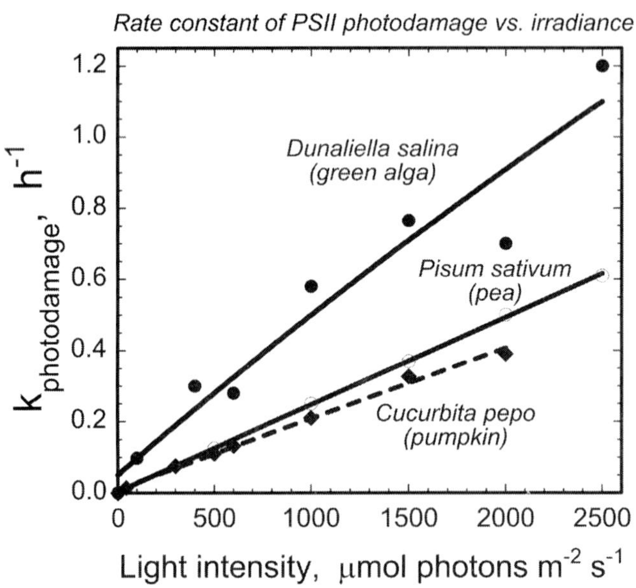

Figure 2: Linear relationship between rate constant of PSII photodamage and as a function of growth irradiance in *Pisum sativum* (pea, Park et al. 1995), *Cucurbita pepo* (pumpkin, Tyystjärvi and Aro 1996), and *Dunaliella salina* (green algae, Baroli and Melis 1996).

Light-harvesting Chl Antenna Size of PSII Modulates the Rate of Photodamage

Since rate constant of PSII photodamage depends strictly on light intensity, it is logical that rate of photon energy absorption by the PSII light-harvesting Chl antenna must modulate the rate of this adverse effect. A number of papers (Cleland and Melis 1987, Park et al. 1997, Melis 1998, 1999) have supported the notion that the rate of photodamage *in vivo* is modulated by the PSII Chl antenna size, including studies in *Dunaliella salina* (Baroli and Melis 1998, Melis et al. 1998). Accordingly, plants with a smaller PSII Chl antenna size are less likely to sustain PSII photodamage. Conversely, plants with a large PSII Chl antenna size have a higher susceptibility to photo-oxidative damage under the same irradiance conditions.

Rate Constant of Photodamage is Modulated by PSII Excitation Pressure

Several lines of evidence also showed that the rate constant of PSII photodamage depends on the redox state of the primary quinone acceptor Q_A of the PSII (Maxwell et al. 1995, Baroli and Melis 1998, Huner et al. 1998). It appears that the likelihood of photodamage to PSII by irradiance is quite different depending on whether Q_A is in the oxidized or reduced state. When Q_A is oxidized and excitation energy is effectively utilized through photochemical electron transport, probability of the PSII being damaged is low. Conversely, there is a significantly greater chance for photodamage if Q_A is in a reduced state and the absorbed excitation is wasted via nonphotochemical dissipation (Baroli and Melis 1998, Huner et al. 1998, Melis 1999). The effect of the two parameters, redox state of Q_A and the flux of photon arrival at the PSII reaction center point to the concept of PSII excitation pressure (Maxwell et al. 1995, Huner et al. 1996, 1998) as a unifying theory by which to explain many biotic conditions that modulate the rate constant of PSII photodamage.

Electron Transport and Photosynthesis Mitigate against Photodamage

One of the bottlenecks for photosynthetic productivity, other than the light-driven electron transport from water to $NADP^+$, is the CO_2 assimilation process in chloroplasts (Stitt 1986). Limitations in the CO_2 assimilation reactions are known to enhance the rate of PSII photo-oxidative damage (Stitt 1986, Long et al. 1994). A slow-down of the CO_2 fixation reactions would result in accumulation of NADPH and reduced ferredoxin, which in turn would restrict the process of linear electron transport. A rigorous study was undertaken with *Dunaliella salina* grown under HL conditions with variation in supply of inorganic carbon source (Baroli and Melis 1998). Limited supply of inorganic carbon, provided by an initial concentration of 25 mM $NaHCO_3$ to the medium, confined the rate of photosynthesis (P_{max} = ~100 pmol O_2 (10^6 cells)$^{-1}$ s^{-1}). Supplementation of inorganic carbon by bubbling of 3% CO_2 in air enhanced the rate of photosynthesis (P_{max} = ~250 pmol O_2 (10^6 cells)$^{-1}$ s^{-1}). *D. salina* cultures with $NaHCO_3$ only as the inorganic carbon source had a rate of PSII photodamage with a half time $t_{1/2}$ of ~40 min while that of CO_2 bubbling cultures is ~70 min. Rate constants of PSII photodamage under these two conditions differed by a factor of ~2, confirming the notion that enhanced rates of electron transport and photosynthesis mitigate against the PSII photo-oxidative damage (Park et al. 1996b, Baroli and Melis 1998).

5. PSII Repair Process in *Dunaliella salina*

Throughout evolution, no organism of the oxygenic photosynthesis was able to avoid the photo-oxidative damage to the PSII reaction center. In order for the damaged PSII to return to the active state, it needs to undergo a repair process. Whenever the repair process cannot keep pace with the rate of PSII photodamage, productivity of photosynthesis declines causing a slow-down in plant growth. Without an effective repair mechanism, it has been estimated that plant productivity would be less than 5% of today's level (Melis 1999). Thus, it is of interest to elucidate the fundamental mechanism of the PSII repair from photo-oxidative damage. Studies in *Dunaliella salina* have contributed considerable information on the mechanism of the PSII damage and repair cycle. At present, knowledge is far from complete concerning the identity and regulation of genes, enzymes or proteins involved in the PSII repair. Nevertheless, for the most part, the PSII repair mechanism appears to be highly conserved and is similar in higher plants and green algae. In general, photoinactivated PSII reaction centers in the grana thylakoids undergo partial disassembly, with the PSII-core complex becoming exposed to the stroma (partial grana unstacking), where it can be accessed by components of the repair apparatus, including proteases, molecular chaperons and chloroplast ribosomes. The photodamaged D1 protein undergoes selective removal by specific proteases and a *de novo* synthesized D1 is inserted into the PSII core complex (Aro et al. 1993, Melis 1999, Yokthongwattana and Melis 2006).

Figure 3 illustrates the temporal sequence of events in the PSII damage and repair cycle in *D. salina*. The cycle is initiated upon the photo-oxidative damage to the PSII D1 reaction center protein. The affected PSII complex undergoes partial disassembly into the peripheral Chl *a-b* light-harvesting complex and the inactive PSII-core complex (PSII Repair Intermediate in Figure 3). Specific proteases, Deg (Haußühl et al. 2001, Kapri-Pardes et al. 2007) and FtsH (Lindahl et al. 2000,

Adam et al. 2006), degrade the photodamaged and inactive D1 protein. It is generally accepted that D1 degradation is the rate-limiting step of the PSII repair process (Sundby et al. 1993, Vasilikiotis and Melis 1994). In *D. salina*, the D1 degradation step occurs with a half time of ~2 h (Kim et al. 1993, Neidhardt et al. 1998). Following degradation and *de novo* synthesis and assembly of the nascent D1, the PSII holocomplex is reassembled, reactivated and firmly incorporated back into the grana partition regions. Thus PSII returns to a functional status again.

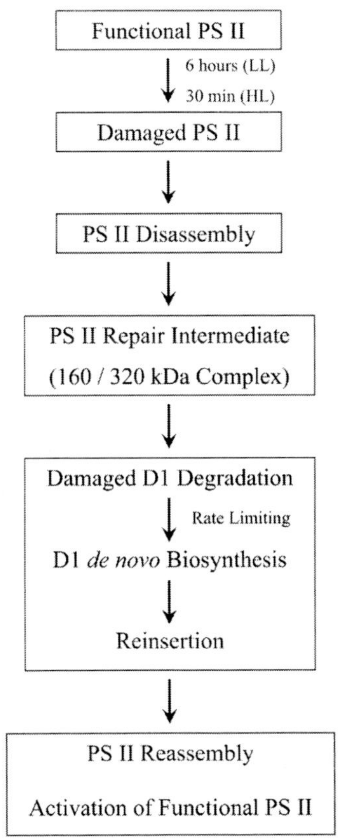

Figure 3: Temporal sequence of events in the PSII damage and repair cycle in *Dunaliella salina* showing of PSII damage, disassembly, D1 degradation and replacement. The rate of photodamage to PSII is directly proportional to the incident light intensity, resulting in variable rates of photodamage, and occurring with halftimes, from 6 hours to 30 minutes. The rate of PSII disassembly is not limiting under a broad range of incident intensities. Direct D1 degradation and *de novo* D1 biosynthesis are required for the repair of PSII from this photo-oxidative damage.

6. Disassembly and Stabilization of Photodamaged PSII in Higher Plants and Green Algae

While most of the steps in the PSII repair process are highly conserved and common among all oxygenic photosynthetic organisms, there is evidence of minor but important differences among higher plants and green algae. As a prerequisite for the PSII repair, the PSII-core complex of the photodamaged PSII must separate from the associated LHC-II, become partially disassembled and exposed to the stromal phase, where specific proteases perform D1 degradation (Melis 1991, Aro et al. 1993, Melis 1998, 1999, Adam and Clarke 2002). Plants have evolved a regulatory mechanism by which they appear to coordinate and control the disassembly process of the photodamaged PSII. In higher plants, several major protein subunits of PSII including D1, D2, PsbH, CP43 and LHC-II are reversibly phosphorylated upon illumination with strong light (Bennett 1991, Michel et al. 1988). This reversible phosphorylation of the PSII subunits is thought to play an important role in the regulation of the D1 protein turnover (Aro et al. 1992, Elich et al. 1992, Koivuniemi et al. 1995, Kruse et al. 1997, Yokthongwattana and Melis 2006). Phosphorylation of the D1 protein does not protect PSII from photodamage but rather prevents a premature degradation of the photodamaged D1 (Koivuniemi et al. 1995, Kruse et al. 1997). As a result, dephosphorylation of the photodamaged D1 is a prerequisite for the repair process to continue (Rintamäki et al. 1996). Thus, reversible phosphorylation of the photodamaged D1 in higher plants is thought to function as a regulatory mechanism to prevent a premature degradation of this protein D1 before it reaches the stroma-exposed region that is accessible by the chloroplast protein translation machinery (Baena-González et al. 1999).

Reversible phosphorylation of the D1 protein in cyanobacteria, red and many green algae (especially in *Dunaliella* spp.), however, has never been observed (Pursiheimo et al. 1998). This raises the prospect of alternative mechanism(s) employed by these organisms to stabilize and regulate the coordinated degradation and replacement of the photodamaged D1, which take the place of the reversible D1 phosphorylation in higher plants. Information on the mechanism employed by green algae in the regulation of the PSII repair process was obtained in studies with the green alga *Dunaliella salina*. In *D. salina*, photodamaged PSII reaction centers disassemble and the corresponding PSII-core complex forms a high-molecular-weight 'repair intermediate' complex of about 160 kD, as evidenced in SDS-PAGE analysis (Kim et al. 1993). Kinetic studies of the 160 kD appearance and decline matched the photodamage, turnover and replacement of D1 protein (Kim et al. 1993, Baroli and Melis 1996). The 160 kD complex PSII repair intermediate was initially found to consist of D1, D2 and of other repair-related proteins of unknown origin (Melis and Nemson 1995). Cross-linked products between D1 and proximal PSII proteins are commonly detected on urea-SDS-

PAGE of photoinhibited thylakoid membranes (Barbato et al. 1992, Ishikawa et al. 1999, Yamamoto 2001). It is likely that cross-linking of these proteins, occurring specifically in photoinhibited thylakoid samples, is a useful artifact, taking place during denaturing solubilization of the photoinhibited thylakoid membranes.

Further analysis by native gel electrophoresis demonstrated that native form of the 160 kD protein repair intermediate appears as a high-molecular-weight 320-kD complex, containing CP47 and the chloroplast-localized heat-shock protein 70B (HSP70B) (Yokthongwattana et al. 2001). This 320 kD complex appeared nearly colorless on native gels prior to staining with Coomassie, supporting the notion that it does not contain peripheral Chl *a/b* light-harvesting antenna proteins. The observation that the molecular chaperone HSP70B associates with the PSII repair intermediate in *D. salina* is consistent with earlier findings that overexpression of the *hsp70B* gene in another green alga, *Chlamydomonas reinhardtii*, allowed the organism to recover from photoinhibition significantly faster than the wild type counterpart (Schroda et al. 1999). Similarly, mutants transformed with an *hsp70B* antisense construct recovered more slowly than the wild type strain. These lines of evidence lead to the hypothesis that HSP70B may function to stabilize the photodamaged and disassembled PSII reaction center prior to and during the D1 degradation and replacement (Schroda et al. 1999, Yokthongwattana et al. 2001, Schroda 2004, Yokthongwattana and Melis 2006). Since many green algae lack the D1 phosphorylation that ostensibly regulates the D1 degradation in higher plants (Aro et al. 1992, Koivuniemi et al. 1995, Baena-González et al. 1999, Baena-González and Aro 2002), such function could possibly be assigned to the chloroplast-localized molecular chaperone HSP70B. According to this hypothesis, photodamage to the D1 protein could lead to a conformational change that triggers a specific interaction with the plastidic HSP70B. Binding of the HSP70B to the photodamaged D1 protein might facilitate the partial disassembly of the holocomplex and, in doing so, form the PSII repair intermediate, consisting of D1, D2, CP47, HSP70B and possibly CP43 (Figure 4, PSII Repair Intermediate) (Yokthongwattana et al. 2001). The repair intermediate must be exposed to the stroma phase where degradation of the inactive D1 takes place. HSP70B may not only stabilize the irreversibly damaged PSII reaction center complex but further facilitate D1 degradation and insertion of the *de novo* synthesized nascent D1 protein (Schroda et al. 2001, Schroda 2004).

Whereas HSP70B plays a role in the stabilization of the disassembled PSII on the stroma-facing side of the complex, it appears that other proteins might be doing the same on the luminal side of the complex. Psb27, a cyanobacterial lipoprotein, is important for reintegration/reassembly of the water-oxidizing components during the PSII repair in *Thermosynechococcus elongatus* (Nowaczyk et al. 2006). Consistent with this interpretation, a *psb27* loss-of-function mutant of *Arabidopsis thaliana* recovers from photoinhibition more slowly than the wild type counterpart (Chen et al. 2006).

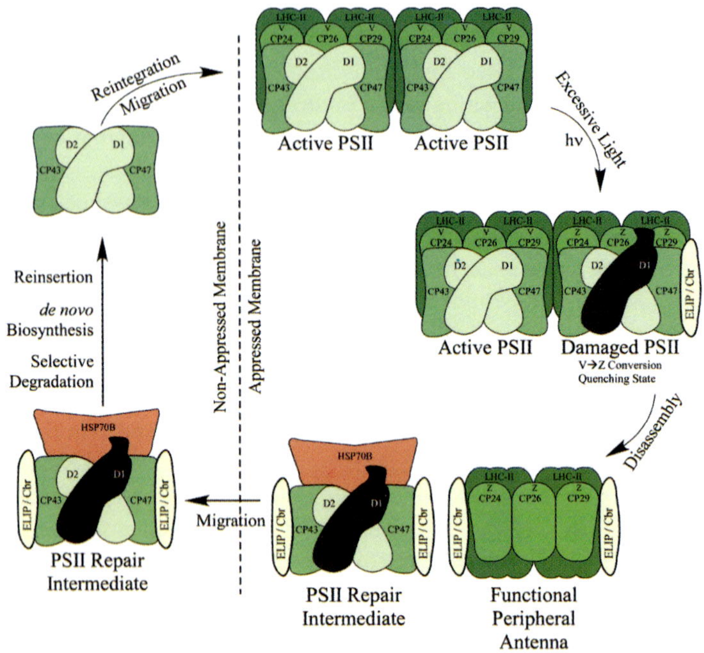

Figure 4: A 'working hypothesis' schematic of the PSII damage and repair cycle in *Dunaliella salina*. In high light, PSII centers undergo irreversible photo-oxidative damaged. Under such conditions, synthesis of the Cbr protein is accelerated and violaxanthin, associated with the light-harvesting antenna, is de-epoxidized to zeaxanthin. Cbr and zeaxanthin accumulate in parallel with photodamaged PSII and could function to protect the remaining unaffected Chl-proteins (antenna proteins, CP43, CP47 and D2) from massive photo-bleaching by quenching the incoming excitation energy. At this stage, the photodamaged PSII could become an efficient excitation energy trap that may also protect nearby active PSII centers occurring in the same pigment bed. Prior to degradation and replacement of photodamaged D1, the PSII holocomplex is partially disassembled. A specific interaction with HSP70B might facilitate this disassembly, forming a distinct PSII repair intermediate.

7. Role of the ELIP/Cbr Proteins and of Zeaxanthin in the PSII Repair Process

Xanthophylls in plants and algae are found in association with the Chl *a-b* light-harvesting proteins. Under physiological (non-stress) growth conditions, lutein and violaxanthin are the most abundant xanthophylls. Under stress conditions, such as excess irradiance or low temperature, however, excitation pressure at the PSII reaction center is thought to trigger the conversion (a de-epoxidation reaction) of violaxanthin to zeaxanthin. Zeaxanthin remains in association with the thylakoid membrane and persists until removal of the condition(s) that caused the plant/algal stress. The relaxation reaction entails epoxidation of zeaxanthin and its conversion back to violaxanthin (Gilmore et al. 1994). Conversion of violaxanthin to zeaxanthin, and vice versa, is referred to as the 'reversible xanthophyll cycle', the function of which is to afford 'photoprotection' to the photosynthetic apparatus (Gilmore et al. 1998, Niyogi 1999, 2000, Demmig-Adams 2003, Demmig-Adams and Adams 2006). Evidence to suggest involvement of zeaxanthin and of Cbr protein, an algal homolog of early-light-inducible protein (ELIP) (Lers et al. 1991, Levy et al. 1992, 1993, Banet et al. 2000), in the PSII repair process came from the study of a photoinhibition-sensitive mutant of *Dunaliella salina* (*dcd1*) (Jin et al. 2001). Under the same irradiance conditions, the *dcd1* mutant displayed an enhanced amplitude of photoinhibition compared to the wild type, manifested as a lowering of the Chl fluorescence Fv/Fm ratio, decline in the amplitude of Q_A photoreduction, and accumulation of 160/320 kD PSII repair intermediates in the thylakoid membranes. In addition to these accepted markers of photoinhibition, accumulation of zeaxanthin and enhanced levels of the Cbr protein were also observed. Although the onset of these changes occurred at different levels of irradiance for the wild type and for the *dcd1* mutant, there was a strict correlation between the amounts Cbr protein, zeaxanthin, and the fraction of photodamaged PSII complexes that accumulated in the thylakoids.

A role for Cbr and zeaxanthin in the PSII damage and repair cycle in *D. salina* was further inferred from studies of the kinetic relationship in the accumulation and relaxation of photodamaged PSII and Cbr/zeaxanthin in wild type and *zea1*, a mutant that constitutively accumulates zeaxanthin under all growth conditions (Jin et al. 2003a, 2003b). Following a LL→HL transition, both wild type and the *zea1* mutant accumulated the Cbr protein in direct proportion to the fraction of photodamaged PSII, and decayed it in tandem with the recovery of the cells from photoinhibition (Jin et al. 2003b). In wild type, zeaxanthin was also found to accumulate in parallel with the accumulation of photodamaged PSII centers and decayed in tandem with the recovery of the alga from photoinhibition (Jin et al. 2003b). However, in the *zea1* mutant, levels of zeaxanthin remained constant and independent of the photoinhibition and recovery in this strain.

EST analysis of *Dunaliella salina* cDNA indicated that Cbr gene transcripts are enhanced under irradiance stress (Park et al. 2006). This observation, together with other reports from work with higher plants (Jahns and Miehe 1996, Demmig-Adams et al. 1998, Depka et al. 1998, Jahns et al. 2000), raised the possibility that Cbr protein accumulation, and even that of zeaxanthin, are a response to photodamage in PSII and might play a role in the protection of the disassembled PSII reaction centers, and of the remaining functional chlorophyll antenna subunits from massive pigment photo-bleaching (see Figure 4). This function may then be essential during the process of D1 degradation and replacement (Jin et al. 2001, 2003b). Such hypothesis is consistent with the report that the kinetics of zeaxanthin epoxidation to violaxanthin resembled those of D1 degradation and replacement during recovery of pea chloroplasts from photoinhibition (Jahns and Miehe 1996). The notion was also supported by a study of an obligate shade species in which the de-epoxidation state of the xanthophyll-cycle carotenoids remained directly proportional to the level of photoinhibition in the leaves and independent of the light-intensity seen by the plant (Demmig-Adams et al. 1998). Further evidence in support of this idea came from studies in overwintering plant species, in which there appears to be interplay between photoprotection, zeaxanthin accumulation and status of the PSII damage and repair cycle (Adams et al. 2004). Recent reports on overwintering coniferous forests also revealed that the degree of photoinhibition correlates with the amount of accumulated zeaxanthin and ELIP proteins, the latter being a higher plant homolog of Cbr (Zarter et al. 2006a, 2006b). Moreover, a study in *Arabidopsis thaliana* also showed that ELIP genes are up-regulated upon light stress and correlate with the degree of PSII photodamage in the leaf (Heddad et al. 2006). All of these findings point in the direction of a photoprotective role for ELIP/Cbr and of zeaxanthin in the PSII repair process.

8. Other Protein Components of the PSII Repair Process

The PSII repair process is a multi-step mechanism entailing function of several enzymatic reactions and a variety of protein components, most of which remain to be discovered. A recent mutagenesis study resulted in the isolation of a PSII repair mutant in *Chlamydomonas reinhardtii*, leading to identification of REP27, a nuclear-encoded chloroplast-localized protein containing two tetratricopeptide repeat (TPR) motifs, apparently essential for the PSII repair process (Park et al. 2007). Results indicated that the REP27 protein is not required for the *de novo* biosynthesis/assembly of the PSII holocomplex but needed for repair-specific D1 biosynthesis and insertion into the existing PSII template. Therefore, the REP27 may function to facilitate translation and/or insertion of the nascent D1 protein in the remaining undegraded PSII components during the course of the repair process (Park et al. 2007).

Also in *Chlamydomonas reinhardtii*, a small PSII subunit, the *PsbT* gene product, has been implicated in the reactivation of Q_A photoreduction and recovery of electron transport activities within the newly repaired PSII complex (Ohnishi et al. 2007).

9. Modulation of the PSII Repair Process

The PSII D1 reaction center protein is undergoing a frequent turnover as part of the PSII damage and repair cycle (Guenther and Melis 1990). Many environmental factors influence this process and thereby modulate the rate of the PSII repair. Foremost among the environmental parameters is irradiance. Although protease degradation of the photodamaged D1 protein is light independent (Prasil et al. 1992, Melis 1999), its *de novo* synthesis is regulated by light (Baena-González and Aro 2002, Zhang and Aro 2002). Another good example of the environmental factors that affect the recovery process of PSII from photoinhibitory damage is nutrient limitation, such as sulfur nutrients (Wykoff et al. 1998, Giordano et al. 2000). In the absence of sulfur, which is an essential component of the amino acids cysteine and methionine, D1 protein biosynthesis is impeded (lack of cysteine and methionine) and the repair cycle is arrested in the PSII Q_B-nonreducing configuration (Wykoff et al. 1998). In consequence, the rate of photosynthesis declines quasi-exponentially in the light as a function of time in S-deprivation with a half time of about 18 h (Wykoff et al. 1998, Cao et al. 2001). This effect is specific to PSII in the thylakoid membrane. In fact, characterization of *rep55*, a repair aberrant DNA insertional transformant of *Chlamydomonas reinhardtii*, revealed that the disrupted *SulP* gene encodes for a chloroplast-targeted sulfate permease (Chen et al. 2003). The function of this sulfate permease is to facilitate sulfate uptake by the chloroplast (Yildiz et al. 1994, Chen et al. 2003), which is apparently critical under high rates of photodamage that require high rates of *de novo* D1 biosynthesis (Melis and Chen 2005). Thus, the supply of inorganic sulfur nutrients to the chloroplast may determine the rate of D1 turnover and may thus represent a significant regulatory step in the PSII repair process (Yokthongwattana and Melis 2006).

Studies in cyanobacteria showed that lack of polyunsaturated fatty acids in the thylakoid membrane lipid bilayer not only makes the PSII complex more susceptible to photodamage but also leads to an impairment of functional D1 protein synthesis (Kanervo et al. 1995, 1997). Independent research has further demonstrated that low temperature, reactive oxygen species (ROS) and salinity stress interfere with the repair of photodamaged PSII by inhibiting the transcription and/or translation of the *psbA* gene in *Synechocystis* PCC 6803 (Nishiyama et al. 2001, 2004, 2006, Allakhverdiev et al. 2002, Allakhverdiev and Murata 2004). It was further demonstrated that disruption of the Calvin cycle greatly affects rates of the PSII repair process at the translation level of the *psbA* gene (Takahashi and Murata 2006a, 2006b).

10. Summary

Oxygen in the vicinity of the PSII reaction center, and irradiance in the form of excitation energy combine to bring about an adverse effect, i.e., a photo-oxidative damage to the PSII D1 reaction center protein in the thylakoid membrane of oxygenic photosynthesis. The D1 protein undergoes a selective degradation and replacement before PSII could regain its useful function. Photodamage specifically brings about impairment in the activity of the D1 reaction center protein, while other PSII reaction center and light-harvesting subunits remain unaffected. It stands to reason that the PSII repair mechanism must also include steps for the stabilization and maintenance of the PSII subunits that are not subject to degradation, and to facilitate the orderly removal and replacement of the centrally-located D1 protein. These steps could include provision of a semblance of order in the thylakoid membrane during the PSII disassembly and D1 removal and replacement, as well as protection of the disassembled PSII-core chlorophyll-protein subunits (mainly CP43, CP47 and D2 and their associated pigments) from irreversible photo-bleaching. ELIP/Cbr accumulation and zeaxanthin formation in *Dunaliella salina* occur in tandem with the advent of photodamage and may play a role in the photoprotection of the disassembled PSII-core subunits. Thus, Cbr and zeaxanthin may be important components of the PSII repair mechanism, acting in the initial stages of the PSII-repair and PSII-repair-intermediate process.

Evidence has accumulated in the literature (Kim et al. 1993, Melis and Nemson 1995, Polle and Melis 1999, Yokthongwattana et al. 2001, Nowaczyk et al. 2006) to indicate the presence and structural configuration of a 'PSII repair intermediate', a protein complex of the partially disassembled PSII-core in *D. salina*, forming in the thylakoid membrane prior to the selective degradation of the photodamaged D1 protein. Thus, important advances in the PSII repair were achieved in research with *D. salina,* summarized in the 'working hypothesis model' of Figure 4. According to this hypothesis, at least a portion of the zeaxanthin pool, forming upon photoinhibitory damage to PSII, serves to protect the disassembled PSII unit. Concomitantly, an accelerated synthesis and accumulation of the ELIP/Cbr protein is observed in the chloroplast thylakoids. The function of the ELIP/Cbr proteins is not clear, although a pigment-binding function can be inferred from the structure of these proteins, which is similar to that of the LHC-II (Adamska 1997, Heddad and Adamska 2000). More research is needed to investigate whether ELIP/Cbr proteins may serve to bind chlorophyll and carotenoids released from the PSII reaction center following photodamage and upon the ensuing PSII disassembly and D1 degradation.

Regarding a PSII repair-specific photoprotective function of zeaxanthin, it could be argued that it serves to dissipate incoming excitation energy, which can no longer be utilized via PSII photochemistry. Non-photochemical energy dissipation by zeaxanthin in *D. salina* may serve to protect other Chl-containing subunits of the inactivated PSII complex from irreversible photo-bleaching. Such photoprotective function of zeaxanthin would last for the duration of the PSII repair and would no longer be needed following insertion of the nascent D1 and functional reassembly of the PSII holocomplex. As a consequence of this zeaxanthin function, the photodamaged PSII complex at this stage of the repair would act as an efficient quencher of incoming excitation energy in the statistical pigment bed of the PSII units (Melis and Duysens 1979). It may thus contribute, at least in part, to the photoinhibition-type of non-photochemical quenching or qI (Krause 1988, Lee et al. 2001, Chow et al. 2002). Consistent with this proposed functional role of zeaxanthin in the thylakoid membrane are reports describing a relationship between zeaxanthin accumulation and the degree of photoinhibition in *Schefflera arboricola* (Demmig-Adams et al. 1998) and *Dictyota dichotoma* (Uhrmacher et al. 1995). It should be noted, however, that this repair-specific photoprotective function of zeaxanthin does not preclude additional and/or different functions for this carotenoid molecule, as described in the literature (Gilmore et al. 1998, Niyogi 1999, 2000, Demmig-Adams 2003, Demmig-Adams and Adams 2006), depending on localization and/or plant species.

Acknowledgement

KY thanks Mahidol University for financial support. Other related work presented in this chapter was supported by USDA National Research Initiative, Competitive Grants Program to AM.

References

Adam, Z. and A.K. Clarke. 2002. Cutting edge of chloroplast proteolysis. Trends Plant Sci. 7: 451-456.

Adam, Z., A. Rudella and K.J. van Wijk. 2006. Recent advances in the study of Clp, FtsH and other proteases located in chloroplasts. Curr. Opin. Plant Biol. 9: 234-240.

Adams, W.W. III., C.R. Zarter, V. Ebbert and B. Demmig-Adams. 2004. Photoprotective strategies of overwintering evergreens. Bioscience 54: 41-49.

Adamska, I. 1997. ELIPs - light-induced stress proteins. Physiol. Plant. 100: 794-805.

Adir, N., H. Zer, S. Schochat and I. Ohad. 2003. Photoinhibition - a historical perspective. Photosynth. Res. 76: 343-370.

Allakhverdiev, S.I., S. Miyairi, N. Inagaki, Y. Nishiyama and N. Murata. 2002. Salt stress inhibits the repair of photodamaged photosystem II by suppressing the transcription and translation of *psbA* gene in *Synechocystis*. Plant Physiol. 130: 1443-1453.

Allakhverdiev, S.I. and N. Murata. 2004. Environmental stress inhibits the synthesis de novo of proteins involved in the photodamage–repair cycle of Photosystem II in *Synechocystis* sp. PCC 6803. Biochim. Biophys. Acta 1657: 23-32.

Anderson, J.M. 1986. Photoregulation of the composition, function, and structure of thylakoid membranes. Annu. Rev. Plant Physiol. 37: 93-136.

Anderson, J.M., I.-Y. Park and W.S. Chow. 1997. Photoinactivation and photoprotection of photosystem II in nature. Physiol. Plant. 100: 214-223.

Andersson, B. and E.-M. Aro. 2002. Photodamage and D1 protein turnover in photosystem II. pp. 377-393. In Regulation of photosynthesis. Advances in Photosynthesis and Respiration Series, Volume 11. E.-M. Aro and B. Andersson (eds) Kluwer Academic Publishers, The Netherlands.

Aro, E.-M., R. Kettunen and E. Tyystjärvi. 1992. ATP and light regulate D1 protein modification and degradation: role of D1* in photoinhibition. FEBS Lett. 287: 29-33.

Aro, E.-M., I. Virgin and B. Andersson. 1993. Photoinhibition of photosystem II. Inactivation, protein damage and turnover. Biochim. Biophys. Acta 1143: 113-134.

Baena-González, E. and E.-M. Aro. 2002. Biogenesis, assembly and turnover of photosystem II units. Phil. Trans. R. Soc. Lond. B 357: 1451-1460.

Baena-González, E., R. Barbato and E.-M. Aro. 1999. Role of phosphorylation in the repair cycle and oligomeric structure of photosystem II. Planta 208: 196-204.

Banet, G., U. Pick and A. Zamir. 2000. Light-harvesting complex II pigments and proteins in association with Cbr, a homolog of higher-plant early light-inducible proteins in the unicellular green alga *Dunaliella*. Planta 210: 947-955.

Barbato, R., G. Friso, F. Rigoni, A. Frizzo and G.M. Giacometti. 1992. Characterization of a 41 kD photoinhibition adduct in isolated photosystem II reaction centres. FEBS Lett. 309: 165-169.

Barber, J. and B. Andersson. 1992. Too much of a good thing: light can be bad for photosynthesis. Trends Biochem. Sci. 17: 61-66.

Baroli, I. and A. Melis. 1996. Photoinhibition and repair in *Dunaliella salina* acclimated to different growth irradiances. Planta 198: 640-646.

Baroli, I. and A. Melis. 1998. Photoinhibitory damage is modulated by the rate of photosynthesis and by the photosystem II light-harvesting chlorophyll antenna size. Planta 205: 288-296.

Bennett, J. 1991. Protein phosphorylation in green plant chloroplasts. Annu. Rev. Plant Physiol. Plant Mol. Biol. 42: 281-311.

Cao, H., L. Zhang and A. Melis. 2001. Bioenergetic and metabolic processes for the survival of sulfur-deprived *Dunaliella salina* (Chlorophyta). J. Appl. Phycol. 13: 25-34.

Chen, H.-C., K. Yokthongwattana, A.J. Newton and A. Melis. 2003. *SulP1*, a nuclear gene encoding a putative chloroplast sulfate permease in *Chlamydomonas reinhardtii*. Planta 218: 98-106.

Chen, H., D. Zhang, J. Guo, H. Wu, M. Jin, Q. Lu, C. Lu and L. Zhang. 2006. A Psb27 homologue in Arabidopsis thaliana is required for efficient repair of photodamaged photosystem II. Plant Mol. Biol. 61: 567-575.

Chow, W.S., H.-Y. Lee, Y.-I. Park, Y.-M. Park, Y.-N. Hong and J.M. Anderson. 2002. The role of inactive photosystem-II-mediated quenching in a last ditch community defence against high light stress *in vivo*. Phil. Trans. R. Soc. Lond. B 357: 1441-1450.

Cleland, R.E. and A. Melis. 1987. Probing the events of photoinhibition by altering electron-transport activity and light-harvesting capacity in chloroplast thylakoids. Plant Cell Environ. 10: 747-752.

Demmig-Adams, B. 2003. Linking xanthophyll cycle with thermal energy dissipation. Photosynth. Res. 76: 73-80.

Demmig-Adams, B. and W.W. III. Adams. 2006. Photoprotection in an ecological context: the remarkable complexity of thermal energy dissipation. New Phytol. 172: 11-21.

Demmig-Adams, B., D.L. Moeller, B.A. Logan and W.W.III. Adams. 1998. Positive correlation between levels of retained zeaxanthin + antheraxanthin and degree of photoinhibition in shade leaves of *Schefflera arboricola* (Hayata) Merrill. Planta 205: 367-374.

Depka, B., P. Jahns and A. Trebst. 1998. beta-carotene to zeaxanthin conversion in the rapid turnover of the D1 protein of photosystem II. FEBS Lett. 424: 267-270.

Edelman, M. and A.K. Mattoo. 2006. The D1 protein: past and future perspectives. pp. 23-38. In Photoprotection, photoinhibition, gene regulation and environment. Advances in Photosynthesis and Respiration Series, Volume 21. B. Demmig-Adams, W.W.III Adams and A. Matto (eds) Springer, Dordrecht, The Netherlands.

Elich, T.D., M. Edelman and A.K. Mattoo. 1992. Identification, characterization, and resolution of the in vivo phosphorylated form of D1 photosystem II reaction center protein. J. Biol. Chem. 267: 3523-3529.

Geider, R.J. and B.A. Osborne. 1992. The photosynthesis - light response curve. pp. 156-191. In Algal photosynthesis. R.J. Geider and B.A. Osborne (eds). Chapman and Hill, New York, USA.

Gilmore, A.M., N. Mohanty and H.Y. Yamamoto. 1994. Epoxidation of zeaxanthin and antheraxanthin reverses non-photochemical quenching of photosystem II chlorophyll a fluorescence in the presence of trans-thylakoid delta pH. FEBS Lett. 350: 271-274.

Gilmore, A.M., V.P. Shinkarev, T.L. Hazlett and Govindjee. 1998. Quantitative analysis of the effects of intrathhylakoid pH and xanthophyll cycle pigments on chlorophyll lifetime distributions and intensity in thylakoids. Biochemistry 37: 13582-13593.

Giordano, M., V. Pezzoni and R. Hell. 2000. Strategies for the allocation of resources under sulfur limitation in the green alga *Dunaliella salina*. Plant Physiol. 124: 857-864.

Glick, R.E. and A. Melis. 1988. Minimum photosynthetic unit size in system I and system II of barley chloroplasts. Biochim. Biophys. Acta 934: 151-155.

Golbeck, J.H. [ed.] 2006. Photosystem I: the light-driven plastocyanin, ferredoxin oxidoreductase. Advances in Photosynthesis and Respiration Series, Volume 24, Springer, Dordrecht, The Netherlands.

Greer, D.M., J.A. Berry and O. Björkman. 1986. Photoinhibition of photosynthesis in intact bean leaves: role of light and temperature, and requirement for chloroplast-encoded protein synthesis during recovery. Planta 168: 253-260.

Guenther, J.E. and A. Melis. 1990. Dynamics of photosystem II heterogeneity in *Dunaliella salina* (green algae). Photosynth. Res. 23: 195-203.

Guenther, J.E., J.A. Nemson and A. Melis. 1988. Photosystem stoichiometry and chlorophyll antenna size in *Dunaliella salina* (green algae). Biochim. Biophys. Acta 934: 108-117.

Hankamer, B., J. Barber and E.J. Boekema. 1997. Structure and membrane organization of photosystem II in green plants. Annu. Rev. Plant Physiol. Plant Mol. Biol. 48: 641-671.

Haußühl, K., B. Andersson and I. Adamska. 2001. A chloroplast DegP2 protease performs the primary cleavage of the photodamaged D1 protein in plant photosystem II. EMBO J. 20: 713-722.

Havaux, M. 1992. Stress tolerance of photosystem II *in vivo*: antagonistic effects of water, heat, and photoinhibition stresses. Plant Physiol. 100: 424-432.

Heddad, M. and I. Adamska. 2000. Light stress-regulated two-helix proteins in Arabidopsis thaliana related to the chlorophyll a/b-binding gene family. Proc. Natl. Acad. Sci. USA 97: 3741-3746.

Heddad, M., H. Norén, V. Reiser, M. Dunaeva, B. Andersson and I. Adamska. 2006. Differential expression and localization of early light-induced proteins in *Arabidopsis thaliana*. Plant Physiol. 142: 75-87.

Huner, N.P.A., D.P. Maxwell, G.R. Gray and L.V. Savitch. 1996. Sensing environmental temperature change through imbalances between energy supply and energy consumption - redox state of photosystem II. Physiol. Plant. 98: 358-364.

Huner, N.P.A., G. Öquist and F. Sarhan. 1998. Energy balance and acclimation to light and cold. Trends Plant Sci. 3: 224-230.

Im, C.-S., Z. Zhang, J. Shrager, C.-W. Chang and A.R. Grossman. 2003. Analysis of light and CO_2 regulation in *Chlamydomonas reinhardtii* using genome-wide approaches. Photosynth Res. 75: 111–125.

Ishikawa, Y., E. Nakatani, T. Henmi, A. Ferjani, Y. Harada, N. Tamura and Y. Yamamoto. 1999. Turnover of the aggregates and cross-linked products of the D1 protein generated by acceptor-side photoinhibition of photosystem II. Biochim. Biophys. Acta 1413: 147-158.

Jahns, P. and B. Miehe. 1996. Kinetic correlation of recovery from photoinhibition and zeaxanthin epoxidation. Planta 198: 202-210.

Jahns, P., B. Depka and A. Trebst. 2000. Xanthophyll cycle mutants from *Chlamydomonas reinhardtii* indicate a role for zeaxanthin in the D1 protein turnover. Plant Physiol. Biochem. 38: 373-376.

Jin, E., J.E.W. Polle and A. Melis. 2001. Involvement of zeaxanthin and of the Cbr protein in the repair of photosystem II from photoinhibition in the green alga *Dunaliella salina*. Biochim. Biophys. Acta 1506: 1389-1397.

Jin, E., B. Feth and A. Melis. 2003a. A mutant of the green alga *Dunaliella salina* constitutively accumulates zeaxanthin under all growth conditions. Biotechnol. Bioeng. 81: 115-124.

Jin, E., K. Yokthongwattana, J.E.W. Polle and A. Melis. 2003b. Role of the reversible xanthophyll cycle in the photosystem-II damage and repair cycle in *Dunaliella salina* (green alga). Plant Physiol. 132: 352-364.

Kanervo, E., E.-M. Aro and N. Murata. 1995. Low unsaturation level of thylakoid membrane lipids limits turnover of the D1 protein of photosystem II at high irradiance. FEBS Lett. 364: 239-242.

Kanervo, E., Y. Tasaka, N. Murata and E.-M. Aro. 1997. Membrane lipid unsaturation modulates processing of the photosystem II reaction-center protein D1 at low temperatures. Plant Physiol. 114: 841-849.

Kapri-Pardes, E., L. Naveh and Z. Adam. 2007. The thylakoid lumen protease Deg1 is involved in the repair of photosystem II from photoinhibition in *Arabidopsis*. Plant Cell 19: 1039-1047.

Kim, J.H., J.A. Nemson and A. Melis. 1993. Photosystem II reaction center damage and repair in *Dunaliella salina* (green alga): analysis under physiological and irradiance-stress conditions. Plant Physiol. 103: 181-189.

Koivuniemi, A., E.-M. Aro and B. Andersson. 1995. Degradation of the D1- and D2-proteins of photosystem II in higher plants is regulated by reversible phosphorylation. Biochemistry 34: 16022-16029.

Kok, B. 1956. On the inhibition of photosynthesis by intense light. Biochim. Biophys. Acta 21: 234-244.

Krause, G.H. 1988. Photoinhibition of photosynthesis. An evaluation of damaging and protective mechanisms. Physiol. Plant. 74: 566-574.

Król, M., D.P. Maxwell and N.P.A. Huner. 1997. Exposure of *Dunaliella salina* to low temperature mimics the high light-induced accumulation of carotenoids and the carotenoid binding protein (Cbr). Plant Cell Physiol. 38: 213-216.

Kruse, O., D. Zheleva and J. Barber. 1997. Stabilization of photosystem two dimers by phosphorylation: implication for the regulation of the turnover of D1 protein. FEBS Lett. 408: 276-280.

Ledford, H.K., I. Baroli, J.W. Shin, B.B. Fischer, R.I. Eggen and K.K. Nyogi. 2004. Comparative profiling of lipid-soluble antioxidants and transcripts reveals two phases of photo-oxidative stress in a xanthophyll-deficient mutant of *Chlamydomonas reinhardtii*. Mol. Genet. Genomics 272: 470-479.

Lee, H.-Y., Y.-N. Hong and W.S. Chow. 2001. Photoinactivation of photosystem II complexes and photoprotection by non-functional neighbours in *Capsicum annuum* L. leaves. Planta 212: 332-342.

Lers, A., H. Levy and A. Zamir. 1991. Co-regulation of a gene homologous to early light-induced genes in higher plants and b-carotene biosynthesis in the alga *Dunaliella baradawil*. J. Biol. Chem. 266: 13698-13705.

Levy, H., I. Gokhman and A. Zamir. 1992. Regulation and light harvesting complex II association of *Dunaliella* protein homologous to early light-induced proteins in higher plants. J. Biol. Chem. 267: 18831-18836.

Levy, H., T. Tal, A. Shaish and A. Zamir. 1993. Cbr, an algal homolog of plant early light-induced proteins, is a putative zeaxanthin binding protein. J. Biol. Chem. 268: 20892-20896.

Lindahl, M., C. Spetea, T. Hundal, A.B. Oppenheim, Z. Adam and B. Andersson. 2000. The thylakoid FtsH protease plays a role in the light-induced turnover of the photosystem II D1 protein. Plant Cell 12: 419-431.

Long, S.P., S. Humphries and P.G. Falkowski. 1994. Photoinhibition of photosynthesis in nature. Annu. Rev. Plant Physiol. Plant Mol. Biol. 45: 633-662.

Masuda, T., J.E.W. Polle and A. Melis. 2002. Biosynthesis and distribution of chlorophyll among the photosystems during recovery of the green alga *Dunaliella salina* from irradiance stress. Plant Physiol. 128: 603-614.

Maxwell, D.P., S. Falk and N.P.A. Huner. 1995. Photosystem II excitation pressure and development of resistance to photoinhibition. I. Light-harvesting complex II abundance and zeaxanthin content in *Chlorella vulgaris*. Plant Physiol. 107: 687-694.

Melis, A. 1991. Dynamics of photosynthetic membrane composition and function. Biochim. Biophys. Acta 1058: 87-106.

Melis, A.1996. Excitation energy transfer: functional and dynamic aspects of *Lhc* (*cab*) proteins. pp. 523-538. In Oxygenic photosynthesis: the light reactions. Advances in Photosynthesis and Respiration Series, Volume 4. D.R. Ort and C.F. Yocum (eds) Kluwer Academic Publishers, Dordrecht, The Netherlands.

Melis, A. 1998. Photostasis in plants: mechanism and regulation. pp. 207-221. In Photostasis and related phenomena.T.P. Williams and A.B. Thistle (eds). Plenum Press, New York, USA.

Melis, A. 1999. Photosystem-II damage and repair cycle in chloroplasts: what modulates the rate of photodamage *in vivo*? Trends Plant Sci. 4: 130-135.

Melis, A. and L.N.M. Duysens. 1979. Biphasic energy conversion kinetics and absorbance difference spectra of photosystem II of chloroplasts. Evidence for two different photosystem II reaction centers. Photochem. Photobiol. 29: 373-382.

Melis, A. and J.A. Nemson. 1995. Characterization of a 160 kD photosystem II reaction center complex isolated from photoinhibited *Dunaliella salina* thylakoids. Photosynth. Res. 46: 207-211.

Melis, A. and H.-C. Chen. 2005. Chloroplast sulfate transport in green algae — genes, proteins and effects. Photosynth. Res. 86: 299-307.

Melis, A., J. Neidhardt and J.R. Benemann. 1998. *Dunaliella salina* (Chlorophyta) with small chlorophyll antenna sizes exhibit higher photosynthetic productivities and photon use efficiencies than normally pigmented cells. *J. Appl. Phycol.* 10: 515-525.

Michel, H., D.F. Hunt, J. Shabanowitz and J. Bennett. 1988. Tandem mass spectrometry reveals that three photosystem II proteins of spinach chloroplasts contain N-acetyl-o-phosphothreonine at their NH_2 termini. J. Biol. Chem. 263: 1123-1130.

Naus, J. and A. Melis. 1991. Changes of photosystem stoichiometry during cell growth in *Dunaliella salina* cultures. Plant Cell Physiol. 32: 569-575.

Neidhardt, J., J.R. Benemann, L. Zhang and A. Melis. 1998. Photosystem II repair and chloroplast recovery from irradiance stress: relationship between chronic photoinhibition, light-harvesting chlorophyll antenna size and photosynthetic productivity in *Dunaliella salina* (green algae). Photosynth. Res. 56: 175-184.

Nishiyama, Y., H. Yamamoto, S.I. Allakhverdiev, M. Inaba, A. Yokota and N. Murata. 2001. Oxidative stress inhibits the repair of photodamage to the photosynthetic machinery. EMBO J. 20: 5587-5594.

Nishiyama, Y., S.I. Allakhverdiev, H. Yamamoto, H. Hayashi and N. Murata. 2004. Singlet oxygen inhibits the repair of photosystem II by suppressing the translation elongation of the D1 protein in *Synechocystis* sp. PCC 6803. Biochemistry 43: 11321-11330.

Nishiyama, Y., S.I. Allakhverdiev and N. Murata. 2006. Regulation by environmental conditions of the repair of photosystem II in cyanobacteria. pp. 193-203. In Photoprotection, photoinhibition, gene regulation and environment. Advances in Photosynthesis and Respiration Series, Volume 21. B. Demmig-Adams, W.W.III Adams and A Mattoo (eds) Springer, Dordrecht, The Netherlands.

Niyogi, K.K. 1999. Photoprotection revisited: genetic and molecular approaches. Annu. Rev. Plant Physiol. Plant Mol. Biol. 50: 333-359.

Niyogi, K.K. 2000. Safety valves for photosynthesis. Curr. Opin. Plant Biol. 3: 455-460.

Nowaczyk, M.M., R. Hebeler, E. Schlodder, H.E. Meyer, B. Warscheid and M. Rögner. 2006. Psb27, a cyanobacterial lipoprotein, is involved in the repair cycle of photosystem II. Plant Cell 18: 3121-3131.

Ohnishi, N., Y. Kashino, K. Satoh, S. Ozawa and Y. Takahashi. 2007. Chloroplast-encoded polypeptide PsbT is involved in the repair of primary electron acceptor Q_A of photosystem II during photoinhibition in *Chlamydomonas reinhardtii*. J. Biol. Chem. 282: 7107-7115.

Osmond, B. and B. Förster. 2006. Photoinhibition: then and now. pp. 11-22. In Photoprotection, photoinhibition, gene regulation and environment. Advances in Photosynthesis and Respiration Series, Volume 21. B.Demmig-Adams, W.W.III Adams and A. Mattoo (eds) Springer, Dordrecht, The Netherlands.

Park, S., J.E.W. Polle, A. Melis, T.K. Lee and E. Jin. 2006. Up-regulation of photoprotection and PSII-repair gene expression by irradiance in the unicellular green alga *Dunaliella salina*. Mar. Biotechnol. 8: 120-128.

Park, S., P. Khamai, J.G. Garcia-Cerdan and A. Melis. 2007. REP27, a tetratricopeptide repeat nuclear-encoded and chloroplast-localized protein, functions in D1/32-kD reaction center protein turnover and photosystem II repair from photodamage. Plant Physiol. 143: 1547-1560.

Park, Y.-I., W.S. Chow and J.M. Anderson. 1995. Light inactivation of functional PSII in leaves of pea grown in moderate light depends on photon exposure. Planta 196: 401-411.

Park, Y.-I., J.M. Anderson and W.S. Chow. 1996a. Photoinactivation of functional PSII and D1 protein synthesis in vivo is independent of the modulation of the photosynthetic apparatus by growth irradiance. Planta 198: 300-309.

Park, Y.-I., W.S. Chow, C.B. Osmond and J.M. Anderson. 1996b. Electron transport to oxygen mitigates against the photoinactivation of photosystem II *in vivo*. Photosynth. Res. 50: 23-32.

Park, Y.-I., W.S. Chow and J.M. Anderson. 1997. Antenna size dependency of photoinactivation of photosystem II in light-acclimated pea leaves. Plant Physiol. 115: 151-157.

Polle, J.E.W. and A. Melis. 1999. Recovery of the photosynthetic apparatus from photoinhibition during dark incubation of the green alga *Dunaliella salina*. Aust. J. Plant Physiol. 26: 679-686.

Powles, S. 1984. Photoinhibition of photosynthesis induced by visible light. Annu. Rev. Plant Physiol. 35: 15-44.

Prasil, O., N. Adir and I. Ohad. 1992. Dynamics of photosystem II: mechanism of photoinhibition and recovery processes. pp. 295-384. In The Photosystems: structure, function and molecular biology. J. Barber (ed) Elsevier, Amsterdam, The Netherlands.

Pursiheimo, S., E. Rintamäki, E. Baena-Gonzalez and E.-M. Aro. 1998. Thylakoid protein phosphorylation in evolutionally divergent species with oxygenic photosynthesis. FEBS Lett. 423: 178-182.

Rintamäki, E., R. Kettunen and E.-M. Aro. 1996. Differential D1 phosphorylation in functional and photodamaged photosystem II centers: dephosphorylation is a prerequisite for degradation of damaged D1. J. Biol. Chem. 271: 14870-14875.

Schroda, M. 2004. The *Chlamydomonas* genome reveals its secrets: chaperone genes and the potential roles of their gene products in the chloroplast. Photosynth. Res. 82: 221-240.

Schroda, M., O. Vallon, F.-A. Wollman and C.F. Beck. 1999. A chloroplast-targeted heat shock protein70 (HSP70) contributes to the photoprotection and repair of photosystem II during and after photoinhibition. Plant Cell 11: 1165-1178.

Schroda, M., J. Kropat, U. Oster, W. Rüdiger, O. Vallon, F.-A. Wollman and C.F. Beck. 2001. Possible role for molecular chaperones in assembly and repair of photosystem II. Biochem. Soc. Trans. 29: 413-418.

Smith, B.M., P.J. Morrissey, J.E. Guenther, J.A. Nemson, M.A. Harrison, J.F. Allen and A. Melis. 1990. Response of the photosynthetic apparatus in *Dunaliella salina* (green algae) to irradiance stress. Plant Physiol. 93: 1433-1440.

Stitt, M. 1986. Limitation of photosynthesis by carbon metabolism. 1. Evidence for excess electron transport capacity in leaves carrying out photosynthesis in saturating light and CO_2. Plant Physiol. 81: 1115-1122.

Sukenik, A., J. Bennett and P.G. Falkowski. 1987a. Light-saturated photosynthesis: limitation by electron transport or carbon fixation. Biochim. Biophys. Acta 891: 205-215.

Sukenik, A., K.D. Wyman, J. Bennett and P.G. Falkowski. 1987b. A novel mechanism for regulating the excitation of photosystem II in a green alga. Nature 327: 704-707.

Sundby, C., S. Mccaffery and J.M. Anderson. 1993. Turnover of the photosystem-II D1-protein in higher-plants under photoinhibitory and nonphotoinhibitory irradiance. J. Biol. Chem. 268: 25476-25482.

Swiatek, M., R. Kuras, A. Sokolenko, D. Higgs, J. Olive, G. Cinque, B. Müller, L.A. Eichacker, D.B. Stern, R. Bassi, R.G. Herrmann and F.-A. Wollman. 2001. The chloroplast gene *ycf9* encodes a photosystem II (PSII) core subunit, PsbZ, that participates in PSII supramolecular architecture. Plant Cell 13: 1347-1367.

Takahashi, S. and N. Murata. 2006a. Interruption of the Calvin cycle inhibits the repair of Photosystem II from photodamage. Biochim. Biophys. Acta 1708: 352-361.

Takahashi, S. and N. Murata. 2006b. Glycerate-3-phosphate, produced by CO2 fixation in the Calvin cycle, is critical for the synthesis of the D1 protein of photosystem II. Biochim. Biophys. Acta 1757: 198-205.

Tanaka, A. and A. Melis. 1997. Irradiance-dependent changes in the size and composition of the chlorophyll *a-b* light-harvesting complex in the green alga *Dunaliella salina*. Plant Cell Physiol. 38: 17-24.

Tyystjärvi, E. and E.-M. Aro. 1996. The rate constant of photoinhibition, measured in licomycin-treated leaves, is directly proportional to light intensity. Proc. Natl. Acad. Sci. USA 93: 2213-2218.

Uhrmacher, S., D. Hanelt and W. Nultsch. 1995. Zeaxanthin content and the degree of photoinhibition are linearly correlated in the brown alga *Dictyota dichotoma*. Mar. Biol. 123: 159-165.

Vasilikiotis, C. and A. Melis. 1994. Photosystem II reaction center damage and repair cycle: chloroplast acclimation strategy to irradiance stress. Proc. Natl. Acad. Sci. USA 91: 7222-7226.

Vass, I., S. Styring, T. Hundal, A. Koivuniemi, E.-M. Aro and B. Andersson. 1992. Reversible and irreversible intermediates during photoinhibition of photosystem II. Stable reduced Q_A species promote chlorophyll triplet formation. Proc. Natl. Acad. Sci. USA 89: 1408-1412.

Webb, M.R. and A. Melis. 1995. Chloroplast response in *Dunaliella salina* to irradiance stress: effect on thylakoid membrane protein assembly and function. Plant Physiol. 107: 885-893.

Wydrzynski, T.J. and K. Satoh. [eds] 2005. Photosystem II: the light-driven water: plastoquinone oxidoreductase. Advances in Photosynthesis and Respiration Series, Volume 22, Springer, Dordrecht, The Netherlands.

Wykoff, D.D., J.P. Davies, A. Melis and A.R. Grossman. 1998. The regulation of photosynthetic electron transport during nutrient deprivation in *Chlamydomonas reinhardtii*. Plant Physiol. 117: 129-139.

Yamamoto, Y. 2001. Quality control of photosystem II. Plant Cell Physiol. 42: 121-128.

Yildiz, F.H., J.P. Davies and A.R. Grossman. 1994. Characterization of sulfate transport in *Chlamydomonas reinhardtii* during sulfur-limited and sulfur-sufficient growth. Plant Physiol. 104: 981-987.

Yokthongwattana, K. and A. Melis. 2006. Photoinhibition and recovery in oxygenic photosynthesis: mechanism of a photosystem-II damage and repair cycle. pp. 175-191. In Photoprotection, photoinhibition, gene regulation and environment. Advances in Photosynthesis and Respiration Series, Volume 21. B. Demmig-Adams, W.W.III Adams and A. Mattoo (eds) Springer, Dordrecht, The Netherlands.

Yokthongwattana, K., B. Chrost, S. Behrman, C. Casper-Lindley and A. Melis. 2001. Photosystem II damage and repair cycle in the green alga *Dunaliella salina*: involvement of a chloroplast-localized HSP70. Plant Cell Physiol. 42: 1389-1397.

Zarter, C.R., W.W. III. Adams, V. Ebbert, D.J. Cuthbertson, I. Adamska and B. Demmig-Adams. 2006a. Winter down-regulation of intrinsic photosynthetic capacity coupled with up-regulation of Elip-like proteins and persistent energy dissipation in a subalpine forest. New Phytol. 172: 272-282.

Zarter, C.R., B. Demmig-Adams, V. Ebbert, I. Adamska and W.W.III Adams. 2006b. Photosynthetic capacity and light harvesting efficiency during the winter-to-spring transition in subalpine conifers. New Phytol. 172: 283-292.

Zhang, L. and E.-M. Aro. 2002. Synthesis, membrane insertation and assembly of the chloroplast-encoded D1 protein into photosystem II. FEBS Lett. 512: 13-18.

12 Regulation of Photoacclimation in *Dunaliella*

Denis P. Maxwell

Abstract

In response to changes in light quantity and quality the green algae *Dunaliella salina* Teod and *Dunaliella tertiolecta* Butcher can dramatically remodel their photosynthetic processes. Under low light conditions alterations occur that maximize light capture and utilization, while upon exposure to excess light, mechanisms are put in place that attempt to protect the photosynthetic apparatus from damage. Research with both *D. salina* and *D. tertiolecta* has revealed that these alterations can be detected within minutes of a change in irradiance and include adjustments in pigment and protein composition and abundance that are most often the result of changes in nuclear and chloroplastic gene expression. All photosynthetic organisms attempt to maintain a balance between the light energy absorbed by the photosystems with the capacity to use that light energy in metabolism. Evidence from work on *Dunaliella* indicates that this balance, or photostasis, is sensed by the photosynthetic electron transport chain, specifically the redox state of the plastoquinone pool. Acting as a sensor, the redox state of plastoquinone can modulate a wide range of activities including gene expression in the nucleus. Changes to the structure and function of the photosynthetic apparatus occur whenever this redox balance is upset which not only occurs in response to changes in irradiance, but also in response to environmental factors such as a change in temperature or nutrient availability that would alter the cell's capacity to utilize the light energy absorbed.

Introduction

The transformation of solar energy into chemical potential energy occurs within the biosphere through the process of photosynthesis, which in plants and algae takes

[1] Department of Biology, University of Western Ontario, North Campus Building, Room 406, London, Ontario, N6A 5B7, Canada, Email: dmaxwell@uwo.ca

place within the chloroplast. Like many biochemical processes photosynthesis is tightly controlled, however it has unusual complexity due to the fact that protein components involved in the energy-transducing reactions of photosynthesis are encoded by two different genomes – the chloroplast and the nucleus. This spatial separation requires the coordination of events in both compartments that necessitates intra-organellar communication (Rodermel 2001). In addition, a unique feature of photosynthesis is that it involves the capture of light energy and while light is a requirement of photoautotrophic growth, the absorption of excess light is potentially very destructive to the photosynthetic apparatus (Long et al. 1994).

Following a brief overview of photosynthesis, the molecular and physiological mechanisms that photosynthetic organisms use to adjust to changing light conditions through a process termed photoacclimation will be described. This review focuses on research conducted on green algae of the genus *Dunaliella,* specifically *D. salina* and *D. tertiolecta.*

Overview of Photosynthesis

What follows is brief overview of photosynthesis. For a more thorough treatment readers are encouraged to peruse two recent and excellent monographs: Falkowski and Raven (2007) and Blankenship (2002).

Photochemistry and Electron Transport

The primary light-harvesting processes and photochemical reactions of photosynthesis occur within two distinctly different photosystems, PSII and PSI, both of which are multi-protein complexes that span the thylakoid membrane of the chloroplast (Figure 1). The absorption of light energy is facilitated by chlorophyll and carotenoid molecules which are bound to various integral membrane proteins that surround the reaction centre of each of the photosystems. The majority of the pigments are bound to the major peripheral light harvesting antenna complexes of PSII and PSI, denoted as LHCII and LHCI, respectively. Each of these antenna complexes is composed of varying combinations of polypeptides belonging to the Lhca (PSI) and Lhcb (PSII) protein families (van Amerongen and Dekker 2003). Lesser amounts of pigment are associated with the PSII core antenna polypeptides CP47, CP43, the PSII reaction centre polypeptides D1, D2 and the PSI reaction centre polypeptides PsaA and PsaB (Figure 1) (van Amerongen and Dekker 2003, Green and Durnford 1996).

The photochemical reactions which take place within PSI and PSII are dependent upon the absorption of photons of light by pigment molecules within

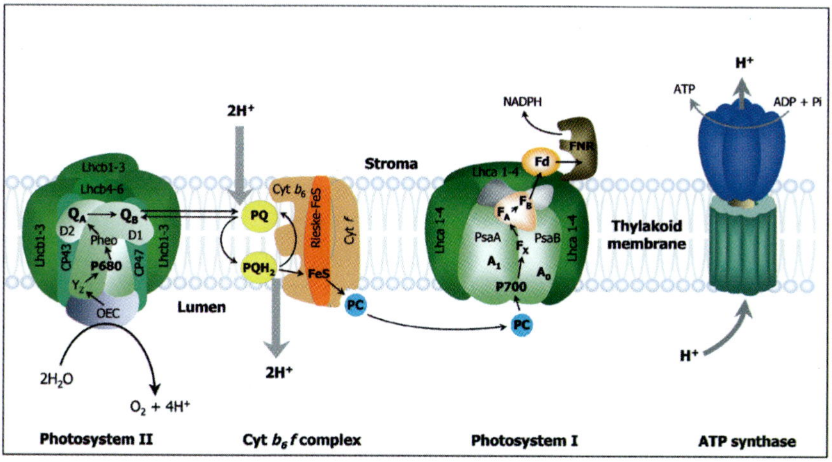

Figure 1: A model of the thylakoid membrane of the chloroplast showing the major polypeptides and components responsible for electron transport and ATP synthesis. The transfer of electrons from H_2O to $NADP^+$ is shown by the solid arrows. PSII: The reaction centre is comprised of two polypeptides D1 and D2 that bind the photochemical chlorophyll P680 as well as pheophytin, and the quinones Q_A and Q_B. The major (distal) light-harvesting polypeptides (Lhcb1-3) and minor (proximal) light-harvesting polypeptides (Lhcb4-6) surround each reaction centre and bind chlorophyll *a*, chlorophyll *b* and a range of carotenoids. Energy transfer from the antenna proteins to the reaction centre is facilitated by the chlorophyll *a* containing CP43 and CP47 polypeptides. The oxygen-evolving complex (OEC) is located on the luminal side of PSII and catalyzes the oxidation of water. Electrons are shuttled from PSII to the cytochrome b_6f-complex by the reduction/oxidation of a pool of plastoquinone (PQ) that migrates within the lipid bilayer. The cytochrome b_6f complex consists of a number of redox active components that function to transfer electrons between PQ and plastocyanin (PC) a mobile carrier that ferries electrons to PSI. The reaction center heterodimer of PSI (PsaA and PsaB) bind the special photochemical chlorophyll pair P700 and two redox active 4Fe-4S clusters (F_A, F_B) that facilitate electron transfer from p700 to ferredoxin (Fd). Reduction of $NADP^+$ to NADPH is catalyzed by ferredoxin NADPH reductase (FNR). Concurrently with electron transport, protons are pumped across the thylakoid membrane (grey arrows) during the oxidation/reduction of PQ. This contributes to the acidification of the thylakoid lumen and represents potential energy that is used for the synthesis of ATP by the ATP synthase complex.

the antenna complexes and the subsequent migration of excitation energy to the specific reaction centre chlorophyll molecules, P680 and P700 associated with PSII and PSI, respectively. When excitation energy reaches an '*open*' PSII reaction centre [Y_Z P680 Pheo Q_A Q_B] the photo-oxidation of P680 results in reduction of the first stable electron acceptor Q_A, [Y_Z $P680^+$ Pheo Q_A^- Q_B]. In this state PSII is considered closed and cannot process another quantum of excitation energy. To reopen a closed PSII, $P680^+$ is reduced by donation of an electron from H_2O. The oxidation of H_2O occurs at the oxygen evolving complex that resides on the luminal side of PSII. The electron is first transferred to a specific tyrosine (Y_Z) residue on the D1 protein before reducing $P680^+$. Concurrently with the reduction of $P680^+$ by H_2O, Q_B is reduced by Q_A^- that yields an 'open' reaction centre [Y_Z P680 Pheo Q_A Q_B^-]. Following two photochemical events the formation of doubly reduced Q_B yields plastoquinol (PQH_2) with the 2 H^+ originating from the stromal side of the thylakoid (Figure 1). PQH_2 diffuses in the thylakoid membrane and is oxidized by the cytochrome b_6f complex regenerating PQ and releasing the 2 H^+ into the thylakoid lumen. This shuttling of PQH_2 from PSII to the cytochrome b_6f complex is considered the rate limiting step of electron transport (Haehnel 1984). Absorption of light by PSI results in photo-oxidation of P700, yielding $P700^+$, which is reduced by oxidation of the cytochrome b_6f complex via the mobile electon carrier plastocyanin (PC). The reduction of $NADP^+$ requires two photo-oxidation reactions of P700 with the electrons being initially donated to ferridoxin (Fd) and the enzyme ferredoxin NADPH reductase (Figure 1). During photosynthetic electron transport, the oxidation of H_2O on the luminal side of PSII as well as the oxidation of PQH_2 result in release of protons into the thylakoid lumen, lowering the pH (Figure 1). The establishment of this proton gradient across the thylakoid membrane represents potential energy that is consumed by the ATP synthase to synthesize ATP.

The Calvin Cycle

The reducing power (NADPH and ferredoxin) and ATP generated by photosynthetic electron transport is used to reduce CO_2 into carbohydrates by the Calvin cycle. Localized to the chloroplast stroma, the eleven enzyme-catalyzed steps the Calvin cycle can conceptually be divided into three phases: Carbon fixation, reduction, and regeneration. The initial carbon fixation reaction that incorporates CO_2 into a five-carbon sugar ribulose bisphosphate (RuBP) is catalyzed by the enzyme ribulose-bisphosphate carboxylase oxygenase (Rubsco) resulting in the formation of two molecules of 3-phosphoglycerate (PGA). In subsequent steps, ATP and NADPH are used to reduce each molecule of PGA to glyceraldehydes-3-phospphate (G3P).

Stoichiometrically for every three CO_2 incorporated 6 molecules of G3P are synthesized. Five of these molecules are used to regenerate 3 molecules of the substrate RuBP while the remaining G3P molecule is used for the synthesis of starch in the chloroplast or exported for sucrose synthesis in the cytosol.

For a unicellular alga like *Dunaliella* approximately 95% of the NADPH and more than 60% of the photosynthetically generated ATP is used for carbon fixation (Falkowski and Raven 2007). The remainder is used primarily for nitrogen and sulfur reduction. It should be noted here that unlike many unicellular algae *Dunaliella* species are halotolerate and this trait requires the synthesis of the 3-carbon polyalcohol, glycerol (Chitlaru and Pick 1991). This is synthesized from three carbon intermediates of the Calvin cycle as glycerol-3 phosphate that is subsequently exported to the cytsol were it gets dephosphorylated.

The Calvin cycle is tightly regulated and responsive to photosynthetic electron transport. Light activates a number of Calvin cycle enzymes through increases in stromal pH, as well as decreases in Mg^{2+} (Malkin and Niyogi 2000). Independent of these, many Calvin cycle enzymes including fructose-1,6-bisphosphatase, sedoheptulose-1,7-bisphosphatase as well as phosphoglycerate kinase are regulated by covalent modification (Jacquot et al. 2002). This involves a ferredoxin-thioredoxin system that links reduction of ferredoxin by the oxidation of PSI to enzyme activation by the oxidation of specific thiol groups (Malkin and Niyogi 2000). Rubisco is activated in the light through a number of unique mechanisms including carbomylation of the active site that is facilitated by the enzyme Rubisco activase (Malkin and Niyogi 2000). In *D. tertiolecta* Rubisco activity has been shown to decline rapidly in darkness within 9-18 minutes as well as being rapidly activated within 3-4 minutes upon re-illumination (MacIntrye et al. 1997).

Photoacclimation

As discussed by Hüner et al. (1998), a sudden change in any environmental factor (e.g. light, temperature, nutrient status) brings about an initial stress response that can be defined as a series of transient, physiological, biochemical and molecular perturbations. If the change in the environment is sustained, the stress response leads to acclimation that can be defined as phenotypic changes to the structure and function of the organism that are stable (Hüner et al. 1998), and are usually the result of changes in gene expression and protein abundance. While such changes are readily reversible, adaptation, which occurs over the course of generations, results from changes in genotype, are not. Typically, it is assumed that acclimation serves to increase growth rate under suboptimal conditions above that which could be attained if the physiology of the cell did not change (see Geider et al. 1998). The term photoacclimation has been formally defined as reversible phenotypic changes that compensate for changes in spectral irradiance, independent of plant developmental properties (see Falkowski and LaRoche 1991, Durnford and Falkowski 1997). While irradiance is strictly defined as the quantity of light (photons per m^{-2} s^{-1}) changes in light quality (altered spectral composition) can also lead to changes to the photosynthetic apparatus and thus should also be considered to be involved in photoacclimation.

Photoacclimation to Changes in Irradiance

The rates of the processes involved in light absorption, energy transfer and photochemistry occur on a time scale of femtoseconds to nanoseconds (10^{-15} -10^{-9} s). In comparison, metabolism and ultimately growth, occur on a time scale of seconds to minutes. Due to this huge rate differential, the photosynthetic apparatus is very susceptible to light-dependent damage. The component most at risk is PSII, in part because its oxidation is dependent on the much slower downstream processes of electron transport and metabolism. Damage to PSII involves the photo-oxidation of D1, the reaction centre polypeptide. This damage is met by a high rate of D1 synthesis that is part of an elaborate system of PSII repair (Melis 1999). However if the damage to PSII exceeds the capacity to carry out repair, the extent of damage can decrease photosynthetic efficiency and under severe conditions decrease the rate of CO_2 fixation and even limit growth (Long et al. 1994).

To maintain high photosynthetic efficiency while attempting to protect the photosynthetic apparatus, all photoautotrophs acclimate to changes in irradiance. Species of unicellular green alga of the genus *Dunaliella* have become a model system for the study of photoacclimation in part because they are easy to grow and maintain in laboratory culture, but also because they respond so obviously to changes in irradiance – cultures turn from a deep green to almost yellow within about 24 hours of being shifted from low light (LL: 50-150 µmol m^{-2} s^{-1}) to high light (HL: 700-2000 µmol m^{-2} s^{-1}) (Durnford and Falkowski 1997, Smith et al. 1990). In addition, an advantage to using *Dunaliella* is that, by being single-celled, it is far more amenable to the study of photoacclimation that in plants is confounded by the slow rate of cell turnover and the developmental and structural complexity inherent in the leaf. Interestingly, while plants do exhibit the capacity to photoacclimate (Anderson et al. 1995), they do not undergo the dramatic changes in pigmentation seen in *Dunaliella* and other green algae. This difference may reflect the limited plasticity of plants, or alternatively, may indicate that plants invoke other strategies besides adjusting pigmentation during photoacclimation (Ensminger et al. 2006).

Adjustments to light-harvesting capacity - Early work by Falkowski and co-workers (Falkowski and Owens 1980, Falkowski 1984) using *D. tertiolecta* showed the now classic response of acclimation to high irradiance - a decrease in total chlorophyll per cell with a concomitant increase in the chlorophyll *a/b* ratio. Changes in the chlorophyll *a/b* ratio are used to gauge light-harvesting antenna size. Since chlorophyll *b* is only associated with antenna proteins and not with proteins of the reaction centres, an increase in the chlorophyll *a*/*b* ratio indicates a decrease in overall LHC abundance. Approximately 40-60% of the total cell chlorophyll is bound to the major LHC surrounding PSII, LHCII (Boardman et al. 1978) and using polyclonal antibodies, Sukenik et al. (1988) were able to identify four distinct LHCII apoproteins in *D. tertiolecta* with molecular masses of 31, 30, 28.5 and 24.5 kDa. Similarly, four LHCII apoproteins have been identified in *D. salina* (denoted LHCII-1 – LHCII-4) with similar, if not identical, electrophoretic mobilities (Tanaka and Melis 1997).

In *D. tertiolecta* the increase in the chlorophyll *a*/*b* ratio seen during a LL to HL shift is associated with a specific 75% decrease in the abundance of the 31 and 30 kDa LHCII proteins (Sukenik et al. 1988). Similarly, in *D. salina* acclimation to HL resulted in an almost complete loss of LHC-II-1 and a substantial reduction in LHC-II-2 (Tanaka and Melis 1997). While the stoichiometry of chlorophyll and carotenoids bound to individual LHCII polypeptides seems to be static in plants (Anderson et al. 1995), it has been shown to be variable in *D. tertiolecta*. Sukenik et al. (1987) showed that in addition to a general decrease in the abundance of LHCII polypeptides, acclimation to HL results in a decrease in chlorophyll *b* associated with individual proteins and an increased binding of the energy dissipating carotenoid lutein, resulting in a decrease in the efficiency of energy transfer to chlorophyll *a* (Sukenik et al. 1987).

A detailed kinetic analysis of the alterations to the photosynthetic apparatus following a change in irradiance has been conduced in *D. tertiolecta* following a 10-fold decrease in growth irradiance from 700 to 70 µmol m^{-2} s^{-1} (Sukenik et al. 1990). The sudden decrease in energy absorbed by the photosynthetic apparatus resulted in a three-fold decrease in growth rate within the first 24 hours after the shift. Over the same 24-hour period, synthesis of chlorophyll *a* and chlorophyll *b* increased approximately 5 and 7-fold respectively, resulting in a decrease in chlorophyll *a*/*b* from about 11 to 7 (Sukenik et al. 1990). Concomitant with the increase in chlorophyll there was rapid synthesis of the four major polypeptides that comprise LHCII, the result being an effective increase in the absorptive cross-section of PSII. The increase in LHCII abundance is followed by a general increase in the concentration of PSII reaction centres that is accompanied by synthesis of thylakoid membrane lipids and increased membrane stacking (Sukenik et al. 1990).

The rapid increase in both chlorophyll and LHCII apoprotein content following a shift from HL to LL requires tight coordination as normally no excess pigment is synthesized that is not immediately bound to nascent LHCII apoproteins, and no significant amounts of protein accumulate in the absence of chlorophyll. The influence of chlorophyll biosynthesis on LHCII stability and accumulation has been studied initially in *D. tertiolecta* (Mortain-Bertrand et al. 1990), and subsequently in *D. salina* (Masuda et al. 2002) through the use of gabaculine an inhibitor of chlorophyll synthesis. A finding of both studies is that while chlorophyll synthesis is not required for accumulation of LHCII apoproteins, stabilization of the apoproteins by chlorophyll is required for the increase in LHCII abundance during acclimation to low light. In *D. salina*, a shift from HL to LL in the absence of gabaculine resulted in an increase in the number of chlorophylls associated with LHC proteins from about 55 to 300 molecules, while in cells shited to LL in the presence of gabaculine, the antenna size increased to only about 130 chlorophyll molecules (Masuda et al. 2002). This difference reflects a limited ability of gabuculine-treated cells to assemble functional LHCII complexes. Gabuculine had no effect on the transcription of genes encoding LHCII polypeptides or a key enzyme involved in chlorophyll biosynthesis (Masuda et al. 2002). The overall finding from these studies (Masuda et al. 2002, Mortain-Bertrand et al. 1990) is that the actual size of the chlorophyll antenna in *Dunaliella* requires post-translational stabilization that is controlled by chlorophyll availability.

When cells acclimate to low irradiance the increase in pigment composition and associated increase in thylakoid membrane structure actually decreases the optical absorption cross section of the cell on a per chlorophyll basis. Effectively, as cells accumulate more chlorophyll, each chlorophyll molecule becomes less effective in absorbing light. This phenomenon, that has been described in *D. tertiolecta* (Berner et al. 1989), is called the 'package' effect and has important implications when considering the 'costs' associated with building the photosynthetic apparatus because there comes a point at which increasing pigment composition confer no advantage in light absorption (Falkowski and Raven 2007).

Light induced changes in gene expression - Changes in the abundance of structural proteins as well as enzymes in response to photoacclimation most often involve changes to gene expression. Interestingly, while the chloroplast possesses its own genome that encodes for protein components critical to the electron transport chain, the majority of photosynthetic genes reside in the nucleus, with translation taking place in the cytosol prior to the protein being imported into the chloroplast. Included in this group is *Lhc* (previously known as *cab*) a small family of genes that encode the LHC proteins of PSII and PSI. In *D. tertiolecta*, LHCII is encoded by three to five genes, one of which, denoted *Lhcb1* (*cab1*), has been cloned and used as a probe to assess changes in overall *Lhcb* gene expression (LaRoche et al. 1990). Following a shift from 700 to 70 μmol m^{-2} s^{-1}, *Lhcb1* expression was rapidly induced with changes detected as early as 1.5 hours following the shift and an overall 4-fold increase seen after 9 hours (LaRoche et al. 1991), which was similar to fully LL acclimated cells. Treatment of cells with actinomycin D, an inhibitor of transcription, blocked the increase in *Lhcb1* expression and prevented the normal photoacclimatory increases in both LHCII and chlorophyll indicating that transcription is required for photoacclimation (LaRoche et al. 1991).

Lhcb1 expression is also induced transiently when cultures are shifted from HL into total darkness (LaRoche et al. 1991). However, starting at about 24 hours in darkness transcript abundance starts to rapidly decline to the point where by 36 hours *Lhcb1* abundance is below that of HL acclimated cells. This rapid decline in *Lhcb1* abundance is probably not linked to photoacclimation but rather to cell death processes that have been shown to be activated in *D. tertiolecta* in response to prolonged exposure to darkness (Segovia et al. 2003).

While *Lhcb1* is the most widely studied light-responsive gene, it is just one of many that have been shown to be induced by HL in *Dunaliella*. Other genes identified include one encoding the chlorophyll *a* oxygenase, that catalyzes the conversion of chlorophyll *a* to chlorophyll *b* (Masuda et al. 2002) as well as a number isolated from a *D. salina* HL cDNA library (Park et al. 2006), including genes involved in oxidative stress such as superoxide dismutatase and ascorbate peroxidase, as well as a Clp protease (Park et al. 2006).

In both plants and green algae changes in *Lhcb* expression is transcriptionally regulated by changes in binding to promoter elements of specific transcription factors (Argüello-Astorga and Herrera-Estrella 1998) - and while the *trans*-acting factors and *cis*-binding elements involved in phytochrome-mediated *Lhcb* expression in plants are well characterized (Surpin et al. 2002) there is little information to date on the regulation of *Lhcb* expression during photoacclimation (see Chen et al. 2004). This being said, in *D. tertiolecta,* three protein complexes have been identified that display DNA-binding activity specific for *Lhcb1*, with the extent of binding correlated with increased transcript abundance (Chen et al. 2004). *In silico* analysis of the *Lhcb1* promoter has identified a number of possible binding sites for these complexes (Chen et al. 2004). Similarly to *Chlamydomonas reinhardtii* (Hahn and Kück 1999) these *cis*-elements lack any similarity to the conserved GT-1 and G-boxes that seem to be conserved in the *Lhcb* genes of plants (Argüello-Astorga and Herera-Estrella 1998). While there are some regions of conservation between plants and algae, the general lack of consensus would suggest that regulation of the transcription, and perhaps the mechanism of intracellular signalling is fundamentally different (Chen et al. 2004). Besides regulating transcription there is evidence from *D. salina* (Masuda et al. 2003), *D. tertiolecta* (Escoubas et al. 1995), as well as *C. reinhardtii* (Durnford et al. 2003) that a shift from LL to HL results in increased rate of mRNA decay of *Lhcb* transcripts, which in *D. salina* has been shown to significantly decrease the transcript half-life (Masuda et al. 2003).

Xanthophyll cycle - In addition to chlorophyll, the protein complexes of PSII and PSI also bind a range of carotenoids that serve as accessory pigments by absorbing wavelengths not captured by chlorophyll. Carotenoids also have a fundamental role in preventing photo-oxidative damage to the photosynthetic apparatus by serving as a conduit for the de-excitation of both singlet oxygen ($^1O_2$) and the triplet excited state of chlorophyll (Young 1991). The most widely studied carotenoids implicated in photo-protection are three xanthophylls that collectively constitute the xanthophyll cycle: violaxanthin, antheraxanthin, and zeaxanthin (Demmig-Adams et al. 1996, Niyogi 1999). While violaxanthin serves as an accessory pigment, donating excitation energy to chlorophyll *a,* under conditions of high irradiance, violaxanthin is converted into antheraxanthin and subsequently zeaxanthin the two energy-dissipating pigments of the cycle.

Increased synthesis of zeaxanthin results in the diversion of excitation energy away from the reaction centres, and since this energy is lost as heat rather than being used for photochemistry, the xanthophyll cycle is a major component of what is called non-photochemical quenching (Müller et al. 2001).

The operation of the xanthophyll cycle is elegantly regulated by two pH sensitive enzymes, a violaxanthin de-epoxidase and a zeaxanthin epoxidase localized to the thylakoid lumen and tentatively to the stroma respectively (Demmig-Adams et al. 1996). Under conditions of HL the over-reduction of electron transport results in acidification of the thylakoid lumen that activates the violaxanthin depoxidase leading to the rapid conversion of violaxanthin into zeaxanthin (Demmig-Adams et al. 1996). Under conditions of LL when the pH of the lumen increases, zeaxanthin is rapidly converted back into violaxanthin by zeaxanthin epoxidase.

In *D. salina*, acclimation to HL results in a 6-fold increase in xanthophyll cycle pigments on a chlorophyll basis, with an approximate 37-fold increase in zeaxanthin (Jin et al. 2001). The epoxidation state of the xanthophyll cycle, which is often used as a measure of its energy dissipation engagement, is calculated by determining the concentrations of antheraxanthin and zeaxanthin relative to the total amounts of all three pigments [(A+Z)/(A+Z+V)]. Acclimation to HL in *D. salina* resulted in an increase in epoxidation state from 0.31 in LL cells to near 1 upon shifting to HL (Jin et al. 2001). Although xanthophylls are thought to have defined binding sites on LHCII, the question arises if the abundance of LHCII decreases upon acclimation to HL, where does the zeaxanthin bind? On a total dry weight basis, HL-acclimated *D. salina* cells contain about 6.5 times more zeaxanthin than LL-acclimated cells (Jin et al. 2001). It has previously been shown that Cbr, a protein belonging to the LHCII superfamily binds zeaxanthin in *Dunaliella*, and is strongly induced in HL-acclimated cells (Levy et al. 1992, Jin et al. 2001). More recently, Banet et al. (2000) showed that Cbr co-purifies with LHCIIb and that the majority of zeaxanthin, as well as lutein, does seem to bind to LHCIIb. They hypothesize that Cbr may serve to help facilitate or stabilize the binding of zeaxanthin, as well as lutein, to LHCII proteins in HL-acclimated cells (Banet et al. 2000).

Photoacclimation to Changes in Light Quality

Besides changes in irradiance, the photosynthetic apparatus can acclimate to changes in light quality as differences in the spectral composition of incident light lead to unequal excitation of the two photosystems. While the pigment-protein complexes of PSI preferentially absorb longer wavelengths of light, PSII absorbs light of slightly shorter wavelengths. The ability to acclimate in response to changing light quality is ecologically important as many habitats including plants growing under a forest canopy or aquatic photoautotrophs are exposed to light that is altered in its spectral composition. The light reaching a forest floor is depleted in blue-green wavelengths while water strongly attenuates red light (Falkowski and LaRoche 1991, Melis 1998).

State transitions - Acclimation to changes in light quality has been shown to result in changes to the thylakoid membrane composition, ultrastructure, as well as molecular organization (Humbeck et al. 1988, Sense and Senger 1990). The most widely studied response, which is found in almost all photosynthetic organisms, including *Dunaliella* (Harrison and Allen 1993), are what are called state transitions. Usually LHCII is closely associated with PSII (State 1), however under conditions where cells are exposed to light that preferentially excites PSII, a mobile pool of LHCII can migrate away from PSII and increase excitation delivery to PSI (State 2). Migration is triggered by reversible phosphorylation of LHCII that results in its detachment from PSII. Potential over-excitation of PSI would result in the dephosphorylation of LHCII and its migration back to PSII. By maintaining equal excitation of PSII and PSI the efficiency of linear electron transport is kept high.

Changes in photosystem stoichiometry – Another mechanism to ensure equal rates of electron flow through PSI and PSII, which operates over a longer time frame, is the adjustment in the relative content of the two photosystems (Allen and Pfannschmidt 2000). The relative abundance of PSII and PSI has also been shown to change in *D.* salina in response to changes in growth irradiance (Smith et al. 1990, Naus and Melis 1991).

Photoacclimation and Photoreceptors

In both *D. salina* and *D. tertiolecta* changes in growth irradiance lead to dramatic changes in the pigment and protein composition of the photosynthetic apparatus that are triggered by changes in nuclear and chloroplastic gene expression. A question that arises from this work is how does the cell perceive the change in irradiance that, in turn, activates the signalling pathway(s) leading to the change in phenotype? The widely studied process of photomorphogenesis in plants involves defined photoreceptors such as phytochrome and cryptochrome (Fankhauser and Chory 1997) that detect light of specific wavelengths, and through the isolation and study of photomorphogenic mutants in *Arabidopsis*, the role of specific photoreceptors and their associated signal transduction pathways are becoming well characterized. However, photoacclimation to changes in irradiance does not involve changes in the spectral composition of light and it occurs during steady state growth, not during a specific developmental process (Durnford and Falkowski 1997). That photoacclimation is a distinctly different process is supported by Walters et al. (1999) who used the availability of a wide range of photoreceptor mutants of *Arabidopsis* to show that, while they were all impaired in photomorphogenesis, they retained their ability to photoacclimate. In support of this, no phytochrome sequences have been detected in a genomic library of

D. tertiolecta and the phytochrome chromophore has never been isolated from any green alga (Durnford and Falkowski 1997).

Redox Control of Photoacclimation

In contrast to the photoreceptor model of light sensing, early work by Beale and Appleman (1971) implicated a role for the chloroplast electron transport chain in photoacclimation by showing that in the green alga *Chlorella*, cultures grown in the presence of low concentrations of DCMU (3-(3,4-dichlorophenyl)-1,1-dimethyl urea), which inhibits electron transport at the acceptor-side of PSII, resulted in an increase in cellular chlorophyll content. This led to the hypothesis that photoacclimation was not driven by changes in irradiance *per se* but rather by an imbalance between energy availability and that capacity to utilize the energy through metabolism (Falkowski and Owens 1980, Melis et al. 1985).

A significant advancement in our understanding of how the photosynthetic electron transport chain may regulate photoacclimation was achieved by research undertaken by the Hüner laboratory (see Hüner et al.1998, Wilson et al. 2006), which ironically, was studying photosynthetic acclimation to low temperature. Initial experiments using *Chlorella vulgaris* showed that, compared to cells grown at 27°C and 150 μmol m^{-2} s^{-1}, cultures acclimated to 5°C and the same irradiance displayed characteristics of high-light acclimation that included a decrease in LHCII abundance, increase chlorophyll *a*/*b* ratio and increased carotenoid/chlorophyll ratio (Maxwell et al. 1994). So why would acclimation to low temperature mimic the effects of acclimation to high light? It was hypothesized that both conditions result in an imbalance whereby the photosynthetic apparatus absorbs more light than it can effectively dissipate through metabolism. It is easy to see how a 10-fold increase in growth irradiance would cause an imbalance but the same effect can be obtained without changing irradiance at all but by simply lowering the growth temperature. Since the enzyme-catalyzed rates of the Calvin cycle and other metabolic sinks, and ultimately growth rate, would be much slower at 5°C, it was reasoned that that the cell's demands for light energy would be much lower (Maxwell et al. 1994).

Measuring the Redox State of Photosynthetic Electron Transport

One can assess photosynthetic energy balance non-invasively in intact cells by determining the redox state of PSII through the use of chlorophyll *a* fluorometry (see Krause and Weis 1991). While typically less than 3% of absorbed light is re-emitted as fluorescence (Krause and Weis 1991), the amount varies depending upon the redox state of Q_A, the primary electron acceptor of PSII. The redox state of Q_A reflects the overall redox state of the electron transport chain (Krause and Weis 1991), and it has been shown that chlorophyll *a* fluorescence can be used to monitor changes in electron transport and overall photosynthesis (Schreiber et al. 1994).

The relative reduction state of Q_A, which is $[Q_A^-]/[Q_A] + [Q_A^-]$, can be estimated *in vivo* by the pulse amplitude modulated (PAM) chlorophyll *a* florescence quenching parameter 1-qP (Schreiber et al. 1994). An increase in 1-qP has come to be referred to as the 'excitation pressure' on PSII and reflects the balance between the rate of reduction of PSII reaction centres through light absorption and the rate of their oxidation through intersystem electron transport. While excitation pressure (1-qP) is zero in the dark, it reaches a maximum (a value of 1) when all PSII centres are closed, which is achieved experimentally by exposing cells to very high irradiance (2,000 – 6,000 μmol m^{-2} s^{-1}).

Redox Control of Photoacclimation in *Dunaliella*

The findings using *C. vulgaris* that acclimation to high irradiance can be mimicked by low temperature was also found to be true for *D. salina*. Compared to cells grown at 30°C and 150 μmol m^{-2} s^{-1} (30/150), *D. salina* grown at low temperature (13°C) and the same irradiance (13/150) displayed a decrease in chlorophyll/cell from 1.96 to 0.24 pg/cell and an increase in chlorophyll *a/b* from 4.5 to 13.7 (Maxwell et al. 1995a and n). The surprising finding was that these low temperature-acclimated cultures where virtually indistinguishable from cultures grown at 30°C but at a 16-fold greater irradiance (2500 μmol m^{-2} s^{-1}) with respect to pigmentation (chlorophyll and carotenoid), LHCII and Cbr protein content (Figure 2). The similarity was explained by the fact that cultures grown at either 13/150 or 30/2500 were exposed to comparably high excitation pressures (Maxwell et al. 1995a and b) (Figure 2). In contrast, cultures grown at 30/150 had a excitation pressure (1-*qP*) of 0.10 (indicating that 90% of PSII centres are open) that would be achieved at 13°C but required cells to be grown at a 8-fold lower irradiance (20 μmol m^{-2} s^{-1}) (Maxwell et al. 1995a and b).

Redox Modulation of Nuclear Gene Expression

Changes in excitation pressure have far reaching consequences that are not limited to the chloroplast. As mentioned previously, LHCII is encoded by a family of nuclear genes, and as shown in Figure 2C, the *Lhcb* transcript abundance is comparably high in cultures of *D. salina* grown at either 30/150 or 13/20 and comparably much lower in the two conditions that elicit high excitation pressure, either 13/150 or 30/2500.

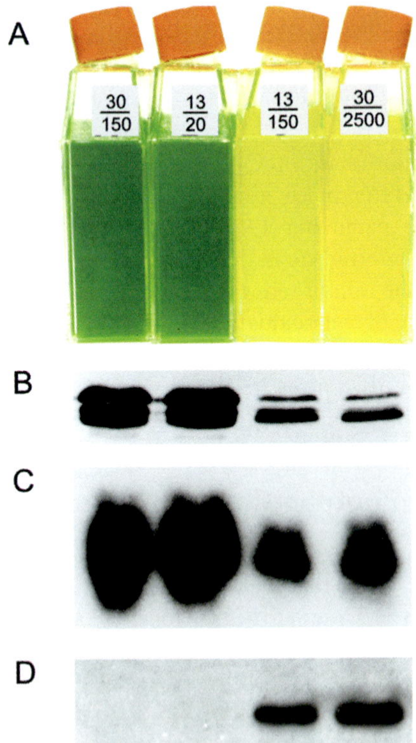

Figure 2: The effects of growth regime on *D. salina*. (A) pigmentation of cultures; (B) immunoblot showing the accumulation of light-harvesting polypeptides (Lhcb) (C) RNA blot showing *Lhcb* mRNA accumulation; and (D) immunoblot showing the accumulation of the carotenoid-binding protein, Cbr. Cells grown at either 30°C and 2500 μmol m^{-2} s^{-1} (30/2500) or 13/150 were exposed to comparable high PSII excitation pressure whereas cells grown at either 30/150 or 13/20 were exposed to comparable low PSII excitation pressure. The different effect of high or low PSII excitation pressure are very distinct.

Concurrent with the work by the Hüner laboratory, Falkowski and co-workers (Escoubas et al. 1995) used a pharmacological approach to modulate the redox state of intersystem electron transport. The group employed two inhibitors of photosynthetic electron transport: DCMU which inhibits the oxidation of Q_A, and DBMIB (2,5-dibromo-3-methyl-6-isopropyl-p-benzoquinone) which binds to the cytochrome b_6f complex. Untreated cells of *D. tertiolecta* shifted from HL to LL resulted in an increase in chlorophyll/cell as well as a 3-fold increase in *Lhcb* gene expression as measured by run-on transcription (Escoubas et al. 1995). A similar increase in chlorophyll as well as a 2-fold increase in *Lhcb* expression could be attained by simply treating HL grown cells with DCMU, which by blocking electron transport at PSII, resulted in the complete oxidation of downstream electron transport components. Thus DCMU mimics the effects of a decrease in irradiance. Similarly, when LL acclimated cells were incubated with DBMIB cellular chlorophyll levels dropped by 25% and there was a clear reduction in *Lhcb* gene expression (Escoubas et al. 1995). By blocking electron flow at cytochrome b_6f complex DBMIB treatment results in over-reduction of the plastoquinone (PQ) pool that would also occur if cells were shifted from LL to HL conditions. These data provided evidence that it is the redox state of intersystem electron transport, specifically the plastoquinone pool, that acts as a sensor that triggers photoacclimation.

Photostasis

The research employing both *D. salina* or *D. tertiolecta* suggest that a fundamental aspect of photoautotrophic growth is the maintenance of balance between energy absorption by the photosystems with the capacity to utilize the trapped energy through metabolism. This balance is refereed to as photostasis (Ensminger et al. 2006) and can be formally described using the following equality (Falkowski and Chen 2003) (Figure 3),

$$\sigma_{PSII} \cdot E_k = \tau^{-1}$$

where σ_{PSII} is the functional absorption cross-section of PSII, E_k is the irradiance (I) at which the maximum photosynthetic quantum yield balances photosynthetic capacity, and τ^{-1} is the rate at which photosynthetic electrons are consumed in metabolism. The product, $\sigma_{PSII} \cdot E_k$ represents the rate at which photons are effectively harvested by PSII and electrons injected into the photosynthetic electron transport chain (Falkowski and Chen 2003), and is synonymous with excitation pressure (Maxwell et al. 1995).

Exposure of *Dunaliella* to an irradiance that exceeds E_k results in a departure from photostasis that is characterized by $\sigma_{PSII} \cdot E_k > \tau^{-1}$. This occurs because the rate at which energy absorbed by PSII and the rate of electron transport exceed the capacity of the metabolic sinks. This has the effect of increasing the excitation pressure (1-qP) on PSII. It is suggested by the model that it is the corresponding over-reduction of the electron transport chain that acts as the sensor which initiates photoacclimatory mechanisms that attempt to regain photostasis ($\sigma_{PSII} \cdot E_k = \tau^{-1}$) (Figure 3). This is achieved, in part, by reducing σ_{PSII} through decreasing LHCII and chlorophyll abundance as well as increased non-photochemical quenching mechanisms such as activation of the xanthophyll cycle (Falkowski and Chen 2003).

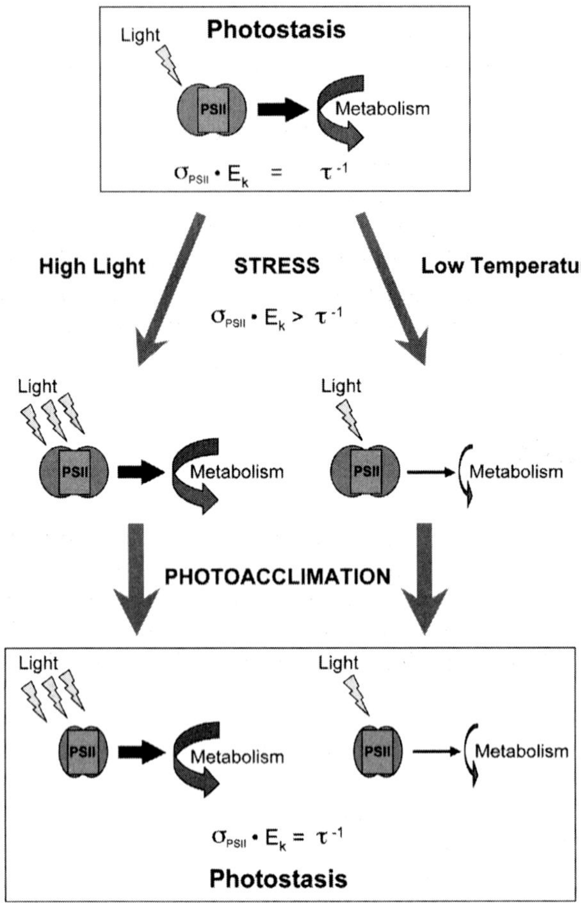

Figure 3: Schematic illustration of photostasis as define by the equation $\sigma_{PSII} \cdot E_k = \tau^{-1}$ (Falkowski and Chen 2003), where σ_{PSII} is the effective absorption cross-section of PSII, E_k is the irradiance at which the maximum photosynthetic quantum yield balances photosynthetic capacity, and τ^{-1} is the rate at which photosynthetic electrons are consumed by metabolism. Photostasis exists when the rate of light absorption ($\sigma_{PSII} \cdot E_k$) is balanced by the rate of energy utilization (τ^{-1}). High light stress upsets photostasis by increasing ($\sigma_{PSII} \cdot E_k$), while low temperature stress upsets photostasis by decreasing (τ^{-1}). A return to photostasis is driven by photoacclimatory changes to the photosynthetic apparatus. Regardless of the stress, in *Dunaliella* the response is to decrease $\sigma_{PSII} \cdot E_k$ through: decreased LHCII size, decreased chlorophyll/cell and increased xanthophylls cycle pigments.

The inequality $\sigma_{PSII} \cdot E_k > \tau^{-1}$ can also be arrived at without changing irradiance but by decreasing the temperature. While light harvesting and energy transfer, and thus, $\sigma_{PSII} \cdot E_k$, are essentially insensitive to temperature in the range relevant to biology, the rate of enzyme-catalyzed reactions, decrease appreciably with temperature, which is reflected in a decrease in τ^{-1}. The Q_{10} for most biological processes is approximately 2 (a two-fold change in reaction rate for every 10°C change in temperature). In *C. vulgaris* for example, a 10°C decrease in temperature results in a suppression of growth rate by a factor of 2.3 (Wilson and Hüner 2000). Thus, at low temperature even moderate irradiance (100-200 μmol m^{-2} s^{-1}) can be considered 'high light' since the metabolic demand for the energy derived from light harvesting is restricted. This explains why *D. salina* exposed to low temperature mimics exposure to high light - both result in an inequality and in both cases the cell attempts to regain photostasis by decreasing $\sigma_{PSII} \cdot E_k$.

Photostasis in Algae versus Plants

Research using *Dunaliella* was critical in elucidation of the concept of photostasis in photoacclimation because, unlike plants, the phenotype of *Dunaliella* changes so dramatically when photostasis is disrupted. Compared to *Dunaliella* however, plants, after exposure to high excitation pressure ($\sigma_{PSII} \cdot E_k > \tau^{-1}$), return to photostasis using a fundamentally different approach. Instead of adjusting components and processes involved in light absorption plants seem to regain photostasis by upregulating carbon metabolism (increasing τ^{-1}). This is achieved by increasing the levels of key enzymes of photosynthetic carbon metabolism including Rubisco, stromal and cytosolic fructose bisphosphatase and sucrose phosphate synthase (see Ensminger et al. 2006). Species of green algae such as *Dunaliella* and *Chlorella* seem unable to adjust the capacity of electron consuming sinks (τ^{-1}) in response to either excess light or low temperature (Savitch et al. 1996) and thus must rely on adjusting σ_{PSII} in order to regain photostasis. Interestingly, the model plant *Arabidopsis thalina* seems to possess an intermediate acclimation mechanism. Compared to other plant species examined, *A. thalina* can adjust τ^{-1} only moderately, and thus also relies on increases in non-photochemical quenching via the xanthophylls cycle to reduce σ_{PSII} (Savitch et al. 2001).

Redox Sensing

The redox state of the plastoquinone pool has been implicated in regulating a wide range of photoacclimatory processes. In a seminal piece of work it was Allen et al. (1981) who showed that state transitions were regulated not specifically by changes to the spectral composition of light but rather the redox state of plastoquinone.

Light that preferentially excites PSII over reduces the plastoquinone pool leading to migration of some LHCII proteins to PSI while light that preferentially excites PSI would lead to oxidation of the plastoquinone pool and migration of LHCII components to PSII. This has recently been followed by the identification of the kinase that phosphoyrlates the LHCII pool regulating its migration (Snyder and Kohorn 1999, 2001). Related to this, work by Pfannschmidt et al. (2001) has shown that changes in photosystem stoichiometry are regulated by the plastoquinone redox state by its ability to modulate the expression of the chloroplastic genes that encode the D1 protein (*psbA*) and the two polypeptides that make up the PSI core (*psaAB*).

While the redox state of the plastoquinone pool seems to regulate many acclamatory processes, evidence suggests that it is not the sole regulator of photoacclimation. For example, the trans-thylakoid pH gradient, which is strongly correlated with excitation pressure on PSII, has been shown to be involved in non-photochemical quenching by regulating the key enzymes involved in the xanthophyll cycle (Wilson and Hüner 2000, Li et al. 2004). In addition, the trans-thylakoid pH gradient is implicated in having an important, complementary role, in regulating *Lhcb1* expression in *D. tertiolecta* (Chen et al. 2004).

Signal Transduction

A major avenue of research with regards to redox control of photoacclimation is the elucidation of the components of the signal transduction pathway that links redox changes in the chloroplast with alterations in gene expression in the nucleus? In *D. tertiolecta*, Escoubas et al. (1995) demonstrated that the increase in chlorophyll/cell characteristic of a HL to LL shift could be inhibited by between 33-56% by pre-treating cells with low concentrations of a protein phosphatase inhibitor such as okadaic acid. This finding indicated that protein phosphorylation, which is involved in a wide range of intracellular signalling pathways (see Karin and Hunter 1995), may be critical in conveying the redox signal from the thylakoid membrane to the nucleus. Escoubas et al. (1995) proposed a working model for this pathway suggesting that reduction of the PQ pool induces the phosphorylation of a chloroplastic phosphoprotein that in turn is exported from the chloroplast to the cytosol where it activates a protein kinase. This kinase in turn regulates *Lhcb* transcription through activation of a specific transcription factor.

The molecular genetic dissection of signal transduction using *Dunaliella* has been limited by the lack of genome-wide sequence data in combination with the lack of a reliable genetic transformation system. While nuclear integration and expression of transgenes is now routine in *Chlamydomonas reinhardtii* (Grossman et al. 2003), transformation of *Dunaliella* has been found to be particularly recalcitrant (Walker et al. 2005a), for reasons that include strong intrinsic antibiotic resistance as well as problematic cell wall and membrane properties (Walker et al. 2005b). However, recent reports of successful nuclear transformation for *D. tertiolecta* (Walker et al. 2005a) and *D. salina* (Jin et al. 2001, Geng et al. 2003) suggest that a reliable and stable method of transformation may be arrived at shortly. This, combined with the initiation of a U.S. Department of Energy-funded *D. salina* sequencing project (J. Polle, personal communication), suggest that *Dunaliella* may soon prove to be a useful molecular system, as well as an excellent physiological system, for studying photoacclimation.

The most powerful method to identify signal transduction components involved in the chloroplast-to-nucleus pathway requires the use of genetic screens in order to identify mutants and subsequently isolate genes involved in the pathway. This method has proven successful in understanding a wide-range of intracellular signalling pathways particularly in model systems such as yeast and *Arabidopsis*. Research using mutants of *Arabidopsis* indicate that intermediates in chlorophyll biosynthesis, specifically the H-unit of Mg-chelatase and Mg-protoporphyrin IX may play a role in intracellular communication between the chloroplast and nucleus (Kropat et al. 1997). However, work in *C. vulgaris* suggest that Mg-protoporphyrin IX does not seem to play a role in regulating *Lhcb1* expression (Wilson et al. 2003). Alternatively, there is evidence that the redox changes in the chloroplast may be signalling to the nucleus through alterations in the production of reactive oxygen species (ROS) (Karpinski et al. 1999). In plants, numerous genes have been seen to be regulated either directly by ROS formation (Mittler et al. 2004) or by components that constitute part of the antioxidant defense system, most notably ascorbate peroxidase (Karpinski et al.1997). A possible role for ROS involvement in *Lhcb* regulation has recently been shown in *C. vulgaris* where ROS formation is positively correlated with excitation pressure on PSII (Wilson et al. 2003). It is interesting to note that there is evidence that ROS is involved as an intracellular signalling molecule linking mitochondrial dysfunction to changes in nuclear gene expression (Maxwell et al. 2002).

Photostasis as a Sensor of Environmental Stress

As photostasis reflects overall cellular energy balance ($\sigma_{PSII} \cdot E_k > \tau^{-1}$), any environmental stress will cause an energy imbalance usually because it will restrict the ability of the organism to utilize the absorbed energy. Besides light and temperature stress this would also include a range of other factors including nutrient deprivation, low water availability, and low CO_2 concentration. There are a number of examples in literature of stress factors other than light and temperature that result in an acclimation response that may, in part, be driven by an over-reduction of the plastoquinone pool (Wykoff et al. 1998; Moseley et al. 2002). For example, in *D. tertiolecta,* transfer of cells to low nitrogen conditions results in a decrease in total chlorophyll, increase in the chlorophyll a/b ratio, decrease in photochemical efficiency, and increase in carotenoids/chlorophyll ratio (Young and Beardall 2003). This modulation of the photosynthetic apparatus is strikingly similar to that seen when *D. tertiolecta* is exposed to high irradiance. The potential importance of the maintenance of photostasis in the acclimation response to nutrient limitation has been examined by Geider et al. (1998). In an in-depth study with *D. tertiolecta*, the adjustments to the photosynthetic apparatus in response to phosphorus and nitrogen limitation were found to decrease reductant supply (light absorption and photochemistry) such that it matched the low reductant demand (metabolism) imposed by the nutrient stress. Interestingly, the reestablishment of photostasis in nitrogen limited versus phosphorus limited cells, was achieved by different mechanisms (Geider et al. 1998).

The study of stress responses in photosynthetic organisms either in the laboratory or under natural field conditions will undoubtedly involve changes brought about by changes in excitation pressure and not only by the specific stress being studied. Therefore, studies purporting that molecular, physiological and biochemical changes are due to a specific environmental stress factor need to account for the fact that a common component of the response to any stress factor will be to respond to the increase in excitation pressure, and movement away from photostasis. Due to this, additional control experiments are needed to separate excitation pressure effects from those due specifically to the stress being studied. For example, plants grown under iron stress result in a decrease in chlorophyll and LHCII abundance. However these changes were found to be completely reversible if iron-stressed plants were grown under a lower irradiance (Abadia et al. 2000). The identification of so-called 'nitrogen-stress induced genes' or 'salt-induced processes' as examples, needs to be confirmed as not the result of an energy imbalance by the undertaking of low light control experiments. Such experiments have already been done using microarrays to separate out changes in gene expression caused by cold and high-light from those induced by increased excitation pressure (NDong et al. 2001). While more costly and time-consuming, such additional experiments are required if one is ever going to parse the complex interactions that exist in photosynthetic organisms exposed to environmental stress.

Conclusion—Photoacclimation Redefined

Historically, photoacclimation has focused on the alterations to the photosynthetic apparatus that occur in response to a change in irradiance. However, research using species of *Dunaliella* has shown that photoacclimation is driven not by changes in absolute irradiance *per se* but rather to changes in the redox state of photosynthetic electron transport. By utilizing electron transport as the 'light sensor', photosynthetic organisms are able to readily integrate the light conditions under which they grow with a myriad of other factors including ambient temperature and nutrient status, which influence energy utilization. Thus, the term photoacclimation should be redefined as the molecular, physiological and biochemical alterations that occur upon the disruption of photostasis.

References

Abadia, J., F. Morales and A. Abadia. 2000. Photosystem II efficiency in low chlorophyll, iron deficient leaves. Plant and Soil 215: 183-192.

Allen, J.F., J. Bennett. K.E. Steinback. and C.J. Arntzen. 1981. Chloroplast protein phosphorylation couples plastoquinone redox state to distribution of excitation energy between photosystems. Nature 291: 21-25.

Allen, J.F. and T. Pfannschmidt. 2000. Balancing the two photosystems: photosynthetic electron transfer governs transcription of reaction centre genes in chloroplasts. Phil. Trans. R. Soc. London. B. 355: 1351-1359.

Anderson, J.M., W.S. Chow and Y.I. Park. 1995. The grand design of photosynthesis: acclimation of the photosynthetic apparatus to environmental cues. Photosynth. Res. 46: 129–139.

Argüello-Astorga, G. and L. Herrera-Estrella. 1998. Evolution of light-regulated plant promoters. Annu. Rev. Plant Physiol. Plant Mol. Biol. 49: 525–555.

Banet, G., U. Pick and A. Zamir. 2000. Light-harvesting complex II pigments and proteins in association with Cbr, a homolog of higher-plant early light-inducible protein in the unicellular green alga *Dunaliella*. Planta 210: 947-955.

Beale, S.I. and D. Appleman. 1971. Chlorophyll synthesis in *chlorella*. Plant Physiol. 47: 230–235.

Berner, T., K. Wyman, Z. Dubinsky and P.G. Falkowski. 1989. Photo-adaptation and the "package" effect in *Dunaliella tertiolecta* (Chlorophyceae). J. Phycol 25: 70-78

Blankenship, R.E. 2002. Molecular Mechanisms of Photosynthesis. Blackwell Publishing Limited, UK.

Boardman, N.K., J.M. Anderson and D.J. Goodchild.1978. Chlorophyll-protein complexes and structure of mature and developing chloroplasts. pp 35-109. In Current Topics in Bioenergetics. Vol 8.: Academic Press. New York, USA.

Chen, Y-B., D.G. Durnford, M. Koblizek and P.G. Falkowski. 2004. Plastid regulation of *Lhcb1* transcription in the chlorophyte alga *Dunaliella tertiolecta*. Plant Physiol. 136: 3737-3750.

Chitlaru, E. and U. Pick. 1991. Regulation of glycerol synthesis in response to osmotic changes in *Dunaliella*. Plant Physiol. 96: 50-60.

Demmig-Adams, B., I. Adams and W. William. 1996. The role of xanthophyll cycle carotenoids in the protection of photosynthesis. Trends Plant Sci. 1: 21–26.

Durnford, D.G. and P.G. Falkoskwi. 1997. Chloroplast redox regulation of nuclear gene transcription during photoacclimation. Photosynth. Res. 53: 229–241.

Durnford, D.G., J.A. Price, S.M. McKim and M.L. Sarchfield. 2003. Light-harvesting complex gene expression is controlled by both transcriptional and post-transcriptional mechanisms during photoacclimation in *Chlamydomonas reinhardtii*. Physiol. Plant. 118: 193–205.

Ensminger, I., F. Busch and N.P.A. Hüner. 2006. Photostasis and cold acclimation: sensing low temperature through photosynthesis. Physiol. Plant. 126: 28-44.

Escoubas, J.M., M. Lomas, J. LaRoche and P.G. Falkowski. 1995. Light intensity regulation of cab gene transcription is signaled by redox state of the plastoquinone pool. Proc. Natl. Acad. Sci. USA. 92: 10237-10241.

Falkowski, P.G. 1984. Kinetics of light intensity adaptation in *Dunaliella tertiolecta*, a marine plankton chlorophyte. Photosyn. Res. 18: 62-68.

Falkowski, P.G. and T.G. Owens. 1980. Light-shade adaptation: two strategies in maine phytoplankton. Plant Physiol. 66: 592-595.

Falkowski, P.G. and J. LaRoche. 1991. Acclimation to spectral irradiance in algae. J. Phycol. 27, 8-14.

Falkowski, P.G. and Y-B. Chen. 2003. Photoacclimation of light harvesting systems in eukaryotic algae. pp. 423-447. In Light-Harvesting Antennas in Photosynthesis. B.R. Green and W.W. Parson (eds). Kluwer Academic Publishers, Dordrecht, The Netherlands.

Falkowski, P.G. and J.A. Raven. 2007. Aquatic Photosynthesis. Princeton University Press, Princeton, New Jersey, USA.

Fankhauser, C. and J. Chory. 1997. Light control of plant development. Annu. Rev. Cell and Dev. Biol. 13: 203-229.

Geng, D., Y. Wang, P. Wang, W. Li and Y. Sun. 2003. Stable expression of hepatitis B surface antigen gene in *Dunaliella salina.* (chlorophyllorophyta). J. Appl. Phycol. 15:451–6.

Geider, R.J., H.L. MacIntyre, L.M. Graziano and R.M.L. McKay. 1998. Responses of the photosynthetic apparatus of *Dunaliella tertiolecta* (Chlorophyceae) to nitrogen and phosphorus limitation. Eur. J. Phycol. 33: 315-332.

Green, B.R. and D.G. Durnford. 1997. The chlorophyll-carotenoid proteins of oxygenic photosynthesis. Annu. Rev. Plant Physiol. Plant Mol. Biol. 47: 685-714.

Grossman, A.R., E.E. Harris, C. Hauser, P.A. Lefebvre, D. Martinez, D. Rokhsar, J. Shrager, C.D. Silflow and D. Stern. 2003. *Chlamydomonas reinhardtii* at the crossroads of genomics. Euk. Cell 2: 1137-1150.

Haehnel, W. 1984. Photosynthetic electron transport in high plants. Annu. Rev. Plant Physiol. 35: 659-693.

Hahn, D. and U. Kück. 1999. Identification of DNA sequences controlling light- and chloroplast-dependent expression of the *lhcb1* gene from *Chlamydomonas reinhardtii.* Curr. Genet. 34: 459–466.

Harrison, M.A. and J.F. Allen. 1993. Differential phosphorylation of individual LHC-II polypeptides during short-term acclimation to light regime in the green alga *Dunaliella salina*. Biochim. Biophys. Acta 1141: 37-44.

Humbeck, K., B. Hoffman and H. Senzer. 1988. Influence of energy flux and quality of light on the molecular organization of the photosynthetic apparatus in *Scenedesmus.* Planta 173: 205–12.

Hüner, N.P.A., G. Oquist and F. Sarhan. 1998. Energy balance and acclimation to light and cold. Trends Plant Sci. 3: 224-230.

Jacquot, J.P., N. Rouhier and E. Gelhaye. 2002. Redox control by dithiol-disulfide exchange in plants: I. the chloroplastic systems. Ann. N.Y. Acad. Sci. 973: 508-519.

Jin, E.S., J.E.W. Polle and A. Melis. 2001. Involvement of zeaxanthin and of the Cbr protein in the repair of photosystem II from photoinhibition in the green alga *Dunaliella salina.* Biochem. Biophys. Acta 1506: 244-259.

Karin, M. and T. Hunter. 1995. Transcriptional control by protein phosphorylation: signal transmission form the cell surface to the nucleus. Curr. Biol. 5: 747-757.

Karpinski, S., C. Escobar, B. Karpinska, G. Creissen. and P.M. Mulllineaux. 1997. Photosynthetic electron transport regulates the expression of cytosolic ascorbate peroxidase genes in Arabidopsis during excess light stress. Plant Cell 9: 627–640.

Karpinski, S., H. Reynolds, B. Karpinska, G. Creissen and P.M. Mulllineaux. 1999. Systemic signaling and acclimation in response to excess excitation energy in Arabidopsis. Science 284: 654–657.

Krause, G.H. and E. Weis. 1991. Chlorophyll fluorescence and photosynthesis: the basics. Ann. Rev. Plant Physiol. Plant Mol. Biol. 42: 313-349.

Kropat, J., U. Oster, W. Rudiger and C.F. Beck. 1997. Chlorophyll precursors are signals of chloroplast origin involved induction of nuclear heat-shock genes. Proc. Natl. Acad. Sci. U.S.A. 14168-14172.

LaRoche, J., J. Bennett and P.G. Falkowski. 1990. Characterization of a cDNA encoding for the 28.5-kDa LHCII apoprotein from the unicellular marine chlorophyte, *Dunaliella tertiolecta*. Gene 95: 165-171.

LaRoche, J., A. Mortain-Bertrand and P.G. Falkowski. 1991. Light intensity-induced changes in cab mRNA and light harvesting complex II apoprotein levels in the unicellular chlorophyte *Dunaliella tertiolecta*. Plant Physiol. 97: 147-153.

Levy, H., I. Gokman and A. Zamir. 1992. Regulation and light-harvesting complex II association of a *Dunaliella* protein homologous to early light-induced proteins in higher plants. J. Biol. Chem. 267: 18831-18836.

Li, X-P., A.M. Gilmore, S. Caffarri, R. Bassi, T. Golan, D. Kramer and K.K. Niyogi. 2004. Regulation of Photosynthetic Light Harvesting Involves Intrathylakoid Lumen pH Sensing by the PsbS Protein. J. Biol. Chem. 279: 22866-22874.

Long, S.P., S. Humphries and P.G. Falkowski. 1994. Photoinhibition of photosynthesis in nature. Annu. Rev. Plant Physiol. Plant Mol. Biol. 45: 633-662.

MacIntryre, H.L., T.D. Sharkey and R.J. Geider. 1997. Activation and Deactivation of Ribulose-1,5-bisphosphate carboxylase/oxygenase (Rubisco) in three marine microalgae. Photosynth. Res. 51: 93-106.

Malkin, S. and K. Niyogi. 2000. Photosynthesis. pp. 568-628. In Biochemistry and Molecular Biology of Plants. B. Buchanan, W. Gruissem and R.L. Jones (eds) American Society of Plant Physiologists, Rockville Maryland, USA.

Masuda, T., J.E.W. Polle and A. Melis. 2002. Biosynthesis and distribution of chlorophyll among the photosystems during recovery of the green alga *Dunaliella salina* from irradiance stress. Plant Physiol. 128: 603–614.

Masuda, T., A. Tanaka and A. Melis. 2003. Chlorophyll antenna size adjustments by irradiance in *Dunaliella salina* involve coordinate regulation of chlorophyll a oxygenase (CAO) and Lhcb gene expression. Plant Mol. Biol. 51: 757–771.

Maxwell, D.P., S. Falk C.C. Trick and N.P.A. Hüner. 1994. Growth at low temperature mimics high-light acclimation in *Chlorella vulgaris*. Plant Physiol. 105: 535–543.

Maxwell, D.P., D.E. Laudenbach and N.P.A. Hüner. 1995. Redox regulation of light-harvesting complex II and *cab* mRNA abundance in *Dunaliella salina*. Plant Physiol. 109: 787–795.

Maxwell, D.P., R. Nickels and L. McIntosh. 2002. Evidence of mitochondrial involvement in the transduction of signals required for the induction of genes associated with pathogen attack and senescence. Plant J. 29: 452-458.

Melis, A. 1998. Photostasis in plants: mechanisms and regulation. pp. 207-21. In Photostasis and Related Phenomena. W. Thistle (ed) Lenum Press, New York, USA.

Melis, A. 1999. Photosystem-II damage and repair cycle in chloroplasts: What modulates the rate of photodamage? Trends Plant Sci. 4: 130-135.

Melis, A., A. Manodoria, R.E. Glick, M.L. Ghirardi, S.W.McCauley and P.J. Neale. 1985. The mechanism of photosynthetic membrane adaptation to environmental stress conditions: a hypothesis on the role of electron-transport capacity and of ATP/NADPH pool in the regulation of thylakoid membrane organization and function. Physiol. Vég. 23: 757-765.

Mortain-Bertrand, A., J. Bennett, P.G. Falkowski. 1990. Photoregulation of the light-harvesting chlorophyll protein complex associated with photosystem II in *Dunaliella tertiolecta*. Plant Physiol. 94: 304-311.

Moseley, J.L., T. Allinger, S. Herzog, P. Hoeth, E. Wehinger, S. Merchant and M. Hippler. 2002. Adaptation to Fe-deficiency requires remodeling of the photosynthetic apparatus. EMBO J. 21: 6709-6720.

Müller, P., X-P. Li and K.K. Niyog. 2001. Non-photochemical quenching. A response to excess light energy. Plant Physiol. 125: 1558-1566.

Naus, J. and A. Melis. 1991. Changes of Photosystem Stoichiometry during Cell Growth in *Dunaiella salina* Cultures. Plant and Cell Physiol. 32: 569-575.

NDong, C., J. Danyluk and N.P.A. Hüner. 2001. Survey of gene expression in winter rye during changes in either growth temperature, irradiance or excitation pressure. Plant Mol. Biol. 45: 691–703.

Niyogi, K.K. 1999. Photoprotection revisited: genetic and molecular approaches. Annu. Rev. Plant Physiol. Plant Mol. Biol. 50: 333-359.

Nishigaki, A., S. Ohshima and K. Nakayama. 2000. Characterization of three forms of light-harvesting chlorophyll a/b-protein complexes of photosystem II isolated from the green alga *Dunaliella salina*. Plant Cell Physiol. 41: 591-599.

Park, S., J.E.W. Polle, A. Melis, T.K. Lee and E. Jin. 2006. Up-regulation of photoprotection and PSII-repair gene expression by irradiance in the unicellular green alga *Dunaliella salina* Marine Biotech. 8: 120-128.

Pfannschmidt, T., A. Nilsson and J.F. Allen. 1999. Photosynthetic control of chloroplast gene expression. Nature 397: 625–628.

Pfannschmidt, T., K.Schutze, M.Brost and R. Oelmüller. 2001. A novel mechanism of nuclear photosynthesis gene regulation by redox signals from the chloroplast during photosystem stoichiometry adjustment. J. Biol. Chem. 276: 36125–36130.

Rodermel, S. 2001. Pathways of plastid-to-nucleus signaling. Trends Plant Sci. 6: 471–478.

Savitch, L.V., D.P. Maxwell and N.P.A. Hüner. 1996. Photosystem II excitation pressure and photosynthetic carbon metabolism in *Chlorella vulgaris*. Plant Physiol. 111: 127-136.

Savitch, L.V., J. Barker-Åstrom, A.G. Ivanov, V. Hurry, G. Öquist, N.P.A. Hüner and P. Gardeström. 2001. Cold Acclimation of *Arabidopsis thaliana* results in incomplete recovery of photosynthetic capacity, associated with an increased reduction of the chloroplast stroma. Planta 214: 295-303.

Segovia, M., L. Haramaty, J.A. Berges and P.G. Falkowski. 2003. Cell death in the unicellular chlorophyte *Dunaliella tertiolecta*. Plant Physiol. 132: 99-105

Schreiber, U., W. Bilger and C. Neubauer. 1994. Chlorophyll fluorescence as a nonintrusive indicator for rapid assessment of in vivo photosynthesis. pp. 49-70. In Ecophysiology of Photosynthesis. E-D. Schulze and M.M. Caldwell (eds) Springer-Verlag.

Sense, M. and H. Senger. 1990. Functional changes in the photosynthetic apparatus during light adaptation of the green alga *Chlorella fusca*. J. Photochem. Photobiol. B Biol. 8: 63–71.

Smith, B.M., P.F. Morrissey, J.E. Guenther, J.A. Nemson, M.A. Harrison and A. Melis. 1990. Response of the photosynthetic apparatus in *Dunaliella salina* to irradiance stress. Plant Physiol. 93: 1433-1440.

Snyders, S. and B. Kohorn. 1999. TAK's, Thylakoid Membrane Kinases associated with energy transduction. J. Biol. Chem. 274: 9137-9140.

Snyders, S. and B. Kohorn. 2001. Disruption f thylakoid kinase TAK1 leads to alteration of light energy transduction in *Arabidopsis*. J. Biol. Chem. 276: 32169-32176.

Sukenik, A., K.D. Wyman, J. Bennett and P.G. Falkowski. 1987. A novel mechanism for regulating the excitation of photosystem II in a green alga. Nature 327: 704-707.

Sukenik, A., J. Bennett and P.G. Falkowski. 1988. Changes in the abundance of individual apoproteins of light-harvesting chlorophyll a/b-protein complexes of photosystem I and II with growth irradiance in the marine chlorophyte *Dunaliella tertiolecta*. Biochim. Biophys. Acta. 932: 206–215.

Sukenik, A., J. Bennett, A. Mortain-Bertrand and P.G. Falkowski. 1990. Adaptation of the photosynthetic apparatus to irradiance in *Dunaliella tertiolecta*. A kinetic study. Plant Physiol. 92: 891-898.

Surpin, M., R.M. Larkin and J. Chory. 2002. Signal transduction between the chloroplast and the nucleus. Plant Cell (Suppl) 14: s327–s338.

Tanaka, A. and A. Melis. 1997. Irradiance-dependent changes in the size and composition of the chlorophyll a-b light-harvesting pigment-protein complex in the green alga *Dunaliella salina*. Plant Cell Physiol. 38: 17-24.

van Amerongen, H. and J.P. Dekker. 2003. Light-harvesting in Photosystem II. pp 219-251. In Advances in Photosynthesis and Respiration: Light Harvesting Antennas in Photosynthesis, Vol. 13. B.R. Green and W.W. Parsons (eds) Kluwer Academic Publishers. Dordrecht. The Netherlands

Walker, T.L., D.K. Becker, J.L. Dale and C. Collet. 2005a. Towards the development of a nuclear transformation system for *Dunaliella tertiolecta*. J. App. Phycol. 17: 363-368.

Walker, T.L., C. Collet and S. Purton. 2005b. Algal transgenics in the genomic era. J. Phycol. 41: 1077-1093.

Walters, R.G., J.J.M. Rogers, F. Shephard and P. Horton. 1999. Acclimation of *Arabidopsis thaliana* to the light environment: the role of photoreceptors. Planta 209: 517-527.

Wilson, K.E. and N.P.A. Hüner. 2000 The role of growth rate, redox-state of the plastoquinone pool and the trans-thylakoid ΔpH in photoacclimation of *Chlorella vulgaris* to growth irradiance and temperature. Planta 212: 93-102.

Wilson, K.E., S.M. Sieger and N.P.A. Hüner. 2003. The temperature-dependent accumulation of Mg-protoporphyrin IX and reactive oxygen species in *Chlorella vulgaris*. Physiol. Plant. 119: 126–136.

Wilson, K.E., A.G. Ivanov, G. Öquist, B. Grodzinski, F. Sarhan and N.P.A. Hüner. 2006. Energy balance, organellar redox status, and acclimation to environmental stress. Can. J. Bot. 84: 1355-1370.

Wykoff, D.D., J.P. Davies, A. Melis and A.R. Grossman. 1998. The regulation of photosynthetic electron transport during nutrient deprivation in *Chlamydomonas reinhardtii*. Plant Physiol. 117: 129-139.

Young, A.J. 1991. The photoprotective role of carotenoids in higher plants. Physiol. Plant. 83: 702-708.

Young, E.B. and J. Beardall. 2003. Photosynthetic Function in *Dunaliella tertiolecta* (Chlorophyta) During a Nitrogen Starvation and Recovery Cycle. J. Phycol. 39: 897-905.

13 Selected Functional and Molecular Adaptations Contributing to the Exceptional Stress Tolerance of *Dunaliella*

Ada Zamir

Abstract

Members of the genus *Dunaliella* are exceptional among marine algae, and photosynthetic eukaryotes in general, in their ability to adapt to broadly varying environmental conditions, ranging from the mild to the extreme. This chapter describes mechanisms underlying the extraordinary adaptive capabilities of *D. bardawil and D. salina* to excessive light fluxes or multimolar salinities as revealed in differential scans for genes/proteins specifically upregulated/accumulated in cells challenged with these stresses. The responses to light stress include the accumulation of Cbr, a protein identified in LHCII protein-pigment complexes enriched in zeaxanthin and lutein and assigned a role in a photoprotective mechanism operating independently of a trans-thylakoid pH gradient. Related mechanisms have now been described also in higher plants. The salt-induced responses included the induction of fatty acid elongase and corresponding modifications of fatty acid composition of microsomal membranes, a unique transferrin protein involved in iron uptake and two salt-inducible plasma membrane α-type carbonic anhydrases (dCA I and dCA II) implicated in alleviating salt-imposed limitations on CO_2 availability. The *Dunaliella* carbonic anhydrases differ drastically from animal and *Chlamydomonas reinhardtii* counterparts, that are sensitive to inhibition by halide and other anions, in retaining activity in zero to at least 3.0 M NaCl. The striking functional singularity of the *Dunaliella* carbonic anhydrases drew our attention to a largely neglected aspect of the salt-adaptability of *Dunaliella,* i.e., the role played by protein molecular adaptation. Critical insights into the structural basis of the exceptional anion resistance/salt tolerance of dCAs were gained from the crystal structure determined for dCA I and dCA II (the first structures determined for *Dunaliella* proteins), that differed from all previously-determined CA structures in exhibiting a negative surface electrostatic

Department of Biological Chemistry, Weizmann Institute of Science, Rehovot 76100, Israel, Email: ada.zamir@weizmann.ac.il

potential, intermediate between typical potentials of salt-sensitive and halophilic proteins. The significance of these electrostatic features in dCAs halotolerance were supported by the prediction and biochemical confirmation of the unanticipated halotolerance of a mammalian CA, the murine CA XIV.

Introduction

Algae of the genus *Dunaliella,* such as *D. salina* and *D. bardawil*, show extraordinary adaptability to highly variable environmental conditions, ranging from the ambient to harsh extremes. The outstanding tolerance to high salinities (Avron, 1992) and excessive light intensities as well as prospects of biotechnological applications (Ben-Amotz and Avron, 1983) have motivated extensive ecological, physiological and biochemical studies toward the characterization of major stress-adaptive responses of the alga.

While well-established transformation protocols and complete genomic data are still in the making for *Dunaliella*, the application of the tools of molecular genetics and molecular and structural biology have enabled the discovery of entirely new aspects of the acclimation capacities of *Dunaliella* (Zamir, 1992) and the unraveling of their mechanistic and structural basis. The studies summarized here used as a starting point the screening for genes that are differentially activated during the transition from ambient to stressful conditions, specifically focusing on two types of stress: light intensities exceeding the photosynthetic capacity and multimolar salinity levels.

The elucidation of previously unrecognized functions, processes and unique molecular adaptations provided essential insights into the exceptional resilience of *Dunaliella* and, in some cases, even foreshadowed later findings in plants and even mammals.

Photoprotective mechanisms

The Cbr Protein and its *in Vivo* Complexes

Dunaliella bardawil responds to high light, nutrient deprivation and several other types of stress by massive accumulation of β-carotene ((Lers et al., 1990; Jin and Polle, this volume). The onset of massive β-carotene accumulation in *D. bardawil* could therefore help to delineate the conditions and time frame under which the induction of additional, hitherto unrecognized responses to stressful light conditions is likely to take place (Lers et al., 1990). A scan for genes upregulated in correspondence with β-carotene accumulation led to the cloning of a gene, *cbr*, encoding a protein belonging to the chlorophyll a/b binding proteins family (Lers et al., 1991) and most closely related to early light-induced proteins (Elips), originally identified in greening pea seedlings and consequently assigned a role in chloroplast development. Subsequently, Elips, like Cbr, were assigned a photoprotective role against photo-oxidative damage in higher plants (Adamska et al., 1992).

The upregulation of *cbr* expression in *Dunaliella* was, however, not tightly coupled to massive β-carotene accumulation as it occurred not only in *D. bardawil*, a β-carotene over-producer, but also in a non-overproducing strain of *D. salina* (Levy et al., 1992). The association of Cbr with light-harvesting II (LHCII) complexes, first revealed in low resolution subcellular analyses (Levy et al., 1992), suggested that Cbr was engaged in the dissipation of excessive excitation of the light-capturing antennae. By its homology to chlorophyll a/b binding proteins and subcellular co-localization, Cbr could be envisaged as a pigment-binding protein that, by the circumstances of its induction, could be functionally linked to the operation of the light-stress induced xanthophyll cycle in higher plants and more recently in algae. The cycle consists of the de-epoxidation of the xanthophyll (oxo-carotenoid) violaxanthin to zeaxanthin under light stress and the reversal of this reaction on the restoration of normal conditions (Demmig-Adams and Adams III, 2006). The operation of the xanthophyll cycle was demonstrated in *D. salina* cells exposed to high light or starved for sulfate under moderate light (Levy et al., 1993; Banet et al., 1999). In these cells, violaxanthin to zeaxanthin conversion starts earlier and proceeds faster than Cbr protein accumulation. High resolution analyses of pigment-protein complexes from detergent-solubilized thylakoid membranes from light-stressed *D. salina* indicated preferred associations among the four LHCIIb polypeptides themselves as well as with Cbr. Specifically, subcomplexes including Cbr contained one or two of the more acidic of the four *D. salina* LHCIIb polypeptides together with exceptionally large proportions of lutein and zeaxanthin relative to chlorophyll a/b as compared to normal light controls. The LHCIIb subcomplexes containing Cbr were highly enriched in zeaxanthin and lutein (Banet et al., 2000).

Possible Role of Cbr in Nonphotochemical Quenching (NPQ) of Chlorophyll Fluorescence

In higher plants photosynthesis, light absorption by the light-harvesting pigment-protein complexes is followed by the transfer of excitation energy to reaction centers that perform the primary photochemistry. Modulation of the photosynthetic apparatus allows it to adjust to variations in light quality, intensity and duration. When the rate of photon absorption in the light-harvesting complexes exceeds the maximum rate of energy utilization by the dark reactions of photosynthesis, the excess excitation energy, funneled to the generation of reactive oxygen species, will damage the photosynthetic machinery. A photoprotective mechanism coming into play under light stress conditions enhances the thermal dissipation of the excessive energy and balances light absorption and photosynthesis (Demmig-Adams and Adams III, 2006). The operation of this mechanism (nonphotochemical quenching, NPQ) in plants is monitored indirectly by measuring the reversible decline in chlorophyll fluorescence reflecting the development of a temporary nonphotochemical energy-quenching mechanism. Until the mid '90s, it was generally accepted that reversible NPQ in higher plants was invariably dependent on the existence of a trans-thylakoid pH gradient. Based on the extraordinary tolerance of *Dunaliella* to excessive light intensities, the possibility was raised that additional NPQ mechanisms may be operating in *Dunaliella*. Using *D. bardawil* cells subjected to sulfate starvation under variable light intensities in combination of the uncoupler SF-6847, that abolished the trans-thylakoid pH-gradient, it was shown that during exposure to excessive light, the initial trans-thylakoid ΔpH-dependent Cbr-independent mode of NPQ was transformed into a ΔpH-independent and Cbr-requiring NPQ (Braun et al., 1996). These observations were among the first to provide concrete evidence for ΔpH-independent energy-dissipating mechanisms (sustained quenching) that have since been described in higher plants, particularly in evergreens. The role of Elip family proteins, and the intensively studied PsbS, another member of the light-harvesting protein family (Li et al., 2002), in sustained NPQ mechanisms is still not fully clarified (Demmig-Adams and Adams III, 2006). These developments stress the significance of *Dunaliella* as an organism where a 'sustained' mode of photo-protection was early demonstrated.

Functional and Molecular Adaptations Related to the Salt-tolerance of *Dunaliella*

General

D. bardawil and *D. salina* are among the *Dunaliella* strains able to proliferate over practically the entire range of NaCl concentrations (from 0.1 M to near saturation) while maintaining a relatively low internal ionic concentration. The algal cells achieve osmotic balance by accumulating up to multimolar concentrations of glycerol, generally considered as a fully compatible osmolyte (Avron, 1986, 1992). Consequently, intracellular proteins and other components of *Dunaliella* are not expected to encounter high salinities but rather high concentrations of glycerol.

In contrast, proteins with extracellular essential domains are expected to retain adequate structural and functional properties within the broad range of salinities potentially encountered by the algae.

The osmo-regulatory mechanism operating in *Dunaliella* fundamentally differs from that operating in extreme halophilic archaea, e.g., *Haloarcula marismortui,* where osmotic adjustment to external high salinity is achieved by intracellular accumulation of potassium salts. Consequently, the entire biochemical machinery in extreme halophiles is adapted to function in high salt concentrations. Thus, halophilic enzymes, while similar to their non-halophilic counterparts in catalytic function, exhibit an obligatory requirement for 1-4 M salt for proper folding and solubility and, consequently, for biological activity (Lanyi, 1974; Madern et al., 2000; Mevarech et al., 2000). The strategy used by *Dunaliella* to maintain the salt tolerance of proteins while avoiding salt dependence poses one of the most intriguing questions raised by the physiological adaptability of these algae. As a whole, the unbiased screening for genes/proteins that are preferentially expressed/accumulated under high salinities has helped to unravel distinct functions and molecular modifications contributing to the outstanding salt tolerance of *Dunaliella*.

Salt-Induction of Fatty Acid Elongase (β-ketoacyl-CoA Synthase) and Microsomal Membrane Lipid Modifications

A salt-inducible cDNA cloned from *D. salina* cells transferred from low (0.5 M) to high (3.5 M) concentrations of NaCl, encoded a ~70 kDa protein closely resembling plant β-ketoacyl CoA synthases (Kcs) (Azachi et al., 2002), microsomal enzymes that act in fatty acid elongation by condensing malonyl-CoA with acyl-CoA (acyl chain lengths >C12). This reaction is the first, and rate-limiting, of four reactions that include reduction and dehydration steps, which lead to fatty acids elongation by C2 units. The close similarity between the *Dumaliella* and plant homologs extends along most of their length including two predicted membrane spanning domains. The expression of *kcs* in *D. salina* was stimulated to a far greater extent in response to salt than to iso-osmotic glycerol levels indicating that *kcs* induction mainly reflected a salt-specific rather than a general osmotic response.

Analyses for Kcs activity localized the enzyme in a microsomal fraction similar to its location in higher plants. As substrates, acyl-CoAs with saturated acyls were preferred (in the order C14:0<C16:0>C18:0) over the mono-unsaturated acyls C16:1 and C18:1.

Recombinant Kcs synthesized in *Esherichia coli* transformed with a plasmid-cloned *kcs* cDNA was localized to a membrane fraction. When solubilized, this fraction exhibited Kcs activity with substrate preferences resembling those of the native enzyme. These results confirmed that the cloned gene indeed encoded a functional Kcs enzyme.

The finding that a lipid-modifying enzyme such as Kcs was upregulated in response to salt raised the intriguing possibility that salt adaptation could entail the elongation of fatty acids in membrane lipids. Analysis of lipids isolated from membrane fractions (including microsomes and plasma membranes) from low- or high-salt grown *D. salina* indicated a very significant decrease of in the proportion of C16 fatty acids which is mainly balanced by an increase in C18 fatty acids and to a much lesser degree in C14 and C22 fatty acids. In contrast, purified plasma membranes maintained the same proportions of C14, C16 and C18 fatty acids in cells grown in high or low salt. Based on these findings, salt-induced modifications in fatty acid chain length were confined to the microsomal membranes.

Fatty acids of chain lengths C16 and C18 are the most abundant in the microsomal and plasma membranes of *Dunaliella*. While C16 fatty acids are represented almost exclusively by the fully saturated palmitic acid, C18 fatty acids consist mostly of several unsaturated species. Comparison of the various C18 fatty acids in a fraction containing both microsomal and plasma membranes indicated a large rise in the proportion of 18:1 and 18:2 fatty acids in high-compared to low-salt grown cells. For purified plasma membranes, the most pronounced salt-related difference was a large increase in the proportion of 18:2 and a decline in 18:3 and 18:4. These analyses indicate that in response to salt, plasma membrane C18 fatty acids undergo changes in desaturation pattern while microsomal fatty acids are modified both in chain length and degree of desaturation.

While these observations led to the conclusion that microsomal fatty acid elongation, as well as desaturation, contribute to the salt tolerance of *D. salina*, the elucidation of the mechanistic consequences of these modifications is made difficult by the fact that homeostatic mechanisms maintain the intracellular ionic concentration at a rather low level, regardless of the external salinity. Consequently, the endoplasmic reticulum or Golgi apparatus are not exposed to high ionic concentrations, but to high levels of glycerol (in excess of 4.0 M) that osmotically balance the external high salinity. Glycerol is generally thought to be fully compatible with the stability and function of cellular components. Still, the possibility that some cellular components may not operate optimally in the presence of such high levels of glycerol cannot be dismissed and, hence, the need for adaptive modifications exemplified by the present observations.

Whereas the targets potentially affected by glycerol in *Dunaliella* remain elusive, a plausible candidate is the intracellular secretory pathway responsible for transport of proteins and lipids between the endoplasmic reticulum and Golgi apparatus by secretory vesicles which bud from one compartment and fuse with another. The budding and fusion of such vesicles, and the recognition of cargo and targeting molecules, require specific protein-membrane and membrane-membrane interactions and are affected by membrane lipid composition. As some of these interactions may be rendered less effective in high intracellular glycerol concentrations, modifications in lipid composition as catalyzed by the salt-induced Kcs and additional activities in *Dunaliella* may be required to modify membranes of the endoplasmic reticulum and/or Golgi apparatus so as to optimize vesicular transport in cells grown in high salinity. Membrane modifications of this sort may not be unique to *Dunaliella,* as intracellular accumulation of inorganic or organic solutes is a ubiquitous salt-adaptive, osmo-regulatory response in taxonomically-varied organisms.

Extracellular Carbonic Anhydrases

As indicated above, *Dunaliella* achieves osmotic balance by accumulating iso-osmotic concentrations of glycerol. However, *Dunaliella* proteins that are extracellularly-oriented are expected to remain active over the broad range of salinities encountered in the medium and therefore display far greater salt tolerance than their homologs in other organisms that, for the most part, are active only in low to medium salinities. Two plasma membrane proteins from *D. salina* identified in our studies (Fisher et al., 1996; Premkumar et al., 2003) fully conform to this expectation (Bageshwar et al., 2004). Both proteins belong to the α-type carbonic anhydrase (CA) family. The first discovered enzyme, dCA I, is a 60 kDa protein consisting of two, ~52% identical in sequence, tandemly-arranged α-type carbonic anhydrase domains. The later discovered dCA II is a ~30 kDa, single-domain α-type carbonic anhydrase exhibiting ~55% sequence identity to each of the dCA I domains (Premkumar et al., 2003).

Carbonic anhydrases constitute a family of Zn enzymes present in all kingdoms of life. The α-type members of the family are found predominantly in animals, but also occur in some bacteria, green algae and higher plants. By reversibly converting CO_2 and water to bicarbonate and proton ions, carbonic anhydrases fulfil essential roles in a variety of physiological processes including the transport of CO_2 and bicarbonate, acid-base regulation and water and electrolyte balance. In photosynthetic organisms, carbonic anhydrases perform crucial roles in CO_2 uptake and concentrating mechanisms conducive to efficient photosynthetic CO_2 fixation. Carbonic anhydrases are typically subject to inhibition by a variety of monovalent anions, including chloride and bromide, that displace the catalytically-essential Zn-bound water/hydroxyl and disrupt hydrogen-bonded networks at the active site, as indicated in crystal structures of CA-anion adducts (Liljas et al., 1994; Lindskog and Silverman, 2000). In this context, the capability of the two dCAs to act in up to multimolar anion concentrations poses a particularly intriguing problem.

Assays of the *Dunaliella* CAs for three different CA activities, CO_2 hydration, bicarbonate dehydration and ester hydrolysis, revealed an astounding degree of tolerance to salt up to 4.0 M NaCl (Bageshwar et al., 2004). In sharp contrast, homologs such as human CA isozymes and the *Chlamydomonas reinhardtii* periplasmic CA are strongly inhibited by Cl^- in all three CA activities, *e.g.*, the I_{50} for Cl^- inhibition of CO_2 hydration activity did not exceed 0.2 M NaCl. Thus, dCAs are exceptional in their salt tolerant, and manifestly immune to anion inhibition (Bageshwar et al., 2004). The *Dunaliella* CAs also differ strikingly from proteins from halophilic *Archaea* that generally require 2-4 M KCl to maintain the native conformation, solubility and activity (Frolow et al., 1996; Mevarech et al., 2000).

Structural Basis for the Unique Salt Tolerance of *D. salina* Extracellular Carbonic Anhydrases

Crystal Structure of dCA II, the First to be Determined for a Dunaliella Protein

To explore the structural basis of the exceptional salt tolerance of the *Dunaliella* extracellular carbonic anhydrases, the crystal structure of dCA II was determined and used for homology modeling of the separate dCA I repeats. These structures are the first to be determined for any *Dunaliella* protein and for salt tolerant proteins in general (Premkumar et al., 2005).

The X-ray structure of the recombinant dCA II was determined at 1.86 Å resolution. The crystal structure shared the global fold of salt-sensitive homologs in containing a central anti-parallel ten-stranded β-sheet, two α-helices and a catalytic Zn^{2+}. Distinctive features of dCA II include the extensions of the two α-helices, that make them significantly longer than those in salt-sensitive CAs, and a unique insertion forming a sodium binding loop (Figure 1).

The core structure of dCA II is well conserved. Specifically, all the amino acid residues located within ~8 Å from the catalytic Zn are superimposable with those in the mesophilic human CA isozyme II. These include the catalytic Zn ligands, active site H-bonding residues and the substrate binding residues. However, the solvent-accessible surface of dCA II significantly differs from that of the other CAs in possessing a higher ratio of acidic over basic amino acid residues, achieved by a lowered proportion of basic residues against a conserved proportion of acidic residues. Determination of surface electrostatic potentials revealed that dCA II exhibited a predominantly negative potential as compared with the uneven surface distribution of neutral, negative and positive potentials displayed by other CAs (Figure 2). The uniform surface negative potential is reminiscent of halophilic proteins in which the typically negative surfaces were assigned salt protective roles such as the formation of hydrated cation networks (Dym et al., 1994; Frolow et al., 1996; Madern et al., 2000). The halophilic proteins differ however in that

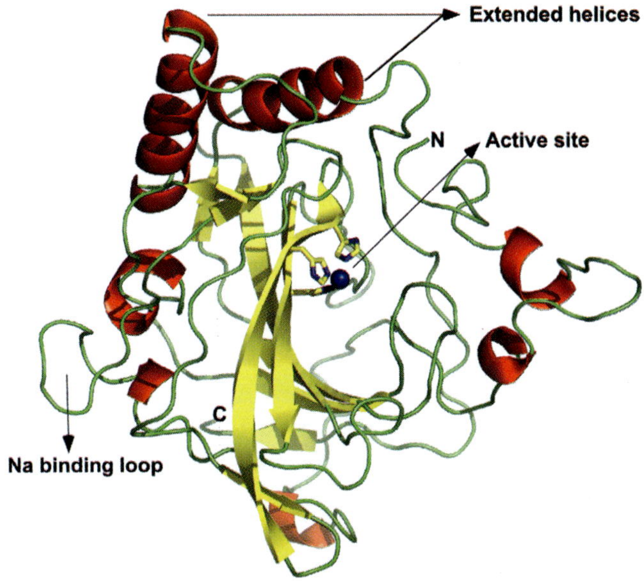

Figure 1: Ribbon diagram of the crystal structure of dCA II. Unique backbone features include a Na-binding loop and two extended α-helices.

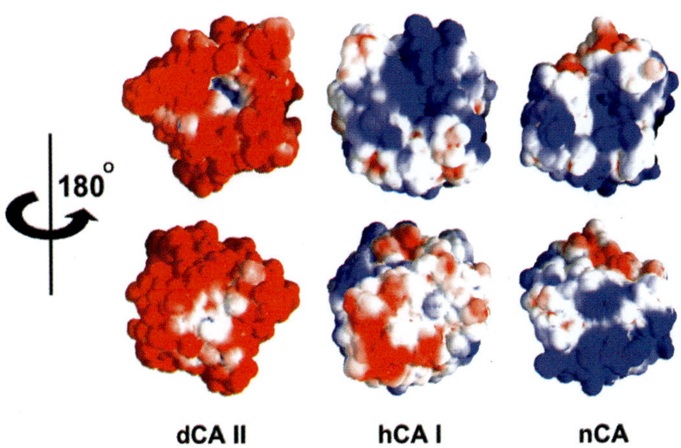

Figure 2: Surface electrostatic potential of dCA II and other CAs. Blue, negative potential; red, positive potential, white, neutral. Upper row, images looking down the active site; lower row, 180° rotated view. dCA II, *D. salina* carbonic anhydrase II, hCA I, human carbonic anhydrase isozyme I; nCA, *Neisseria gonorrhoeae* carbonic anhydrase.

their negative surface potential arises both from a decrease in basic residues and a marked increase in surface acidic residues. It is this difference in charge density that is likely to differentiate between the salt-requiring halophilic proteins from the salt-independent, but remarkably salt-tolerant dCAs (Bageshwar et al., 2004).

As dCA II is ~55% identical in sequence to each of the dCA I domains (Fisher et al., 1996; Bageshwar et al., 2004), it was possible to use the crystal structure of dCA II as a template to build homology models for the N-terminal and C-terminal CA repeats of dCA I.

The models show that the amino acid sequence insertions that extend the two conserved α-helices and form the Na-binding loop in dCA II are also present in the models for the N and C terminal repeats. Similar to dCA II, the solvent-accessible surfaces of the dCA I repeats display a high ratio of acidic over basic amino acid residues compared to mesophilic CAs and a predominantly negative surface potential. The similarities between dCA I repeats and dCA II suggest a common evolutionary origin. The presumed gene duplication yielding the internally repeated structure of dCA I afforded additional stability, as indicated by comparing the activity of isolated C-terminal repeat (the N-terminal repeat showed only rudimentary activity) with that of the native, duplicated form (Bageshwar et al., 2004). The two forms may also exhibit specialized physiological functions still to be discovered. Gene duplication events that create tandemly-fused, internally duplicated polypeptides in *Dunaliella* are not unique for dCA I but are also evident in TTf, an extracellular transferrin-like protein from *D. salina*. This iron-carrying protein, found mostly in animals, consists normally of two homologous domains (α β), but in *Dunaliella* has assumed an internally triplicated form conforming to an α_2 β structure (Fisher et al., 1997).

It is intriguing to note that the electrostatic features proposed to underlie the halotolerance of the dCAs allowed us to predict, and biochemically confirm, the unanticipated halotolerance of a mammalian CA, the murine CA XIV. It is thus intriguing to speculate that convergent evolution has arrived at similar modes of molecular solutions to achieve a high degree of salt tolerance.

Conclusion

Although far from genomic scale coverage, the unbiased screening for stress inducible genes/proteins in *Dunaliella* has yielded critical novel insights regarding essential processes and molecular details enabling the astounding stress tolerance of these algae. The detailed multi-tiered regulatory, functional, biochemical and structural analyses characterized a photo-protective mechanism involving a light stress-induced LHCIIb-associated protein of the chlorophyll a/b binding family, identified protein components compensating for salt-imposed iron and inorganic carbon deficiencies, outlined molecular adjustments of microsomal membranes, thought to facilitate intracellular trafficking in the presence of multimolar concentrations of glycerol, and provided the molecular basis for the unique salt-tolerance of extracellular enzymes, such as the two □-type carbonic anhydrases, discovered in characterized in depth in these studies.

Acknowledgements

The author acknowledges with deep appreciation and gratitude the essential contributions of members of her lab and collaborating colleagues. Without them this journey could not have taken place. The continuous support of Nikken-Sohonsha Corporation, Hashima City, Japan, Nature Beta Technologies, Eilat, Israel and the Magnet Program of the Ministry of Commerce and Industry, State of Israel, is gratefully acknowledged.

References

Adamska, I., I. Ohad and K. Kloppstech. 1992. Synthesis of the early light-inducible protein is controlled by blue light and related to light stress. Proc. Natl. Acad. Sci. USA 89: 2610-2613.

Avron, M. 1986. The osmotic components of halotolerant algae. Trends in Biochem. Sci. 11: 5-6.

Avron, M. 1992. Osmoregulation. pp. 135-164. In *Dunaliella*: Physiology, Biochemistry and Biotechnology. M. Avron and A. Ben-Amotz (eds) CRC Press. Boca Raton, FL, USA.

Azachi, M., A. Sadka, M. Fisher, P. Goldshlag, I. Gokhman and A. Zamir. 2002. Salt induction of fatty acid elongase and membrane lipid modifications in the extreme halotolerant alga *Dunaliella salina*. Plant Physiol. 129: 1320-9. Erratum. 2003. Plant Physiol. 132: 111.

Bageshwar, U.K., L. Premkumar, I. Gokhman, T. Savchenko, J.L. Sussman and A. Zamir. 2004. Natural protein engineering: a uniquely salt-tolerant, but not halophilic, α-type carbonic anhydrase from algae proliferating in low- to hyper-saline environments. Protein Eng. Des. Sel. 17: 191-200.

Banet, G., U. Pick, S. Malkin and A. Zamir. 1999. Differential responses to different light spectral ranges of violaxanthin deepoxidation and accumulation of Cbr, an algal homolog of plant early light inducible proteins, in two strains of *Dunaliella*. Plant Physiol. 37: 875-879.

Banet, G. U. Pick and A. Zamir. 2000. Association of large amounts of xanthophylls and an algal early light induced protein with light harvesting complex IIb in the alga *Dunaliella*. Planta 210: 947-955.

Ben-Amotz, A. and M. Avron. 1983. On the factors that determine massive β-carotene accumulation in the halotolerant alga *Dunaliella bardawil*. Plant Physiol. 72: 593-597.

Braun, P., G. Banet, T. Tal, S. Malkin and A. Zamir. 1996. Possible role of Cbr, an algal early light induced protein, in non-photochemical quenching of chlorophyll fluorescence. Plant Physiol. 110: 1405-1411.

Demmig-Adams, B. and W. W. Adams III. 2006. Photoprotection in an ecological context: the remarkable complexity of thermal energy dissipation. New Phytologist 172: 11-21.

Dym, O., M. Mevarech and J. L. Sussman. 1994. Structural features that stabilize halophilic malate dehydrogenase from an Archaebacterium. Science 267: 1344-1346.

Fisher, M., I. Gokhman, U. Pick and A. Zamir. 1996. A salt-resistant plasma-membrane carbonic anhydrase is induced by salt in *Dunaliella salina*. J. Biol. Chem. 271: 17718-17723.

Fisher, M., I. Gokhman, U. Pick and A. Zamir. 1997. A structurally novel transferrin-like protein accumulates in the plasma membrane of the unicellular green alga *Dunaliella salina* grown in high salinities. J. Biol. Chem. 272: 1565-1570.

Frolow, F., M. Harel, J.L. Sussman, M. Mevarech and M. Shoham. 1996. Insights into protein adaptation to a saturated salt environment from the crystal structure of a halophilic 2Fe-2S ferredoxin. Nat. Struct. Biol. 3: 452-458.

Lanyi, J.K. 1974. Salt-dependent properties of proteins from extremely halophilic bacteria. Bacteriol. Rev. 38: 272-290.

Lers, A., Y. Biener and A. Zamir. 1990. Photoinduction of massive β-carotene accumulation by the alga *Dunaliella bardawil*. Plant Physiol. 93: 389-395.

Lers, A., H. Levy and A. Zamir. 1991. Co-regulation of a gene homologous to early light-induced genes in higher plants and β-carotene biosynthesis in the alga *Dunaliella bardawil*. J. Biol. Chem. 266: 13698-13705.

Levy, H., I. Gokhman and A. Zamir. 1992. Regulation and light-harvesting complex II association of a *Dunaliella* protein homologous to early light induced proteins in higher plants. J. Biol. Chem. 267: 18831-18836.

Levy, H., T. Tal, A. Shaish and A. Zamir, 1993. Cbr, an algal homolog of plant Elips is a putative zeaxanthin binding protein. J. Biol. Chem. 268: 20892-20896.

Li, X-P., P. Muller-Moule, A.M. Gilmore and K.K. Niyogi. 2002. PbS-dependent enhancement of feedback de-excitation protects photosystem II from photoinhibition. Proc. Natl. Acad. Sci. USA 99: 15222-15227.

Liljas, A., K. Hakansson, B.H. Jonsson and Y. Xue. 1994. Inhibition and catalysis of carbonic anhydrase: recent crystallographic analyses. Eur. J. Biochem. 219: 1-10.

Lindskog, S. and D.N. Silverman. 2000. The catalytic mechanism of mammalian carbonic anhydrases. pp 175-195. In The Carbonic Anhyrases. W.R. Chegwidden, N.D. Carter and Y.H. Edwards (eds) New Horizons. Brikhauser Verlag, Basel, Switzerland.

Madern, D., C. Ebel and G. Zaccai. 2000. Halophilic adaptation of enzymes. Extremophiles 4: 91-98.

Mevarech, M., F. Frolow and L. Gloss. 2000. Halophilic enzymes: proteins with a grain of salt. Biophys. Chem. 86: 155-164.

Premkumar, L., H.M.Greenblatt, U.K. Bageshwar, T. Savchenko, I. Gokhman, A. Zamir and J.L. Sussman. 2003. Identification, cDNA cloning, expression, crystallization and preliminary X-ray analysis of an exceptionally halotolerant carbonic anhydrase from *Dunaliella salina*. Acta Crystallogr. D Biol. Crystallogr. 59: 1084-1086.

Premkumar, L., H.M. Greenblatt, U.K. Bageshwar, T. Savchenko, I. Gokhman, J.L. Sussman and A. Zamir. 2005. Three-dimensional structure of a halotolerant algal carbonic anhydrase predicts halotolerance of a mammalian homolog. Proc. Natl. Acad. Sci. USA 102: 7493-7498.

Zamir, A. 1992. Molecular Biology of *Dunaliella*. pp. 195-203. In *Dunaliella*: Physiology, Biochemistry and Biotechnology. M. Avron and A. Ben-Amotz (eds) CRC Press, Boca Raton, FL. USA.

14

Salinity Tolerance and Iron Deprivation Resistance Mechanisms Revealed by Proteomic Analyzes in *Dunaliella salina*

Adriana Katz[1], *Yakov Paz*[1] and *Uri Pick*[1]*

Abstract

The halotolerant alga *Dunaliella salina* can adapt to practically the entire range of salt concentrations. This capacity is achieved primarily by massive accumulation of glycerol and by efficient elimination of sodium ions, which require major metabolic investment.

To clarify the molecular mechanisms of salinity tolerance in *D. salina,* we performed a proteomic differential analysis aimed to identify salt-regulated proteins in different sub-cellular fractions: Soluble proteins were identified by 2D IEF/SDS-PAGE combined with MALDI-TOF MS/MS, whereas plasma membrane proteins were biotin-tagged, separated by blue-native/SDS-PAGE, and identified by nano-LC/MS-MS.

High salt up-regulated plasma membrane carbonic anhydrases, which mediate bicarbonate acquisition, key enzymes in Calvin cycle, starch mobilization and redox energy production. These results indicated that *D. salina* enhances photosynthetic CO_2 assimilation and diverts its principle carbon and energy resources for synthesis of glycerol.

In the plasma membrane we identified: (i) bacterial-type surface coat proteins (peptidoglycan-associated lipoprotein) and tubulin, which probably function in stabilizing the membrane against osmotic lysis; (ii) small GTP-binding proteins which may be involved in signal-transduction in response to salt/osmotic stress; (iii) lipid metabolizing enzymes, possibly associated with osmotic sensing; (iv) chaperones and proteolytic enzymes probably involved in enhancing the turnover and stabilization of membrane proteins at high salinity; (v) ion transporters for protons, iron, nitrate, ammonium and possibly sodium.

[1] Department of Biological Chemistry, Weizmann Institute of Science, Rehovot, 76100, Israel.

* Author for correspondence, Email: uri.pick@weizmann.ac.il.

Taken together, these results suggest that concerted changes in multiple pathways contribute to unique ability of *D. salina* to withstand high salinity.

A surprising observation was the finding that high salinity induced iron deficiency stress in *D. salina*. We discovered that *D. salina* evolved special strategies to cope with iron limitation. It utilizes a unique mechanism for iron acquisition, via membrane-associated transferrins, that bind and internalize ferric ions into acidic vacuoles. In the chloroplast, iron deprivation induced one major protein, identified as a PS-I chlorophyll a/b-binding protein, which largely increased the size of PS-I units. Interestingly, iron-deprived cyanobacteria, accumulate in PS-I a different type of chlorophyll-binding protein. This may represent a general strategy of photosynthetic organisms to adapt to iron deprivation.

Salt-regulated Proteins in *Dunaliella*- Early Observations

Dunaliella is generally recognized as a photosynthetic organism that thrives in the most hypersaline environments known on earth, such as the Dead Sea in Israel. However, what is striking in this alga, is the fact that it can grow equally well also at low salinity without apparent morphological changes (Pick, 2002). This remarkable adaptation capacity is achieved primarily by a robust osmoregulatory mechanism, namely, massive synthesis of glycerol (Avron, 1992) and by efficient Na^+ homeostasis, which maintains a fairly constant cytoplasmic Na^+ concentration throughout a wide salt concentration range (Pick et al., 1986). However, exposure to hypersaline solutions also affects the immediate vicinity of the cell surface, namely, it decreases protein hydration and perturbs salt bridges and decreases the availability of essential minerals, such as CO_2. Therefore, successful adaptation to hypersalinity would depend on changes in the plasma membrane surface elements, on accumulation of selected transporters and on upregulation of cellular metabolism. Such changes should be manifested by changes in the proteomic profile of *Dunaliella* at high salinity.

Indeed, earlier studies indicated that high salinity induces in *Dunaliella* accumulation of major proteins in the plasma membrane as well as up-regulation of a few other proteins and enzymes. The first identified salt-induced protein was a plasma membrane 150 kD glycoprotein, which turned out to be a transferrin (Sadka et al., 1991; Fisher et al., 1997). Two other major plasma membrane proteins were identified as surface carbonic anhydrases (Fisher et al., 1996; Premkumar et al., 2003). These proteins turned out to mediate Fe and CO_2 uptake, suggesting that these elements are rate limiting for survival in hypersaline conditions. Two plasma membrane secondary sodium transporters are upregulated at high salt concentrations: a Na^+/H^+ antiporter, activity detected in plasma membrane preparations (Katz et al., 1986; Katz et al., 1992; Katz et al., 1994) and a sodium-dependent phosphate transporter (Li et al., 2006). These proteins probably function in pH homeostasis and in mediating phosphate uptake, by utilizing the Na^+ electrochemical gradient across the plasma membrane at high salinity (Katz and Pick, 2001; Weiss et al., 2001). A special isoform of dihydroxyacetone phosphate reductase, which is induced at high salinity in *Dunaliella* chloroplasts, was identified and characterized, suggesting that this enzyme, which catalyzes glycerol synthesis, controls massive glycerol accumulation at high salinity (Gee et al., 1993; Ghoshal et al., 2002). A fatty acid elongase (FAE1), which is transcriptionally upregulated at high salinity in the endoplasmic reticulum, was cloned and characterized (Azachi et al., 2002). It was suggested that this FAE1 functions in modifying plasma membrane lipid composition at high salinity. These few examples provided preliminary evidence that gross changes in the proteome of *Dunaliella* take place during adaptation to high salinity. In this chapter we summarize a more comprehensive proteomic analysis of salt-regulated proteins in *D. salina*, carried out during 2000-2007 by collaboration between our research team at the WIS and that of Dr. A. Shevchenko at the MPI in Dresden.

Application of Proteomics for Identification of Salt-induced Proteins in *D. salina*

Differential proteomics has become a major and powerful tool for broad characterization of stress responses in diverse organisms, which complements and extends differential gene expression analyses. However, proteome analyses in *Dunaliella* is hindered by two major limitations: First, the absence of a genome or a large EST database, hampered efficient proteome characterization by conventional data base searching. Second, the most interesting proteins, which are expected to respond to salt stress, are plasma membrane proteins, which are typically under-represented by conventional 2D gel methodologies, due to poor solubilization or precipitation on isoelectric focusing gels. To overcome these problems, we fractionated cellular proteins into soluble and plasma membrane fractions and resolved them on 2D gels by different methodologies: the soluble (cytoplasmic and chloroplastic) fraction was resolved by conventional isoelectric focusing/SDS PAGE (Liska et al., 2004), whereas plasma membrane proteins were first tagged with biotin, than resolved by Blue Native (BN) gel electrophoresis/SDS-PAGE and identified by Western analysis with avidin. Extracted protein digests from soluble fractions were analyzed by MALDI-TOF MS/MS and the plasma-membrane proteins by LC-MS/MS sequencing, followed by sequence-similarity database searching, Mascot and MS BLAST (Katz et al., 2007). In order to improve protein identification, we employed advanced sequence-similarity database searching algorithm, MS BLAST, pioneered by Dr. Shevchenko (Shevchenko et al., 2001). Application of these complementary methodologies and expertise, enabled to obtain for the first time a comprehensive overview of salt-induced proteins in a halotolerant specialized organism whose genome is not yet resolved.

The Experimental Strategies and their Rationale

The experimental design that we chose to analyze the proteome of *D. salina* included the following steps:

Sub-cellular fractionation: A notable advantage of *Dunaliella* in comparison to most algae for proteomic analysis is the fact that it does not have a rigid cell wall and therefore can be lysed very gently by an osmotic shock (Katz et al., 1986). This enables to obtain enriched sub-cellular fractions of soluble cytoplasmic, soluble chloroplastic, chloroplast membrane and plasma membrane proteins. We took advantage of this virtue to compare by 2D gels the proteome compositions of soluble cytoplasmic and chloroplastic and of plasma membrane protein fractions derived from low salt (0.5M) and from high salt (3M) *D. salina* cells.

Biotin tagging of plasma membrane proteins: As mentioned above, analysis of plasma membrane proteome by conventional proteomic methodologies is problematic. A major problem that we encountered is in purification: plasma membrane preparations are usually contaminated by other major proteins from different cellular compartments. Conversely, extensive purification often eliminates major surface-associated proteins from the preparation. To circumvent these problems, we labeled intact cells with a membrane-impermeable biotin tag, which can be sensitively detected by Western analysis with avidin.

As shown in Figure 1A, most of the separated proteins on 2D gel are soluble proteins that were localized in the cytoplasm or in the chloroplast (Liska et al., 2004). Only a small number of plasma membrane proteins are separated on the 2D, as shown by the western analysis of the biotin tagged proteins (Figure 1B). An interesting observation is that most of the plasma membrane proteins are very acidic. This observation has been documented earlier both in halophilic bacteria and for some proteins from *Dunaliella* (Dym et al., 1995; Premkumar et al., 2005). The acidic nature of plasma membrane proteins may be an important adaptation in stabilization of the membrane at high salinity: the increase in negative surface charge of proteins is believed to compensate for the decreased hydration in hypersaline solutions and for perturbation of salt bridges which loosen protein structure (Madern et al., 2000). In order to get a good separation of the membrane proteins we utilized a different type of separation, as described below.

Resolution on Blue-Native (BN) gels: Blue-Native gel electrophoresis (BN/SDS-PAGE), was originally developed for resolution of protein complexes from mitochondria and chloroplast membranes (Schagger and Vonjagow, 1991; Kugler et al., 1997). It offers several advantages for plasma membrane proteome analysis:

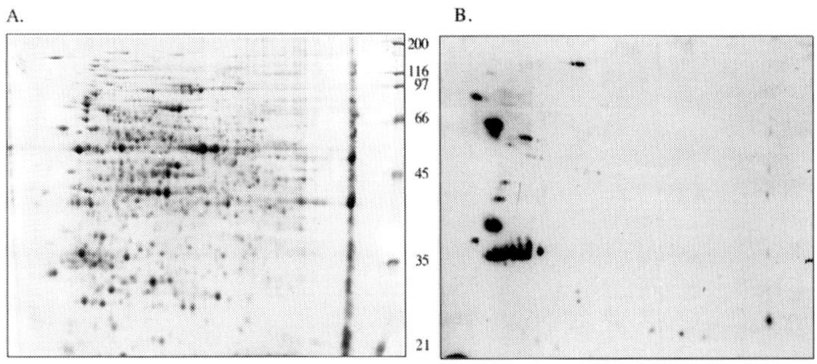

Figure 1: Two dimensional gel electrophoresis of *Dunaliella* total cell protein extract. *Dunaliella* cells grown in 3M NaCl medium were first biotinylated by the impermeable analog LC-NHS-biotin, washed and than total proteins were extracted by phenol extraction as described before (Liska et al., 2004). A. Coomasie Brilliant Blue R-250 staining of 250 ug protein separated on 18 cm IPG dry strips, pH 3-10NL, and then applied on 11% SDS-PAGE for second dimension . B. Western analysis of cell surface biotinylated proteins, 100 ug protein separated on 2D gels as in A, and probed with streptavidin.

it enables a complete solubilization and resolution of membrane proteins, it identifies components of protein complexes and it allows isolating functionally intact complex units. Even though it does not have the high resolution of conventional IEF/SDS-PAGE, it circumvents most limitations of separation and quantification of membrane proteins by conventional 2D gel electrophoresis.

A representative 2D BN/SDS-PAGE map of *D. salina* plasma membrane proteins reveals more than 50 spots (Figure 2). Most protein spots were labeled with biotin, indicating that the preparation indeed contains mostly plasma membrane proteins. A comparison of protein maps from the low salt (0.5 M NaCl) and high salt (3 M NaCl) membranes, shows that 20 spots were reproducibly increased by high salt by more than 2-fold, whereas only 2 were almost completely suppressed. The major biotinylated spots corresponded to the transferrin TTf (150 kDa) and the two carbonic anhydrases, dCAI (60 kDa) and dCAII (30 kDa), but many uncharacterized proteins were also labeled, suggesting that they are localized at the plasma membrane surface. Several pairs of proteins co-migrated on BN gels, indicating that they may interact with each other. One notable example is the transferrin, TTf with the heat-shock protein HSP-90 (Katz et al., 2007).

Mass spectrometry and advanced bioinformatics sequence analysis: A critical element in the successful proteomic analysis was the choice of comprehensive mass spectrometry and multiple DB searching techniques. We used the conventional DB-searching method, Mascot, and a sequence-similarity database searching, MS BLAST, to probe protein and EST databases. From a

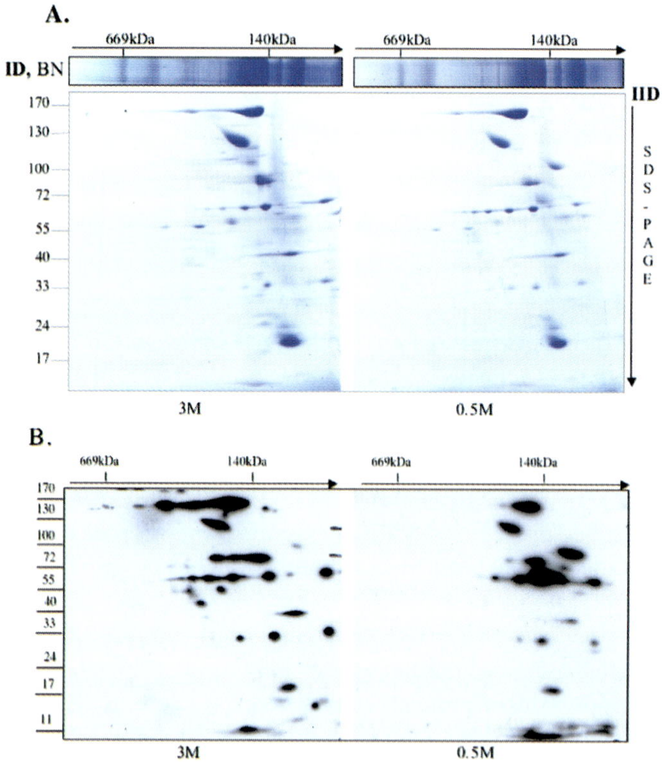

Figure 2: Proteomic map of membrane proteins from 0.5M NaCl and 3M NaCl, separated on BN/SDS-PAGE. Plasma membranes were prepared from biotinylated Dunaliella cells. Plasma membrane proteins were solubilized with 2% DM, separated on BN gel for first dimension, followed by 6-18% acrylamide SDS-PAGE for second dimension (according to (Katz et al., 2007).
A. Coomasie staining of 100 ug plasma membrane protein extract separated on 2D BN/SDS gels. B. Western analysis of biotynilated 10 ug plasma membrane proteins separated on 2D BN/SDS gels, probed with streptavidin.
BN separated plasma membrane proteins from 0.5M (right) and 3M (left) cells were run in parallel on the same SDS gel for comparison (adapted from (Katz et al., 2007)

total of 113 spots that were upregulated by more than 2-fold at high salinity, we were able to identify about 80% of the selected proteins. Table 1 shows the power of MS BLAST in identification of proteins, it identified 2 fold more spots than Mascot.

Table 1: Mass Spectrometry-driven database searching strategy. Spots were excised, digested with trypsin and analyzed by peptide mass fingerprinting, PMF. All the remaining unidentified spots were analyzed by nanoESI tandem mass spectrometry and DB searching with Mascot protein, Mascot EST and MS BLAST (according to (Liska et al., 2004)). In total 78% of spots were identified and 22% of the spots remained unidentified.

Step No.	Analytical technique-DB searching method	Database type	Identified spots
1	MS-peptide mass fingerprinting, DB searching	Protein	11%
2	MS/MS-Mascot	Protein+EST	30%
3	MS/MS-MS BLAST	Protein	67%

The identified proteins were divided into eight categories, according to their functional annotation. Based on these identifications, we propose an integrated metabolic scheme of salinity tolerance in *D. salina*, illustrated in Figure 4.

Classification of Salt-regulated Proteins and their Predicted Functions

Functional categories: Classification of the salt-induced proteins into functional groups revealed that the largest number of upregulated proteins were associated with central metabolic networks, namely: photosynthesis, respiration, carbon metabolism and protein synthesis or degradation. Smaller functional categories were surface coat proteins, membrane transporters, signaling and cellular trafficking and general stress proteins (Figure 3).

Carbon Assimilation and Energy Production

We found that photosynthetic activity in *D. salina* is greatly enhanced at high salinity (Liska et al., 2004). This observation was very surprising because in plants and most other photosynthetic organisms high salinity severely inhibits photosynthesis. The proteomic analysis revealed that the stimulation of photosynthesis results from upregulation of key enzyme in carbon uptake and assimilation. One major limitation for photosynthesis at high salinity is the reduced availability of CO_2 (Booth and Beardall, 1991). To overcome this limitation, *Dunaliella* accumulates three different carbonic anhydrases (CA) at the plasma membrane surface (Fisher

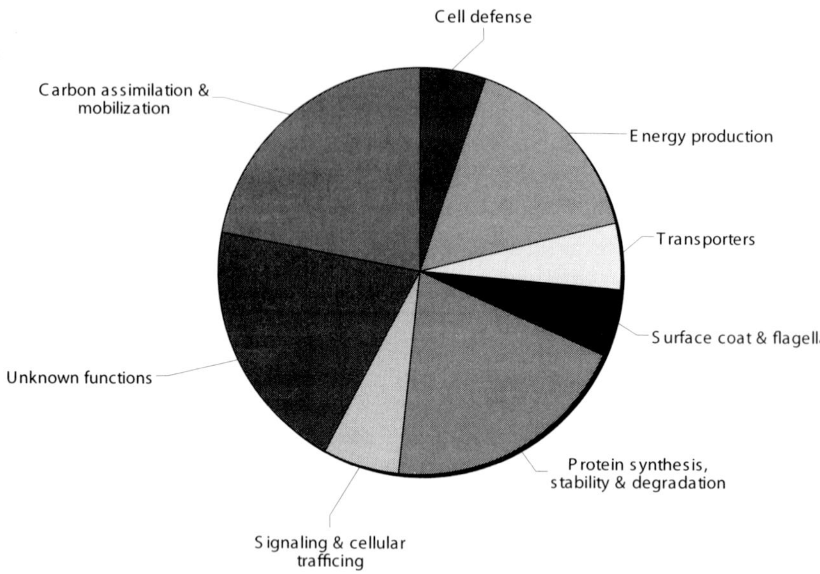

Figure 3: Functional classification of identified proteins.

et al., 1996; Katz et al., 2007). Two of these CA, dCAI and dCAII, have been extensively characterized and shown to be salt-resistant and highly acidic proteins (Fisher et al., 1996; Premkumar et al., 2003). In addition, five central enzymes in the Calvin cycle, mediating CO_2 assimilation were upregulated at high salinity, including Rubisco and Rubisco activase, universally recognized as the major rate-limiting step in CO_2 assimilation. Three major enzymes in the pentose phosphate pathway accumulate at high salinity. This metabolic pathway has a central role in mobilization of carbon reserves, primarily starch, and in providing reducing power in the form of NADPH for different biosynthetic activities. We hypothesized that the enhanced photosynthetic activity and starch mobilization at high salinity, serve for massive synthesis of glycerol, the osmotic element in *D. salina* (Liska et al., 2004).

Protein synthesis, degradation and stabilization

The remarkable changes in protein composition and abundance at high salinity should depend on active synthesis and degradation of proteins and on chaperones, which stabilize and mobilize newly synthesized proteins. Indeed, we observed upregulation of several classes of proteins that function in protein synthesis, stabilization and degradation. They include regulatory factors in protein synthesis initiation and elongation, such as EF-Tu and the dual-function eIF3, whose homologs in plants and animals are associated with stress signaling (Chen et al., 1995; Jiang and Clouse, 2001); chaperons (HSP-70, HSP-90) associated with protein trafficking and stabilization and ubiquitin, zinc metaloproteases and 26S proteasome regulatory protein involved in protein degradation.

These results suggest that high salt induces in *D. salina* a dynamic remodeling of protein composition in different cellular compartments.

Transporters

In the plasma membrane we identified different types of transport proteins that were all tagged by biotin. Some of these proteins have been previously cloned and characterized: Two types of carbonic anhydrases mentioned above, the transferrin-like protein TTf, which mediates Fe uptake (see next section), and two P-type ATPases, an H^+-ATPase (Weiss and Pick, 1996) and second P-type ATPase which accumulates at high salinity (Katz et al., 2007). Interestingly, strong functional evidence for a Na^+-translocating P-type ATPase in plasma membranes of *D. maritima* has been reported (Popova et al., 2005). It is conceivable that the salt-induced P-type ATPase is the Na^+-ATPase since it is expected to be over-expressed at high salinity to enhance Na^+ elimination. Two other transport proteins that were also identified are a nitrate and an ammonium transporter.

Sensing and signaling

Very little is known about the mechanism of sensing and signaling osmotic stress in *Dunaliella*. There is evidence that the MAP kinase-signaling cascade may be involved in the process (Jimenez et al., 2004), yet, no proteins have been conclusively identified. We identified in *D. salina* plasma membrane abundance of small GTP-binding proteins, yptV2/Rab8 and yptC6/Rab11, subgroups of the Ras superfamily, which function in animal and in plant cells as molecular switches in signal transduction cascades, in response to external signals (Berken et al., 2005; Jaffe and Hall, 2005). It is conceivable that these G proteins are involved in sensing and signaling salt/osmotic stress in *D. salina*.

Another factor that may control the osmotic response in *Dunaliella* are changes in the lipid composition and organization (Zelazny et al., 1995). In accordance with this idea, we found that the expression of three lipid metabolizing enzymes, at the plasma membrane was affected by salinity. Notably, lysophospholipase A was found to be strongly downregulated at high salinity. Previous reports in bacteria and in fission yeast indicated that lysophosphlipases are directly implicated in triggering stress response, presumably by mediating changes in membrane fluidity (Yang et al., 2003; Baysse et al., 2005). It is conceivable that the lower activity of lysophospholipase A in the plasma membrane of *Dunaliella*, would slow-down hydrolysis of phospholipids to lysophospholipids and to free fatty acids, which is expected to decrease the plasma membrane fluidity at high salinity.

Membrane structure stabilization

The proteomic study provides clues that *D. salina* cells may have some kind of protective outer layer to prevent lysis upon osmotic changes. We identified two surface coat proteins that are involved in stabilization of cell envelope structure in bacterial cells, peptidoglycan-associated and murein lipoprotein, which were not previously reported in plants or algae. In bacteria these proteins are localized at the extracellular cell envelope and help to maintain the cell integrity (RodriguezHerva et al., 1996; Fadl et al., 2005). Notably, it has been reported that the surface coat of *D. parva* is sensitive to lysozyme treatment, suggesting that peptidoglycans at the cell surface are essential for maintaining cell integrity in saline solutions (Hatanaka et al., 1998).

Figure 4 schematically illustrates the major metabolic networks, transporters and structural elements that are upregulated at high salinity in *D. salina* (marked red). They include the enhanced carbon assimilation and starch mobilization which stimulate enhancing to glycerol biosynthesis, protein synthesis and degradation, antioxidative activity, ion transport, plasma membrane lipid metabolism and reorganization which trigger osmotic signaling and surface coat stabilization. These responses most probably evolved to enable *Dunaliella* to withstand high salinity.

Iron Deficiency Induced Proteins and their Function in Dunaliella

Iron deficiency is a common limitation for plant and algae proliferation because of its low availability in aerobic aqueous solutions. Therefore, all photosynthetic organisms developed efficient mechanisms for uptake and storage of iron.

D. salina can adapt very well to iron deprivation, as manifested by its proliferation and by maintenance of efficient photosynthetic activity under iron deprivation (Varsano et al., 2003). In earlier studies in our group we found that *D. salina* has an unusual mechanism for Fe uptake, mediated by the transferrin-like protein, TTf, mentioned above (Fisher et al., 1997; Fisher et al., 1998). More recent and comprehensive proteomic analyses revealed that iron deprivation induces in *D. salina* the accumulation of five major protein bands (Figure 5A) in different cellular compartments: three in the plasma membrane, one in thylakoid membranes and one soluble protein.

The soluble 27 kD protein that is induced under Fe deprivation was identified by immuno cross-reactivity as flavodoxin (Fdx), a protein that substitutes ferredoxins in respiratory and photosynthetic electron transport systems in bacteria, cyanobacteria and diatoms. Flavodoxin has been previously identified in Fe-deprived *D. tertiolecta* (La Roche et al., 1993) and was suggested as a general indicator for Fe deprivation in photosynthetic microorganisms (La Roche et al., 1996).

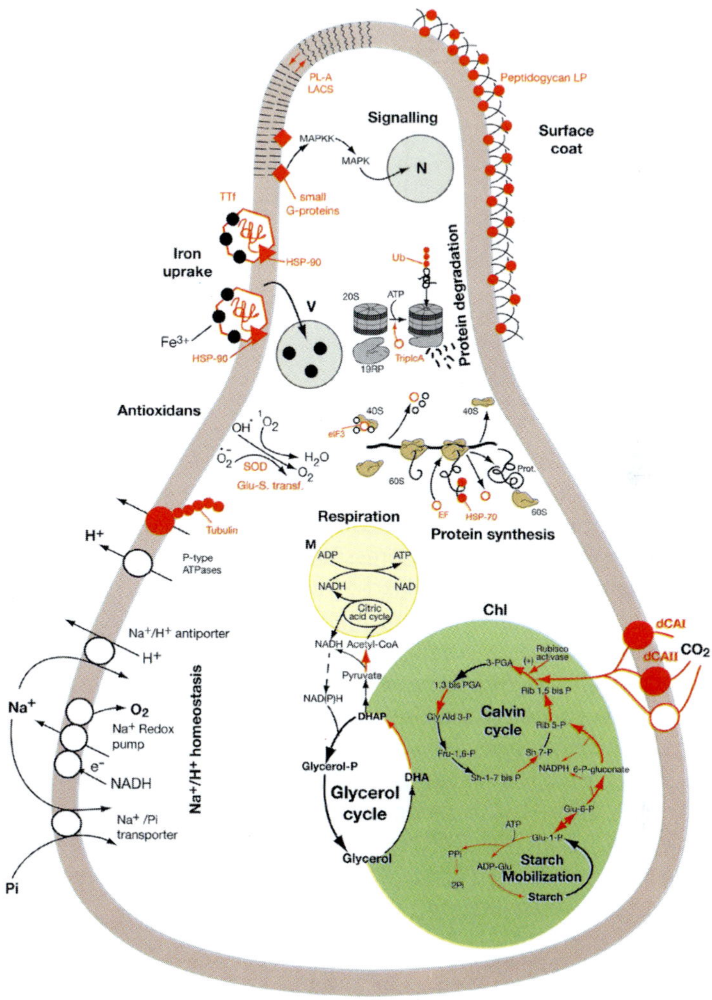

Figure 4: Schematic presentation of salt upregulated metabolic networks and proteins in *Dunaliella*. Red arrows indicate the salt-induced enzymes or proteins identified in the proteomic analysis. Chl, chloroplast; PM, plasma membrane; V, vacuole; N, nucleu; M, mitochondria.

The 40 kD protein that is induced in chloroplast thylakoids under Fe deprivation was identified as a novel chlorophyll a/b binding (CAB) protein and was termed Tidi, for thylakoid iron deficiency induced protein (Varsano et al., 2006). Tidi was localized in photosystem I (PS-I) light harvesting antenna and was found to increase significantly the size of PS-I units. These findings are particularly

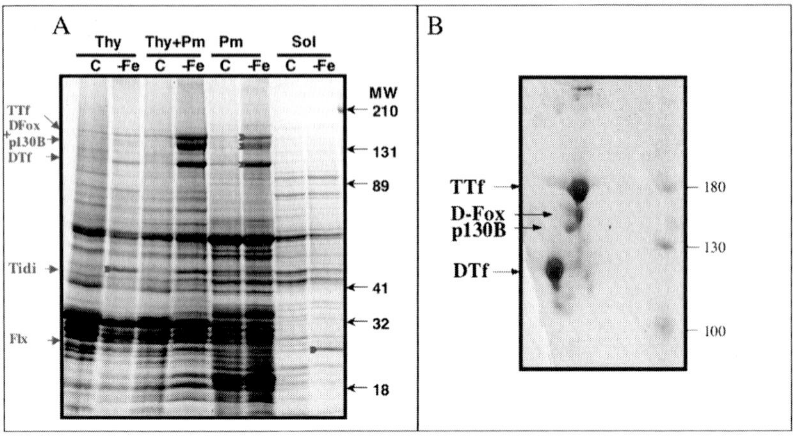

Figure 5: Iron deficiency induced proteins in *Dunaliella*.
A. Subcellular fractionation: *Dunaliella salina* cells were cultured for 48 h in the presence (C) or absence (-Fe) of 1.5 μM Fe-EDTA. Cells were lysed with a Yeda-press pressure cell and fractionated into soluble proteins (Sol), crude thylakoid fraction (Thy + Pm), purified thylakoid fraction (Thy) and plasma membrane-enriched fraction (Pm). Red arrows indicate iron deficiency-induced proteins (from (Varsano et al., 2003), with permission). B. 2D resolution of plasma membrane iron deficiency-induced proteins. Plasma membrane proteins (350 μg) derived from iron-deficient cells, were solubilized in IEF buffer and loaded on a non-linear (pH 3.5-5.0) IPG strip. Second dimension was on 7% SDS-PAGE. A section of the stained 2D gel is shown (from (Paz et al., 2007a)). TTf, DTf: transferrins, D-Fox: multi-copper ferroxidase, p130B: unknown glycoprotein, Tidi: thylakoid iron deficiency induced protein, Flx: flavodoxin.

important because PS-I is recognized as the most sensitive photosynthetic unit to Fe deprivation (Straus, 1994; Moseley et al., 2002; Varsano et al., 2006). This is probably due to the fact that PS-I is the photosynthetic unit with the highest Fe content of 12-13 Fe ions. It may be that *Dunaliella* evolved the capability to stabilize PS-I by increasing its antenna size with TiDi. Interestingly, several species of cyanobacteria evolved a similar strategy, but with a different protein: In a series of elegant studies, the research groups of Barber and Dekker demonstrated that CP43', a protein that resembles the PS-II chlorophyll binding protein CP43, which is induced in cyanobacteria under Fe deprivation, is organized around trimeric PS-I units in 18 units per trimer, and increase the light harvesting capacity of PS-I antenna (Bibby et al., 2001; Andrizhiyevskaya et al., 2002; Varsano et al., 2006). Thus, Tidi and CP43' may represent a common solution to fortify PS-I units under Fe deprivation.

The proteins that are induced in the plasma membrane were resolved on 2D gels into 4 different proteins (Figure 5B), which were all cloned and sequenced. The two major proteins were identified as transferrins, and they are in fact the first transferrins ever identified in plants. The above-mentioned TTf is an internally triplicated transferrin, it is constitutively produced in *D. salina*, but its level is significantly elevated both at high salinity and under Fe deprivation (Sadka et al., 1991; Fisher et al., 1997). TTf is a high-affinity iron binding protein, which resembles mammalian transferrins in its iron binding characteristics (Fisher et al., 1998; Schwarz et al., 2003). The three other proteins are induced coordinately in parallel with a large increase Fe binding activity. These proteins were identified as: DTf, a second transferrin (Schwarz et al., 2003), D-Fox, a multicopper ferroxidase (Paz et al., 2007a), which closely resembles Fox1 from *C. reinhardtii*, a protein that is involved in high-affinity Fe uptake by a Redox-driven mechanism (La Fontaine et al., 2002), and an unknown glycoprotein, p130B (Paz et al., 2007a). These proteins were found to associate with TTf creating a large Fe-binding complex at the plasma membrane outer surface (Scheme in Figure 6). These results led to identification of a novel mechanism for Fe uptake, that differs from other known mechanisms in algae in the following characteristics: First, the basic mechanism involves Fe binding and internalization rather than carrier mediated uptake; Second,

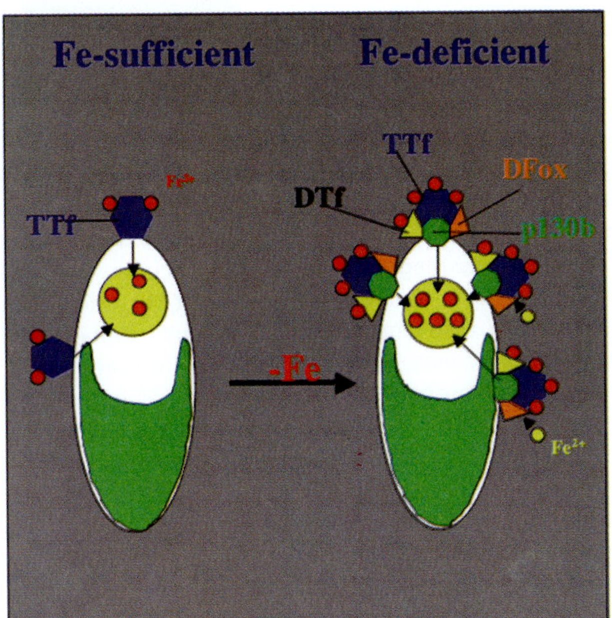

Figure 6: Schematic representation of the proposed organization of iron binding proteins in Fe-sufficient and in Fe-deficient Dunaliella.

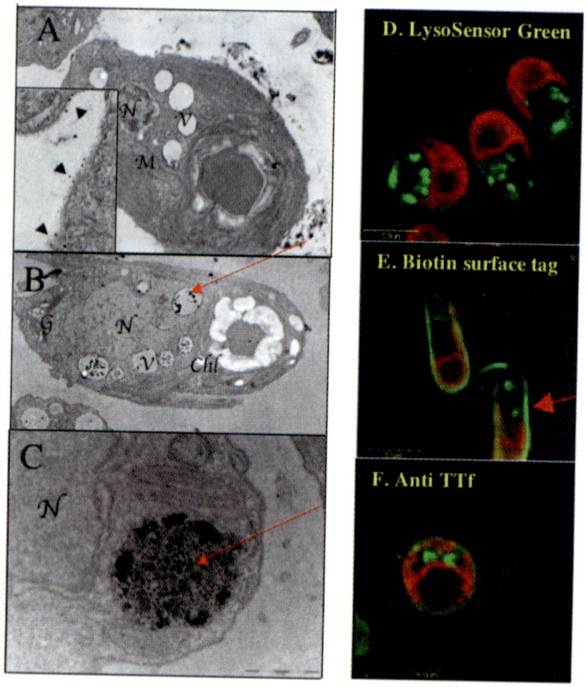

Figure 7: Internalization of iron and of iron-binding proteins into acidic vacuoles.

A-C: Iron localization with acid ferrocyanide (Prussian Blue) staining. A, Cells before internalization- notice empty vacuoles. Inset- magnified outer surface section. Notice extracellular bound iron (black arrows). B, Cells after internalization. Notice filled vacuoles (red arrow). C, Magnified Fe-loaded vacuole. D. Acidic vacuoles localization in a confocal microscope (CM): Acidic vacuoles were stained with LysoSensor Green (Molecular Probes) and shown in green fluorescence. Red fluorescence indicates chlorophyll and marks the chloroplast. E. Internalization of biotin-tagged surface proteins viewed by CM. *D. salina* cells were tagged with a membrane-impermeable biotin-derivative, treated to bind and internalize iron, fixed, permeabilized and incubated with fluorescein avidin and view in CM. Notice internalized biotin in vacuolar-like structures. F. Immunolocalization of internalized TTf visualized by CM. *D. salina* cells were treated to internalize iron, fixed, permeabilized and incubated first with primary polyclonal rabbit antibodies against TTf and next with fluorescein goat anti-rabbit antibodies. Notice internalized TTf in two vacuolar-like structures (modified from (Paz et al., 2007b) with permission).

the internalized iron was localized within acidic vacuoles and not in the cytoplasm (Figure 7, (Paz et al., 2007b); Third, iron deprivation does not significantly enhance the rate of Fe uptake but rather elevates the level of Fe binding to about 1-2 x 10^6 Fe ions per cell, an amount sufficient to satisfy the whole iron budget of the cells. We proposed that *Dunaliella* evolved this unusual strategy for iron accumulation as part of its adaptation to hypersaline solutions: transferrins are generally salt-resistant proteins, which specifically bind ferric ions with high affinity. Therefore, transferrins are very suitable for mediating Fe acquisition in hypersaline solutions and we postulated that *Dunaliella* has acquired transferrins early in its evolution by a gene transfer event from a primitive mammalian ancestor cell. The existence of a multicopper ferroxidase in *Dunaliella* suggests that it may be a relict from an earlier redox-driven Fe uptake mechanism, whose function has been adapted to promote Fe binding by association with transferrins (Paz et al., 2007a).

Acknowledgments

The work described here was supported by the Avron Minerva Center for Photosynthesis and by the Charles and Louise Gartner fund to UP.

References

Andrizhiyevskaya, E.G., T.M.E. Schwabe, M. Germano, S.D. Haene, J. Kruip, R. van Grondelle and J.P. Dekker. 2002. Spectroscopic properties of PSI-IsiA supercomplexes from the cyanobacterium Synechococcus PCC 7942. Biochim. Biolphys. Acta 1556: 265-272.

Avron, M. 1992. Osmoregulation. pp. 135-164. In *Dunaliella*: Physiology, Biochemistry, and Biotechnology. M. Avron and A. Ben-Amotz (eds) CRC Press Inc., Boca Raton, FL, USA.

Azachi, M., A. Sadka, M. Fisher, P. Goldshlag, I. Gokhman and A. Zamir. 2002. Salt induction of fatty acid elongase and membrane lipid modifications in the extreme halotolerant alga *Dunaliella salina*. Plant Physiol. 129: 1320-1329.

Baysse, C., M. Cullinane, V. Denervaud, E. Burrowes, J.M. Dow, J.P. Morrissey, L. Tam, J.T. Trevors and F. O'Gara. 2005. Modulation of quorum sensing in *Pseudomonas aeruginosa* through alteration of membrane properties. Microbiology-Sgm 151: 2529-2542.

Berken, A., C. Thomas and A. Wittinghofer. 2005. A new family of RhoGEFs activates the Rop molecular switch in plants. Nature 436: 1176-1180.

Bibby, T.S., J. Nield and J. Barber. 2001. Iron deficiency induces the formation of an antenna ring around trimeric photosystem I in cyanobacteria. Nature 412: 743-745.

Booth, W.A. and J. Beardall. 1991. Effects of salinity on inorganic carbon utilization and carbonic-anhydrase activity in the halotolerant alga *Dunaliella salina* (Chlorophyta). Phycologia 30: 220-225.

Chen, R.H., P.J. Miettinen, E.M. Maruoka, L. Choy and R. Derynck. 1995. A WD-domain protein that is associated with and phosphorylated by the type II TGF-beta receptor. Nature 377: 548-552.

Dym, O., M. Mevarech and J. Sussman. 1995. Structural features that stabilize halophilic malate dehydrogenase from an archaebacterium. Science 267: 1344-1346.

Fadl, A.A., J. Sha, G.R. Klimpel, J.P. Olano, D.W. Niesel and A.K. Chopra. 2005. Murein lipoprotein is a critical outer membrane component involved in *Salmonella enterica serovar typhimurium* systemic infection. Infection and Immunity 73: 1081-1096.

Fisher, M., I. Gokhman, U. Pick and A. Zamir. 1996 A salt-resistant plasma membrane carbonic anhydrase is induced by salt in *Dunaliella salina*. J. Biol. Chem. 271: 17718-17723.

Fisher, M., I. Gokhman, U. Pick and A. Zamir. 1997. A structurally novel transferrin-like protein accumulates in the plasma membrane of the unicellular green alga *Dunaliella salina* grown in high salinities. J. Biol. Chem. 272: 1565-1570.

Fisher, M., A. Zamir and U. Pick. 1998. Iron uptake by the halotolerant alga Dunaliella is mediated by a plasma membrane transferrin. J. Biol. Chem. 273: 17553-17558.

Gee, R., A. Goyal, R.U. Byerrum and N.E. Tolbert. 1993 Two isoforms of dihydroxyacetone phosphate reductase from the chloroplasts of *Dunaliella tertiolecta*. Plant Physiol. 103: 243-249.

Ghoshal, D. Mach, M. Agarwal and A. Goyal. 2002. Osmoregulatory isoform of dihydroxyacetone phosphate reductase from *Dunaliella tertiolecta*: Purification and characterization. Protein Expression and Purification 24: 404-411.

Hatanaka, Y., K. Inaoka, O. Kobayashi, M. Higashihara and K. Hiyama. 1998. Sensitivity of the surface coat of the halotolerant green alga *Dunaliella parva* (Volvocales, Chlorophyceae) to lysozyme. Phytological Research 46: 147-153.

Jaffe, A.B. and A. Hall 2005. Rho GTPases: Biochemistry and biology. Annual Review of Cell and Developmental Biology 21: 247-269.

Jiang, J. and S.D. Clouse. 2001. Expression of a plant gene with sequence similarity to animal TGF-beta receptor interacting protein is regulated by brassinosteroids and required for normal plant development. Plant J. 26: 35-45

Jimenez, C., T. Berl, C.J. Rivard, C.L. Edelstein and J.M. Capasso. 2004. Phosphorylation of MAP kinase-like proteins mediate the response of the halotolerant alga *Dunaliella viridis* to hypertonic shock. Biochim. Biophys. Acta 1644: 61-69.

Katz, A. and U. Pick. 2001. Plasma membrane electron transport coupled to Na^+ extrusion in the halotolerant alga *Dunaliella*. Biochim.Biophys. Acta-Bioenergetics 1504: 423-431.

Katz, A., H. Kaback and M. Avron. 1986. Na^+/H^+ antiport in isolated plasma membrane vesicles from the halotolerant alga *Dunaliella salina*. FEBS Lett. 202: 141-144.

Katz, A., U. Pick and M. Avron. 1992. Modulation of Na^+/H^+ antiporter activity by extreme pH and salt in the halotolerant alga *Dunaliella salina*. Plant Physiol. 100: 1224-1229.

Katz, A., T.R. Kleyman and U. Pick. 1994. Utilization of amiloride analogs for characterization and labeling of the plasma-membrane Na^+/H^+ antiporter from *Dunaliella salina*. Biochemistry 33: 2389-2393.

Katz, A., P. Waridel, A. Shevchenko and U. Pick. 2007. Salt-induced changes in the plasma membrane proteome of the halotolerant alga *Dunaliella salina* as revealed by blue-native gel electrophoresis and nano-LC-MS/MS analysis. Molecular and Cellular Proteomics 6: 1459-1472.

Kugler, M., L. Jansch, V. Kruft, U.K. Schmitz and H.P. Braun. 1997. Analysis of the chloroplast protein complexes by blue-native polyacrylamide gel electrophoresis (BN-PAGE). Photosynthesis Research 53: 35-44.

La Fontaine, S., J.M. Quinn, S.S. Nakamoto, M.D. Page, V. Gohre, J.L. Mosely, J. Kropat and S. Merchant. 2002. Copper-dependent iron assimilation pathway in the model photosynthetic eukaryote *Chlamydomonas reinhardtii*. Eukaryotic Cell 1: 736-757.

La Roche, J., R.J. Geider, L.M. Graziano, H. Murray and K. Lewis. 1993. Induction of specific proteins in eukaryotic algae grown under iron-deficient, phosphorus-deficient, or nitrogen-deficient conditions. Journal of Phycology 29: 767-777.

La Roche, J., P.W. Boyd, R.M.L. McKay and R.J. Geider. 1996. Flavodoxin as an in situ marker for iron stress in phytoplankton. Nature 382: 802-805.

Li, Q., X. Gao, Y. Sun, Q. Zhang, R. Song and Z. Xu. 2006 Isolation and characterization of a sodium-dependent phosphate transporter gene in *Dunaliella viridis*. Biochem. Biophys. Res. Commun. 340: 95-104.

Liska, A.J., A. Shevchenko, U. Pick and A. Katz. 2004. Enhanced photosynthesis and redox energy production contribute to salinity tolerance in *Dunaliella* as revealed by homology-based proteomics. Plant Physiol. 136: 2806-2817.

Madern, D., C. Ebel and G. Zaccai. 2000. Halophilic adaptation of enzymes. Extremophiles 4: 91-98.

Moseley, J.L., T. Allinger, S. Herzog, P. Hoerth, E. Wehinger, S. Merchant and M. Hippler. 2002. Adaptation to Fe-deficiency requires remodeling of the photosynthetic apparatus. EMBO J 21: 6709-6720.

Paz, Y., A. Katz and U. Pick. 2007a. Multicopper ferroxidase involved in iron binding to transferrins in *Dunaliella salina* plasma membranes. J. Biol. Chem. 282: 8658-8666.

Paz, Y., E. Shimoni, M. Weiss and U. Pick. 2007b Effects of iron deficiency on iron binding and internalization into acidic vacuoles in *Dunaliella salina*. Plant Physiol. 144: 1407-1415.

Pick, U. 2002. Adaptation of the halotolerant alga Dunaliella to high salinity. pp. 97-112. In Salinity: Environment-Plants-Molecules. A. Läuchli and U. Lüettge (eds) Kluwer Acad. Dordrecht, The Netherlands.

Pick, U., L. Karni and M. Avron. 1986. Determination of ion content and ion fluxes in the halotolerant alga *Dunaliella salina*. Plant Physiol. 81: 92-96.

Popova, L., G. Shumkova, I. Andreev and Y. Balnokin. 2005. Functional identification of electrogenic Na^+-translocating ATPase in the plasma membrane of the halotolerant microalga *Dunaliella maritima*. FEBS Letters 579: 5002-5006.

Premkumar, L., U.K. Bageshwar, I. Gokhman, A. Zamir and J.L. Sussman. 2003. An unusual halotolerant alpha-type carbonic anhydrase from the alga *Dunaliella salina* functionally expressed in *Escherichia coli*. Protein Expression and Purification 28: 151-157.

Premkumar, L., H. Greenblatt, U.T.S. Bageshwar, I. Gokhman, J. Sussman and A. Zamir. 2005. Free in PMC Three-dimensional structure of a halotolerant algal carbonic anhydrase predicts halotolerance of a mammalian homolog. Proc. Natl. Acad. Sci. USA. 102: 7493-7498.

RodriguezHerva, J.J., M.I. Ramos Gonzalez and J.L. Ramos. 1996. The *Pseudomonas putida* peptidoglycan-associated outer membrane lipoprotein is involved in maintenance of the integrity of the cell envelope. Journal of Bacteriology 178: 1699-1706.

Sadka, A., S. Himmmelhoch and A. Zamir. 1991. A 150 kilodalton cell-surface protein is induced by salt in the halotolerant green-alga *Dunaliella salina*. Plant Physiol 95: 822-831.

Schagger, H. and G. Vonjagow. 1991. Blue-native electrophoresis for isolation of membrane protein complexes in enzymatically active form. Analytical Biochemistry 199: 223-231.

Schwarz, M., N. Sal-Man N, A. Zamir and U. Pick. 2003. A transferrin-like protein that does not bind iron is induced by iron deficiency in the alga *Dunaliella salina*. Biochim. Biophy. Acta 1649: 190-200.

Schwarz, M., A. Zamir and U. Pick. 2003. Iron-binding properties of TTf, a salt-induced transferrin from the alga *Dunaliella salina*. Journal of Plant Nutrition 26: 2081-2091.

Shevchenko, A., S. Sunyaev, A. Loboda, A. Shevchenko, P. Bork, W. Ens and K.G. Standing. 2001. Charting the proteomes of organisms with unsequenced genomes by MALDI-quadrupole time-of-flight mass spectrometry and BLAST homology searching. Anal. Chem. 73: 1917-1926.

Straus, N. 1994. Iron Deprivation: Physiology and Gene Regulation. Kluwer Academic Publishers Group, Dordrecht, The Netherlands.

Varsano, T., D. Kaftan and U. Pick. 2003. Effects of iron deficiency on thylakoid membrane structure and composition in the alga *Dunaliella salina*. J. Plant Nutri. 26: 2197-2210.

Varsano, T., S.G. Wolf and U. Pick. 2006. A chlorophyll a/b-binding protein homolog that is induced by iron deficiency is associated with enlarged photosystem I units in the eucaryotic alga *Dunaliella salina*. J. Biol. Chem. 281: 10305-10315.

Weiss, M. and U. Pick. 1996. Primary structure and effect of pH on the expression of the plasma membrane H^+-ATPase from *Dunaliella acidophila* and *Dunaliella salina*. Plant Physiol. 112: 1693-1702.

Weiss, M., G. Haimovich and U. Pick. 2001. Phosphate and sulfate uptake in the halotolerant alga Dunaliella are driven by Na^+-symport mechanism. J. Plant Physiol. 158: 1519-1525.

Yang, P.R., H.Y. Du, C. Hoffman and S. Marcus. 2003. The phospholipase B homolog, Plb1, is a mediator of osmotic stress response and nutrient-dependent repression of sexual differentiation in the fission yeast, *Schizosaccharomyces pombe*. Yeast 20: S208-S208.

Zelazny, A.M., A. Shaish and U. Pick. 1995. Plasma-membrane sterols are essential for sensing osmotic changes in the halotolerant alga *Dunaliella*. Plant Physiol. 109: 1395-1403.

15 Carbon Dioxide Fixation by *Dunaliella* spp. and the Possible Use of this Genus in Carbon Dioxide Mitigation and Waste Reduction

John A. Raven

Abstract

Dunaliella spp. live in a wide range of habitats with significant variations in inorganic carbon availability. As far as has been investigated, *Dunaleilla* spp. have the normal range of carboxylases for green algae and high plants, including Form IB Rubisco. Tracer carbon studies show that the biochemical pathway of CO_2 assimilation in *Dunaliella* involves the photosynthetic carbon reduction cycle with no prior carboxylation-decarboxylation (C_4-like) cycle except, perhaps, in stationary phase cells. All *Dunaliella* spp. so far examined, except perhaps *D. acidophila*, have inorganic carbon concentrating mechanisms (CCMs) based, apparently, on active transport of inorganic carbon (or protons) across one or more membranes. These CCMs, by accumulating CO_2 at the site of Rubisco, increases the affinity of cells for inorganic carbon relative to what would be expected from the kinetics of Rubisco from relatives of *Dunalialla*. CCMs decrease, but do not eliminate, the occurrence of the oxygenase activity of Rubisco and the subsequent glycolate excretion and/or metabolism. Both CO_2 and HCO_3^- can enter both whole cells, and isolated, photosynthetically active, chloroplasts, of *Dunaliella*. The mechanism(s) of the CCM(s) of *Dunalialla* needs much further investigation. To the extent that microalgae can be used in carbon dioxide mitigation, *Dunaliella* spp. have the advantage of using water (saline or acidic) which would not be usable to grow most other algae (or for most other human uses), and of having high-value products such as β-carotene in addition to *Dunalialla* biomass that can be used as a fuel.

Division of Plant Sciences, University of Dundee at SCRI, Scottish Crop Research Institute, Invergowrie, Dundee DD2 5DA, UK, Email: j.a.raven@dundee.ac.uk

Introduction

Dunaliella spp. and the Inorganic Carbon Chemistry of their Natural Habitats

In considering the means by which *Dunaliella* spp. acquire inorganic carbon and convert it into organic carbon, and the possibilities that the genus can be used in carbon mitigation procedures, it is essential to circumscribe the genus so that we know what organisms we are dealing with. It is also important to know their native habitats, and especially the inorganic carbon availability in those habitats and extent to which inorganic carbon availability determines growth rate.

González et al. (2001) used nuclear ITS rDNA sequence studies to study the molecular phylogeny of 15 strains of *Dunaliella*. The results showed that the 14 strains representing the halophilic subgenus *Dunaliella* formed a clade, while the only representative (*Dunaliella lateralis*) of the freshwater subgenus *Pascheria* was genetically almost as far from the subgenus *Dunaliella* as was the outgroup *Chlamydomonas reinhardtii*. On the basis of ultrastructural features the species of subgenus *Pascheria* that is closest to subgenus *Dunaliella* is the acidophilic *Dunaliella acidophila* (formerly *Spermatozopsis acidophila*): Albertano et al. (1981); Melkonian and Preisig (1984); González et al. (2001). Importantly for considerations of inorganic carbon acquisition, *D. acidophila* is one of the two species out of the five species in subgenus *Pascheria* that has a pyrenoid. In this Chapter *D. acidophila* will be considered as a member of the genus *Dunaliella*.

As for the habitats, some species of the subgenus *Dunaliella* can grow in saturated salt (predominantly NaCl) solutions, while *D. acidophila* can grow at a pH as low as pH 1, equivalent to a concentration of ions of not less than 20% of seawater (or 1% of saturated brine). Implications of salinities (practical salinity scale) as high as 330 for the solubility of CO_2 and the speciation of the dissolved inorganic carbon are a (non-linear) continuation of the decrease in solubility of CO_2, and in the pK_{a1} and especially pK_{a2}, seen in going from freshwater to seawater at 35 (Sass and Ben-Yaakov 1977; Raven and Richardson 1986). Thus, the Dead Sea has (at 30°C) a pK_{a1} of 5.09 and a pK_{a2} of 6.23, as compared to 5.9 and 9.1 respectively for seawater (Sass and Ben-Yaakov 1977). With a pH of not more than 6.45 at the surface of the Dead Sea (Ben-Yaakov and Sass 1977), the ratios of the three major inorganic forms are 38 CO_3^{2-} and 23 HCO_3^- for each 1 CO_2.

The solubility of CO_2 does not necessarily dictate the CO_2 concentration in a common *Dunaliella* habitat, inland saline waters. One reason is that CO_2 is generated in these habitats (as in most inland waters) by chemo-organotrophs from dissolved (DOC) and particulate (POC) organic carbon. These two forms of organic carbon are derived from atmospheric CO_2 by terrestrial photosynthesis in the catchment and are delivered to the water body in the influent rivers (Einsele et al. 2001). The influent rivers also generally have CO_2 concentrations in excess of air equilibrium as a result of respiration in the soil of organic matter generated from atmospheric CO_2 by vegetation in the catchment (Raven and Maberly 2008). Finally, the portion of the CO_2 from respiration in the soil which is not lost to rivers in groundwater or by gas-phase diffusion to the atmosphere is consumed in chemical (biologically assisted) weathering of carbonates to bicarbonates and of silicates to silicic acid and bicarbonates. These soluble compounds, with the corresponding cations, move via groundwater to rivers and hence to the water body, with in dissolved ionic inorganic carbon plus dissolved CO_2 termed dissolved inorganic carbon (DIC). In the water body there is frequently precipitation of solid carbonates (particulate organic carbon or POC) from the bicarbonates, with production of CO_2 (Barkan et al. 2001; Einsele et al. 2001).

In summary, carbon from atmospheric CO_2 assimilated by terrestrial vegetation is transferred, as DIC, DOC and POC to the water body, where there is further CO_2 production in metabolism of POC and DOC and in the conversion of DIC to PIC. The CO_2 production is in excess of the CO_2 assimilated in the water body by *Dunaliella* and other primary producers, as a result of limitations by 'bottom up' factors (supply of other inorganic nutrients, limitation by light supply if there is deep mixing, and limitation by extremely high salinity: Oren et al. 1995) and 'top down' factors (e.g. grazers and parasitoids).

Having set the scene phylogenetically and environmentally, we can put into context the large body of data on how *Dunaliella* spp. acquire and assimilate inorganic carbon.

Biochemistry of Inorganic Carbon Assimilation into Organic Carbon in *Dunaliella*

Enzyme assays, assays with antibodies, and genome sequence have shown the occurrence of a range of carboxylases, and related enzymes, in *Dunaliella* spp.

Rubisco (ribulose-1.5-bisphophosphate carboxylase oxygenase) is the core enzyme of CO_2 incorporation into organic matter in oxygen-producing photolithotrophs, and many other autotrophs. Enzyme activity assays (starting with Beardall et al. 1976 and Mukerji et al. 1978), enzyme protein assays (Sukenik et al. 1987; Giordano and Bowes 1997; Lin and Carpenter 1997) and genomic and proteomic studies (Liska et al. 2004) show the occurrence of Form IB Rubisco in *Dunaliella* spp. Strangely, there is no reason nine years on to change the conclusion of Badger et al. (1998) that there are no data for the kinetics (half-saturation concentrations, maximum reaction rates) of the carboxylase and oxygenase activities of *Dunaliella* Rubisco. The activation and deactivation of the *D. tertiolecta* Rubisco resembles that of the (Form IB) Rubisco of vascular plants (MacIntyre et al. 1997) and genomic and proteomic evidence for the occurrence, as in the other eukaryotes with Form IB Rubisco that have been examined, of Rubisco activase (Liska et al. 2004).

Rubisco is mainly located in the pyrenoids of the chloroplast stroma (Markelova et al. 1990 for *D. salina*; delRio et al. 1996, and Lin and Carpenter 1997 for *D. tertiolecta*), with more of the Rubisco protein in the pyrenoid for cells grown at higher photon flux densities (Lin and Carpenter 1997). There is a greater content of Rubisco protein per cell when *D. tertiolecta* is grown at higher photon flux densities (Sukenik et al. 1987). Rubisco activity and protein content are higher on a cell basis when *D. salina* is grown at higher rather than lower CO_2 levels, and with ammonium rather than nitrate as the nitrogen source for growth (Giordano and Bowes 1997). CO_2 is the inorganic carbon substrate for RUBISCO.

PEPC (phosphoenolpyruvate carboxylase) is the major anaplerotic carboxylase involved in supplying C_4 dicarboxylic acids to replenish the tricarboxylic acid cycle as carbon skeletons are removed from it for biosynthesis of amino acids, tetrapyrrols and pyrimidines in green (and many other) algae and embryophytic plants. This enzyme also acts as the initial carboxylase in almost all variants of C_4 photosynthesis (Granum et al. 2005). Enzyme assays have shown the occurrence of PEPC in *Dunaliella* spp. (Beardall et al. 1976, Mukerji 1978, Mukerji et al. 1978, Giordano and Bowes 1997, Giordano 2001). PEPc activity in *Dunaliella* is higher when cells are grown at low rather than higher CO_2 concentrations (Aizawa et al. 1985: *D. tertiolecta*; Giordano and Bowes 1997; Giordano 2001: *D. salina*), and with NH_4^+ rather than NO_3^- as the nitrogen source at high available nitrogen concentrations but no difference between the two nitrogen sources when grown at low available nitrogen concentrations (Giordano and Bowes 1997; Giordano 2001; Norici et al. 2002: *D. salina*; Giordano et al. 2003: *D. parva*). The activity of PEPc is decreased when *D. salina* cells are grown with limiting SO_4^{2-} concentrations for growth (Giordano et al. 2000). PEPC presumably accounts for most of the light-independent inorganic carbon fixation in *Dunaliella* (Giordano and Bowes 1997; Giordano 2001: *D. salina*). HCO_3^- is the inorganic carbon substrate for PEPC.

PEPCK (phosphoenolpyruvate carboxykinase) can behave as a (C_3+C_1) carboxylase, or as a (C_4-C_1) decarboxylase, with CO_2 as the substrate for carboxylation and the product of decarboxylation. PEPCK can, among other roles, act as a decarboxylase in some variants of C_4 photosynthesis (Granum et al. 2005). This enzyme has been assayed in *D. salina* by Giordano and Bowes (1997), who found that the activity of PEPCK was 1/3 – 1/2 that of PEPC. Giordano and Bowes (1997) also showed that PEPCK activity in *D. salina* was higher in cells grown on ammonium rather than nitrate as nitrogen source, but is independent of the CO_2 concentration used for growth.

NADPME (NADP malic enzyme) activity was reported by Mukerji (1978) as being higher than that of PEPC in *D. tertiolecta*. Among other roles, NADPME is the decarboxylase in some variants of C_4 photosynthesis (Granum et al. 2005). CO_2 is the inorganic species involved in both the carboxylation and decarboxylation reactions. There seem to be no measurements of NADME (NAD malic enzyme) activity in *Dunaliella*.

ACC (acetyl CoA carboxylase) is involved in the addition of acetyl units in the synthesis of fatty acids and isoprenoids; this use is catalytic, in that the CO_2 is released when the malonyl CoA produced by ACC is used to add two carbons to the lipid compound. Activity of this ACC in *Dunaliella* was demonstrated by Rabbani et al. (1998). HCO_3^- is the inorganic carbon substrate of ACC.

PPDK (pyruvate-orthophosphate dikinase), an enzyme very generally involved in C_4 photosynthesis (Granum et al. 2005), has a relatively lower activity in *Dunaliella* than in the two diatoms tested (Mukerji et al. 1978).

CA (carbonic anhydrase) activity in *Dunaliella* has been detected in the periplasm (extracellular), and inside cells, with evidence for two distinct carbonic anhydrases in plastids (Aizawa and Miyachi 1984; Aizawa et al. 1986; Amoroso et al. 1998). There are also genomic and proteomic studies on CAs in *Dunaliella* (Liska et al. 2004). The activity of CAs is increased at lower inorganic carbon concentrations at constant salinity (Aizawa and Miyachi 1984: *D. tertiolecta*) and increased salinity at constant gas-phase CO_2 (Booth and Beardall 1991; Fisher et al. 1997: *D. salina*). The role of these CAs has been mainly probed by treatments that inhibit the external CA only, e.g. proteolysis of periplasmic proteins, including CA, by subtilisin, and membrane-impermeant CA inhibitors such as DBS (dextran-bound sulphonamide) or, less certainly impermeable, AZA (acetazolamide), and a membrane-permeant inhibitor that inhibit both external and internal CAs, EZA (ethoxyzolamide) (Aizawa and Miyachi 1984; Palmqvist et al. 1991). To the extent that these CA inhibition treatments can be trusted, it appears that the external CA acts in catalysing the conversion of external HCO_3^- to CO_2, with subsequent uptake of the CO_2 to a much greater extent in air-grown cells than in cells grown in at higher CO_2 concentrations, but even in air-grown cells the inhibition of photosynthesis is not complete. Inhibition of the extracellular CA decreases the photosynthetic affinity for inorganic carbon in air-grown cells. One or both of the intracellular CAs is required for photosynthesis in cells grown at both high and low CO_2 levels. Giordano and Bowes (1997) showed, in a unique (for *Dunaliella*) measurement of the CA activity in units of the rate substrate transformation per cell, that extracellular CA is much greater than the intracellular CA activity in air-grown *D. tertiolecta*. Activity of the extracellular CA of *D. salina* is not affected by the SO_4^{2-} concentration used for growth, although S-limited cells have lower inorganic carbon-saturated growth rates and lower affinities for inorganic carbon in photosynthesis (Giordano et al. 2000).

Isotopic Labelling Studies on Early Inorganic Carbon Fixation Products in *Dunaliella*

Determination of the earliest organic products of inorganic carbon fixation by isotopic (^{14}C) labelling involves the use of the shortest possible time between supplying isotopically enriched inorganic carbon to the cell suspension and killing the cells for extraction of soluble, acid-stable (i.e. not inorganic carbon) compounds and their separation and quantitation of radioactivity in each compound. Labelling in 3-phosphoglycerate and other sugar phosphates at the shortest labelling time suggest C_3 biochemistry, while labelling in malate and /or aspartate suggests C_4 biochemistry. 1-2 seconds is technically the shortest practicable time of labelling.

The first such experiments on an unspecified species *Dunaliella* (Wegmann 1969) had 15 s as the shortest labelling time; this is too long to give a firm indication of the earliest labelled product(s), so the observed labelling after 15 s in sugar phosphates alone cannot be taken as a clear indication of C_3 biochemistry. Beardall et al. (1976) had 2 s as the shortest labelling time, so the observation that all of the acid-stable label in 2 s is in 3-phosphoglycerate is a clear indication of C_3 photosynthetic biochemistry in *D. tertiolecta*. At longer labelling times the fraction of the label in 3-phosphoglycerate decreases and label occurs in sugar phosphates, then in glycerol and amino acids. These data come from cells growing in low inorganic carbon concentrations during the exponential phase.

Mukerji et al. (1978) compared the short-term labelling products of *D. tertiolecta* in exponentially growing and stationary phase cultures. The results from the exponential phase cells agree with those of Beardall et al. (1976). By contrast, the stationary phase cultures only had 10% of the label after 2s in 3-phosphoglycerate and sugar phosphates, 43% in aspartate, and 24% in tricarboxylic acid cycle intermediates with very little in malate. This labelling pattern suggests C_4 photosynthesis. Mukerji et al. (1978) point out that the observed labelling of the C_4 compounds cannot be quantitatively accounted for by the anaplerotic role C_3+C_1 carboxylation (see also Giordano and Bowes 1997; Granum et al. 2005). The C_4-like photosynthetic biochemistry found in stationary stage cells is correlated with lower ratios of Rubisco activity to PEPC activity than occurs in the exponentially growing cells with their C_3 biochemistry.

Presence of a CCM in *Dunaliella* spp.

A CCM (inorganic carbon concentrating mechanism), such as C_4 biochemistry of photosynthesis, or active transport of one or more of CO_2, HCO_3^- or H^+ across one or more membranes, brings about a CO_2 concentration around Rubisco in steady-state photosynthesis that is higher than the CO_2 concentration in the medium (Giordano et al. 2005).

A priori evidence for the occurrence of a CCM, based on the kinetics of whole-cell photosynthesis and of Rubisco, is that the ratio maximum rate on a cell basis to the half-saturation value for CO_2 of photosynthesis *in vivo*, is lower than the ratio of the maximum rate on a cell basis to the half-saturation value for CO_2 for Rubisco *in vitro* (see Badger et al. 1998). Diffusive entry of CO_2 requires that the former ratio exceeds the latter to allow for the restrictions on CO_2 diffusion between the bulk medium and Rubisco. When the former ratio is less than the latter there must be a CCM which provides a higher steady-state concentration of CO_2 around Rubisco than occurs in the bulk medium. While there is evidence correlating Rubisco content with photosynthetic rate for *Dunaliella* cells grown at a range of photon flux densities (Sukenik et al. 1987), there are no data on the kinetics of *Dunaliella* Rubisco with respect to CO_2 and O_2 concentrations (Badger et al. 1998; see above). Thus the suggestion by Badger et al. (1998), based on the arguments used above, that *Dunaliella* has a CCM was based (as Badger et al. 1998 point out) on the Rubisco kinetics of closely related algae such as *Chlamydomonas*. Badger et al. (1998) assumed a $K_{1/2}$ for CO_2 of 30 μM for Rubisco from *Dunaliella* spp, which is more than an order of magnitude higher than all but one of the $K_{1/2}$ values for whole-cell photosynthesis in air-grown cells of halophilic species of *Dunaliella* (Table 1). However, the values for Rubisco are much more similar to the value for *in vivo* photosynthesis in the acidophilic *D. acidophila* (Table 1) which probably lacks a CCM (Gimmler and Slovick 1995). Some of the values (e.g. Gordillo et al. 2003) for the $K_{1/2}$ for inorganic carbon, expressed as CO_2, that can be estimated from measurements of the growth of halophilic *Dunaliella* are higher than most of the values for photosynthesis in Table 1. However, it must be remembered that it is difficult to ensure equilibrium between the gas stream and the culture medium, and any lack of equilibrium would under-estimate the $K_{1/2}$ for CO_2 in growth.

Strong evidence for a CCM in *Dunaliella* comes from work showing higher internal than external CO_2 concentrations during photosynthesis, based on measurements of internal inorganic carbon concentrations after supply of external supply of ^{14}C inorganic carbon. The CO_2 concentration is computed from measurements of internal pH. Such measurements of intracellular inorganic carbon have been carried out on halophilic species of *Dunaliella* grown at low external inorganic concentrations by Zenvirth and Kaplan (1981), Burns and Beardall (1987), Young et al. (2001) and Beardall et al. (2002), and showed a higher internal than external inorganic carbon concentration. The intracellular pH measurements of Burns and Beardall (1987), Gimmler et al. (1988), Kuchitsu et al. (1989) and Young et al. (2001) show that the results cannot be explained by diffusive CO_2 entry and an 'alkaline trap' of such CO_2 as HCO_3^- (Werdan et al. 1972; Heldt et al. 1973): there is a higher internal than external CO_2 concentration during photosynthesis, i.e. a CCM is operating. These experiments give average values for the intracellular concentration of inorganic carbon and, using the mean pH of the cell contents, the average CO_2 concentration inside the cells; they do not directly indicate where in the cell CO_2 is accumulated. If the compartment in which CO_2 is concentrated is less than the whole cell (the smallest compartment proposed is the pyrenoid), then the CO_2 concentration in that compartment relative to that in the rest of the cell, and to the medium, will be correspondingly higher.

The CCM at the whole cell level in *Dunaliella* is light-dependent, at least in part because of the need for energy from photosynthesis to power the pump(s)

Table 1: Photosynthetic half-saturation value for inorganic carbon, expressed as CO_2, as a function of the species of *Dunaliella* and the conditions for growth.

***Dunaliella* sp.**	**Growth conditions**	**$K_{1/2}CO_2$ mmol m^{-3}**	**Reference**
D. acidophila	pH 1, HC	60.4	Gimmler & Slovick (1995)
D. acidophila	pH 1, LC	21.0	Gimmler & Slovick (1995)
D. bioculata	HC	2.1	Aizawa et al. (1985)
D. bioculata	LC	0.56	Aizawa et al. (1985)
D. parva	HC	166	Gimmler & Slovick (1995)
D. parva	LC	2.6	Gimmler & Slovick (1995)
D. salina	LC, 210 mol m^{-3} NaCl	13.4	Booth & Beardall (1991)
D. salina	LC, 2.2 kmol m^{-3} NaCl	0.15	Booth & Beardall (1991)
D. salina	LC $NO_3^{-\#}$	0.02(0.21)	Giordano & Bowes (1997)
D. salina	LC $NH_4^{+\#}$	0.12(0.33)	Giordano & Bowes (1997)
D. salina	LC, 1.5 kmol m^{-3} NaCl HS	0.3	Giordano et al. (2000)
D. salina	LC, 1.5 kmol m^{-3} NaCl LS	1.0	Giordano et al. (2000)
D. tertiolecta	LC, +eCA	0.51	Aizawa & Miyachi (1984)
D. tertiolecta	LC, -eCA	1.12.	Aizawa & Miyachi (1984)
D. tertiolecta	LC, -eCA, +CA	0.4	Aizawa & Miyachi (1984)
D .tertiolecta	HC	1.25	Aizawa et al. (1985)
D. tertiolecta	LC	0.48	Aizawa et al. (1985)
D. tertiolecta	LC	1.35	Burns & Beardall (1987)
D. tertiolecta	HC	13.9	Amoroso et al. (1998)

D. tertiolecta	LC	1.7	Amoroso et al. (1998)
D. tertiolecta	HC, Chlpl	5.8	Amoroso et al. (1998)
D. tertiolecta	LC, Chlpl	0.5	Amoroso et al. (1998)
D. tertiolecta	LC, H PAR*	0.1	Young & Beardall (2005)
D. tertiolecta	LC, L PAR*	0.25	Young & Beardall (2005)
D. tertiolecta	LC, HN	0.33	Young & Beardall (2005)
D. tertiolecta	LC, LN	0.20	Young & Beardall (2005)
D. tertiolecta	LC, H Fe	0.30	Young & Beardall (2005)
D. tertiolecta	LC, L Fe	0.10	Young & Beardall (2005)
D. viridis	HC	1.8	Aizawa et al. (1985)
D. viridis	LC	0.7	Aizawa et al. (1985)

Notes: Unless otherwise stated, seawater or artificial seawater, neutral to alkaline pH, grown with saturating PAR, NO_3^-, Fe, extracellular CA (eCA) not inhibited, $K_{1/2}$ based on inorganic carbon dependence of O_2 evolution

$K_{1/2}CO_2$ concentrations as given in the cited papers, or calculated from the total inorganic carbon or HCO_3^- concentrations given using the given salinity and temperature (and hence pK_a) values and pH

HC = CO_2 at least 10 times the present air level; LC = CO_2 at present air level

+eCA = with eCA unmodified, -eCA = eCA inhibited by AZA or subtilisin. –eCA+CA = eCA inhibited by subitilisin, subitilisin removed, CA added

Chlpl = measurements made on isolated chloroplasts.

#Values in parentheses relate to estimates from ^{14}C-inorganic carbon assimilation.

HS = saturating SO_4^{2-} for growth, LS = limiting SO_4^{2-} for growth, H PAR = saturating PAR for growth, L PAR = limiting PAR for growth, HN = saturating NO_3^- for growth, LN = limiting NO_3^- for growth, H Fe = saturating Fe for growth, L Fe = limiting Fe for growth

*Significant difference between saturating and limiting PAR for batch cultures (shown) but not for turbidostat cultures

(Burns and Beardall 1987). The kinetics of the CCM, including the affinity for inorganic carbon, are altered by many environmental conditions (Table 1). Booth and Beardall (1991) found that the affinity for inorganic carbon in *D. salina* increased with increasing salinity of the growth medium, in parallel with increased expression of an extracellular CA (see Fisher et al. 1997). Sültemeyer (1997) found that there was variation in CCM through the cell cycle of *D. tertiolecta* (cf. Naus and Melis 1991). Giordano et al. (2000) showed that sulfur deficiency decreased both the maximum rate of photosynthesis and the affinity for inorganic carbon in *D. salina*. Surprisingly, oxygen inhibits the CCM in *D. tertiolecta* (Ghoshal and Goyal 2001); the mechanism of this inhibition is not clear. Beardall et al. (2002) found that the CCM in *D. tertiolecta* is unaffected by fluxes of UVB that significantly inhibit photosynthetic CO_2 assimilation. Young and Beardall (2005) showed that growth of *D. tertiolecta* at limiting fluxes of photosynthetically active radiation decreased the activity of the CCM, as indicated by the affinity of cells for inorganic carbon (the reciprocal of the half-saturating inorganic carbon concentration). The activity of the CCM in *D. tertiolecta* was, by the same measure, increased when growth was limited by the availability of nitrogen or iron (Young and Beardall 2005; cf. Lin et al. 2001). These findings are very interesting ecologically, but tell us little about the mechanisms of the CCM.

Goyal and Tolbert (1989) examined uptake of inorganic carbon by isolated, photosynthetically functional chloroplasts of the halophilic *D. tertiolecta*, and showed a 4-fold accumulation of inorganic carbon by illuminated chloroplasts from air-grown cells, but negligible accumulation in chloroplasts from cells grown in 5% CO_2 in air or in chloroplasts from air-grown cells in the dark. Could this accumulation be a result of alkalinisation of the stroma in the light and operation of an 'alkaline trap' mechanism (Werdan et al. 1972; Heldt et al. 1973)? Such a mechanism would in principal account for the decreased accumulation seen in the presence of metabolic inhibitors (Goyal and Tolbert 1989) which might limit the stromal alkalisation. However, such an accumulation would need a stromal pH in the light of 8.1, whereas the highest estimated stromal pH of intact cells of *D. parva* in the light is only 7.5 (Gimmler et al. 1988). Accordingly, a CCM acting at the chloroplast level, as suggested by Goyal and Tolbert (1989) seems to be more likely.

Modelling of photosynthetic CO_2 supply on air-grown cells of the acidophilic *D. acidophila* (Gimmler and Slovick 1995; see Gimmler et al. 1988 for intracellular pH values and Gimmler et al. 1989 for the transplasmalemma electrical potential difference) suggests that photosynthesis may be accounted for on the basis of catalysed diffusion of CO_2; this mechanism clearly operates for CO_2 transport in cells grown on 3% CO_2 in air. The possibility of diffusive entry of CO_2 for growth of *D. acidophila* can be seen from the lower affinity for CO_2 of air-grown cells of this species than for the halophilic/neutrophilic – alkalophilic speces of the genus (Table 1). Not all acidophilic algae lack CCMs: Zenvirth et al. (1985) showed than the cyanidophyte red alga *Cyanidioschyzon merolae* showed accumulation of inorganic carbon within the cells during photosynthesis, and had an affinity for CO_2 that was an order of magnitude greater than that of *D. acidophila*.

Gimmler and Slovik showed, using the same modelling technique that they employed for *D. acidophila*, and in agreement with earlier work, that a CCM is needed for halophilic *D. parva* cells grown in either air or 3% CO_2.

Occurrence of Photorespiration in *Dunaliella* spp.

Rubisco oxygenase activity, and thus glycolate excretion and photorespiratory glycolate metabolism, would be expected to be inhibited by the CCM operating in at least the halophilic *Dunaliella* spp. examined when grown with air unenriched CO_2. This is because the CCM increases the CO_2/O_2 ratio around Rubisco. In agreement with this Hellebust (1965) and Huntsman (1972) found that glycolate was not more than 2.2% of the organic carbon excreted by illuminated *D. tertiolecta*. Furthermore, Giordano et al. (1994) found no stimulation of organic carbon excretion when cells were transferred from high CO_2 to lower CO_2 conditions that would be expected to increase glycolate excretion before CCM and glycolate metabolism enzymes could be upregulated. Such experiments do not, however, address the metabolism of glycolate: a low rate of glycolate excretion could be related to a high rate of glycolate production if there was effective metabolism of glycolate by the photorespiratory carbon oxidation cycle (PCOC) or its equivalent. That suppression of Rubisco oxygenase by the CCM is not always complete is shown by studies (Leboulanger et al. 1998) of the effect of light-dark transients on the intracellular pools of two intermediates, glycine and serine, of the PCOC, and on production of extracellular glycolate, in air-grown *D. tertiolecta*. This significant carbon flux into, and through, glycolate even for air-grown cells could help to explain the inhibition by oxygen of ^{14}C-inorganic carbon assimilation in the light in *D. tertiolecta* with normal seawater inorganic carbon concentrations.

The mean inhibition in six experiments on *D. tertiolecta* was by 17% with 21% (i.e. air level) oxygen, and by 38% with 100%, relative to the rate in the nominal absence of oxygen (Beardall et al. 1976). Part of any photorespiration-related inhibition would relate to loss of previously fixed ^{14}C as $^{14}CO_2$ in the conversion of glycine to serine in the PCOC, and thus not measured in the experimental procedure (Beardall and Morris 1975), although any ^{14}C glycolate excreted by the cells would have been measured and so would not contribute to the inhibition. However, a major component would have come from oxygen inhibition of Rubisco carboxylase activity.

Such an inhibition of photosynthetic $^{14}CO_2$ fixation not associated with oxygen uptake is at least qualitatively consistent with the observations of Bate et al. (1988) who measured the rate of $^{18}O_2$ uptake in the light and the dark by *D. tertiolecta* in a medium (0.2 mM inorganic carbon at pH 7, i.e. the same CO_2 concentration as air-equilibrated seawater at pH 8) that saturated photosynthesis with respect to inorganic carbon. They found that labelled oxygen uptake decreased from the dark rate with increasing photon flux density. This means that the sum of oxygen-uptake processes in the light (Rubisco oxygenase, oxidative steps of glycolate metabolism, the Mehler peroxidase reaction, and mitochondrial respiration) is less than the sum of the processes in the dark, i.e. mitochondrial respiration and chlororespiration. The data of Bate et al. (1988) is also consistent with the observations of Ghoshal and Goyal (2001) that the CCM in *D. tertiolecta* is inhibited by oxygen, presumably by a process involving little or no oxygen uptake.

Turning to the reactions involved in the intracellular processing of glycolate, members of the Chlorophyceae typically have a mitochondrial glycolate dehydrogenase rather than the peroxisomal glycolate oxidase found in embryophytic plants and their green algal ancestors, the Charophyceae (Stabenau et al. 1993; Stabenau and Winkler 2005). Stabenau et al. (1993) found mitochondrial glycolate dehydrogenase in the two *Dunaliella* spp. (*D. primolecta, D. salina*) that they examined. However, Stabenau et al. (1993) could find no evidence of the peroxisomal glycolate oxidase reported, with a mitochondrial glycolate dehydogenase, for *D. primolecta* and *D. salina* by Nimer et al. (1990). They could not detect glycolate dehydrogenase in the thylakoid membranes of the chloroplasts of these two species, although this activity was earlier reported by Sallal et al. (1987) for *D. salina* and later by Goyal and Tolbert (1996) and Goyal (2002) for *D. tertiolecta*. Goyal (2002), on the basis of the mitochondrial and chloroplast enzyme activities, and inhibitor sensitivities, that the chloroplast enzyme is predominant in glycolate metabolism. Clearly there is no consensus on the location or nature of the glycolate-oxidising enzyme(s) of *Dunaliella* spp. Apparently, the only other enzyme of the PCOC which has been assayed in *Dunaliella* spp.is hydroxypyruvate reductase; this was found in mitochondria of *D. primolecta* and *D. salina* (Stabenau et al. 1993).

Species of Inorganic C Crossing Membranes of *Dunaliella* spp.

The species considered here are CO_2 and HCO_3^-; CO_3^{2-} is widely held not to be important in membrane transport, at least in the context of photosynthesis. Another reason for not considering CO_3^{2-}, at least in the first instance, is that it is technically much more difficult to distinguish fluxes of CO_3^{2-} from those of HCO_3^- than to distinguish those of CO_2 from HCO_3^- (and CO_3^{2-}). This is because the two ionized species equilibrate very rapidly whereas, in the absence of carbonic anhydrase activity in the compartment from which the fluxes are being measured, the half-time of equilibration of CO_2 and HCO_3^- is from seconds to tens of seconds, depending on temperature and ionic strength (Johnson 1982; Matsuda et al. 2001).

Thus, in the absence of carbonic anhydrase activity (natural, by mutation or by inhibition, noting that carbonic anhydrase inhibitors can inhibit at least certain bicarbonate transporters: Pollock and Colman 2001) in the relevant compartment, three techniques can be used to distinguish CO_2 fluxes from HCO_3^- fluxes. One of these, the demonstration that the rate of photosynthesis exceeds the uncatalysed rate of conversion of HCO_3^- to CO_2, shows that the observed rate of photosynthesis either can, or cannot, be accounted for by CO_2 fluxes alone. The other two can potentially (isotope disequilibrium) and invariably (membrane inlet mass spectrometry) allow estimation of fluxes of the both of the inorganic carbon species. All three have been applied to *Dunaliella* spp. (Mukerji et al. 1978; Aizawa et al. 1986; Burns and Beardall 1987; Amoroso et al. 1998).

Mukerji et al. (1978; isotope disequilibrium) showed that while CO_2 was the major inorganic carbon species used by exponentially growing *D. tertiolecta*, HCO_3^- predominated in stationary phase cells. The term 'used' rather than 'entering' is used here because no attempt was made to measure and, if present, inhibit extracellular CA activity. Aizawa et al. (1986, isotope disequilibrium) also used *D. tertiolecta* and found that cells in which the extracellular CA had not been inhibited used predominantly HCO_3^-, while CO_2 was the main inorganic C form entering cells in which the external CA was inhibited. Aizawa et al. (1986) suggested that the findings of Mukerji et al. (1978) resulted from the presence of high CA activity on stationary phase cells but not on exponentially growing cells, with CO_2 entering the cells in both cases. Burns and Beardall (1987) showed that the photosynthetic rate of *D. tertiolecta* was greater than the rate of uncatalysed HCO_3^- conversion in the medium, but also demonstrated extracellular CA activity and did not add an inhibitor. Accordingly, they wisely did not say that HCO_3^- was necessarily entering the cells. Young et al. (2001) showed that, with extracellular CA inhibited, photosynthesis was faster than the uncatalysed rate of conversion of HCO_3^- to CO_2 so that there must have been direct HCO_3^- uptake. This uptake was not inhibited by the anion transport inhibitor 4,4'-diisothiocyanostilbene-2,2'-disulphonic acid (DIDS) (Young et al. 2001).

Amoroso et al. (1998) used the membrane inlet mass spectrometer technique on whole cells and isolated chloroplasts of *D. tertiolecta* in the presence of an (allegedly) membrane-impermeant CA inhibitor, and showed that HCO_3^- was the dominant species of inorganic carbon taken up by whole cells, while both species were taken up by chloroplasts. It is not clear how the fluxes across the chloroplast envelope relate to polypeptides in the inner envelope membrane that are induced by growth at low CO_2 (Thielmann et al. 1992; Goyal et al. 1998).

These results for halophilic *D. tertiolecta* show that whole cells can take up both CO_2 and HCO_3^-. When extracellular CA is not inhibited, and especially at pH values below 9 or so, the uptake of CO_2 produced from HCO_3^- is a very significant mode of inorganic C entry, while with CA inhibited, or at higher pH values, direct HCO_3^- uptake predominates. For isolated chloroplasts of *D. tertiolecta* in the absence of external CA both inorganic C species are taken up.

The acidophilic *D. acidophila* at an external pH of 1.0 has essentially no HCO_3^-: the equilibrium HCO_3^-:CO_2 ratio is less than 10^{-5}, and entry of CO_2 is the obvious mechanisms of inorganic carbon uptake by this alga (Gimmer and Slovik 1995). There is, however, the question of the occurrence of extracellular CA in this alga (Geib et al.1996): although the activity per cell is lower than in halophilic species, it shows the same increase in activity with acclimation from growth in air enriched in CO_2 to growth at air levels of CO_2 (Gimmler and Slovick 1995; Geib et al. 1996).

Mechanism of the CCM in *Dunaliella* spp.

We do not known where the active transport step(s) for CO_2, HCO_3^- and/or H^+ are located in halophilic *Dunaliella* spp. The variant of the CCM based on active H^+ transport that is most likely to occur in *Dunaliella* spp. involves acidification of the thylakoid lumen; this mechanism originated with Pronina et al. (1981), Pronina and Semenenko (1984, 1992) and Pronina and Borodin (1993), and was developed and tested by Raven (1997), von Hunnik and Sültemeyer (2002), Hanson et al. (2003) and Mitra et al. (2005); see also Funke et al. (1997).

Assuming for the moment that the CCM is based on the transport of inorganic carbon species rather than H^+ at the thylakoid membrane, we know that inorganic carbon can be accumulated in isolated chloroplasts of *D. tertiolecta* (Goyal and Tolbert 1989), and that transport of CO_2 and HCO_3^- into isolated chloroplasts involve relatively high-affinity transporters (Amoroso et al. 1998). However, it is not clear whether this accumulation, taking into account the fraction of whole cell occupied by the chloroplast, is adequate to account for the whole-cell accumulation of inorganic carbon under normal external inorganic carbon concentrations. One problem is that inorganic carbon accumulation experiments on whole cells are normally performed, for practical reasons, with external inorganic carbon concentrations lower than those found in the normal growth medium. It is not even possible to argue that HCO_3^- influx at the plasmalemma must be by active (energized) transport. The electrical potential across the plasmalemma of halophilic (neutrophilic - alkalophilic) *Dunaliella* spp. is inside-negative, in *D. bardawil*, *D. parva* and *D. salina* by 40 – 100 mV (Oren-Shamir et al. 1990; Remis et al. 1992), in part a function of the activity of an electrogenic Na^+-ATPase (and H^+-ATPase) in the plasmalemma (Sekler and Pick 1993; Remis et al. 1994; Weiss and Pick 1996; Popova et al. 2005). The inside-negative electrical potential means that the influx of HCO_3^- ions to a cytosol with the same HCO_3^- concentration as in the medium would need an energy input, depending on the potential difference, of at least 4 - 10 kJ per mole HCO_3^- (from the Nernst Equation). However, considerations of acid-base balance show that there is an efflux of about 1 OH^- (or influx of 1 H^+) for each HCO_3^- entering the cells and CO_2 assimilated (Raven and Smith 1974; Goldman and Brewer 1980), so that the overall electroneutral reaction can be energetically neutral (see Raven et al. 2005).

Turning to the proposed CCM involving the H^+ electrochemical difference across the thylakoid membrane, it is possible that the active H^+ transport across the thylakoid membrane driven by photochemical coupled redox reactions could energize the complete CCM (Raven 1997), including the observed levels of inorganic carbon accumulation averaged over the whole cell (Raven et al. 2005). In other words, none of the inorganic carbon fluxes across membranes need be energized.

Two features mentioned above that need to be (and can) be incorporated into all of the possible modes of CCM operation are the universal occurrence of pyrenoids in *Dunaliella* spp., and the occurrence (where tested) of most or all of the cell content of Rubisco in the pyrenoid. This is consistent with models in which CO_2 is supplied to Rubisco within the pyrenoid by a spatially closely related carbonic anhydrase acting on HCO_3^- that has either been accumulated in the stroma (Badger et al. 1998), or is in the acidic lumen of thylakoids penetrating or near lumen where the equilibrium CO_2 is higher than it is in the medium (Raven 1997). Another feature mentioned above that can also be accommodated by all models is the need for carbonic anhydrase(s) in some compartment(s) of the chloroplast.

Allocation of Carbon Downstream of Photosynthesis in *Dunaliella* spp.

Some of the organic carbon from gross photosynthesis does not contribute to particulate organic carbon. Of this, some losses are stoichiometrically unavoidable. Growth requires some carbon skeletons that are only produced in a 'dark' respiratory process, the tricarboxylic acid cycle. By contrast, carbon skeletons produced in the oxidative pentose phosphate pathway can be produced from the reductive pentose phosphate cycle (photosynthetic carbon reduction cycle), or the reversible reactions of these cycles in the absence of redox reactions. Thus, even in continuous light, where it is possible that the NADPH and ATP needed for growth using the reduced carbon products of photosynthesis come from thylakoid reactions, there is still a need for the tricarboxylic acid cycle reactions (Falkowski and Raven 2007). The other requirement for respiration is in powering maintenance processes, which, again, can be powered by photosynthetic processes in the light (Falkowski and Raven 2007). It is not clear whether the rate of dark respiration in *Dunaliella* spp. exceeds what can be explained in terms of the requirement for growth and maintenance processes. There is also the question of enhanced post-illumination respiration which has been demonstrated in *Dunaliella* spp. (Beardall et al. 1994; Heraud and Beardall 2002). It is not clear how this latter process relates to a carry-over of photorespiration or any light-stimulated tricarboxylic acid cycle involved in biosyntheses. More work is needed on the location of respiratory processes other than those found in mitochondria (Kombrink and Wöber 1980a,b)..

Organic carbon efflux is another aspect of photosynthate that does not become particulate organic matter. Glycolate excretion occurs in the light in relation to photorespiration (Leboulanger et al. 1989). Organic carbon excretion from *D. salina* was measured by Giordano et al. (1994) as a function of salinity, nitrogen and inorganic carbon supply. Organic carbon efflux was 0.8 – 11% of photosynthesis, with the highest values during acclimation to NH_4^+ as N source, caused by lower photosynthesis rather than higher rates per cell. Gordillo et al. (2003) investigated the effects on inorganic carbon and inorganic nitrogen supply on dissolved organic carbon excretion from lag-phase, log-phase and stationary phase cultures of *Dunaliella viridis*, and found that excretion was a much higher percentage of photosynthesis, from 30% to 80%.

Another possible aspect of soluble organic carbon loss by *Dunaliella* spp. is the leakage of the compatible osmolyte glycerol which, especially in cells growing in the highest salinities, has a very large concentration difference across the plasmalemma which, depending on the permeability of the plasmalemma, could give significant glycerol effluxes (Raven 1984). The argument is that the low costs of making the osmolyte (glycerol is the energetically cheapest osmolyte to manufacture) is in part offset by the higher permeability of lipid bilayers to smaller polyhydroxy compounds than to larger, and more expensive to make, polyhydroxy or zwitterionic compatible solutes. The role of glycerol in osmoregulation (volume regulation) involves synthesis of glycerol with increased external osmolarity, and intracellular removal (not loss to the medium) with decreased external osmolarity (Ben-Amotz and Avron 1973).

Use of *Dunaliella* spp. in Mitigation of Carbon Dioxide?

The use of microalgae as a means of removing CO_2 from industrial effluents has attracted much attention, with a frequent emphasis on tying the microalgal production to a higher value product than bulk algal biomass production which could be used in (for example) fermentation to methane, or as a solid fuel. We first address general aspects of the use of microalgae, then the potential for the use of *Dunaliella* spp. in particular.

Microalgal growth on industrial effluent has been addressed by, for example, Benemann (1993), Hanagata et al. (1992), Maeda et al. (1995), Matsumoto et al. (1995), Ono and Cuello (2004), Doucha et al. (2005) and Ono and Cuello (2006), usually with a (freshwater) species of *Chlorella* as the microalga. The rationale for such schemes seems to be as follows. Most microalgae in nature grow with CO_2

in solution at an average concentration similar to air equilibrium to a few times greater than, air equilibrium (see Introduction). In more potentially productive waters (higher nutrient concentrations, relatively shallow epilimnion/upper mixed layer) the rate of photosynthesis can draw down the CO_2 in the surface water. Resupply of CO_2 from the atmosphere in these conditions is limited by the physics and physical chemistry of the air-water interface, and inorganic carbon can limit productivity, even though most of the algae have CCMs and high affinities for CO_2, especially when grown at low CO_2 (Table 1; Giordano et al. 2005). In algal cultures where the actual CO_2 concentration in solution is of less concern than the requirement to produce large biomass densities rapidly, bubbling with air enriched in CO_2 is commonplace. This means that bubbling with air enriched in CO_2 stimulates algal productivity while removing some of the CO_2 from the gas stream. It also means that the properties, and composition, of the algae used in CO_2 mitigation technology resemble those of high CO_2 grown algal cultures (low inorganic carbon affinity; low expression of CCM, high carbohydrate or lipid content) than air-grown laboratory cultures or algae in nature.

Benemann (1993) emphasised the need to attain high productivities in the algal cultures and that, if these could be attained, overall costs would compare favourably with physical-chemical methods of CO_2 mitigation from the effluent of fossil fuel-fired power plants, with the resulting algae used to produce gaseous or liquid fuels and/or to produce higher value products such as dietary supplements (Benemann 1989). Benemann (1993) comments that the overall costs would be higher than for a system based on terrestrial plants; he also points out that a land plant system could not use flue gases directly. Microalgal cultures can also be of advantage when the available water is too saline to be used in agriculture without desalination.

Hanagata et al. (1992) screened five species of freshwater green microalgae, and concluded that, of these, *Chlorella* sp. was the most suitable, followed by *Scenedesmus* sp. Maeda et al. (1995) selected for a strain of *Chlorella* that was tolerant of high temperatures and higher CO_2 concentrations, as well as of SO_x and NO_x, and that could be used in fixing CO_2 from coal-fired thermal power plants. Matsumoto et al. (1995) also focussed on gaseous effluent from thermal power plants; their results suggested that the area of outdoor raceway algal cultures required less land area per CO_2 assimilated than alternatives such as tree farms, and that the resulting algal biomass could be used a solid fuel. Doucha et al. (2005) also used *Chlorella*, and found that their outdoor cultures removed 10-50% (with, as expected, a smaller percentage removed when the injection rate was higher) of the CO_2 from the combustion of natural gas in a boiler. A scheme was suggested in which organic agricultural waste was anaerobically digested to produce biogas which, by combustion, yields the CO_2-rich gas injected into the algal cultures; the algal biomass produced could be used in animal feed (Doucha et al. 2005). The scheme was quantitatively assessed (Doucha et al. 2005).

Other work has concentrated on the combination of such microalgal culture methodology with solar concentration mechanisms in which uniform light distribution within the algal culture was regarded as a major desirable attribute of the apparatus (Ono and Cuello 2004, 2006). In their analysis of the economics of such a scheme, Ono and Cuello (2006) showed that the generation of (a) commercially valuable product(s) (listed in their Table 1) was essential if the cost was to be lower than the target CO_2 mitigation price of 30 US$ per tonne of CO_2. Ono and Cuello (2006) also comment that the microalgal cultures can (with additional expenditure) lose much less water vapour to the atmosphere than terrestrial plants performing the same CO_2 mitigation task; this is an important consideration in a world that is increasingly concerned about water availability. However, some water is lost as vapour in the process of passing the effluent gas through the algal culture: the gas approaches water vapour equilibrium with the liquid water at the relevant temperature and osmolarity in the approach to equilibrium of the carbon dioxide in the gas phase with that in solution which is consumed by the algae.

These tests of concept, and economic analyses, suggest that microalgal cultures might be commercially viable options for CO_2 mitigation. Where *Dunaliella* (Cortinas et al. 1984; Suzuki et al. 1995) might be the favoured alga? Halophilic *Dunaliella* spp. might be the favoured microalgae when only saline (including hypersaline) water is available (Table 2); such locations are also those where it is least likely that areas used for microalgal growth are least likely to compete with other commercial uses of land area, e.g. agriculture, horticulture and sylviculture. Furthermore, water loss from the culture in gas bubbling, thereby concentrating the solutes in the water, might have a smaller impact on halophilic *Dunaliella* spp. than on freshwater algae. Less likely is the case where only acidic water is available; in such a case *D. acidophila* could be used (Table 2). Another reason for using at least the halophilic *Dunaliella* spp. is where the generation of a high-value algal end product, in this case β-carotene, is an important economic driver (Borowitzka et al. 1984; Borowitzka and Borowitzka 1988; Benemann 1989 and Benneman 1993; Ono and Cuello 2006) (Table 2). At the moment, commercially viable production of β-carotene from *Dunaliella* spp. does not involve artificial aeration of the outdoor cultures. Another case where *Dunaliella*, as *D. acidophila* (or a related acidophilic alga), might be useful is where very acidic waters, some of which are produced by industrial process, are available for algal culture (Gross 2000; Messerli et al. 2005). *D. acidophila* does not seem to have been investigated for the possibility of β-carotene production. It should be noted that there is a debate in medical literature about the health benefits, or otherwise, of nutritional supplements, including β-carotene (Chiu and Taylor 2007; Greenwald et al. 2007; Maserejian et al. 2007; Prentice 2007; Riccioni et al. 2007). Time will tell whether this debate will have implications for the sales of β-carotene produced by *Dunaliella* (Table 2).

An advantage of using extremophiles (halophiles, acidophiles) in outdoor cultures for CO_2 mitigation combined with production of a high-value end product is the reduced likelihood of successful invasion by other organisms which do not

Table 2: Advantages and disadvantages of *Dunaliella* spp. as the microalgae used in CO_2 mitigation.

Characteristic of *Dunaliella* spp.	Advantages	Disadvantages
Use of very saline or acidic water	Use of water not usable for growth of most algae, or any economically significant vascularl and plants Limits invasion by non-target microalgae	The industry producing CO_2 may not be located at places where very saline or acidic water is available
Production of β–carotene	Economics of mitigation may be improved by production of a high-value dietary supplement	Dependence of the economics of mitigation on β–carotene involves continued consumer faith in the dietary supplement

make the desired end-product, since few photosynthetic organisms can tolerate the culture conditions (Table 2). Thus far β-carotene, but not glycerol produced in abundance by halophilic *Dunaliella* spp., is the high-value end-product of *Dunaliella* culture.

Conclusions

Much is known about the physiology and biochemistry of inorganic carbon assimilation by *Dunaliella* spp., though there are significant gaps, e.g. our ignorance of the dependence of the activity of *Dunaliella* Rubisco *in vitro* on the concentrations of CO_2 and O_2, separately and together. Further molecular genetic data would also be very useful.

As for the use of *Dunaliella* spp. in removing CO_2 from industrial effluent, the use of species from this extremophile genus would be particularly useful where only saline (including hypersaline) or acidic water was available for the growth of photosynthetic organisms in CO_2 mitigation, and in the use of land area not readily usable for the commercial growth of other photolithotrophs. There is also the added commercial advantage of the high-value product, β-carotene.

Acknowledgements

Colleagues past and present, especially John Beardall and Mario Giordano, have greatly influenced my views on the content of this chapter. My research on inorganic carbon acquisition by microalgae is supported by the Natural Environment Research Council UK. The University of Dundee is a registered Scottish Charity, No:SC015096.

References

Aizawa, K. and S. Miyachi. 1984. Carbonic anhydrase located on cell-surface increases the affinity for inorganic carbon in photosynthesis of *Dunaliella tertiolecta.* FEBS Letters 173: 41-44.

Aizawa, K., K. Nakamura and S. Miyachi. 1985. Variations in phosphoenolpyruvate carboxylase activity in *Dunaliella* associated with changes in atmospheric CO_2 concentration. Plant and Cell Physiology 26: 1199-1203.

Aizawa, K., M. Tsuzuki and S. Miyachi. 1986. Kinetic studies on the active species of inorganic carbon absorbed by the cells of *Dunaliella tertiolecta.* Plant and Cell Physiology 27: 37-47.

Albertano, P., G. Pinto, S. Santisi and T. Taddei. 1981. *Spermatozopsis acidophila* Kalina (Chlorophyta, Volvocales), a little known alga from highly acidic environments. Giornalo Botanica Italiano 115: 65-76.

Amoroso, S., D. Sültemeyer, C. Thyssen and H.P. Fock. 1998. Uptake of HCO_3^- and CO_2 in cells and chloroplasts from the microalgae *Chlamydomonas reinhardtii* and *Dunaliella tertiolecta.* Plant Physiology 116: 193-201.

Amoroso, G., C. Weber, D. Sültemeyer and H.P. Fock. 1996. Intracellular carbonic anhydrase activities in *Dunaliella tertiolecta* (Butcher) and *Chlamydomonas reinhardtii* (Dangeard) in relation to inorganic carbon concentration during growth: Further evidence for the existence of two distinct carbonic anhydrases with the chloroplasts. Planta 199: 177-184.

Badger, M.R., T.J. Andrews, S.M. Whitney, M. Ludwig, D.C. Yellowlees, W. Leggat and G.D. Price. 1998. The diversity and coevolution of Rubisco, plastids, pyrenoids, and chloroplast-based CO_2-concentrating mechanisms in algae. Canadian Journal of Botany 76: 1052-1071.

Barkan, E., B. Luz and B. Lazar. 2001. Dynamics of the carbon dioxide system in the Dead Sea. Geochmica and Cosmochimica Acta 65: 355-368.

Bate, G.C., D.F. Sültemeyer and H.P. Fock. 1988. $^{18}O_2/^{16}O_2$ analysis of oxygen exchange in *Dunaliella tertiolecta.* Evidence for the inhibition of mitochondrial respiration in the light. Photosynthesis Research 16: 219-231.

Beardall, J. and I. Morris. 1975. Effects of environmental factors on photosynthesis patters in *Phaeodactylum tricornutum* (Bacillariophyceae). II. Effect of oxygen. Journal of Phycology 11: 430-434.

Beardall, J., D. Mukerji, H.E. Glover and I. Morris. 1976. The path of carbon in photosynthesis by marine phytoplankton. Journal of Phycology 12: 409-417.

Beardall, J., T. Burgerwiersma, M. Rijkboer, A. Suenki, J. Lemalle, Z. Dubinsky and D. Forntrielle. 1994. Studies on enhances post-illumination respiration in microalgae. Journal of Plankton Research 16: 1401-1410.

Beardall, J., P. Heraud, S. Roberts, K. Shelly and S. Stojkovic. 2002. The effect of UV-B radiation on inorganic carbon acquisition by the marine microalga *Dunaliella tertiolecta* (Chlorophyceae). Phycologia 41: 268-272.

Ben-Amotz, A. and M. Avron. 1973. The role of glycerol regulation of the halophilic alga *Dunaliella parva.* Plant Physiology 51: 875-878.

Ben-Yaakov, S. and E. Sass. 1977. Independent estimate of pH of Dead Sea brines. Limnology and Oceanography 22: 374-376.

Benemann, J.R. 1989. The future of microalgal biotechnology. pp. 317-337. In Algal and Cyanobacterial Biotechnology. C. Cresswell, T.A.V. Rees and N. Shah (eds) Longman Scientific and Technical, Essex, UK.

Benemann, J.R. 1993. Utilization of carbon dioxide from fossil fuel-burning plants with biological systems. Energy Conversion and Management 34: 999-1004.

Booth, W.A. and J. Beardall. 1991. Effects of salinity on inorganic carbon utilization and carbonic anhydrase activity in the halotolerant green alga *Dunaliella salina* (Chlorophyta). Phycologia 30: 220-225.

Borowitzka, L.J., M.A. Borowitzka and T.P. Moulton. 1984. The mass-culture of *Dunaliella salina* for fine chemicals - from laboratory to pilot-plant. Hydrobiologia 116: 115-121.

Borowitzka, M.A, and L.J. Borowitzka. 1988. *Dunaliella.* pp. 26-58. In Micro-algal Biotechnology. M.A. Borowitzka and L.J. Borowitzka (eds) Cambridge University Press, Cambridge, UK.

Burns, B.D. and J. Beardall. 1987. Utilization of inorganic carbon by marine microalgae. Journal of Experimental Marine Biology and Ecology 107: 75-86.

Chiu, C.J. and A. Taylor. 2007. Nutritional antioxidants and age-related cataract and maculopathy. Experimental Eye Research 84: 229-245.

Cortinas, T.I., H.J. Silva and R.J. Ertola. 1984. Photosynthetic production of *Dunaliella* biomass from natural salt-water and carbon dioxide. Journal of Chemical Technology and Biotechnology B – Biotechnology 34: 291-295.

delRio, M.J., G. GarciaReina and Z. Ramazanov. 1996. The ultrastucture and polypeptide composition of the pyrenoid from *Dunaliella tertiolecta.* Scientica Marina Supplement 1 60: 155-160.

Doucha, J., F. Straka and K. Livansky. 2005. Utilization of flue gas for cultivation of microalgae (*Chlorella* sp.) in an outdoor open thin-layer bioreactor. Journal of Applied Phycology 17: 403-412.

Einsele, G., J. Yan and M. Hinderer. 2001. Atmospheric carbon burial in modern lake basins and its significance for the global carbon budget. Global and Planetary Change 30: 167-195.

Falkowski, P.G. and J.A. Raven. 2007. Aquatic Photosynthesis, Second Ed. Princeton University Press, Princeton, USA.

Fisher, M., I. Gokham, U. Pick and A. Zamir. 1997. A salt-resistant plasma membrane carbonic anhydrase is induced by salt in *Dunaliella salina.* Journal of Biological Chemistry 272: 1565-1570.

Funke, R.F., J.L. Kovar and D.P. Weeks. 1997. Intracellular carbonic anhydrase is essential to photosynthesis in *Chlamydomonas reinhardtii* at atmospheric levels of CO_2. Demonstration via genomic complementation of the high-CO_2-requiring mutant *ca-1.* Plant Physiology 114: 237-244.

Geib, K., D. Golldak and H. Gimmler. 1996. Is there a requirement for an external carbonic anhydrase in the extremely acid-resistant green alga *Dunaliella acidophila*? European Journal of Phycology 31: 272-284.

Ghoshal, D. and A. Goyal. 2001. Oxygen inhibition of dissolved inorganic carbon uptake in unicellular green algae. Phycological Research 49: 319-324.

Gimmler, H. and S. Slovick. 1995. The mode of uptake of dissolved inorganic carbon in the extremely acid-resistant green alga *Dunaliella acidophila* – theoretical considerations and experimental observations. Plant Physiology and Biochemistry 33: 655-664.

Gimmler, H., H. Kugel, D. Leibfritz and A. Mayer. 1988. Cytoplasmic pH of *Dunaliella parva* and *Dunaliella acidophila* as monitored by *in vivo* P-31-NMR spectroscopy and the DMO method. Physiologia Plantarum 74: 521-530.

Gimmler, H., U. Weis, U. Weiss, H. Kugel and B. Treffny. 1989. *Dunaliella acidophila* (Kalina) Masyuk – an alga with a positive membrane-potential. New Phytologist 113: 175-184.

Giordano, M. 2001. Interactions between C and N metabolism in *Dunaliella salina* cells cultured at elevated CO_2 high N concentrations. Journal of Plant Physiology 158: 577-581.

Giordano, M., and G. Bowes. 1997. Gas exchange and C allocation in *Dunaliella salina* cells in response to the N source and CO_2 concentration used for growth. Plant Physiology 115: 1049-1056.

Giordano, M., J.S. Davis and G. Bowes. 1994. Organic carbon release by *Dunaliella salina* (Chlorophyta) under different conditions of CO_2, nitrogen and salinity. Journal of Phycology 30: 249-257.

Giordano, M., V. Pezzoni and R. Hell. 2000. Strategies for the allocation of resources under sulphur limitation in the green alga *Dunaliella salina*. Plant Physiology 124: 857-864.

Giordano, M., A. Norici and D.J. Gilmour. 2003. Influence of the nitrogen source and metabolites on the V_{max} of phosphoenolpyruvate carboxylase from the unicellular green alga *Dunaliella parva* CCAP 19/9 (Volvocales, Chlorophyceae). Phycologia 42: 133-137.

Giordano, M., J. Beardall and J.A. Raven. 2005. CO_2 concentrating mechanisms in algae: Mechanisms, modulation, and evolution. Annual Review of Plant Biology 56: 99-131.

Goldman, J.C. and P.G. Brewer. 1980. Effects of nitrogen source and growth rate on phytoplankton-mediated changes in alkalinity. Limnology and Oceanography 25: 352-327.

González, M.A., A.W Coleman, P.I. Gómez and R. Montoya. 2001. Phylogenetic relationship among various strains of *Dunaliella* (Chlorophyceae) based on nuclear ITS rDNA sequences. Journal of Phycology 37: 604-611.

Gordillo, F.J.L., C. Jiminez, F.L. Figueroa and F.X. Niell. 2003. Influence of elevated CO_2 and nitrogen supply on the carbon assimilation performance and cell composition of the unicellular alga *Dunaliella viridis*. Physiologia Plantarum 19: 513-518.

Goyal, A. 2002. Glycolate metabolism in algal chloroplasts: inhibition by salicylhydroxamic acid (SHAM). Physiologia Plantarum 116: 264-270.

Goyal, A. and N.E. Tolbert. 1989. Uptake of inorganic carbon by isolated chloroplasts from air-adapted *Dunaliella*. Plant Physiology 89: 1264-1269.

Goyal, A. and N.E. Tolbert. 1996. Association of glycolate oxidation with photosynthetic electron transport in plant and algal chloroplasts. Proceedings of the National Academy of Sciences of the USA 93: 3319-3324.

Goyal, A., J. Thielmann and N.E. Tolbert. 1998. Isolation of chloroplast envelopes from *Dunaliella salina*. Canadian Journal of Botany 76: 1146-1152.

Granum, E., J.A. Raven and R. Leegood. 2005. How do marine diatoms fix 10 billion tonnes of inorganic carbon per year? Canadian Journal of Botany 83: 989908.

Greenwald, P., D. Anderson, S.A. Nelson and P.R. Taylor. 2007. Clinical trials of vitamin and mineral supplements for cancer prevention. American Journal of Clinical Nutrition 85: 3145-3175.

Gross, W. 2000. Ecophysiology of algae living in highly acidic environments. Hydrobiologia 433: 31-37.

Hanagata, N., T. Takeuchi, Y. Fukuju, D.J. Barnes and I. Karube. 1992. Tolerance of microalgae to high CO_2 and high temperature. Phytochemistry 31: 3345-3348.

Hanson, D.T., L.A. Franklin, G. Samuelsson and M.R. Badger. 2003. The *Chlamydomonas reinhardtii Cia3* mutant lacking a thylakoid-localized carbonic anhydrase is limited by CO_2 utilization by Rubisco and not PSII function *in vivo.* Plant Physiology 132: 2267-2275.

Heldt, H.W., K. Werdan, M. Milovank and G. Geller. 1973. Alkalization of chloroplast stroma caused by light-dependent proton flux into thylakoid space. Biochimica et Biophysica Acta 314: 224-241.

Hellebust, J.A. 1965. Excretion of some organic compounds by phytoplankton. Limnology and Oceanography 10: 192-206.

Heraud, P. and J. Beardall. 2002. Ultraviolet radiation has no effect on respiratory oxygen consumption and enhanced post-illumination respiration in three species of microalgae. Journal of Photochemistry and Photobiology B – Biology 68: 109-116.

Huntsman, S.A. 1972. Organic excretion in *Dunaliella tertiolecta.* Journal of Phycology 8: 56-63.

Johnson, K.S. 1982. Carbon dioxide hydration and dehydration kinetics in seawater. Limnology and Oceanography 27: 849-855.

Kombrink, E. and G. Wöber. 1980a. Preparation of intact chloroplasrts by chemically induced lysis of the green alga *Dunaliella marina.* Planta 149: 123-129.

Kombrink, E. and G. Wöber. 1980b. Identification and subcellular localization of starch-metabolizing enzymes in the green alga *Dunaliella marina.* Planta 149: 130-137.

Kuchitsu, K., M. Katsuhara and S. Miyachi. 1989. Rapid cytoplasmic alkalinization and dynamics of interacellular compartmentation of inorganic phosphate during adaptation against salt stress in a halotolerant unicellular green alga *Dunaliella tertiolecta*: 31P-Nuclear Magnetic resonance study. Plant Cell Physiology 30: 407-414.

Leboulanger, C., V. Martin-Jezequel, C. Descolas-Gros. A. Sciandra and H.J. Jupin. 1998. Photorespiration in continuous cultures of *Dunaliella tertiolecta* (Chlorophyta): relationships between serine, glycine and extracellular glycolate. Journal of Phycology 34: 651-654.

Lin, S. and E.J. Carpenter. 1997. Rubisco of *Dunaliella tertiolecta* is redistributed between the pyrenoid and the stroma as a light/shade response. Marine Biology 127: 521-529.

Lin, S., C.J. Gobler and E.J. Carpenter. 2001. Cytological and biochemical responses of *Dunaliella tertiolecta* (Volvocales, Chlorophyta) to iron stress. Phycologia 40: 403-410.

Liska, A.J., A. Shevchenko, U. Pick and A. Katz. 2004. Enhanced photosynthesis and redox energy production contribute to salinity tolerance in *Dunaliella* as revealed by homology-based proteomics. Plant Physiology 136: 2806-2817.

MacIntyre, H.L., T.D. Sharkey and R.J. Geider. 1997. Activation and deactivation of ribulose-1,5-bisphosphate carboxylase-oxygenase (Rubisco) in three marine microalgae. Photosynthesis Research 51: 93-106.

Maeda, K., M. Owada, N. Kimura, K. Omata and I. Karube. 1995. CO_2 fixation from the flue in coal-fired thermal power-plant by microalgae. Energy Conversion and Management 36: 717-720.

Markelova, A.G., M.G. Vladirova and V.E. Semenenko. 1990. Ultrastructural localization of RUBISCO in algal cells. Soviet Plant Physiology 37: 675-679.

Maserejian, M.N., E. Giovannucci, B. Rosner and K. Joshipura. 2007. Prospective study of Vitamins C, E and A and carotenoids and risk of oral premalignant lesions in men. International Journal of Cancer 120: 970-977.

Masserli, M.A., L.A. Amaral-Zettler, E. Zettler, S.K. Jung, P.J.S. Smith and M.L. Sogin. 2005. Life at acidic pH imposes an increased energetic cost for a eukaryotic acidophile. Journal of Experimental Biology 208: 2569-2579.

Matsuda, Y., T. Hara and B. Colman. 2001. Regulation of the induction of bicarbonate uptake by dissolved CO_2 in the marine diatom *Phaeodactylum tricornutum*. Plant, Cell and Environment 24: 611-620.

Matsumoto, H., N. Shioiji, A. Hamasaki, Y. Ikuta, Y. Fukuda, M. Sato, N. Endo and T. Tsukamoto. 1995. Carbon dioxide fixation by microalgae photosynthesis using actual flue-gas discharged from a boiler. Applied Biochemistry and Biotechnology 51: 681-692.

Melkonian, M. and H.R. Priesig. 1984. An ultrastructural comparison between *Spermatopsis* and *Dunaliella* (Chlorophyceae). Plant Systematics and Evolution 164: 1724-1726.

Messerli, M.A., L.A. Amaral-Zettler, E. Zettler, S-K. Jung, P.J.S. Smith and M.L. Sogin. 2005. Life at acidic pH imposes an increased energetic cost for a eukaryotic acidophile. Journal of Experimental Biology 208: 2569-2579.

Mitra, M., C.B. Mason, Y. Xiao, R.A. Ynalvez, S.M. Lato and J.V. Moroney. 2005. The carbonic anhydrase family of *Chlamydomonas reinhardtii*. Canadian Journal of Botany 83: 780-795.

Mukerji, D. 1978. Pyruvate-orthophosphate dikinase and malic enzyme: their assays and significance in the photosynthetic carbon assimilation in some marine algae. Photosynthetica 14: 517-522.

Mukerji, D., H.E. Glover and I. Morris. 1978. Diversity in the mechanism of carbon dioxide fixation in *Dunaliella tertiolecta* (Chlorophyceae). Journal of Phycology 14: 137-142.

Naus, J. and A. Melis. 1991. Changes in photosystem stoichiometry during cell growth in *Dunaliella salina* cultures. Plant and Cell Physiology 32: 569-575.

Nimer, M.A., A.K.J. Sallal, R.H. Ahasan and M.J. Merrett. 1990. The presence of glycolate oxidase and dehydrogenase in *Dunaliella primolecta*. Planta 181: 374-377.

Norici, A., A. Dalpass and M.Giordano. 2002. The role of phosphoenolpyruvate carboxylase in anaplerosis in the green microalga *Dunaliella salina* cultures under different nitrogen regimes. Physiologia Plantarum 116: 186-191.

Ono, E. and J.L. Cuello. 2004. Design parameters of solar concentrating systems for CO_2-mitigating algal photobioreactors. Energy 59: 1651-1657.

Ono, E. and J.L. Cuello. 2006. Feasibility assessment of microalgal carbon dioxide sequestration technology with photobioreactor and solar collector. Biosystems Engineering 95: 597-606.

Oren, A., P. Gurevich, D.A. Anati, E. Barkan and B. Luz. 1995. A bloom of *Dunaliella parva* in the Dead Sea in 1992 – biological and biogeochemical aspects. Hydrobiologia 297: 173-185.

Oren-Shamir, M., U. Pick and M. Avron. 1990. Plasma membrane potential of the alga *Dunaliella*, and its relation to osmoregulation. Plant Physiology 93: 403-408.

Palmqvist, K., Z.M. Ramazanov and G. Samuelsson. 1991. Role of carbonic anhydrase in assimilation of inorganic carbon by *Dunaliella salina*. Soviet Plant Physiology 38: 339-345.

Pollock, S.V. and B. Colman. 2001. The inhibition of the carbon concentrating mechanism of the green alga *Chorella saccharophila* by acetazolamide. Physiologia Plantarum 111: 527-532.

Popova, L.G., G.A. Shumkova, I.M. Andreev and Y.V. Balnokin. 2005. Functional identification of electrogenic Na^+-translocating ATPase in the plasma membrane of the halotolerant microalga *Dunaliella maritima*. FEBS Letters 579: 5002-5006.

Prentice, R.L. 2007. Clinical trials and observational studies to assess the chronic disease benefits and risks of multivitamin-multimineral supplements. American Journal of Clinical Nutrition 85: 3085-3135.

Pronina, N.A. and V.E. Semeneko. 1984. Localization of membrane bound and soluble carbonic anhydrase in *Chlorella* cells. Fiziologia Rastenii 31: 241-251

Pronina, N.A. and V.E. Semenenko. 1992. Role of the pyrenoid in concentration, generation and fixation of CO_2 in the chloroplast of microalgae. Soviet Plant Physiology 39: 494-502.

Pronina, N.A. and V.V. Borodin. 1993. CO_2 stress and CO_2 concentration mechanism: investigation by means of photosystem-deficient and carbonic anhydrase-deficient mutants of *Chlamydomonas reinhardtii*. Photosynthetica 28: 515-522.

Pronina, N.A., S. Avramova, D. Gieogiev and V.E. Semenenko. 1981. A pattern of carbonic anhydrase activity in *Chlorella* and *Scenedesmus* cells on cell adaptation to high level light intensity and low CO_2 concentration. Fiziologia Rastenii 28: 43-52

Rabbani, S., P. Beyer, J. Von Lintig, P. Hugueney and H. Kleinig. 1998. Induced β–carotene synthesis driven by triacylglycerol deposition in the unicellular alga *Dunaliella bardawil*. Plant Physiology 116: 1239-1248.

Raven, J.A. 1984. Energetics and Transport in Aquatic Plants. A.R. Liss, New York, USA.

Raven, J.A. 1997. CO_2 concentrating mechanisms: a role for thylakoid lumen acidification? Plant, Cell and Environment 20: 147-154.

Raven, J.A. and F.A. Smith. 1974. Significance of hydrogen ion transport in plant cells. Canadian Journal of Botany 52: 1035-1048.

Raven, J.A. and K. Richardson.1986. Photosynthesis in marine environments. pp. 337-398. In Photosynthesis in Specific Environments.N.R.Baker and S.P. Long (eds) Elsevier, Amsterdam.The Netherlands

Raven, J.A., L.A. Ball, J. Beardall, M. Giordano and S.C. Maberly. 2005. Algae lacking carbon concentrating mechanisms. Canadian Journal of Botany 83: 879-890

Raven, J.A. and S.C. Maberly. Phytoplankton nutrition and related mixotrophy, in press. In 2008. Encyclopedia of Inland Waters. G.E. Likens (ed) Elsevier, Amsterdam. The Netherlands

Remis, D., W. Simonis and H. Gimmler. 1992. Measurement of the transmembrane electrical potential of *Dunaliella acidophila* by microelectrodes. Archives of Microbiology 158: 3450-355.

Remis, D., B. Treffny and H. Gimmler. 1994. Light-induced H^+ transport across the plasma-membrane of the acid-resistant green alga *Dunaliella acidophila*. Plant Physiology and Biochemistry 32: 75-84.

Riccioni, G., T. Buccarelli, B. Mancini, F. Corradi, C. Di Ilio, P.A. Mattei and N. D'Orazio. 2007. Antioxidant supplementation in cardiovascular diseases. Annals of Clinical and Laboratory Studies 37: 89-95.

Sallal, A.K.J., R.H. Alhasan and M.A. Nimer. 1987. Localization of glycolate dehydrogenase in *Dunaliella salina*. Planta 171: 429-432.

Sass, E. and S. Ben-Yaakov. 1977. The carbonate system in hypersaline solutions: Dead Sea brines. Marine Chemistry 5: 183-199.

Sekler, I., and U. Pick. 1993. Purification and properties of a plasma-membrane H^+-ATPase from the extremely acidophilic alga *Dunaliella acidophila*. Plant Physiology 101: 1055-1061.

Stabenau, H. and U. Winkler. 2005. Glycolate metabolism in green algae. Physiologia Plantarum 123: 235-245.

Stabenau, H., U. Winkler and W. Saftel. 1993. Localization of glycolate dehydrogenase in 2 species of *Dunaliella*. Planta 191: 362-364.

Sukenik, A., J. Bennett and P.G. Falkowski. 1987. Light-saturated photosynthesis-limitation by electron transport or carbon fixation? Biochimica et Biophysica Acta 891: 205-215.

Sültemeyer, D. 1997. Changes in the CO_2 concentrating mechanism during the cell cycle in *Dunaliella tertiolecta*. Botanica Marina 110: 55-61.

Suzuki, T., T. Matsuo, K. Ohtaguchi and K. Koide. 1995. Gas-sparged bioreactors for CO_2 fixation by *Dunaliella tertiolecta*. Journal of Chemical Technology and Biotechnology 33: 225-234.

Thielmann, J., A. Goyal and N.E. Tolbert. 1992. Polypetides in the inner chloroplast envelope of *Dunaliella tertiolecta* induced by low CO_2. Plant Physiology 100: 2113-2115.

van Hunnik, E., and D. Sültemeyer. 2002. A possible role for carbonic anhydrase in the lumen of thylakoids in green algae. Functional Plant Biology 29: 243-249.

Wegmann, K. 1969. On the pathways of CO_2 fixation in *Dunaliella*. pp. 1559-1564. In International Union of Biological Sciences. H. Metzner (ed) 1996.Tubingen.Germany.

Weiss, M. and U. Pick. 1996. Primary structure and effect of pH on the expression of the plasma membrane H^+-ATPase from *Dunaliella acidophila* and *Dunaliella salina*. Plant Physiology 112: 1693-1702.

Werdan, K., H.W. Heldt and G. Geller. 1972. Accumulation of bicarbonate in intact chloroplasts following a pH gradient. Biochimica et Biophysica Acta 283: 430-441.

Young, E.B. and J. Beardall. 2005. Modulation of photosynthesis and inorganic carbon acquisition in a marine microalga by nitrogen, iron and light availability. Canadian Journal of Botany 83: 917-928.

Young, E., J. Beardall and M. Giordano. 2001. Inorganic carbon acquisition by *Dunaliella salina* (Chlorophyta) involved external carbonic anhydrase and direct HCO_3^- utilization insensitive to the anion exchanger DIDS. European Journal of Phycology 36: 81-88.

Zenvirth, D. and A. Kaplan. 1981. Uptake and efflux of inorganic C in *Dunaliella salina*. Planta 152: 8-12.

Zenvirth, D., M. Volokita and A. Kaplan. 1985. Photosynthesis and inorganic carbon accumulation in the acidophilic alga *Cyanidioschyzon merolae*. Plant Physiology 77: 237-239.

16

Paratransgenic Approaches to the Control of Infections of Shrimp: Prospects for Use of *Dunaliella*

Ravi Durvasula[1], Bobban Subhadra[1], Scott Matthews[1], Ivy Hurwitz[1] and D.V. Subba Rao[2]*

Abstract

Dunaliella, a unicellular green alga, offers enormous potential benefits to the commercial mariculture industry. Tolerant of very wide ranges of temperature, salinity and pH, this organism accumulates high levels of nutrients such as β-carotenoids, making it a favorable source of live feed. Advances in the genetics and transformation of *Dunaliella* with foreign genes are relatively recent and much remains to be understood. In this chapter, the prospects for using *Dunaliella* as a tool for paratransgenic control of infectious diseases of farmed shrimp are discussed. Paratransgenesis is a strategy that employs commensal or symbiotic organisms to express molecules that interfere with transmission cycles of infectious pathogens. Application of paratransgenesis to viral and bacterial diseases of farmed shrimp has been described with genetically modified cyanobacteria as the delivery agents. Here, a transformation system for *Dunaliella* and a framework for application of engineered *Dunaliella* in commercial mariculture are presented.

Introduction

Dunaliella is a unicellular, bi-flagellated green alga that belongs to the class *Chlorophyceae*. These organisms were recently placed into their own order of *Dunalillales* and family *Dunaliellaceae* (Gonzalez et al. 2001). This genus is further subdivided into two subgenera, *Pascheria* (freshwater) and *Dunaliella* (marine). Morphologically, *Dunaliella* is very similar to *Chlamydomonas*. Both organisms have

[1] Division of Infectious Diseases, Department of Internal Medicine, School of Medicine, MSC 10 5550, School of Medicine, 1 University of New Mexico, NM 87131.
[2] Bedford Institute of Oceanography, Dartmouth, NS, Canada, B2Y 4A2.
[*] Corresponding author: Email: Ravi.Durvasula@med.va.gov

complex life cycles that encompass, in addition to division of motile vegetative cells, the possibility of sexual reproduction. These organisms are both photosynthetic, and relatively easy to maintain in a laboratory setting. Unlike *Chlamydomonas*, the genetics of *Dunaliella* are poorly understood. *Dunaliella* is by far one of the most salt-tolerant eukaryotic organisms (Ben-Amotz and Avron 1990). Furthermore, it is highly resistant to stresses such as high light intensity and dramatic pH and temperature changes. Although there is an increasing interest in the mechanisms that allow such physiological versatility, research in this area is still in its infancy. To date, few of these stressed-induced genes have been cloned from *Dunaliella* (Fisher et al. 1996; Fisher et al. 1997; Sanchez-Estudillo et al. 2006), and the information that is available has shed little light on the genomic organization or the biological significance of some of the unique sequence features that have been identified (Sun et al. 2006).

One of the stress-induced responses in *Dunaliella* is the production and accumulation of the carotenoid, β-carotene. *Dunaliella* is one of the richest natural producers of carotenoid, producing up to 15% of its dry weight under suitable conditions. Interestingly, it is thought that the carotenoid functions as a "sun-screen" to protect chlorophyll and DNA from harmful UV-irradiation (Ben-Amotz et al. 1989). The carotenoids are stored as intracellular oil droplets near the outer membrane and it has been suggested that carotenoid and triacylglycerol biosynthesis are interrelated (Rabbani et al. 1998). The commercial cultivation of *Dunaliella* began in the 1960's once it was realized that their halotolerance allowed for monoculture in large brine ponds. Till today, β-carotene remains the major natural product harvested from *Dunaliella*. Common uses of β-carotene include food coloring, additives to multivitamin preparations, health food products, cosmetics, and animal feed as provitamin A. The ease of maintaining *Dunaliella* in culture – its ability to grow in very high salt concentrations, tolerance to high temperature and to extreme pH changes - makes this species a highly desirable target for exploitation as a biological factory for the large-scale production of foreign proteins.

Algae as Bioreactors

Algae are highly suited as bioreactors for the large-scale production of foreign proteins for several reasons. First, they are relatively easy to culture as they will grow in a laboratory setting, subsisting on an inexpensive medium of simple salts. Second, unlike many cell lines, algae can be grown in continuous culture. Third, the cost for production on this platform was calculated to be approximately $0.002 per liter, compared to $1000-$2000 per gram in cultured mammalian cells and $0.05 per gram in a plant system. Besides the tremendous cost advantage, the generation of initial transformants to production volumes can occur within a short period of time. This system is also highly scalable in that transformed algal lines can be grown in few milliliters to 500,000 liters in a cost effective manner as their growth medium can be recycled. Furthermore, both the chloroplast and nuclear genome of algae can be genetically transformed, opening the possibility of expressing multiple recombinant products in a single organism. This eukaryotic system also offers the advantages of post-translational modifications of expressed protein products. The economics, ease of use and flexibility of this system make it highly desirable for the expression of complex recombinant products.

Transformation of Algae and Diatoms

The green alga *Chlamydomonas reinhardtii* has long served as a model system for photosynthesis and flagellar function. This unicellular green alga will grow on a simple medium of inorganic salts in the light, using a photosynthesis system that is similar to that of higher plants to provide energy. *Chlamydomonas* will also grow in total darkness if an alternate carbon source, usually in the form of acetate, is provided. Both the ~15.8 Kb mitochondrial genome (Genbank accession: NC001638 (Vahrenholz et al. 1993)) and the complete >200 Kb chloroplast genome for this organism are available online (Genbank accession: BK000554 (Maul et al. 2002)). The current assembly of the nuclear genome is available online at http://genome.jgi-psf.org/Chlre3/Chlre3.info.html. The *Chlamydomonas* Center located at www.chlamy.org continues to be an informative resource to the *Chlamydomonas* community.

Over the last two decades, several highly efficient methods for nuclear, chloroplast and mitochondrial transformation have been developed for *C. reinhardtii*. Introduction of foreign DNA into the nuclear genome of *C. reinhardii* was initially performed using bombardment with DNA-coated microparticles, and/or agitation with glass beads or silicon carbide whiskers (Debuchy et al. 1989; Dunahay 1993; Gumpel and Purton 1994). Transformation efficiencies were found to be very low using all these methods. Molecular analysis of transformants revealed predominantly random recombination of transforming DNA into the nuclear genome, resulting in a distribution of expression levels caused by positional effects of the integrating DNA. Although the expression of re-introduced endogenous nuclear genes, particularly those that rescue nutritional auxotropes, has been relatively successful, the expression of heterologous genes transformed into the nucleus of *C. reinhardtii* remains problematic. More recently, electroporation was used to introduce foreign DNA as large as 14 kb into this organism (Brown et al. 1991). Not only is this method quick and simple, it also yielded a transformation efficiency that was 2 orders of magnitude higher than previously described methods (Shimogawara et al. 1998).

Currently, microprojectile particle bombardment appears to be the \most efficient way of introducing DNA into the chloroplast genome of *C. reinhardtii*. Once inside the organelle, foreign DNA will usually integrate into the genome by homologous recombination (Boynton et al. 1988). This highly reproducible protocol had allowed for studies involving specific gene disruptions and site directed mutagenesis on plastid genes. Even with these developments, the expression of recombinant therapeutic proteins, such as antibodies, in green algae is limited. It was not until 2003 when Mayfield et al. elegantly expressed human monoclonal antibodies in transgenic algal chloroplasts. In this work, *C. reinhardtii* chloroplast *attpA* or *rbcL* promoters were used to drive the expression of an engineered large single-chain antibody directed against herpes simplex virus (HSV) glycoprotein D. This antibody accumulated as a functional soluble protein in transgenic chloroplasts, and bound herpes virus proteins, as determined by ELISA assays. This breakthrough serves as the first demonstration of microalgae as an expression platform for complex recombinant proteins, and is currently being utilized by Rincon Pharmaceuticals Inc, a San Diego-based biopharmaceuticals company for expression of monoclonal antibodies for use in cancer therapy.

The bottleneck for genetic transformation of diatoms was resolved in 1995. Dunahay et al. generated lines of transgenic *Cyclotella cryptica* and *Navicula saprophila* with plasmid vectors containing the *E. coli* neomycin phosphtransferase II gene using helium accelerated particle bombardment (Dunahay et al. 1995). This was followed by the successful transformation of *Phaeodactylum tricornutum* (Apt et al. 1996) and *Cylindrotheca fusiformis* (Fisher et al. 1999). A landmark transformation study was demonstrated by Zaslavskaia et al. in 2001. Most diatoms are solely photosynthetic and lack the ability to grow in the absence of light. These investigators successfully engineered *P. tricornutum*, a photosynthetic diatom, to grow on exogenous glucose in the dark by transformation with the glucose transporter gene *Glut1* from human erythrocytes or *Hup1* from the microalga *Chlorella kessleri*. Positive transformants exhibited glucose uptake and grew in the dark in the presence of glucose (Zaslavskaia et al. 2001). The exciting trophic conversion of an obligate photoautotrophic diatom is a critical first step for successful large-scale cultivation using microbial fermentation technology. Commercial benefits from such a system are enormous, ranging from an increase in biomass and productivity to reduced loss from contamination by obligate phototropic microbes.

The multicellular organism *Volvox carteri* represents an ideal model organism to study the transition from unicellularity to multicellularity. Using *C. reinhardtii* as a model, stable nuclear transformation of *V. carteri* was reported in 1994 by Schiedlmeier et al. Elegant studies with the intent of generating selectable markers for gene replacement and gene disruption analysis were subsequently developed (Hallmann and Sumper 1994). One of these studies resulted in a *V. carteri* transformant that carried the *Chlorella* hexose/H^+ symporter that is able to survive in the presence of glucose in the dark (Hallmann and Sumper 1996). As in the case with diatoms, this development will only accelerate the development of commercial expression systems for *V. carteri*.

Transformation of *Dunliella*

Although the genetics of *Dunaliella* are poorly understood, this organism is highly suited as an algal bioreactor. It can be cultured easily, rapidly and inexpensively. Until recently, the use of *Dunaliella* was limited by the absence of an efficient and stable transformation system. The first report of successful manipulation of *D. salina* was by Geng et al. in 2003. Using electroporation, these investigators were able to generate stable transformants carrying the hepatitis B surface antigen. Walker et al. in 2005 reported the isolation and characterization of two *D. tertiolecta* nuclear *RbcS* genes and their corresponding 5' and 3' regulatory sequences. The functionality of these regulatory regions was initially used to drive the expression of a selectable marker in *C. reinhardtii*. Subsequently, this expression cassette was electroporated into *Dunaliella* where both stable and transient transformants expressing the *ble* resistance gene were isolated. Jiang et al. (2005) identified and later used the 5' flanking region of an actin gene from *D. salina* to direct stable expression of the bialaphos resistance gene (*bar*) in *D. salina*. In more recent work, Sun et al. (2006) introduced a functional nitrate reductase gene into a *D. salina* mutant that lacked the gene. This group showed that the introduced gene was able to complement the nitrate reductase defective mutant of *D. viridis*. All the studies described are pivotal to the development of an effective transformation system in *Dunaliella*, opening the door for the use of this alga as a bioreactor for production of recombinant proteins.

Transgenic Disease Control in Shrimp Mariculture

One of the avenues that our laboratory is pursuing is the use of available transgenic techniques in marine organisms, such as cyanobacteria and *Dunaliella*, to control disease of mariculture. World aquaculture production has increased to 59.4 million metric tons (MT) in 2004, with a value of \$70 billion. Of this, farmed shrimp production accounts for 2.4 million MT, representing a value of nearly \$10 billion (FAO 2004). Diseases caused by agents such as White Spot Syndrome Virus (WSSV) and Vibrio species have decimated shrimp farming industries in many parts of Asia and South America, and account for nearly \$3 billion of economic loss annually. Unregulated use of antibiotics in farmed shrimp and fish operations has widely been banned and contributes to the epidemic of drug-resistant bacteria in humans. Intensive practices that involve meticulous water exchange with strict standards of hygiene have been effective in reducing transmission of infectious pathogens in farmed shrimp (Otoshi et al. 2002; Otoshi et al. 2001) but are impractical in many lower-income settings of the world. Usually, appearance of disease is associated with loss of harvest for shrimp farmers and a staggering 30% of global production is lost annually. The economic impact of infectious diseases of mariculture is overshadowed only by their tremendous threat to global food security.

Use of Probiotics in Mariculture

Probiotics are defined as micro-organisms that are beneficial to the health of the host. They are not therapeutic agents but, instead, directly or indirectly alter the composition of the microbial community in the rearing environment or in the gut of the host. Although the mode of action of probiotics is not fully understood, it is likely that they function by competitive elusion, that is, they antagonize the potential pathogen by the production of inhibitory compounds or by competition for nutrients and/or space (Verschuere et al. 2000). It is also likely that probiotics stimulate a humoral and/or cellular response in the host.

The first report citing probiotics as a biological control in aquaculture was in the late 1980's. In this mode of farming, probiotics are usually introduced as part of the feeding regimen or applied directly to the water. A variety of micro-organisms, ranging from aerobic Gram-positive bacteria (*Bacillus* spp. (Raengpipat et al. 1998)), to Gram-negative bacteria (*Vibro* spp. (Irianto and Austin 2002)) and yeast (Scholz et al. 1999) have been utilized successfully to increase the commercial yield of farmed marine animals. Several species of microalgae have also effectively been used for this purpose. Of note, the unicellular alga, *Tetraselmis suecica*, has been used as feed for penaeids and salmonids with significant reduction in the level of bacterial diseases. The probiotic activity of *T. suecica* was first reported in 1990 by Austin and Day. These investigators went on to show that when used as a food supplement, the algal cells inhibited laboratory-induced infections in Atlantic salmon by *Aeromonos hydrophila, A. salmonicida, Serratia liquefaciens, Vibrio anguillaram, V. salmonicida* and *Yersnia ruckeri type I* (Austin et al.1992). Although the precise mode of action of *T. suecica* is unknown it is suspected that unspecified antimicrobial organisms in the algal cells might have contributed to its probiotic activities. Along these lines, Avendano and Riquelme (1999) reported the possibility of incorporating bacteria with the ability to produce inhibitory substances into an axenic culture of *Isochrysis galbana*. Thus, this microalga could serve as a vector for transmitting inhibitory substances into cultures of larval bivalves to antagonize pathogenic bacteria.

Dunaliella was initially cultured for its β-carotene. This compound is involved in the optimal function of the immune system and is known to enhance overall disease resistance and improved health and performance in both aquatic and terrestrial animals. Marques et al. (2005) demonstrated that pathogenic effects caused by two different *Vibrio* strains in *Artemia* were reduced or even disappeared when *Dunaliella* was used as feed. Furthermore, Suppamattaya et al. (2005) showed that when used as a supplement in shrimp diet, *Dunaliella* extract appeared to enhance growth and increase resistance to viral infections. Its ease of culture and transformation, coupled with its probiotic and nutritional value makes *Dunaliella* a model organism for our work as a tool for paratransgenic delivery of immunity to commercial animals of mariculture.

Use of Vaccines in Mariculture

In recent years, vaccines based on recombinant DNA technology appear to be a promising approach to controlling infectious diseases in farmed fish (Biering et al. 2005; Clark and Cassidy-Hanley 2005; Heppel et al. 1998). By intramuscular injection of eukaryotic expression vectors encoding the sequence of a pathogen antigen, DNA vaccines offer a method of immunization that overcomes many of the disadvantages such as risk of infection and high costs of traditional live attenuated, killed or subunit protein-based counterparts. They induce strong and long-lasting humoral and cell mediated immune responses which have made them attractive for the aquaculture industry (Heppel and Davis 2000). DNA vaccination has already been proven to be effective in rainbow trout for infectious haematopoietic necrosis virus (Boudinot et al. 1998; Corbeil et al. 1999; Kim et al. 2000; Kurath et al. 2006; Lorenzen et al. 2001; Lorenzen et al. 1999) and viral haemorrhagic septicemia virus (Lorenzen et al. 2002) as well as channel catfish for ictalurid herpes virus 1 (Nusbaum et al. 2002). After intramuscular injection of plasmid DNA carrying promoter-driven reporter genes, protein expression has been achieved in common carp (Hansen et al. 1991), tilapia (Rahman and Maclean 1992), goldfish (Kanellos et al. 1999), zebrafish (Heppel et al. 1998), Japanese flounder (Takano et al. 2004) and gilthead seabream (Verri et al. 2003).

Although there are several ways to administer vaccines, most young fish continue to be vaccinated by hand. In Norway, for example, over 200 million fish are vaccinated each year. Each fish is removed from the water, anesthetized and vaccinated. This method is highly stressful for the fish, and in some circumstances rather impractical. Another method of vaccination is by dip immersion into a solution containing the vaccine. Dip immersion is usually used in fish stocks that are too young or small for manual handling. Unfortunately, this method alone is not sufficient to achieve a long duration of protection. Thus, the fish are usually subjected to intra-peritoneal re-vaccination injection as soon as their size allows. Oral vaccine delivery systems are by far the most desirable method for immunizing fish. But reports have indicated that this system is ineffective. All these hurdles point to the need for the development of a more user-friendly methodology for vaccine administration (Lin et al. 2005).

Case in Point—White Spot Syndrome Virus (WSSV)

WSSV is the most striking example of shrimp viral disease. This disease has devastated many parts of the world with grave economic consequences and reduction in available food supply. Infection of peneaid shrimp by WSSV can result in up to 100% mortality within 3 to 7 days. The virus is extremely virulent and has a broad host range including other marine invertebrates such as crayfish and crab. The global annual economic loss due to WSSV is estimated to be $3 billion (Hill 2005). In much of the world, there is no effective method to control this disease.

Entry and pathogenesis of WSSV in peneaid shrimp occur either via oral ingestion or water-borne contact (Chou et al. 1998). Work by several investigators has demonstrated that VP28, a structural protein found on the virion envelope, is responsible for viral attachment, penetration and consequently the systemic infection of shrimp (Chappel et al. 2004; van Hulten et al. 2001). Although studies on the shrimp immune response are limited, the presence of viral inhibiting proteins in both experimental and natural survivors of WSSV infections suggests that an adaptive immune response exists (Venegas et al. 2000; Wu et al. 2002). Several approaches using VP28 and another structural envelope protein, VP19, have been used to elicit an immune response in shrimp. Witteveldt et al. (2004) orally vaccinated *P. monodon* and *L. vannamei* (Witteveldt et al. 2006), two of the most important cultured shrimp species, using feed pellets coated with inactivated bacteria that were over-expressing VP28. In both cases, lower mortality was found in test versus control animals up to three weeks post vaccination. In a similar study, crayfish were protected fully from WSSV following injection with fusion VP19 + VP28 polyclonal antiserum (Li et al. 2005). Vaccination trials with VP292, a newly identified envelope protein, also resulted in significant resistance to WSSV for up to 30 days post initial vaccination (Vaseeharan et al. 2006). Using a different strategy, Robalino et al. (2004, 2005) and Tirosophon et al. (2005) demonstrated that the administration of dsRNA specific for WSSV genes induces a potent and virus-specific antiviral response in shrimp. Both studies revealed significant reduction in mortality in the shrimp population protected by vp28 and vp19 dsRNA injections.

These approaches to controlling WSSV involve induction of an immune response to virulence epitopes of WSSV and suggest that this could potentially control this disease. In each approach, however, vaccine delivery constrains implementation. The method used in the studies cited above, individual inoculation of shrimp, is highly impractical under field conditions. Given that a typical shrimp grow-out pond can harbor upwards of 300,000 post-larvae per hectare, labor costs imposed by this method rule out commercial application. The coating of dry feed with inoculum appears logical, but the feeding behavior of shrimp involves the slow nibbling of feed particles. This sluggish behavior can cause substantial losses of inoculum through leaching. It has been demonstrated that within an hour, shrimp feed can lose more than 20% of its crude protein, about 50% of its carbohydrates and 85 to 95% of its vitamin content (Rosenberry 2005). In light of the tremendous global impact of WSSV on shrimp farming and the constraints of high-intensity cultivation, new strategies to impart immunity against WSSV are essential. It is also critical that such a technology be economically viable, scalable to large shrimp farming facilities and be easily delivered to the shrimp.

Paratransgenesis

Our laboratory has developed a novel approach to control of infectious disease transmission, termed paratransgenesis. In this strategy, commensal or symbiotic bacteria found at mucosal sites of pathogen transmission are isolated and genetically altered to elaborate immune peptides or engineered single chain antibody fragments (scFv) that neutralize infectious agents. The transgenic bacteria are then delivered back to mucosal sites where disease transmission occurs. We initially developed this 'Trojan Horse' approach to combat transmission of the Chagas disease parasite, *Trypanosoma cruzi*, by reduviid bug vectors (Beard et al. 2002; Dotson et al. 2003; Durvasula et al. 1997; Durvasula et al. 1999). We have since applied it to commensal bacteria of the human respiratory tree (Sundaram et al. 2008) and sharpshooter vectors of the grape bacterial disease, Pierce's Disease (Bextine et al. 2004).

We have now developed a novel approach to control of infectious diseases of commercial mariculture. Lines of marine cyanobacteria, algae and diatoms - common components of feed for farmed shrimp and fish - can be transformed to produce antibodies that neutralize infectious pathogens such as WSSV and Vibrio. Delivery of these feed organisms, either directly in slurry preparations or via a bioamplification strategy (Figure 1) with *Artemia*, will result in passive immunization of the alimentary tract of farmed marine animals. This is the portal of entry for many infectious agents and the delivery of neutralizing antibodies would either abort the infectious process or delay it sufficiently to permit harvest.

We have demonstrated that a marine cyanobacterium, *Synechococcus bacillarus*, could be genetically transformed to express a functional recombinant antibody (Durvasula et al. 2006). We transformed *S. bacillarus* to produce a murine

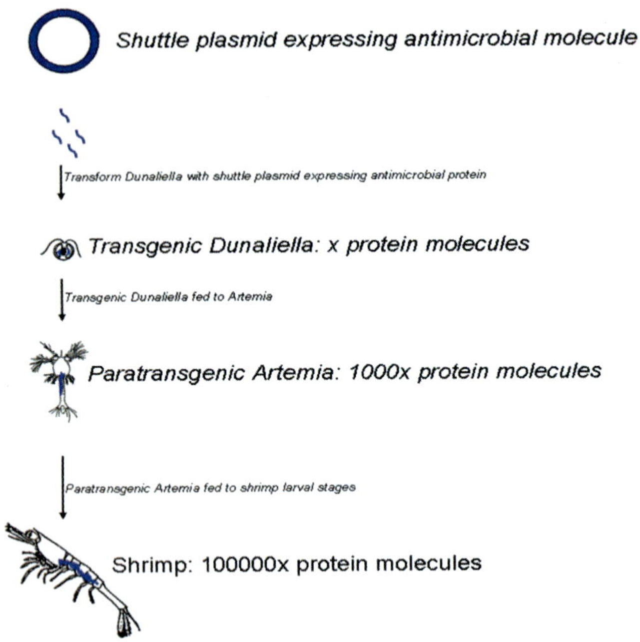

Figure 1: Schematic demonstrating the idealized process of bioamplification. In this strategy, *Dunaliella* is first transformed with a shuttle plasmid expressing an antimicrobial or antiviral molecule. Each transformed *Dunaliella* will express 10 molecules (an arbitrary number) of protective protein molecules. The transformed *Dunaliella* is then fed to *Artemia*. Thus, consumption of 100 transformed organisms would result in (100 *Dunaliella* x 10 molecules protective protein/*Dunaliella*) 1000 molecules of protective protein per *Artemia*. The engorged *Artemia* (now harboring 1000 molecules of protective protein) is fed to the target animal. Ingestion of 100 'paratransgenic' Artemia would therefore result in (100 Artemia x 1000 molecules of protective protein/Artemia) 100000 molecules of protective protein per shrimp. In this manner, the protective molecule is 'bioamplified' as it progresses up the food chain. The paratransgenic *Artemia* will result in passive immunization of the alimentary tract of shrimp thereby aborting the infectious process or delaying it sufficiently to permit harvest. It should be kept in mind that this scheme is highly idealized. A myriad of factors, including gene decay, protein instability, and even feeding rates will affect the efficiency of this delivery system.

antibody (rDB3) against progesterone, using a heterologous expression system. In competitive ELISA studies, the rDB3 antibody bound progesterone in a dose-dependent and specific manner. No cross-reactivity with testosterone, a structurally similar steroid, was detected (Durvasula et al. 2006). This study demonstrated that a transgenic cyanobacterium could elaborate an active recombinant antibody, and serves as a model for future applications of this technology. More recently, we were successful in transforming *D. salina* carrying a construct that contains the chloramphenicol acetyltransferase (CAT) gene as well as the gene that encodes for green fluorescent protein (GFP). Using GFP as a marker, we plan to feed the transformed *Dunaliella* as a slurry to shrimp nauplii and monitor the progression of the GFP within the gut. This study will serve as a proof-of-concept for the delivery of a functional protein from the feed to the target organism. In future work, we will replace this marker gene with genes encoding the single chain antibodies against different capsid proteins of WSSV.

Bioamplification of Foreign Gene Products through Transgenic Microalgae

An alternate strategy for delivery of transgenic *Dunaliella* to the target animal is via bioamplification. In this strategy a feed organism such as *Artemia* initially consumes the transgenic *Dunaliella*. The engorged *Artemia* is then fed to the target animal. In this manner, the supplement is bioamplified as it progresses up the food-chain (Figure 1).

Artemia are non-selective filter feeders and therefore will ingest a wide range of foods. The main criteria for food selection are particle size, digestibility, and nutrient levels (Dobbeleir et al. 1980). Possibly the best foods for *Artemia* are live microalgae such as *Nannochloropsis*, *Tetraselmis*, *Isochrysis*, *Dunaliella* and *Pavlova*. Combinations of live phytoplankton fed to *Artemia* cultures have demonstrated superior enrichment characteristics over feeding single phytoplankton species (D'Agostino 1980). However, not all species of unicellular algae are appropriate for sustaining *Artemia* growth. For example, *Chlorella* and *Stichococcus* have a thick cell wall that cannot be digested by *Artemia*.

The supplementation or bioencapsulation of nutritional components, such as vitamins or calcium, into *Artemia* has been practiced by aquaculture hatcheries for over 10 years. *Artemia* nauplii were initially fed with emulsified fish oils containing highly unsaturated fatty acids (HUFA), to eventually be used as feed for marine finfish and crustacean larvae. Today, live nauplii of the brine shrimp have been used as vectors for delivering compounds of diverse nutritional (Dhert et al. 1999) and/or therapeutic (Campbell et al. 1993) value to larval stages of aquatic animals. Inoculating the digestive tracts of target organisms with probiotic bacteria through bioencapsulation is another alternative use for *Artemia* nauplii. Bacteria with various characteristics have also been incorporated into *Artemia* nauplii prior to oral challenge of turbot larvae with a pathogenic *Vibrio anguillarum* strain (Chair et al. 1994a; Grisez et al. 1996). This route has also been used to vaccinate sea bass fry (Chair et al. 1994b), juvenile carp (Joosten et al. 1995) and fish fry (Campbell et al. 1993).

Summary and Conclusions

Recent advances in molecular techniques have resulted in the successful transformation of *Dunaliella*. These advances coupled with the other attributes of *Dunaliella* (ease of culture, halo- and pH-tolerance) will only contribute to the further use of this species as a biofactory for the large-scale production of proteins.

Our laboratory is most interested in transforming *Dunaliella* to produce antibodies that would neutralize infectious pathogens in mariculture such as WSSV and *Vibrio*. We will then deliver the transformed *Dunaliella* either directly in slurry preparations or via a bioamplification strategy to farmed marine animals. This will result in passive immunization of the alimentary tract of farmed marine animals thereby aborting the infectious process or delaying it sufficiently to permit harvest.

Our ongoing studies suggest that genetically altered *S. bacillarus* can be fed directly to *Artemia* nauplii with accumulation of functional antibodies in the brine shrimp gut (Subhadra et al. manuscript in prep). We were further able to demonstrate accumulation of foreign recombinant proteins in *Artemia* nauplii that had been fed transformed *Dunaliella* (Subhadra, unpublished). These studies support the concept that pre-formed molecules can be delivered via engineered algae or cyanobacteria, and therefore serves as a model for future applications of this technology. We are in the preliminary stages of developing an array of single chain antibodies against different capsid proteins of WSSV for expression in *Dunaliella* and *S. bacillarus*. We are also optimizing the expression of *P. monodon* antiviral proteins (Luo et al. 2003) and antimicrobials (Chiou et al. 2005) in both of these systems. Finally, we are looking at a variety of other microalgal species such as *Isochrysis galbana*, *Phaeodactylum tricornutrum*, and *Chlamydomonas chlorococcum* as the primary players in our bioamplification scheme for combating infectious disease in shrimp.

References

Apt, K. E., P. G. Kroth-Pancic and A. R. Grossman. 1996. Stable nuclear transformation of the diatom *Phaeodactylum tricornutum*. Mol. Gen. Genet. 252:572-579.

Austin, B. and J. G. Day. 1990. Inhibition of prawn pathogenic *Vibro* spp. by a commercial spray-dried preparation of *Tetraselmis suecica*. Aquaculture 90:389-392.

Austin, B., E. Baudet and M.B.C. Stobie. 1992. Inhibition of bacterial fish pathogens by *Tetraselmis suecica*. J.Fish. Diseases 15:55-61.

Avendano, R. E. and C. E. Riquelme. 1999. Establishment of mixed-culture probiotics and microalgae as food for bivalve larvae. Aquaculture Research 30:893-900.

Beard, C. B., C. Cordon-Rosales and R. V. Durvasula. 2002. Bacterial symbionts of the triatominae and their potential use in control of Chagas disease transmission. Annu. Rev. Entomol. 47:123-141.

Ben-Amotz, A. and M. Avron. 1990. The biotechnology of cultivating the halotolerant alga *Dunaliella*. Trends in Biotechnology 8:121-126.

Ben-Amotz, A., A. Shaish and M. Avron. 1989. Mode of action of the passively accumulated beta-carotene of *Dunaliella bardawil* in protecting the alga against damage by excess irradiation. Plant Physiol. 91:1040-1043.

Bextine, B., C. Lauzon, S. Potter, D. Lampe and T. A. M:ller. 2004. Delivery of a genetically marked *Alcaligenes* sp. to the glassy-winged sharpshooter for use in a paratransgenic control strategy. Curr. Microbiol. 48:327-331.

Biering, E., S. Villoing, I. Sommerset and K.E. Christie. 2005. Update on viral vaccines for fish. Dev. Biol. (Basel) 121:97-113.

Boudinot, P., M. Blanco, P. de Kinkelin and A. Benmansour. 1998. Combined DNA immunization with the glycoprotein gene of viral hemorrhagic septicemia virus and infectious hematopoietic necrosis virus induces double-specific protective immunity and nonspecific response in rainbow trout. Virology 249:297-306.

Boynton, J. E., N. W. Gillham, E. H. Harris, J. P. Hosler, A. M. Johnson, A. R. Jones, B. L. Randolph-Anderson, D. Robertson, T. M. Klein, K. B. Shark and et al. 1988. Chloroplast transformation in *Chlamydomonas* with high velocity microprojectiles. Science 240:1534-1538.

Brown, L. E., S. L. Sprecher and L. R. Keller. 1991. Introduction of exogenous DNA into *Chlamydomonas reinhardtii* by electroporation. Mol. Cell Biol. 11:2328-2332.

Campbell, R., A. Adams, M. F. Tatner, M. Chair and P. Sorgeloos. 1993. Uptake of *Vibrio angillarum* vaccine by *Artemia salina* as a potential oral delivery system to fish fry. Fish Shellfish Immunnol. 3:451-459.

Chair, M., M. Dehasque, S. Van Poucke, H. Nelis, P. Sorgeloos and A.P. De Leenheer. 1994a. An oral challenge for turbot larvae with *Vibrio anguillarum*. Aquac Int 2:270-272.

Chair, M., R.S.J. Gapasin, M. Dehasque and P. Sorgeloos. 1994b. Vaccination of European sea bass fry through bioencapsulation of *Artemia* nauplii. Aquac. Int. 2:254-261.

Chappel, J. A., W. O. Rogers, S. L. Hoffman and A. S. Kang. 2004. Molecular dissection of the human antibody response to the structural repeat epitope of *Plasmodium falciparum* sporozoite from a protected donor. Malar. J. 3:28.

Chiou, T. T., J. L. Wu, T. T. Chen and J. K. Lu. 2005. Molecular cloning and characterization of cDNA of penaeidin-like antimicrobial peptide from tiger shrimp (*Penaeus monodon*). Mar. Biotechnol. (NY) 7:119-127.

Chou, H.V., C.Y. Huang, C.F. Lo and G.H. Kou. 1998. Studies on transmission of white spot associated baculovirus (WSBV) in *Penaeus monodon* and *P. japonicus* via water borne contact and oral ingestion. Aquaculture 164:263-276.

Clark, T. G. and D. Cassidy-Hanley. 2005. Recombinant subunit vaccines: potentials and constraints. Dev. Biol. (Basel) 121:151-163.

Corbeil, S., S. E. Lapatra, E. D. Anderson, J. Jones, B. Vincent, Y. L. Hsu and G. Kurath. 1999. Evaluation of the protective immunogenicity of the N, P, M, NV and G proteins of infectious hematopoietic necrosis virus in rainbow trout *oncorhynchus mykiss* using DNA vaccines. Dis. Aquat. Organ. 39:29-36.

D'Agostino, A.S. The vital requirements of *Artemia*: physiology and nutrition. 1980 pp. 55-82. In The Brine Shrimp Artemia G. Personne, P. Sorgeloos, O. Roels and E. Jaspers (eds) Universa Press. Wetteren, Belguim.

Debuchy, R., S. Purton and J. D. Rochaix. 1989. The argininosuccinate lyase gene of *Chlamydomonas reinhardtii*: an important tool for nuclear transformation and for correlating the genetic and molecular maps of the ARG7 locus. EMBO J 8:2803-2809.

Dhert, P., M. Lavens, M. Duray and P. Sorgeloos. 1999. Improved larval survival at metamorphosis of Asian seabass (*Lates calcarifer*) using omega 3-HUFA-enriched live food. Aquaculture 90:63-74.

Dobbeleir, J., N. Adam, E. Bossuyt, E. Bruggeman and P. Sorgeloos. 1980. New asspects of the use of inert diets for high density culturing of brine shrimp. pp. 165 - 174. In The Brine Shrimp Artemia. G. Personne, P. Sorgeloos, O. Roels and E. Jaspers (eds) Universa Press.Wetteren, Belguim.

Dotson, E. M., B. Plikaytis, T. M. Shinnick, R.V. Durvasula and C. B. Beard. 2003. Transformation of *Rhodococcus rhodnii*, a symbiont of the Chagas disease vector *Rhodnius prolixus*, with integrative elements of the L1 mycobacteriophage. Infect. Genet. Evol. 3:103-109.

Dunahay, T. G. 1993. Transformation of *Chlamydomonas reinhardtii* with silicon carbide whiskers. Biotechniques 15:452- 460.

Dunahay, T. G., E. E. Jarvis and P. G. Roessler. 1995. Genetic transformation of the diatoms *Cyclotella cryptica* and *Navicula saprophila*. J. Phycol. 31:1004-1012.

Durvasula, R. V., A. Gumbs, A. Panackal, O. Kruglov, S. Aksoy, R. B. Merrifield, F. F. Richards and C. B. Beard. 1997. Prevention of insect-borne disease: An approach using transgenic symbiotic bacteria. Proc. Natl. Acad. Sci. USA. 94:3274-3278.

Durvasula, R. V., A. Gumbs, A. Panackal, O. Kruglov, J. Taneja, A. S. Kang, C. Cordon-Rosales, F. F. Richards, R. G. Whitham and C. B. Beard. 1999. Expression of a functional antibody fragment in the gut of *Rhodnius prolixus* via transgenic bacterial symbiont *Rhodococcus rhodnii.* Med. Vet. Entomol. 13:115-119.

Durvasula, R.V., R. Sundaram, S. Matthews P. Sundaram and D.V. Subba Rao. 2006. Prospects for paratransgenic applications to commercial mariculture using genetically engineered algae. pp. 865-889. In: Algal cultures, Analogues of blooms and Applications. D.V. Subba Rao (ed) Science Publishers, New Hampshire, USA.

FAO. 2004. The State of World Fisheries and Aquaculture (SOFIA). Food and Agriculture Organization of the United Nations. Rome, Italy.

Fisher, M., I. Gokhman, U. Pick and A. Zamir. 1996. A salt-resistant plasma membrane carbonic anhydrase is induced by salt in *Dunaliella salina*. J. Biol. Chem. 271:17718-17723.

Fisher, M., I. Gokhman, U. Pick and A. Zamir. 1997. A structurally novel transferrin-like protein accumulates in the plasma membrane of the unicellular green alga *Dunaliella salina* grown in high salinities. J. Biol. Chem. 272:1565-1570.

Fisher, H, I. Robl, M. Sumper and N. Kroger. 1999. Targeting and covalent modification of cell wall and membrane proteins heterologously expressed in the diatom *Cylindrotheca fusicormis* (Bacillariophyceae). J. Phycol. 35:113-120.

Geng, D., Y. Wang, P. Wang, W. Li and Y. Sun. 2003. Stable expression of hepatitis B surface antigen gene in *Dunaliella salina*. J. Applied Phycol. 15:451-456.

Gonzalez, M. A., A. W. Coleman, P. I. Gomez and R. Montoya. 2001. Phylogenetic Relationship among various strains of *Dunaliella* (*Chlorophyceae*) based on nuclear 18S rDNA sequences J. Phycol. 37:604-611.

Grisez, L., M. Chair, P. Sorgeloos and F. Ollevier. 1996. Mode of infection and spread of *Vibrio anguillarum* in turbot *Scophthalmus maximus* larvae after oral challenge through live feed. Dis. Aquat. Org. 26:181-187.

Gumpel, N. J. and S. Purton. 1994. Playing tag with *Chlamydomonas*. Trends Cell. Biol. 4: 299-301.

Hallmann, A. and M. Sumper. 1994. Reporter genes and highly regulated promoters as tools for transformation experiments in *Volvox carteri*. Proc. Natl. Acad. Sci. USA. 91: 11562-11566.

Hallmann, A. and M. Sumper. 1996. The *Chlorella* hexose/H+ symporter is a useful selectable marker and biochemical reagent when expressed in Volvox. Proc. Natl. Acad. Sci. U S A. 93:669-673.

Hansen, E., K. Fernandes, G. Goldspink, P. Butterworth, P. K. Umeda and K. C. Chang. 1991. Strong expression of foreign genes following direct injection into fish muscle. FEBS Letters 290:73-76.

Heppel, J. and H.L. Davis. 2000. Application of DNA vaccine technology to aquaculture. Adv Drug Delivery Rev. 43:29-43.

Heppel, J., N. Lorenzen, N. K. Armstrong, T. Wu, E. Lorenzen and K. Einer-Jensen. 1998. Development of DNA vaccines for fish:Vector design, intramuscular injection and antigen expression using viral haemorrhagic septicimia virus genes as model. Fish and Shellfish Immunology 8:271-286.

Hill, B. J. 2005. The need for effective disease control in international aquaculture. pp. 3-12. In Progress in Fish Vaccinology P. J. Midtlyng. (ed) Karger. Bergen, Norway.

Irianto, A. and B. Austin. 2002. Use of probiotics to control furunculosis in Rainbow trout, *Oncorhynchus mykiss*. J. Fish. Diseases 25:1 - 10.

Jiang, G. Z., Y. M. Lu, X. L. Niu and L. X. Xue. 2005. The actin gene promoter-driven bar as a dominant selectable marker for nuclear transformation of *Dunaliella salina*. Yi Chuan Xue Bao 32:424-433.

Joosten, P.H.M., M. Aviles-Trigueros, P. Sorgeloos and J. H. W. M. Rombout. 1995. Oral vaccination of the juvenille carp (*Cyprinus carpio*) and gilthead seabream (*Sparus aurata*) with bioencapsulated *Vibrio anguillarum* bacterin. Fish Shellfish Immunnol. 5:289-299.

Kanellos, T., I. D. Sylvester, C. R. Howard and P. H. Russell. 1999. DNA is as effective as protein at inducing antibody in fish. Vaccine 17:965-972.

Kim, C. H., M. C. Johnson, J. D. Drennan, B. E. Simon, E. Thomann and J. A. Leong. 2000. DNA vaccines encoding viral glycoproteins induce nonspecific immunity and Mx protein synthesis in fish. J. Virol. 74:7048-7054.

Kurath, G., K. A. Garver, S. Corbeil, D. G. Elliott, E. D. Anderson and S. E. LaPatra. 2006. Protective immunity and lack of histopathological damage two years after DNA vaccination against infectious hematopoietic necrosis virus in trout. Vaccine 24:345-354.

Li, H. X., X. L. Meng, J. P. Xu, W. Lu and J. Wang. 2005. Protection of crayfish, *Cambarus clarkii*, from white spot syndrome virus by polyclonal antibodies against a viral envelope fusion protein. J. Fish Dis. 28:285-291.

Lin, J. H., C. C. Yu, C. C. Lin and H. L. Yang. 2005. An oral delivery system for recombinant subunit vaccine to fish. Dev. Biol. (Basel) 121:175 - 180.

Lorenzen, N., E. Lorenzen, K. Einer-Jensen, J. Heppell and H. L. Davis. 1999. Genetic vaccination of rainbow trout against viral haemorrhagic septicaemia virus: small amounts of plasmid DNA protect against a heterologous serotype. Virus Res. 63:19-25.

Lorenzen, N., E. Lorenzen and K. Einer-Jensen. 2001. Immunity to viral haemorrhagic septicaemia (VHS) following DNA vaccination of rainbow trout at an early life-stage. Fish. Shellfish Immunol. 11:585-591.

Lorenzen, N., E. Lorenzen, K. Einer-Jensen and S. E. LaPatra. 2002. Immunity induced shortly after DNA vaccination of rainbow trout against rhabdoviruses protects against heterologous virus but not against bacterial pathogens. Dev. Comp. Immunol. 26:173-179.

Luo, T., X. Zhang, Z. Shao and X. Xu. 2003. PmAV, a novel gene involved in virus resistance of shrimp *Penaeus monodon.* FEBS Letters 551:53-57.

Marques, A., T. Dinh, C. Ioakeimidis, G. Huys, J. Swings, W. Verstraete, J. Dhont, P. Sorgeloos and P. Bossier. 2005. Effects of bacteria on *Artemia franciscana* cultured in different gnotobiotic environments. Appl. Environ. Microbiol. 71:4307-4317.

Maul, J. E., J. W. Lilly, L. Cui, C. W. dePamphilis, W. Miller, E. H. Harris and D. B. Stern. 2002. The *Chlamydomonas reinhardtii* plastid chromosome: islands of genes in a sea of repeats. Plant Cell 14:2659-2679.

Mayfield, S. P., S. E. Franklin and R. A. Lerner. 2003. Expression and assembly of a fully active antibody in algae. Proc. Natl. Acad. Sci. USA. 100:438-42.

Nusbaum, K. E., B. F. Smith, P. DeInnocentes and R. C. Bird. 2002. Protective immunity induced by DNA vaccination of channel catfish with early and late transcripts of the channel catfish herpesvirus (IHV-1). Vet Immunol. Immunopathol. 84:151-168.

Otoshi, C. A, A. D. Montgomery, A. M. Look and S. M. Moss. 2001. Effects of diet and water source on the nursery production of Pacific white shrimp *Litopenaeus vannamei.* J. World Aquaculture Society 32:243-249.

Otoshi, C. A., S. M. Arce and S. M. Moss. 2002. Use of recirculating systems for the production of broodstock shrimp. Panorama Acuícola Magazine 8.

Rabbani, S., P. Beyer, J. Lintig, P. Hugueney and H. Kleinig. 1998. Induced beta-carotene synthesis driven by triacylglycerol deposition in the unicellular alga *Dunaliella bardawil.* Plant Physiol. 116:1239-1248.

Raengpipat, S., W. Phianphak, S. Piyatiratitivorakul and P. Menasveta. 1998. Effects of probiotic bacterium in black tiger shrimp *Penaeus monodon* survival and growth. Aquaculture 167:301-313.

Rahman, A. and N. Maclean. 1992. Fish transgene expression by direct injection into fish muscle. Mol. Mar. Biol. Biotechnol. 1:286-289.

Robalino, J., C. L. Browdy, S. Prior, A. Metz, P. Parnell, P. Gross and G. Warr. 2004. Induction of antiviral immunity by double-stranded RNA in a marine invertebrate. J. Virol. 78:10442-10448.

Robalino, J., T. Bartlett, E. Shepard, S. Prior, G. Jaramillo, E. Scura, R. W. Chapman, P. S. Gross, C. L. Browdy and G. W. Warr. 2005. Double-stranded RNA induces sequence-specific antiviral silencing in addition to nonspecific immunity in a marine shrimp: convergence of RNA interference and innate immunity in the invertebrate antiviral response? J. Virol. 79:13561-13571.

Rosenberry, B. 2005. About Shrimp Farming. Shrimp News.

Sanchez-Estudillo, L., Y. Freile-Pelegrin, R. Rivera-Madrid, D. Robledo and J. A. Narvaez-Zapata. 2006. Regulation of two photosynthetic pigment-related genes during stress-induced pigment formation in the green alga, *Dunaliella salina*. Biotechnol. Lett. 28:787-791.

Schiedlmeier, B., R. Schmitt, W. Muller, M. M. Kirk, H. Gruber, W. Mages and D. L. Kirk. 1994. Nuclear transformation of *Volvox carteri*. Proc. Natl. Acad. Sci. U S A. 91:5080-5084.

Scholz, U., G. Garcia Diaz, D. Ricque, L. E. Cruz Suarez, F. Vargas Albores and J. Latchford. 1999. Enhancement of vibriosis resistance in juvenile *Penaeus vannamei* by supplementation of diets with different yeast products. Aquaculture 176:271-283.

Shimogawara, K., S. Fujiwara, A. Grossman and H. Usuda. 1998. High-efficiency transformation of *Chlamydomonas reinhardtii* by electroporation. Genetics 148:1821-1828.

Sun, X. M., Y. P. Tang, X. Z. Meng, W. W. Zhang, S. Li, Z. R. Deng, Z. K. Xu and R. T. Song. 2006. Sequencing and analysis of a genomic fragment provide an insight into the *Dunaliella viridis* genomic sequence. Acta Biochim. Biophys. Sin. (Shanghai) 38:812-820.

Sundaram, R. K., I. Hurwitz, S. Matthews, E. Hoy, S. Kurapati, C. Crawford, P. Sundaram and R.V. Durvasula. 2008. Expression of a functional single chain antibody in Corynebacterium pseudodiphtheriticum. Eur. J. Clin. Microbiol. Infect. Dis. 27: 617-622.

Sundaram, R. K., I. Hurwitz, S. Matthews, E. Hoy, S. Kurapati, C. Crawford, P. Sundaram and R.V. Durvasula. 2006. Expression of a Functional Single Chain Antibody in *Corynebacterium pseudodiphtheriticum*. Appl. Env. Microbiol (in review).

Suppamattaya, K., S. Kiriratnikom, M. Boonyaratpalin and L. Blorowitzka. 2005. Effect of a Dunaliella extract on growth performance, health condition, immune response and disease resistance in black tiger shrimp (*Penaeus monodon*). Aquaculture 248:207-216.

Takano, T., A. Iwahori, I. Hirono and T. Aoki. 2004. Development of a DNA vaccine against hirame rhabdovirus and analysis of the expression of immune-related genes after vaccination. Fish. Shellfish Immunol. 17:367-374.

Tirasophon, W., Y. Roshorm and S. Panyim. 2005. Silencing of yellow head virus replication in penaeid shrimp cells by dsRNA. Biochem. Biophys. Res. Commun. 334:102-107.

Vahrenholz, C., G. Riemen, E. Pratje, B. Dujon and G. Michaelis. 1993. Mitochondrial DNA of *Chlamydomonas reinhardtii*: the structure of the ends of the linear 15.8-kb genome suggests mechanisms for DNA replication. Curr. Genet. 24:241-247.

van Hulten, M. C., J. Witteveldt, M. Snippe and J. M. Vlak. 2001. White spot syndrome virus envelope protein VP28 is involved in the systemic infection of shrimp. Virology 285:228-233.

Vaseeharan, B., T. Prem Anand, T. Murugan and J. C. Chen. 2006. Shrimp vaccination trials with the VP292 protein of white spot syndrome virus. Lett. Appl Microbiol. 43:137-142.

Venegas, C. A., L. Nonaka, K. Mushiake, T. Nishizawa and K. Murog. 2000. Quasi-immune response of *Penaeus japonicus* to penaeid rod-shaped DNA virus (PRDV). Dis. Aquat. Organ. 42:83-89.

Verri, T., L. Ingrosso, R. Chiloiro, A. Danieli, V. Zonno, P. Alifano, N. Romano, G. Scapigliati, S. Vilella and C. Storelli. 2003. Assessment of DNA vaccine potential for gilthead sea bream (*Sparus aurata*) by intramuscular injection of a reporter gene. Fish Shellfish Immunol. 15:283-295.

Verschuere, L., G. Rombaut, P. Sorgeloos and W. Verstraete. 2000. Probiotic bacteria as biological control agents in aquaculture. Microbiol. Mol. Biol. Rev. 64:655-671.

Walker, T.L., B.K. Douglas, and C.C. Collet. 2005. Characterization of the Dunaliella tertiolecta RbcS genes and their promoter activity in Chlamydomonas reinhardtii. Plant Cell Reports 23: 727-735.

Witteveldt, J., C. C. Cifuentes, J. M. Vlak and M. C. van Hulten. 2004. Protection of *Penaeus monodon* against white spot syndrome virus by oral vaccination. J. Virol. 78:2057-61.

Witteveldt, J., J. M. Vlak and M. C. van Hulten. 2006. Increased tolerance of *Litopenaeus vannamei* to white spot syndrome virus (WSSV) infection after oral application of the viral envelope protein VP28. Dis. Aquat. Organ. 70:167-170.

Wu, J. L., T. Nishioka, K. Mori, T. Nishizawa and K. Muroga. 2002. A time-course study on the resistance of *Penaeus japonicus* induced by artificial infection with white spot syndrome virus. Fish Shellfish Immunol. 13:391-403.

Zaslavskaia, L. A., J. C. Lippmeier, C. Shih, D. Ehrhardt, A. R. Grossman and K. E. Apt. 2001. Trophic conversion of an obligate photoautotrophic organism through metabolic engineering. Science 292:2073-2075.

17 Development of Genetics and Molecular Tool Kits for Species of the Unicellular Green Alga *Dunaliella* (Chlorophyta)

Jürgen E.W. Polle[1] and Song Qin[2]*

Abstract

For over a hundred years, unicellular green algae of the genus *Dunaliella* have been used as model systems in various studies, including salt stress and osmoregulation, photosynthesis, secondary carotenoid metabolism, UV stress, and adaptation to low pH. In particular, *D. salina*, *D. tertiolecta*, and *D. acidophila* have been frequently used by researchers. But in spite of their use as model algae, they have historically presented two major hurdles: 1. Genetic crosses could not be performed on a routine basis, and 2. No transformation system existed. Recent reports, however, have shown that it is indeed possible to reliably perform genetic crosses with many *Dunaliella* species and have also demonstrated the transformation of different species, such as *D. tertiolecta*, *D. salina*, and *D. viridis*. This chapter reviews the life cycle of *Dunaliella* and summarizes the current development status of transformation systems for *Dunaliella* algae.

Introduction

Unicellular green algae of the genus *Dunaliella* are often called 'polyextremophiles' due to their ability to tolerate extreme conditions, such as low temperature, low pH, high salinity, and high light intensities. The term 'polyextremophiles' is nevertheless somewhat incorrect, since researchers are actually working with a number of different *Dunaliella* species, each with its own set of unique characteristics. *Dunaliella acidophila* might be the only true extremophile within the genus *Dunaliella* because

[1] Department of Biology, Brooklyn College of CUNY, 2900 Bedford Ave 200NE, Brooklyn, NY 11210, USA.

[2] Institute of Oceanology, Chinese Academy of Sciences, 266071, P.R. China.

* Corresponding author, Email: jpolle@brooklyn.cuny.edu

it requires low pH for growth. All other species are not strictly extremophiles, for although they can tolerate numerous extreme environmental conditions, they do not grow at maximum rate under those conditions. Still, this ability to tolerate and grow under extreme conditions—albeit at reduced rates—make some species of *Dunaliella* prime model organisms for specific biological studies. Table 1 lists several species of the genus *Dunaliella* that have been used as model organisms (Pick 1998) in various biological investigations. Among the different species, the alga *D. salina* proves to be particularly ideal as a model organism for life in hyper-saline environments. *D. salina* not only adapts to grow in a wide range of hyper-saline conditions, it also tolerates the drastic changes in salinity that result in hypertonic or hypotonic osmotic conditions. Methods of adaptation in order to survive in environments of varying salinity include a special glycerol metabolism (Chitlaru and Pick 1991, Goyal 2007, Thompson 1994), presence of an elastic glycocalyx rather than a rigid cell wall (Oliveira et al. 1980), and induction of secondary carotenoid over-accumulation in the chloroplast (Shaish et al. 1992). In addition, it can be expected that the proteome of the plasma membrane of *D. salina* cells (Katz et al. 2007) differs significantly from that of other green algae, because *Dunaliella* cells have to cope with high salinity.

Table 1: List of exemplary *Dunaliella* species that were used as model organisms for various fields.

Species of *Dunaliella*	**Model organism for**
D. acidophila	Tolerance of low pH
D. salina Teod.	Osmoregulation, photosynthesis, carotenoid over-accumulation, glycerol production, general aquaculture
D. tertiolecta Butcher	Photosynthesis, ecology
D. viridis Teod. / *D. parva* Lerche	Photosynthesis, salt stress responses, aquaculture, ecology of salt lakes, heavy metal tolerance

In the ability of their cells to tolerate high salinity, thereby permitting selective growth conditions, and their property of secondary carotenoid over-accumulation, the species *D. salina* Teod. and *D. bardawil* Avron et Ben-Amotz have become commercially interesting. As a result, these two species are exploited in outdoor mass cultures (Ben-Amotz and Avron 1990, Ben-Amotz et al. 1991, Borowitzka 1991, Borowitzka 1999, Moulton et al. 1987, Raja et al. 2007).

However, despite this introduction of a commercial use for *Dunaliella* algae and despite their physiological uniqueness, more widespread use has not occurred because the absence of a reliable system to perform genetic crosses and lack of a transformation system were always viewed as major drawbacks. But since the early

1990s, when the first book on the alga *Dunaliella* was published (Ben-Amotz and Avron 1992), significant progress has been made in molecular and genetic tool development with algae of the genus *Dunaliella*. This chapter reviews the current state of studies to establish genetic and molecular techniques for the alga *Dunaliella*.

Reproduction

Understanding genetics is crucial to the widespread use of a model organism (Davis 2004). Without the ability to cross strains, for example, it would be impossible to determine linkage of mutations with markers or to map mutations within the genome.

Already in Teodoresco's first publication where he established the genus of *Dunaliella* with the type species *D. salina*, the mode of reproduction for this alga was described (Teodoresco 1905). Under constant growth conditions, reproduction mainly occurs asexually, whereas sexual reproduction appears to be triggered by environmental changes.

Asexual Reproduction

During asexual reproduction, the cells of *Dunaliella* species generally divide longitudinally (Teodoresco 1905, Hamburger 1905, Labbé 1925, Lerche 1937, Liebetanz 1925, Nicolai and Baas-Becking 1935). This process may take up to two hours and had already been shown by Teodoresco (1905, 1906), but it was demonstrated in greater detail with *D. salina* by Hamburger (1905) and Labbé (1925): Early stages of cell division show a crossways-stretched nucleus and two pyrenoids in cells that have one pair of flagella. After the pyrenoid duplicates, the chloroplast and the nucleus divide (Figure 1). At the same time the pair of flagella might duplicate. Cell division then proceeds by apparent simultaneous constriction from anterior and posterior ends, so that in the late stage of division, a plasma-bridge forms between the two daughter cells (Borowitzka and Siva 2007, Figure 1). Division ends with the disappearance of this plasma-bridge as it releases two daughter cells. Both daughter cells may already have two new flagella of equal length (Hamburger 1905, Figure 1), but this is not always the case (Teodoresco 1905, Borowitzka and Siva 2007).

The question of asexual cyst (aplanospore) formation (Borowitzka and Huisman 1993, Hamburger 1905, Liebetanz 1925, Loeblich 1969) remains open. Lerche (1937) reports not to have found any cysts of asexual origin, and she claims that aplanospores were not even produced under stressed conditions. It may be that zygotes had previously been mistaken for aplanospores due to their

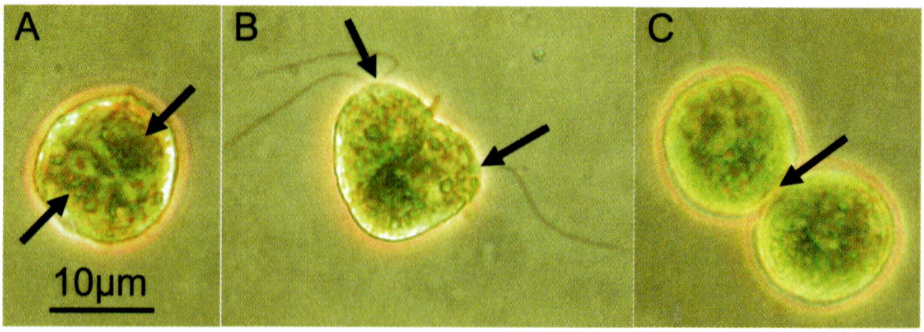

Figure 1: Shown are different stages of dividing cells of *Dunaliella salina* (400x magnification). A) One cell with two chloroplasts indicated by the arrows. B) Two cells after division of the flagellar apparatus. Each new daughter cell already has new full-length flagella (see arrows). C) Two cells still joined through a plasma bridge which is indicated by the arrow.

reported similarity in appearance, and that this contributed to the confusion about the existence of aplanospores. More recently, Borowitzka and Siva (2007) have indicated that not all species and strains within one species have the ability to produce aplanospores, which may explain Lerche's difficulty (1937) in observing aplanospore formation. In contrast to aplanospores, however, reports do agree on the existence of pallmeloid stages for *D. salina* (Labbé 1925, Lerche 1937, Liebetanz 1925, Borowitzka and Siva 2007) and *D. viridis* (Massjuk 1973). One early pallmeloid stage for *D. salina* is shown in Figure 2. In addition, large cells called 'Monsterindividuen' by Hamburger (1905) can often be observed in *Dunaliella* cultures. These 'monster' cells may have multiple pyrenoids, multiple

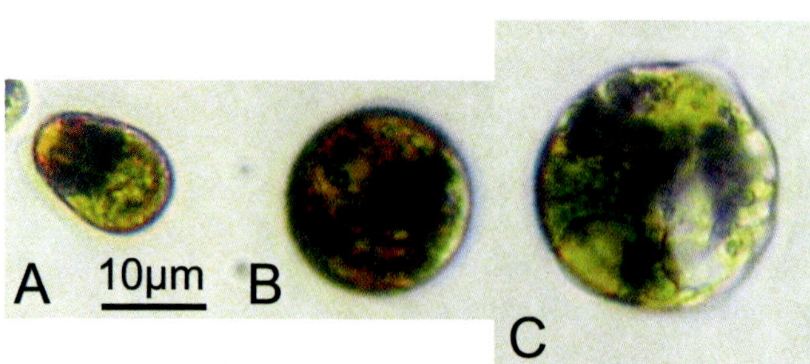

Figure 2: Shown are photographs of *D. salina* cells from 1.0 M NaCl batch cultures (360x magnification). A) Green zoospore. B) Pallmeloid cell. C) 'Monster' cell.

eyespots, and multiple pairs of flagella. It is believed that these large cells represent abnormal asexual division stages. Figure 2 shows a very large cell of *D. salina* that most likely falls into the category of previously described 'monster' cells.

Sexual Reproduction

Isogamic sexual reproduction for *D. salina* was already observed and described by Teodoresco (1905, 1906). Later, various aspects of sexual reproduction were investigated and were reported for different species (Borowitzka et al. 1982, Borowitzka and Siva 2007, Cavara 1906, Gonzalez et al. 2001, Hamel 1931, Liebetanz 1925, Labbé 1925, Leonardi and Cáceres 1997, Lerche 1937, Samanamud 1998). Hamburger (1905) did not observe the full process of conjugation for *D. salina*, but she reported observations of a specific cell type that appeared to be similar to those of two conjugated cells as shown by Teodoresco (1905). The following description of conjugation in *Dunaliella* is mainly based on work by Lerche (1937) who performed very detailed studies on different species. According to Lerche (1937) and Teodoresco (1906), induction of gametogenesis occurs through changes in environmental conditions producing isogametes that look like asexual zoospores (Lerche 1937, Teodoresco 1905, Figure 3). For example, reduction of salinity from 10% to about 3% (Lerche 1937) or simply extreme dilution through the addition of distilled water (Teodoresco 1906) resulted in the induction of gametogenesis and production of isogametes. The procedure described by Lerche (1937) was reproduced in this laboratory, and copulation of gametes was observed after 2-4 days. The reverse treatment, changing from optimal salt conditions to concentrated solutions of $MgSO_4$, also induced gametogenesis, with copulations observed after two days (Teodoresco 1906). Copulation and zygote formation were additionally detected in older cultures, indicating that nutrient deficiency is a factor for the induction of gametogenesis (Lerche 1937, Polle unpublished data, Figures. 3, 4, 5).

According to Lerche (1937), upon reducing the salinity of older cultures, group formation of an average of 60-100 isogametes of *D. salina* was observed after about two days. These groups may stay together for several hours, and from these groups, pairs emerge. Combining gametes of green and red cell types, it was shown that the species of *D. salina* is heterothallic, and two different, specific pheromones were found (Lerche 1937). This technique using red and green cell types of different *D. salina* strains can also be applied to follow the conjugation of gametes from isolates of different origin (Figure 4). The majority of Lerche's (1937) *D. salina* clones were heterothallic. This heterothallic behavior of *D. salina* strains was also reported by Borowitzka and Siva (2007). Lerche (1937) additionally reported that one isolate of *D. salina* had a more or less pronounced tendency to behave homothallically, which was further corroborated by the findings of Gonzalez et al. (2001) that

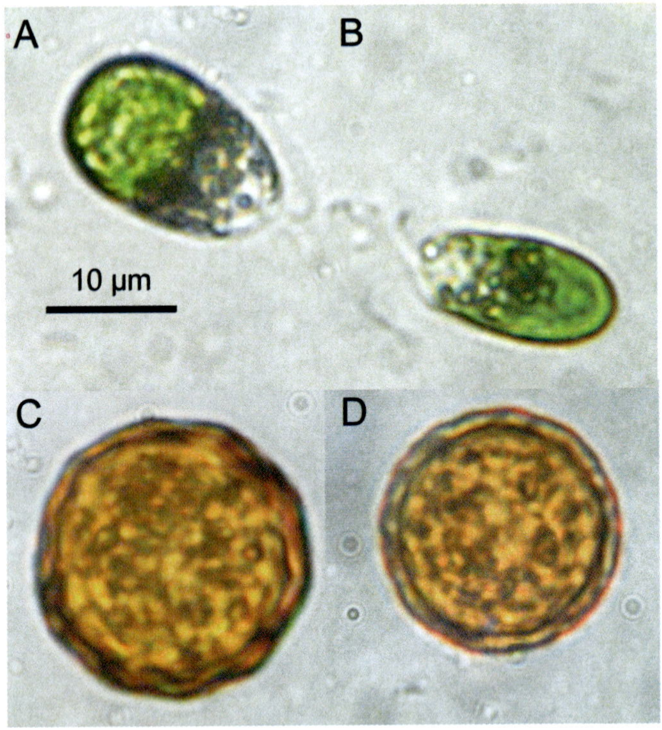

Figure 3: Photographs of cells from an old batch culture of a unialgal strain of *D. salina* (360x magnification). Motile cells in the culture were always green. Orange-red cells of various sizes were always non-motile and represent possible zygotes. A) Example of a green zoospore. B) Possibly a gamete. C) and D) Possibly zygotes, found in old batch cultures of the strain *D. salina*.

some isolates of *D. salina* are homothallic. In contrast to *D. salina*, the sexual reproduction of other species of *Dunaliella* appeared to be clearly homothallic: *D. parva* (Lerche 1937, Gonzalez et al. 2001), *D. peircei* (Lerche 1937), *D. euchlora* (Lerche 1937), *D. minuta* (Lerche 1937), *D. tertiolecta* (Gonzalez et al. 2001). Clone cultures of *D. viridis* Teod. have been observed to form zygotes, thus indicating that this species is also homothallic (Polle, unpublished data).

When pairs of gametes emerge from group formations, they begin to fuse at the anterior, and fusion may then progress along the sides of the two gametes. This process was described in detail for *D. salina* and *D. viridis* by Teodoresco (1905, 1906), Labbé (1925), and Lerche (1937). Such progression of gamete fusion is

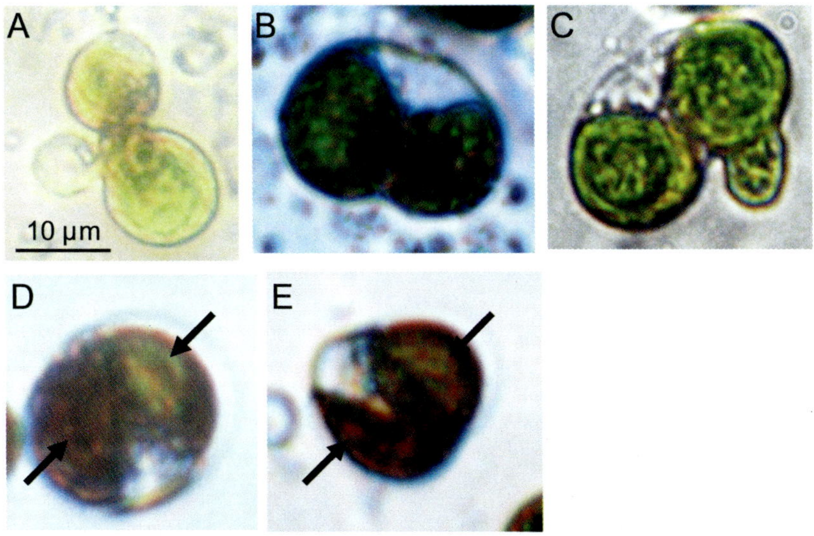

Figure 4: Shown are photographs of gametes of *Dunaliella salina* (360x magnification) fusing from the anterior ends (A & B). However, often numerous abnormal forms of fusing gametes were observed (C). Panels E and F demonstrate that it was possible to mate gametes originating from different strains. Gametes from various strains had different levels of carotenoids resulting in distinguishable colors of their chloroplasts. Arrows point out the two different chloroplasts within the early developing zygote.

similar to that found in algae of the genus *Chlamydomonas*, though it appears to be only one possible mode of fusion with respect to *Dunaliella* algae, since gametes of *Dunaliella* can also fuse from the posterior ends (Teodoresco 1905, 1906). Fusion from the posterior ends appears to occur predominately in older cultures by gametes that have developed one or two flagellum-like extensions that seem to aid in gamete adhesion and subsequent fusion (Polle, unpublished observation). Regardless of the mode, the fusion of gametes of *D. salina* takes at least one hour to complete. By contrast, *D. viridis* fusion may be as fast as 10 minutes although it has been known to take much longer (Teodoresco 1905, Labbé 1925, Lerche 1937). The fusion process of two isogametes is shown in Figures 4 and 5. It is worth noting that numerous abnormal forms of fusing gametes were often observed (Teodoresco 1905, 1906, Lerche 1937, Polle, unpublished results).

After a zygote forms it may begin immediate germination (Teodoresco 1905), as exemplified by *D. spec.* SAG19-5 in Figure 5, or it may continue to grow and go into a resting state with maturation that includes the development of red

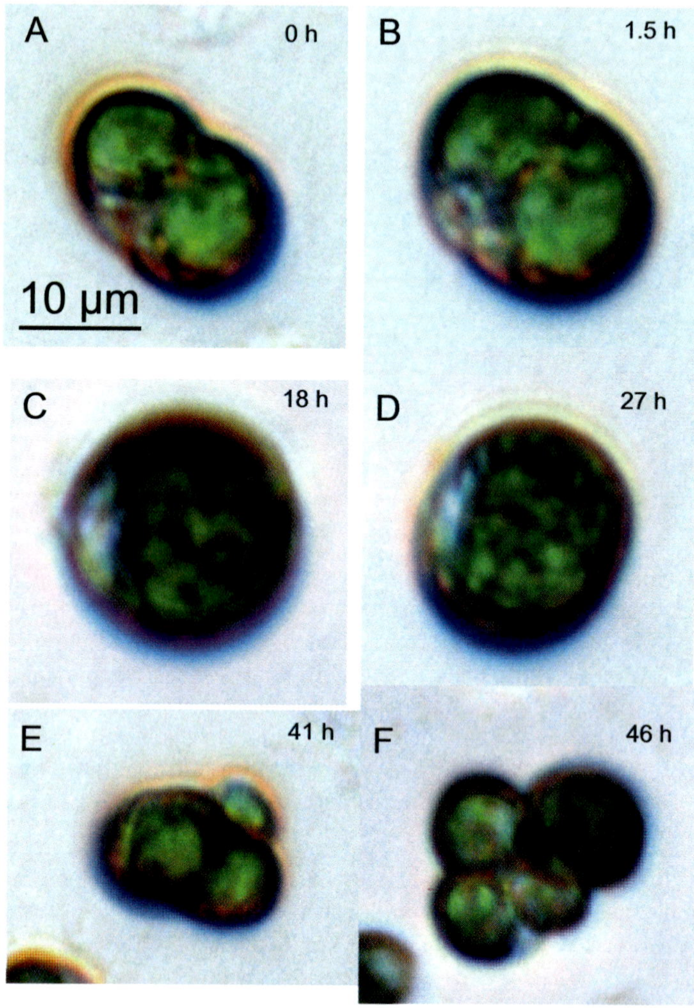

Figure 5: Shown are photographs of cells depicting the sequence of zygote formation and zygote germination for *Dunaliella spec.* SAG19-5 (400x magnification). A) Fusion of two gametes. B) Progressing fusion. C) Zygote. D) Zygote, beginning stage of germination. E) Formation and beginning of release of daughter cells. F) Zygote germination with release of four daughter cells.

coloration (Lerche 1937). All zygotes have a smooth outer endomembrane which may appear deformed due to external factors (Lerche 1937). An example of a possible mature zygote is shown in Figure 3.

In the sexual cell cycle, meiosis happens during zygote germination, which, in the case of *D. salina,* results in up to 16 cells. For other species, the number of germinating progeny varies from 1-32 cells (Lerche 1937). A germination result with fewer than four progeny strongly suggests that meiosis was not necessarily completed by the time germination began.

Recently, using Lerche's (1937) recorded procedure, the complete sexual cycle of *Dunaliella spec.* SAG19-5 was observed including gametogenesis, mating, zygote formation, and of germination (Figure 5). After reducing the salinity of a culture grown under continuous illumination of approximately 50 μmol photons m^{-2} s^{-1} from 1.5 M NaCl to 0.5 M NaCl, conjugation of isogametes was detected within 2–3 days. Figure 5 shows *D. spec.* SAG19-5 and its progression of conjugation, from the fusion of two isogametes to subsequent zygote formation and zygote germination. Similar processes were also observed for strains of *D. salina*, *D. viridis*, and *D. tertiolecta*. Figure 4 shows the formation of zygotes from the fusion of two gametes of different *D. salina* strains. In the early stages of zygote formation, the different gamete origin of each chloroplast can be clearly determined according to the choroplast's color.

Tool Development for Use in Genetics

Further investigation of the nuclear area and number of chromosomes for *Dunaliella* species is under way. Figure 6 shows a living cell of *D. salina* in which DNA was stained with SYBR I Green. Fluorescent dyes that can be used on living cells will improve future studies on chromosome numbers and life cycles of *Dunaliella* species. One question that still remains open concerns the number of chromosomes existing in various *Dunaliella* species. For many species, answers will be provided by new studies using DNA stains. In the particular case of the species *D. salina*, the ongoing genome sequencing project will answer this question once all genes are assigned to independent scaffolds.

4. Development of Transformation Systems

The genetic engineering of microalgae is not trivial, yet it is clear that microalgae will be used in the future as bio-refineries (Franklin and Mayfield 2004, Leon-Banares et al. 2004). In anticipation of this, various groups of researchers have been working since the mid-1990s to establish transformation systems for the alga *Dunaliella*. But as stated in reports on various microalgae species, in order for successful transformation to occur, several factors need to be considered (Walker et al. 2005a, Walker et al. 2005b). Some of the factors are listed below in reference to the alga *Dunaliella*.

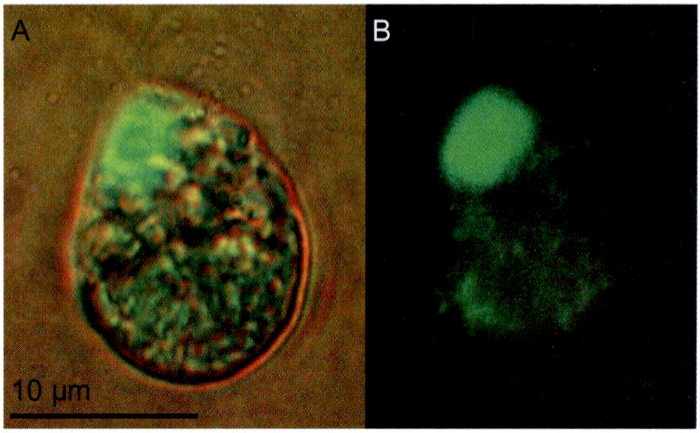

Figure 6: Shown are photographs of A) a living *D. salina* cell, and B) the same cell stained with SYBR Green I (400x magnification). Through green fluorescence the nucleus is visible in the anterior part of the cell. Red autofluorescence originating from chlorophyll indicates the area occupied by chloroplast. Note, that various areas in the chloroplast are fluorescing green indicating nucleoids.

Choice of Species

In contrast to many algae species that have thick and resistant cell walls, euryhaline species of the genus *Dunaliella* are covered by an elastic 'Hautschicht' (Teodoresco 1905, Hamburger 1905) which is made up of proteins (Teodoresco 1905, Ginzburg 1987). As the 'Hautschicht', also called 'glycocalyx', does not contain any cellulose, cells of *Dunaliella* should theoretically be easy objects for delivery of DNA in transformation experiments. Nevertheless, since the first report by Porath et al. (1997), no highly efficient system of transformation has been established for any *Dunaliella* species.

Although numerous reports exist on the transformation of the alga *Dunaliella*, there is a concern with many of the reports about the authenticity of the strains/species used. So far, the species *D. salina*, *D. tertiolecta*, and *D. viridis* have been of primary concern. A second factor to be considered is that these three species have different morphologies (e.g., cell size) and physiologies (e.g., optimal salinities for growth). As a consequence, conditions established for transformation of one species may not be appropriate for other *Dunaliella* species. This chapter reviews the current state of reports on transformation of *Dunaliella* algae.

Methods of Delivery of DNA into Cells

In recent years, three different methods were reported to have been used in the nuclear transformation of *Dunaliella* cells:

1. Glass Bead Method (Jin et al. 2001),
2. Electroporation (Geng et al. 2003, Geng et al. 2004, Sun et al. 2005, Sun et al. 2006, Walker et al. 2005c),
3. Micro-projectile Bombardment [= biolistic] (Jiang et al. 2005, Tan et al. 2005).

To the best knowledge of the authors, only Pan et al. (2004) to date has published information on the development of transformation systems for the plastid genome of *Dunaliella* species. Furthermore, no published report exists at this time on the transformation of mitochondrial genomes for species of the genus *Dunaliella*.

Use of Marker Genes

Three major types of marker genes were reportedly used in attempts to transform cells of different *Dunaliella* species (also see Table 2):

1. Antibiotic resistance:
   Several groups used antibiotic resistance genes originating from bacteria. One plasmid used by several authors (Jin et al. 2001, Porath et al. 1997, Sun et al. 2005) is pSP124S which carries the *Streptoalloteichus hindustanus* gene conferring bleomycin (ble) resistance, thus allowing for selection of transformants. pSP124S was developed and successfully used for the transformation of the unicellular green alga *Chlamydomonas reinhardtii* (Lumbreras et al. 1998). Walker et al. (2005c) used a modified *ble* gene flanked by endogenous *D. tertiolecta RbcS1* 5'and 3'UTR. In addition, another selectable antibiotic marker, the chloramphenicol acetyltransferase (*cat*) gene, was used successfully for transformation (Geng et al. 2004). A number of different antibiotics were also considered and tested for use in transformation systems of *Dunaliella* cells (Geng et al. 2001, Geng et al. 2003, Tan et al. 2005, Walker et al. 2005a,b): Paromomycin, kanamycin, hygromycin B, geneticin (=G4418), streptomycin, spectinomycin, and chlorsulfuron. However, except for bleomycin (= zeocin) and chloramphenicol, the antibiotics tested appeared impractical for use in the transformation of *Dunaliella* species.

Table 2: Marker genes used in transformation of *Dunaliella*.

Gene	Selection for	Plasmid	Reference
Cat	Antibiotic - Chloramphenicol	pUΩHBsAg-CAT	Geng et al. 2004
Ble	Antibiotic - Zeocin	pSP124S	Jin et al. 2001, Porath et al. 1997, Sun et al. 2005, Walker et al. 2005
Bar	Herbicide - Basta (Phosphinothrycin)	p35S-bar	Tan et al. 2005
Nia1	Nitrate Reductase for Functional Complementation -	pDVNR	Sun et al. 2006

2. Herbicide resistance:
   In addition to antibiotics as selectable markers, recently the bialaphos resistance gene (*bar*) from *Streptomyces hygroscopicus* was successfully tested for transformation of *Dunaliella*, allowing use of the herbicide Basta (= phosphinothrycin) for selection (Jiang et al. 2005, Tan et al. 2005). Figure 7 shows three lines of Basta-resistant strains of *Dunaliella spec.* that were generated by biolistic transformation.

3. Functional complementation:
   More recently, Sun et al. (2006) demonstrated use of the endogenous Nitrate Reductase gene (*DvNIA1*) for the transformation of *D. viridis*.

Transient Expression

One step in developing transformation systems is the expression of the gene of interest in the selected organism. To date, various reporter gene systems exist that have a clear and measurable activity. The *GUS* reporter gene system, for example, was used by different groups to demonstrate transient expression of a foreign gene in cells of the alga *Dunaliella* (Geng et al. 2002, Tan et al. 2005). Use of the *GUS* reporter gene system yielded transient expression in cells of *Dunaliella* (Geng et al. 2002, Tan et al. 2005). Figure 7 shows cells of *Dunaliella spec.* that were transiently transformed using the *GUS* reporter gene system. In contrast to the *GUS* system, according to Tan et al. (2005) the *lac Z* reporter gene system (β-galactosidase) and the *EGFP* system (Li et al. 2003) using a fluorescent protein did not prove to be appropriate reporter systems for *Dunaliella*.

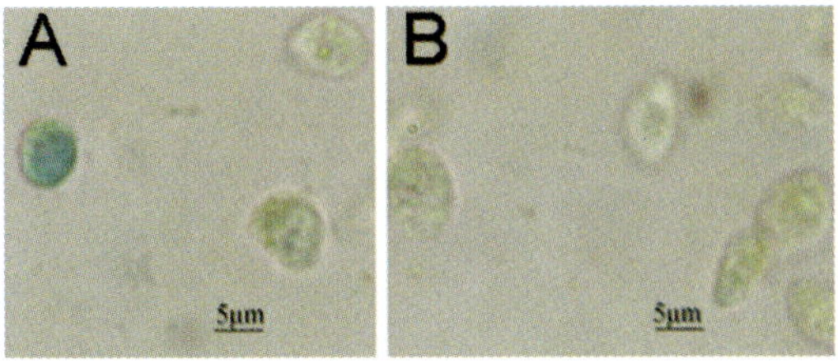

Figure 7: Histochemical staining for GUS activity 48 h after bombardment (magnification: 15 × 40). Reproduced with permission from Tan et al. 2005: (A) Transient GUS expression in transformed *D. salina* cells, blue cells represent the transformants. (B) Negative controls, cells were bombarded with uncoated gold particles.

Transformation Efficiency

In recent years various publications reported successful transformation of the alga *Dunaliella*. Unfortunately, among those reports, the efficiency of transformation was not always calculated (Tan et al. 2005, Sun et al. 2006, Walker et al. 2005c). Sun et al. (2005) does, however, state that using the method of electroporation, 0-2 colonies per 10^7 cells were obtained on agarose plates. A higher transformation efficiency is described by Geng et al. (2003) and Geng et al. (2004) who also used the method of electroporation and claim that 10-60 colonies per $0.4x10^6$ cells were recovered when 3µg of DNA were used. This efficiency is still orders of magnitude lower than what was recently obtained by Shimogawara et al. (1998) where a transformation efficiency of $2x10^5$ transformants per 1µg of DNA was demonstrated for the unicellular green alga *Chlamydomonas reinhardtii*.

Confirmation of Integration and Expression of Foreign DNA

From the 1990s onward, various attempts were made to establish a reliable transformation system for algae of the genus *Dunaliella*. Results from transient expression experiments, for example, clearly demonstrated that foreign DNA was delivered into and successfully expressed by cells (Geng et al. 2002, Tan et al. 2005, Figure 7). Other groups used PCR to demonstrate that products specific to foreign DNA could be amplified from DNA that was isolated from *Dunaliella* cells subsequent to transformation (Tan et al. 2005). Although various reports about successful transformation of *Dunaliella* cells exist, Geng et al. (2004) was the only group demonstrating the integration of foreign DNA into the genome by using the Southern Blot method with genomic DNA. The group was also able to show that PCR products were obtained by amplifying fragments of the HBsAg protein and the CAT protein. In addition, Geng et al. (2004) detected the exogenous protein in *Dunaliella*.

Unraveling the Genomes of *Dunaliella* Species

Only at the start of the new millennium were genomics used to improve understanding of stress responses in algae of the genus *Dunaliella*. Among past and current *Dunaliella* genomics projects, at least three independent research groups have been investigating stress genomics. For example, Expressed Sequence Tags (ESTs) were generated from a cDNA library of light-stressed *Dunaliella spec.* (Park et al. 2006). Out of 1,112 ESTs that were randomly chosen for sequencing, about 75% were unique. Among the unique ESTs, about 25% of the sequences could not be classified into functional gene categories (Park et al. 2006). Another project generated ESTs from salt-stressed *D. salina* (Cushman J.C., NCBI unpublished and http://www.stress-genomics.org). According to data for Stress-Genomics on the PipeOnline database, as of January 2007, about 3,204 clones of *D. salina* have been analyzed which has resulted in 2008 contigs (Ayoubi et al. 2002). Again, about one-third of these contigs could not be assigned to a functional gene category. A third project generated ESTs from *D. salina* exposed to high salinity and low nitrogen, but functional data is not yet available (Baumgartner et al., NCBI unpublished). Taken together, the above results suggest that about 30% of the genome of the alga *D. salina* will contain proteins with no homology to known proteins or functions.

In addition to the above-mentioned EST projects, the genome of *D. salina* strain CCAP19/18 is being sequenced by the Joint Genome Institute in California, USA. As part of the genome sequencing project for *D. salina*, genomic and EST data will be generated. Having cDNA and genomic sequence information available will allow the development of high-density microarrays that can be employed for gene expression profiling (Eberhard et al. 2006). Further, the specific generation of a microarray based on oligonucleotides will allow analysis of the transcriptome of *D. salina* by monitoring global gene expression in response to changes in environmental conditions (Cuming et al. 2007, Müller et al. 2007, Wullschleger and Difazio 2003, Zhang et al. 2004). The combination of transcriptome analysis and other methods such as mutational analysis may further allow identification of genes functioning in unique metabolic pathways necessary for life in hyper-saline environments. In addition, DNA tiling arrays may be constructed to permit analysis of novel transcribed sequences and regulatory elements (Bertone et al. 2005).

Development of a Molecular Tool Kit

Although algae of the genus *Dunaliella* have been used as model organisms to study a variety of physiological responses, currently no repertoire of molecular tool kits exists. As reviewed above, reliable transformation systems for the nuclear genome are still in their infancy. Still lacking, too, are systems for transformation of the plastid and mitochondrial genomes.

With the genome of *D. salina* being sequenced, the development of a variety of molecular tools and resources can be anticipated. The molecular tool kit that has already been developed for the green alga *Chlamydomonas reinhardtii* (Eberhard et al. 2006, Grossman et al. 2003) can be used as an example, not only for what is achievable, but also for techniques that can be adapted for use with *D. salina*. It is necessary to develop, for example:

- A reliable genetic system to perform crosses;
- A genetic map with phenotypic and molecular markers;
- Transformation systems for the chloroplast and mitochondria;
- RNAi techniques;
- A tiled BAC library corresponding to the complete genome;
- Microarrays.

Summary

It may not be appropriate to regard *Dunaliella* as one model alga, but rather to acknowledge several individual species, such as *D. salina*, *D. tertiolecta*, and *D. acidophila*, as separate model systems that are useful in studies concerning acclimation to abiotic stress. Any species used as a model organism, however, needs genetics and molecular tools that can be applied on a routine basis. New studies of genetics involving *Dunaliella* species, for example, have been relatively limited, and since Lerche (1937), only a few reports have been presented on the life cycle of *D. salina* (reviewed in Borowitzka and Siva 2007). So while gametogenesis can be induced and zygotes can be generated, difficulties, such as the induction of zygote germination and the handling of fragile progeny cells, still need to be overcome. And although recently several research groups have demonstrated independently of one another that cells of *Dunaliella* species can be transformed, a stable and highly efficient transformation system still needs to be developed.

The ongoing genome sequencing project for the species *D. salina* in conjunction with the development of reliable transformation techniques should provide an additional basis for the development of new tools to increase the value of *Dunaliella* species as model algae.

Acknowledgements

The authors thank Ms. M. Aksoy for providing the photographs in Figure 1 and Figure 6. The authors thank Mr. A. Arif for technical assistance with *Dunaliella* genetics including Figure 4.

References

Ayoubi, P., X. Jin, S. Leite, X. Liu, J. Martajaja, A. Abduraham, Q. Wan, W. Yan, E. Misawa and R.A. Prade. 2002. PipeOnline 2.0: Automated EST processing and functional data sorting. Nuc. Acids Res. 30 (21): 4761-4769.

Ben-Amotz, A. and M. Avron. 1990. The biotechnology of cultivating the halotolerant alga *Dunaliella*. Trends Biotechnol. 8 (5): 121-126.

Ben-Amotz, A. and M. Avron. 1992. *Dunaliella*: Physiology, Biochemistry, and Biotechnology. CRC Press, Inc. Boca Raton, FL, USA.

Ben-Amotz, A., A. Shaish and M. Avron. 1991. The biotechnology of cultivating *Dunaliella* for production of β-carotene rich algae. Biores. Technol. 38 (2-3): 233-235.

Bertone, P., M. Gerstein and M. Snyder. 2005. Applications of DNA tiling arrays to experimental genome annotation and regulatory pathway discovery. Chr. Res. 13 (3): 259-274.

Borowitzka, L.J. 1991. Development of Western Biotechnology's algal β-carotene plant. Biores. Technol. 38 (2-3): 251-252.

Borowitzka, M.A. 1999. Commercial production of microalgae: ponds, tanks, tubes and fermenters. J. Biotechnol. 70 (1-3): 313-321.

Borowitzka, M.A. and J.M. Huisman. 1993. The ecology of *Dunaliella salina* (Chlorophyceae, Volvocales)-Effect of environmental conditions on aplanospore formation. Bot. Mar. 36: 233–243.

Borowitzka, M.A. and C.J. Siva. 2007. The taxonomy of the genus *Dunaliella* (*Chlorophyta, Dunaliellales*) with emphasis on the marine and halophilic species. J. Appl. Phycol. 19 (5): 567-590.

Borowitzka, M.A., F. Post and L.J. Borowitzka. 1982. The life cycle of *Dunaliella salina*. Austral. Soc. Phycol. Aquatic Bot., Annual Meeting Abstracts. p. 6.

Cavara, F. 1906. Alcune osservazione sulla *Dunaliella salina* (Dun.) Teod. delle saline di Cagliari. Rend. della R. Acad. delle Sc. Fis. e Matemat. di Napoli Ser. 3: 12.

Chitlaru, E. and U. Pick. 1991. Regulation of glycerol synthesis in response to osmotic changes in *Dunaliella*. Plant Physiol. 96 (1): 50–60.

Cuming, A.C., S.H. Cho, Y. Kamisugi, H. Graham and R.S. Quatrano. 2007. Microarray analysis of transcriptional responses to abscisic acid and osmotic, salt, and drought stress in the moss, *Physcomitrella patens*. New Phytol. 176 (2): 275-287.

Davis, R.H. 2004. The age of model organisms. Nat. Rev. Gen. 5 (1): 69-76.

Eberhard, S., M. Jain, CS. Im, S. Pollock, J. Shrager, Y. Lin, A.S. Peek and A. Grossman. 2006. Generation of an oligonucleotide array for analysis of gene expression in *Chlamydomonas reinhardtii*. Curr. Gen. 49: 106-124.

Franklin, S.E. and S.P. Mayfield. 2004. Prospects for molecular farming in the green alga *Chlamydomonas reinhardtii*. Curr. Opinion Plant Biol. 7:159-165.

Geng, D.-G., Y.-Q. Wang, W.-B. Li and Y.-R. Sun. 2001. Study on selective marker of *Dunaliella salina* genetic engineering. Bio/Technol. 11: 1-3.

Geng, D.-G., Y.Q. Wang, W.B. Li and Y.R. Sun. 2002. Transient expression of GUS gene in *Dunaliella salina*. High Tech. Lett. 12: 35-39.

Geng, D.-G., Y.Q. Wang, P. Wang, W.B. Li and Y.R. Sun. 2003. Stable expression of hepatitis B surface antigen gene in *Dunaliella salina*. J. Appl. Phycol. 15: 451-456.

Geng, D.-G, Y. Han, Y. Wang, P. Wang, L. Zhang, W. Li and Y. Sun. 2004. Construction of a system for the stable expression of foreign genes in *Dunaliella salina*. Acta Bot. Sin. 46: 342-346.

Ginzburg, M. 1987. *Dunaliella*: A green alga adapted to salt. Adv. Bot. Res. 14: 93-183.

Gonzalez, M.A., A.W. Coleman, P.I. Gomez and R. Montoya. 2001. Phylogenetic relationship among various strains of *Dunaliella* (Chlorophyceae) based on nuclear ITS rDNA Sequences. J. Phycol. 37: 604-11.

Goyal, A. 2007. Osmoregulation in *Dunaliella*, Part II: Photosynthesis and starch contribute carbon for glycerol synthesis during a salt stress in *Dunaliella tertiolecta*. Plant Physiol. Biochem. 45 (9): 705-710.

Grossman, A.R., E.E. Harris, C. Hauser, P.A. Lefebvre, D. Martinez, D. Rokhsar, J. Shrager, C.D. Silfow, D. Stern, O. Vallon and Z. Zhang. 2003. *Chlamydomonas reinhardtii* at the crossroads of genomics. Eukaryot. Cell 2 (6): 1137-1150.

Hamburger, C. 1905. Zur Kenntnis der *Dunaliella salina* und einer Amöbe aus Salinenwasser von Cagliari. Arch. f. Protistenkd. 6: 111-130.

Hamel, G. 1931 Chlorophycées des côtes françaises. Extr. Rev. Algol. I-V: 4-6.

Jiang, G., Y. Lu, X. Niu and L. Xue. 2005. The actin gene promoter-driven bar as a dominant selectable marker for nuclear transformation of *Dunaliella salina*. Acta Gen. Sin. 32: 424-433.

Jin, E.S., J.E.W. Polle and A. Melis. 2001. Involvement of zeaxanthin and of the Cbr protein in the repair of photosystem II from photoinhibition in the green alga *Dunaliella salina*. Biochim. Biophys. Acta 1506: 244-259.

Katz, A., P. Waridel, A. Shevchenko and U. Pick. 2007. Salt-induced changes in the plasma membrane proteome of the halotolerant alga *Dunaliella salina* as revealed by blue native gel electrophoresis and Nano-LC-MS/MS analysis. Mol. Cell. Prot. 6 (9): 1459-1472.

Labbé, A. 1925. 1er Mémoire: Les cycles biologiques des *Dunaliella*. Arch. d'Anat. Microsc. T. XXI: 314-399.

León-Banares, R., D. González-Ballester, A. Galván and E. Fernández. 2004. Transgenic microalgae as green cell-factories. TRENDS in Biotechnol. 22 (1): 45-52.

Leonardi, P.I. and E.J. Cáceres. 1997. Light and electron microscope observations of the life cycle of *Dunaliella salina* (Polyblepharidaceae, Chlorophyceae). Nova Hedw. 64:621–633.

Lerche, W. 1937. Untersuchungen über Entwicklung und Fortpflanzung in der Gattung *Dunaliella*. Arch. Protistenkd. 88: 236-268.

Li, J., H. Ren, J. Huang, M. Yu and B. Ru. 2003. Transient expression of reporter gene EGFP in transformed *Dunaliella salina*. Chinese J. Biochem. Mol. Biol. 19: 338-342.

Liebetanz, B. 1925. Studja hydrobiologiczne solanek KUJAWSKICH. Bull. Intern. De l'Acad. Polonaise des Sc. Et des Lettres 1-2: 72-94.

Loeblich, L.A. 1969. Aplanospores of *Dunaliella salina* (*Chlorophyta*). J. Protozool. Suppl. 16: 22.

Lumbreras, V., D.R. Stevens and S. Purton. 1998. Efficient foreign gene expression in *Chlamydomonas reinhardtii* mediated by an endogenous intron. The Plant J. 14 (4): 441-447.

Massjuk, N.P. 1973. New taxons from the genus *Dunaliella* Teod., I. Ukr. Bot. Zh. 30: 175.

Moulton, T.P, L.J. Borowitzka and D.J. Vincent. 1987. The mass culture of *Dunaliella salina* for b-carotene: From pilot plant to production plant. Hydrobiol. 151/152: 99-105.

Müller, R., M. Morant, H. Jarmer, L. Nilsson and T.H. Nielsen. 2007. Genome-wide analysis of the arabidopsis leaf transcriptome reveals interaction of phosphate and sugar metabolism. Plant Physiol. 143 (1): 156-171.

Nicolai, E. and L.G.M. Baas-Becking. 1935. Einige Notizen über Salzflagellaten. Arch. Protistenkd. 85: 319-328.

Oliveira, L., Bisalputra, T. and N.J. Antia. 1980. Ultrastructural observation of the surface coat of *Dunaliella tertiolecta* from staining with cationic dyes and enzyme treatments. New Phytol. 85(3): 385-392.

Pan, W., Y. Lu, G. Zhang, X. Niu, G. Hou and L. Xue. 2004. Transcription activities of the atpA promoter of *Chlamydomonas reinhardtii* chloroplast in *Dunaliella salina*. J. Zhengzhou University (Medical Sciences) 39: 35-38.

Park, S., J.E.W. Polle, A. Melis, T.K. Lee and E.S. Jin. 2006. Up-regulation of photoprotection and PSII-Repair gene expression by irradiance in the unicellular green alga *Dunaliella salina*. Mar. Biotechnol. 8 (2): 120-128.

Pick, U. 1998. *Dunaliella* - A model extremophilic alga. Israel J. Plant Sci. 46 (2): 131-139.

Porath, J.B., I. Gokham, A. Lers, H. Levy and A. Zamir. 1997. Developing a transformation system for *Dunaliella*. Phycologia 36 (4 Suppl.): 89.

Raja, R., S. Hemaiswarya and R. Rengasamy. 2007. Exploitation of *Dunaliella* for β-carotene production. Appl. Microbiol. Biotechnol. 74 (3): 517-523.

Samanamud, A. 1998. Crecimiento e historia de vida de *Dunaliella salina* de las salinas de los Chimus, Ancash y de Chilca, Lima, Peru. pp. 309-324. In Anais do IV Congresso Latino-Americano, II Reuniao Iberio-Americana, VII Reunido Brasileira, de Ficologia. Volume II. E.J. de Paula, M. Corediro-Marino, D.P. Santos, E.M. Plastino, M.T. Fujii and N.S. Yokoya (eds). Sociedade Ficologica de America Latina e Caribe and Sociatade Brasiliera de Ficologia, Caxambu, Brazil.

Shaish, A., A. Ben-Amotz, and M Avron. 1992. Biosynthesis of β-carotene in *Dunaliella*. Meth. Enzymol. 213: 439-444.

Shimogawara, K., S. Fujiwara, A. Grossman and H. Usuda. 1998. High-Efficiency Transformation of *Chlamydomonas reinhardtii* by Electroporation. Genetics 148: 1821–1828.

Sun, Y., Z. Yang, X. Gao, Q. Li, Q. Zhang and Z. Xu. 2005. Expression of foreign genes in *Dunaliella* by electroporation. Mol. Biotechnol. 30: 185-192.

Sun, Y., X. Gao, Q. Zhang and Z. Xu. 2006. Functional complementation of a nitrate reductase defective mutant of a green alga *Dunaliella viridis* by introducing the nitrate reductase gene. Gene 377: 140-149.

Tan, C, S. Qin, Q. Zhang, P. Jiang and F. Zhao. 2005. Establishment of a micro-particle bombardment transformation system for *Dunaliella salina*. J. Microbiol. 43: 361-365.

Teodoresco, E.C. 1905. Organization et developpement du *Dunaliella*, nouveau genre de Volvocacee-Polyblepharidee. Beih. Bot. Zentralblatt. Bd.18 Abt.1: 215-232.

Teodoresco, E.C. 1906. Observations morphologiques et biologiquessur le genre *Dunaliella*. Rev. Gén. d. Bot. T18: 353-371.

Thompson Jr., G.A. 1994. Mechanisms of osmoregulation in the green alga *Dunaliella salina*. J. Exp. Zool. 268 (2): 127-132.

Walker, T.L., S. Purton, D.K. Becker and C. Collet. 2005a. Microalgae as bioreactors. J. Plant Cell Reports 24 (11): 629-641.

Walker, T.L., C. Collet and S. Purton. 2005b. Algal transgenics in the genome era. J. Phycol. 41 (6): 1077–1093.

Walker, T.L., D.K. Becker, J.L. Dale and C. Collet. 2005c. Towards the development of a nuclear transformation system for *Dunaliella tertiolecta*. J. Appl. Phycol. 17 (4): 363-368.

Wullschleger, S.D. and S.P. Difazio. 2003. Emerging use of gene expression microarrays in plant physiology. Comp. Funct. Gen. 4 (2): 216-224.

Zhang, Z., J. Shrager, M. Jain, C.W. Chang, O. Vallon and A.R. Grossman. 2004. Insights into the survival of *Chlamydomonas reinhardtii* during sulfur starvation based on microarray analysis of gene expression. Eukaryot. Cell 3: 1331-1348.

18 Application of the Colorless Carotenoids, Phytoene, and Phytofluene in Cosmetics, Wellness, Nutrition, and Therapeutics

Liki von Oppen-Bezalel[1] and Aviv Shaish[2]*

Abstract

Phytoene and phytofluene are the colorless carotenoids precursors in the biosynthetic pathway of colored carotenoids and absorb electromagnetic irradiation in the UV range only. The colorless carotenoids are found in most carotenogenic organisms, including the unicellular alga *Dunaliella*. These molecules have effective and benefiting activities both for health protection and nutritional and beauty applications, mostly in protection against UV and oxidative damage leading to premature aging and other disorders. The colorless carotenoids have been shown to have anti-inflammatory activity along with stabilization, protection, and enhancement of bio-molecules. Effects on skin pigmentation were demonstrated recently.

The following chapter collects and summarizes available information on the use and potential applications of phytoene and phytofluene, as well as the use of *Dunaliella* under defined growth conditions in cosmetics, wellness, nutrition, and therapeutics.

Introduction

Biologically, carotenoids are an important group of compounds with more than 700 members. Carotenoids are found widely throughout nature, but they are synthesized *de novo* only by plants, algae, fungi, and bacteria, where they can aid in the absorption of light and the capture of additional energy, which helps neutralize

[1] IBR, Israeli Biotechnology Research, Ltd., Rehovot, Israel.
[2] The Bert W. Strassburger Lipid Center, The Chaim Sheba Medical Center, Tel-Hashomer, 52621, Israel.
* For correspondence: Likiob@yahoo.com

tissue-damaging free radicals (Bendich and Olson 1989). Chemically, carotenoids are isoprenoid, C-40 molecules that are either linear or cyclized at one or both ends of the molecule. The chemical structure determines their physicochemical properties and their biological activities.

Many dietary carotenoids, both with and without pro-vitamin A activity, are found in the blood and tissues of humans (Bendich and Olson 1989). Carotenoids, especially β-carotene and lycopene, were used as a treatment against photosensitization as early as 1964 (Mathews 1964). Since then, a vast number of studies have shown that carotenoids act as anti-oxidants, anti-inflammatory, and anti-mutagenic agents. Carotenoids are also believed to have the potential to inhibit certain cardiovascular diseases and cancer. In addition, many carotenoids show various beneficial immunological effects. Thus, carotenoids are of interest for protective applications.

Over the years, there has been an increased interest in dietary carotenoids and their contribution to prevention of and protection against age-related diseases, many of which are maturing too early, because of oxidative damage and stress.

Unfortunately, most carotenoids are sensitive to light, a property that considerably limits their use and shortens the shelf life of products that contain them (Krinski et al. 1989). In addition, almost all carotenoids have a distinctive color, which limits their utility for cosmetic and some food applications.

In contrast to all other carotenoids, phytoene and phytofluene (the precursors in the biosynthetic pathway of other, pigmented carotenoids) are colorless and absorb electromagnetic irradiation in the UV range only. Phytoene has been shown to be anti-carcinogenic in the mouse skin cancer model (Mathews-Roth 1982.), and experiments performed with phytoene-producing transgenic mice suggest that phytoene may play a role in cell-to-cell communication with respect to both cancer and general cell communication (Satomi et al. 2004). It is important to note that dietary phytoene and phytofluene, as well as lycopene, are accumulated in human skin (Khachik et al. 2002). The accumulation of these colorless carotenoids can potentially protect the skin by several mechanisms: acting as UV absorbers, as antioxidants, and, as recently suggested, by modulation of gene expression. The distribution of phytoene and phytofluene in human serum, prostate, and skin suggests that phytoene and phytofluene may be responsible for the carotenoid protective effects observed in epidemiological studies.

Although phytoene and phytofluene (also named colorless carotenoids or, in brief, CLC) are part of our diet (tomatoes and the unicellular alga *Dunaliella* sp., for example, accumulate the two carotenoids in addition to lycopene in tomato and β-carotene in *Dunaliella*) and can be detected in plasma, little is known about their tissue distribution and their role in the body.

The aim of this chapter is to describe *Dunaliella* under defined growth conditions leading to excessive synthesis of phytoene and phytofluene and to present available information on the use and potential applications of phytoene and phytofluene in cosmetics, wellness, nutrition, and therapeutics.

Phytoene and Phytofluene: Nomenclature, Biosynthesis, and Structure

Phytoene, 7,7',8,8',11,11',12,12'-octahydro-ψ,ψ-carotene ($C_{40}H_{64}$), is formed from two geranylgeranyl pyrophosphates (GGPP), C-20 compounds derived from four C-5 isoprene units (IPP). Two GGPP molecules are condensed together, followed by removal of diphosphate and proton shift, leading to the formation of phytoene (Figure 1). Phytoene (C40) is the colorless precursor carotenoid in the biosynthesis of all carotenoids and contains three conjugated double bonds in a symmetric molecule. In plants and many carotenogenic organisms, phytoene is accumulated as its 15-*cis* isomer in algae, such as *Dunaliella* sp., it is found to be rich with a mixture of two phytoene stereoisomers (Werman et al. 2002, Orset and Young 2000, Ebenezer and Pattenden 1993).

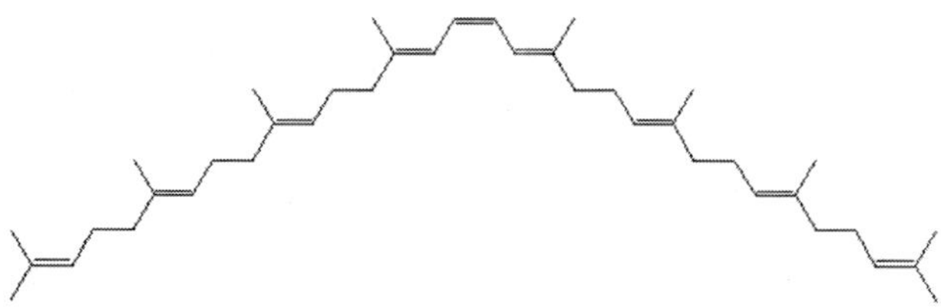

15-cis phytoene
15-*cis*-7,8,11,12,7',8',11',12'-octahydro-ψ,ψ-carotene

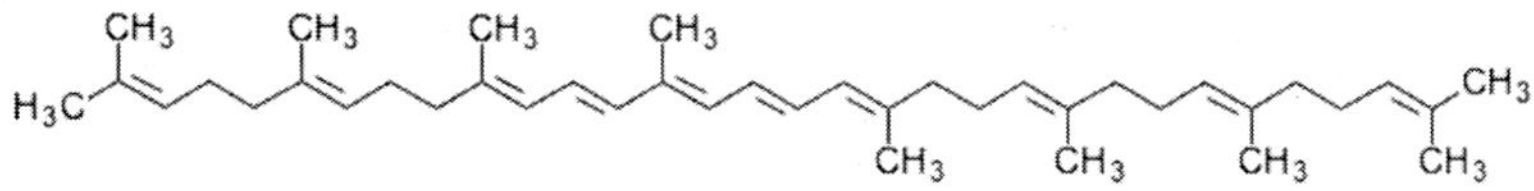

C05414
Phytofluene

Figure 1: Phytoene (a) and phytofluene (b) structure.

Phytofluene, 7,7',8,8',11,12-hexahydro-ψ,ψ–carotene ($C_{40}H_{62}$), is the next carotenoid formed from phytoene in a desaturation reaction, leading to the formation of five conjugated double bonds (Figure 1). In the following step, the addition of carbon-carbon conjugated double bond leads to the formation of ζ-carotene and the appearance of visible color.

Phytoene has typical absorption spectra in the UVB range, with maximum at 286 nm (Figure 2) and with $\varepsilon^{1\%}$ of 915. Phytofluene, with its additional conjugated double bond, has a typical absorption spectra shifted toward longer wave lengths (UVA), with maximal absorption at 348 nm (Figure 2) and with $\varepsilon^{1\%}$ of 1557.

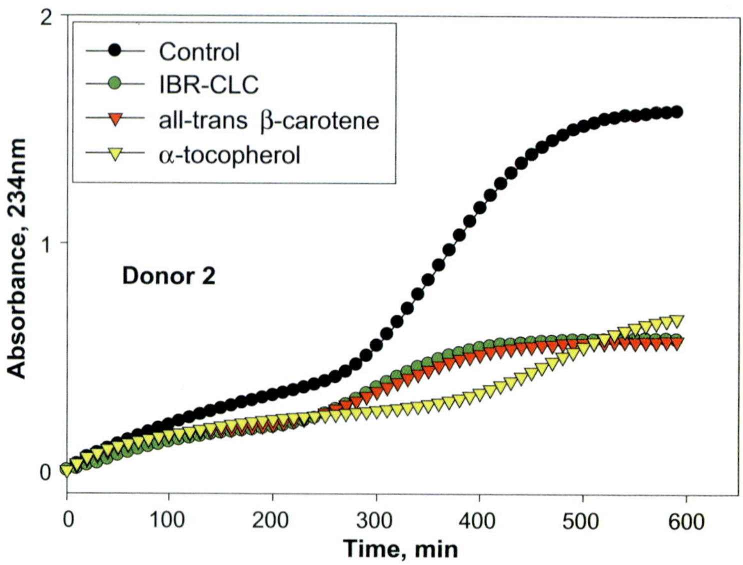

Figure 2: IBR-CLC®, phytoene and phytofluene from *Dunaliella sp.* inhibits Low Density Lipoprotein (LDL) oxidation *in vitro*. The experiment was repeated 3 times, and the results obtained with LDL isolated from donor #2 are presented.

Dietary Sources of Phytoene and Phytofluene

Inasmuch as phytoene, followed by phytofluene, are the basic precursors for all naturally forming carotenoids, phytoene and phytofluene and their various stereoisomers should be found in any carotenogenic organism and in any organism consuming carotenoids containing dietary sources (Khachik et al. 1991). Indeed, Khachik et al. analyzed the distribution of carotenoids in fruits and vegetables and showed that phytoene and phytofluene are found in the vast majority of these fruits and vegetables. Indication to the accumulation of carotenoids in general and phytoene and phytofluene in this case, in carotenoids consuming organisms such as humans one can find with the analysis and findings of carotenoids in skin, serum, plasma, lymphocytes and other organs like prostate (Hata et al. 2000, Campbell et al. 2005, Erdmann 2005, Aust et al. 2005, Bendich and Olson 1989. Ermakov et al. 2004, Porrini et al. 2005, Khachik et al. 1991, Fraser and Bramley 2004).

The algae *Dunaliella* sp. is a rich source for carotenoids as well, mainly β-carotene and its biosynthesis intermediates, such as phytoene, phytofluene, and ζ-carotene.

Phytoene and Phytofluene in *Dunaliella*

In many organisms, inhibitors of enzymes in the biosynthetic pathway of β-carotene lead to the accumulation of intermediates (Britton 1988). These inhibitors are helpful in studying the process of carotenoid biosynthesis. The effects of the herbicide norflurazon, which blocks the conversion of phytoene to phytofluene, were studied in *Dunaliella*.

Increasing norflurazon concentrations caused the accumulation of phytoene to levels of up to 8% of the dry weight and, simultaneously, a gradual decrease in β-carotene content (Ben-Amotz 1988). Similar to β-carotene, phytoene was accumulated in globules in the inter-thylakoids space within the chloroplast, and the ratio between the two phytoene stereoisomers was proportional to the integral irradiance to which they were exposed during a division cycle. From these data, it was concluded that the isomerization reaction, which leads eventually to the production of 9-*cis* and *all-trans* β-carotene stereoisomers, occurs early in the biosynthetic pathway, at or before the formation of phytoene (Ben-Amotz et al. 1988, Ben-Amotz et al. 1987). In further research, norflurazon was used at high and low concentrations. At 0.3 μM, norflurazone inhibited β-carotene accumulation and led to the accumulation of two stereoisomers of phytoene. However, when 0.1 μM was employed, both phytoene and phytofluene were detected. Thus, both the desaturation of phytoene to phytofluene and that of phytofluene to ζ-carotene were inhibited, depending on the concentration of norflurazon employed. In contrast to norflurazon, the inhibitor J-334, when used at high concentration (4,000 nM), led to the accumulation of phytoene, phytofluene and ζ-carotene, suggesting that, at this concentration, J-334 blocks the desaturation reaction of ζ-carotene to neurosporene (Shaish et al. 1990, Yamada et al. 2002).

Another approach to accumulate the intermediates in *Dunaliella* was to induce massive carotenoid production by nitrate or sulfate starvation or by exposure of the algae to high light intensity for a long period of time. Under these conditions, phytoene and phytofluene, as well as other intermediates and β-carotene, were accumulated, and two major peaks of both phytoene and phytofluene were identified after separation on HPLC. The two major stereoisomers of phytoene and phytofluene have not been identified by NMR clearly. However, because of the accumulation of 9-*cis* and *all-trans* β-carotene in *Dunaliella*, phytoene and phytofluene have been suggested to be 9-*cis* and *all-trans* stereoisomers (Werman et al. 2002)

Ebenezer and Pattenden, by NMR spectrum analysis, suggested that *Dunaliella bardawil* contains 15-*cis* phytoene and 9-*cis* phytofluene only (Ebenezer and Pattenden 1993). While Orset and Young identified the 15-cis phytoene only (Orset and Young 2000).

Bioavailability and Tissue Distribution of Phytoene and Phytofluene

Many dietary carotenoids, both with and without pro-vitamin A activity, are found in most human tissues, including in the serum, plasma, and skin (Aust et al. 2005, Bendich and Olson 1989, Ermakov et al. 2004, Porrini et al. 2005, Khachik et al. 1991, Fraser and Bramley 2004).

The distribution of carotenoids throughout the body is not uniform; phytoene and phytofluene are accumulated at higher rates than other carotenoids tested in the skin (Hata et al, 2000) and in other tissues such as lungs, plasma, and lymphocytes (Porrini et al. 2005). However, Campbell et al. (Campbell et al, 2005, Erdmann 2005) showed that lycopene tended to be accepted by the prostate of rats preferentially, and phytofluene and phytoene accumulated in concentrations higher than lycopene in the liver and the adrenal. After an oral dose with phytoene or phytofluene, there were enhanced levels of the respective carotenoids in serum and all tissues measured (Campbell et al. 2005). After dosing, the seminal vesicles and the various lobes of the prostate still remained highest in lycopene, in contrast to most other tissues, which were highest in the dosed phytoene and phytofluene (Campbell et al. 2005).

The effect of 26 days of daily intake of a lycopene-rich carotenoids preparation increased lycopene, phytoene, phytofluene, and β-carotene levels in plasma and lymphocytes. Phytoene underwent the greatest increases in both plasma and lymphocytes. The intake significantly reduced (~42%) oxidative stress-induced DNA damage in lymphocytes (Porrini et al. 2005). It is worth noting that phytoene had the highest accumulation factor in serum and lymphocytes (x 2.0 and x 2.6 respectively) and the most significant protection was observed which seems to be linked to the higher protective rate observed in the experimental group that consumed a mixture of carotenoids with phytoene and phytofluene as well in comparison to a lycopene only group. In an additional study, the plasma concentration of phytoene and phytofluene increased dramatically after the ingestion of tomato juice, oleoresin, or lycopene beadlets (Paetu et al. 1998).

Phytoene and phytofluene, the so-called colorless carotenoids, have the highest accumulation factor in tissues such as skin, plasma, serum, lymphocytes and more, compared to lycopene and β-carotene which could be an indication for better stability against detoxification enzymes and an advantage in phytoene and phytofluene's ability to prevent and protect against oxidation damage *in vivo* (Bartley and Scolnik 1995, Aust et al. 2005, Porrini et al. 2005, Erdman 2005, Campbell et al. 2005). After a rat's carotenoids intake studies, Erdmann concluded that all tomato carotenoids cannot be assumed to be absorbed and metabolized the same. Moreover, their absorption and metabolism is affected by the hormonal status of the host and perhaps the overall redox state of the tissue (Erdman 2005). *Cis* isomers of carotenoids appear to be more bioavailable than *all trans* forms, and the oil carrier contributes to more efficient absorption and distribution (Boileau et al. 2002, Boileau et al. 1999, Clinton et al. 1996, Kaplan et al. 1990, Schmitz et al. 1991).

It has been shown that when carotenoids, such as β-carotene, lycopene, and lutein, are given beyond the optimum levels, further increases in the amount of carotenoids in the cells led to pro-oxidant effects. It has already been demonstrated that related to the concentration and the oxygen tension in the system, carotenoids may act as pro-oxidants *in vitro* (Burton and Ingold 1984, Zhang and Omaye 2001).

With respect to photo-oxidation, ß-carotene enhances the expression of UV-induced heme-oxygenase, a marker of oxidative stress, in cell culture (Obermüller-Jevic et al. 1999). The photo-protective effects of low levels of carotenoids in cell culture are consistent with photo-protective effects of ß-carotene observed in human studies (Stahl et al. 2000, Stahl et al. 2001, Gollnick et al.1996, Mathews-Roth et al. 1972, Lee et al. 2000).

Stahl and Sies suggest that there are optimum levels for protection *in vivo* (Stahl and Sies 2002). The preferential accumulation of phytoene and phytofluene on other carotenoids in tissues, including skin, shows their actual availability in much lower concentrations from dietary sources and supplementation, along with anti-oxidative and photo-protective effects at lower concentrations. This may lead to the conclusion that lower intake levels of phytoene and phytofluene may be favorable over use of pigmented carotenoids that need to be consumed at higher levels to reach an effective dose-activity ratio and that also run the risk of pro-oxidative activity.

Non-invasive Raman spectroscopic and HPLC detection of carotenoids in human skin showed that carotenoid concentration in the skin correlates with the presence or absence of skin cancer. The most prevalent carotenoid was lycopene, followed by phytoene, lutein and zeaxanthin, and phytofluene (Hata 2000). Alaluf et al. have demonstrated that dietary carotenoids contribute to human skin color and UV photosensitivity (Alaluf et al. 2002).

Only one study investigated the bioavailability of phytoene and phytofluene from the alga *Dunaliella*. By using inhibitors of the carotenoid biosynthesis pathway, *Dunaliella* was enriched with a mixture of 9-*cis* and *all-trans* stereoisomers of phytoene and phytofluene. Rats were fed a diet supplemented with *Dunaliella* phytoene-rich spray dried powder. Phytoene accumulated in the liver, and small amounts were found in the adrenals, kidney, and spleen, and in the plasma, as well. The original algal diet content of 9-*cis*-to-*all-trans* ratio of 1:1 was maintained in the plasma and adrenals, but in the liver, spleen, and kidney the ratio was reduced to 1:3. The authors suggested that the preferential accumulation of *all-trans* phytoene over 9-*cis* phytoene in the liver, spleen, and kidney may be interpreted as indicating stronger anti-oxidative effect of 9-*cis* phytoene over the *all-trans* isomer or, alternatively, *in vivo* streoisomerization of 9-*cis* phytoene to the *all-trans* structure (Werman et al. 2002).

In a summary of several intervention studies, Porrini et al. show that bioavailability and plasma concentration of carotenoids, focusing on phytoene and phytofluene, are dependent not only on the initial concentration fed, but also on the source and carrier, with a clear advantage of oil such as oleoresin (Porrini et al. 2005, Paetau et al. 1998, Muller et al. 1999, Richelle et al. 2002, Edwards et al. 2003).

Phytoene and Phytofluene Demonstrated Activities

Hydroxyl Radical's Quenchers

Hydroxyl radical plays a significant role in oxidative and photooxidative damage and is several orders of magnitude more reactive toward cellular constituents than superoxide radicals and hydrogen peroxide. Hydroxyl radicals are short-lived, with a half life of approximately 10^{-9} s, but are the most damaging radicals within the body. This type of free radical can be formed from O_2^- and H_2O_2 via the Harber-Weiss reaction. The interaction of copper or iron and H_2O_2 also produce ·OH, as first observed by Fenton. These reactions are significant, because the substrates are found within the body and could easily interact (Halliwell and Gutteridge 1985).

Sunlight is a source for external/topical generation of hydroxyl radicals, as was demonstrated by Taira et al. This group suggested that sunlight exposure of skin may lead to hydroxyl radical generation and simultaneous lipid peroxidation (Taira et al. 1992). We sought to study the quenching of hydroxyl radicals by phytoene and phytofluene, isolated from the alga *Dunaliella* (von Oppen-Bezalel et al. 2006, Bezalel et al. 1998, Schickler et al. 2000, Soudant et al. 2005). In the assay system, the hydroxyl radical (·OH) was generated by decomposition of hydrogen peroxide by iron [($FeSO_4$) Fenton reaction] in front of spin-trap DMPO. The ·OH is picked up by the DMPO to give an added DMPO-OH, which presents a characteristic signal Electronic Paramagnetic Resonance (EPR). The capacity to trap the hydroxyl radical is detected by decrease of the signal EPR of the added DMPO-OH (for OH). The signal's intensity is proportional to the quantity of ·OH present in the environment.

The capacity of the phytoene and phytofluene mixture to trap the hydroxyl radical is theoretically valued by the difference between the EPR signal value that corresponded prior to the addition of the product and the signal value that resulted post-addition. The EPR signal showed that the phytoene and phytofluene mixture trapped the hydroxyl radical in a dose-dependent manner (Table 1) (von Oppen-Bezalel et al. 2006).

Table 1: Phytoene and phytofluene, the colorless carotenoids quench hydroxyl radicals.

IBR-CLC® final concentration (mg/ml)	Intensity of EPR signal (arbitrary units)	Intensity of the EPR signal after post-addition (arbitrary units)	% of hydroxyl radicals trapping
0	$2.82 \times 10^6 \pm 0.07$		
0.001	$0.34 \times 10^6 \pm 0.04$	$2.19 \times 10^6 \pm 0.13$	66
0.0004	$0.58 \times 10^6 \pm 0.02$	$1.94 \times 10^6 \pm 0.05$	48
0.0001	$1.26 \times 10^6 \pm 0.10$	$2.19 \times 10^6 \pm 0.13$	24

Protection by Phytoene and Phytofluene against UV- and Free Radical Induced Erythema and DNA Damage

Genotoxic agents can be either exogenous (UV irradiation, ionizing radiation, chemical substances) or of endogenous origin, such as free radicals produced by the cellular metabolism. In most cases, damaged DNA is repaired by the enzymatic repair systems of the cell. Occasionally, however, the repair systems fail to function correctly, the damaged DNA will propagate, and the damage generated by the genotoxic agent will cause persistent mutation.

A phytoene and phytofluene mixture, purified from *Dunaliella* extract, was tested *in vitro* for the ability to protect DNA from damage caused by hydroxyl radicals. The activity was measured using repair enzymes and biotinilated DNA base pairs (Salles et al. 1995) after short-term exposure of plasmid DNA to hydroxyl radicals generated in a Fenton reaction.

Plasmid DNA exposed to hydroxyl radicals' damage was protected efficiently by a mixture containing phytoene and phytofluene from the alga *Dunaliella* sp. reaching 50% of inhibiting concentration (IC50) at 11.2 μgr/mL (von Oppen-Bezalel et al. 2006).

These findings are in accord with the intervention study conducted by Porinni et al. with carotenoids from natural tomato extract, containing mostly lycopene, phytoene, phytofluene, and β-carotene. This showed modification of plasma and lymphocyte levels and improved anti-oxidant protection against DNA damage to lymphocytes with increased values of plasma and lymphocyte carotenoids levels, with phytoene at the lead. In a double-blind, cross-over study, 26 healthy subjects consumed 250 ml of tomato drink daily, providing about 6 mg lycopene, 4 mg phytoene, 3 mg phytofluene, 1 mg β-carotene, and 1.8 mg α-tocopherol, or a placebo drink. Treatments were separated by a wash-out period. Plasma and lymphocyte carotenoid and α-tocopherol concentrations were determined by HPLC and DNA damage by the comet assay. The authors concluded that the study supports the hypothesis that a low intake of carotenoids from tomato products improves cell anti-oxidant protection (Porrini et al. 2005).

Carotenoids were used as a treatment against photosensitization as early as 1964 (Mathews 1964). Since then, a vast number of studies have shown that carotenoids act as anti-oxidants, anti-inflammatory, and anti-mutagenic agents. Carotenoids are believed to have the potential to inhibit certain cardiovascular diseases and cancer as well. Several studies showed reduction of UV-induced erythema, because of dietary intake of combinations of carotenoids with significantly increased protection when these combinations contain the phytoene and phytofluene (Cesarini et al. 2003, Aust et al. 2005, Conference Notes, Speaker Sharoni 2003). In most of these studies, it was mentioned that the significant additive protection is probably because phytoene and phytofluene absorb light in the UV range, a property that allows them additional protection against UV generated free radicals (Krinski et al. 1989, Stahl et al. 2001, Lee et al. 2000, Sies and Stahl 2004, Aust et al. 2005, Conference Notes, Speaker Sharoni 2003).

Mathews-Roth demonstrated that carotenoids and phytoene, in particular, have photo-protective capabilities and can prevent UV-induced skin cancer (Mathews-Roth and Krinsky 1987, Mathews-Roth 1990, Mathews-Roth 1982, Mathews-Roth 1987).

Protection against LDL Oxidation

Early atherosclerotic lesions are characterized by lipid-laden foam cells in the arterial intima (Ross 1993). A potential mechanism of foam-cell formation is the unregulated uptake of modified forms of low-density lipoprotein (LDL) by the macrophage scavenger receptors (Goldstein et al. 1979, Henriksen et al. 1981). These receptors recognize LDL modified by products of lipid peroxidation (Haberland et al. 1988, Steinbrecher et al. 1984, Heinecke et al. 1984). Carotenoids, including phytoene and phytofluene, are found in human plasma and are carried by the LDL (Fraser and Bramley 2004). We examined the effect of phytoene and phytofluene mixture, isolated from *Dunaliella*, on LDL oxidation *in vitro* and compared its effect to α-tocopherol and *all trans* β-carotene. Human LDL was isolated and loaded with anti-oxidants, using a modified method for microsome enrichment with anti-oxidants. HPLC-grade purified α-tocopherol, *all trans* β-carotene, and phytoene and phytofluene (500 μM) were incubated with the isolated lipoproteins. The antioxidants-enriched LDL and an untreated control were then incubated in an AAPH-induced LDL oxidation system. Lipoprotein oxidation was monitored as diene conjugation formation at 234 nm (Cortesy of IBR Ltd., personal communication).

Dunaliella phytoene and phytofluene, similar to β-carotene and α-tocopherol, inhibited AAPH-induced LDL oxidation, as measured by a longer lag phase of conjugated diene formation (Figure 2).

Several lines of evidence suggest that low-density lipoprotein (LDL) oxidation plays a role in atherogenesis (more details in the chapter on Application of *Dunaliella* in Atherosclerosis).

Despite the presence of phytoene and phytofluene in human plasma and their anti-oxidative properties, the effect of these carotenoids on atherosclerosis has not been studied so far. The finding that phytoene and phytofluene inhibit LDL oxidation suggests that they have the potential to inhibit and prevent atherogenesis. However, human trials (Dagenais et al. 2000, Brown et al. 2001) showed that anti-oxidant supplementation, including the synthetic, *all-trans* β-carotene, to patients failed to affect either atherosclerosis or cardiovascular diseases and, therefore, the potential to inhibit atherogenesis by antioxidants, and by carotenoids, in particularly β-carotene, is debated.

Phytoene and Phytofluene from *Dunaliella* Protect β-carotene and CoQ10 against Oxidative Damage

One interesting capability of phytoene and phytofluene, the colorless carotenoids, is their ability to protect other compounds used in cosmetics and nutraceuticals from oxidative damage. We demonstrated this capability with β-carotene and CoQ10 (von Oppen-Bezalel et al. 2006, Fuller et al. 2006).

Phytoene and Phytofluene Protect β-carotene against Light-induced Degradation

β-carotene quenches singlet oxygen and free radicals *in vitro* (Burton and Ingold 1984), exhibits radical-trapping anti-oxidant properties at low partial pressure of oxygen (Burton and Ingold 1984), and acts synergistically with a-tocopherol as a radical-trapping antioxidant in membranes (Palozza and Krinsky 1992).

We compared the preservation of β-carotene after exposure to intense light and air with and without the protection of colorless carotenoids (CLC) and BHT (von Oppen-Bezalel et al. 2006). The results clearly showed that the colorless carotenoids, isolated from *Dunaliella*, protected β-carotene against light-induced degradation and that the protection effect is dose dependent. Although in the untreated β-carotene no color was detected following 35 min exposure to sun light and air, the colorless carotenoids (70 μmol/liter) preserved most of the β-carotene and were more efficient than BHT (6800 μmol/liter = 0.15%).

Phytoene and Phytofluene Protected CoQ10 against Oxidative Damage

Coenzyme Q10 (ubiquinone) is a compound that is synthesized in the body. A considerable amount of published data exists on the effects of oral administration of CoQ10 in improving or preventing such varied medical conditions as atherosclerosis, loss of heart muscle contractility, chemotherapy-induced toxicity, and neurodegenerative disorders (including Parkinson's disease, migraine headaches, memory loss, muscle aging, and inflammation) (Sarter 2002, Shults 2003, Linnane 2002). The presence of CoQ10 in the membranes of eukaryotic cells suggests its potential as an antioxidant and scavenger of free radicals, thus preventing the activation of inflammatory signaling pathways.

To study the effect of CLC on CoQ10 stability, CoQ10 was incubated in the presence or absence of hypochlorite (HOCl) for 30 min. at room temperature. Where indicated, reactions contained either lycopene (2 mg/mL) or the colorless carotenoids, phytoene (0.0142%) and phytofluene (0.0012%). Following 30 min. incubation, samples of each reaction were removed and injected onto HPLC for analysis of CoQ10 levels; CoQ10 lost 24% of its content in a control, untreated test tube. HOCl increased degradation to 72.6%. Lycopene (2 mg/mL) stabilized CoQ10 back to 76% (natural unprotected degradation), and colorless carotenoids (0.015 %) retrieved 72% of this amount (Fuller et al., 2006, von Oppen-Bezalel et al. 2006). The phytoene and phytofluene prevented the deterioration of CoQ10 most effectively and in low quantities by hypochloric acid.

Phytoene and Phytofluene from *Dunaliella* Demonstrated Anti-inflammatory Activity

Increased production of IL-6 leads to excessive growth and activation of immune cells and has been linked to arthritis and psoriasis. The phytoene and phytofluene mixture exhibits strong anti-inflammatory activity in phytohemagglutinin-(PHA) activated human Periferal Blood Lymphocytes (PBLs) against IL-6 and IL-12 (data not shown). The reduction of IL-6 was also observed in human dermal fibroblasts. The mixture also reduced inflammation expressed as edema *in vivo*, demonstrated by the oxazalone-induced mouse ear edema model, resulting in 64% inhibition at 0.1% (data not shown). The reduction of IL-6 observed in cell-culture experiments and IL-6's anti-inflammatory activity in an animal model may be an indication of potential activity of the colorless carotenoids in relieving inflamed skin and tissues.

Phytoene and Phytofluene Reduced the Production of Inflammatory Mediator's Markers in Dermal Human Fibroblasts

Prostaglandin E2 (PGE-2) is one of the most significant inflammatory mediators produced in the body. It is made in response to a variety of stimuli in a wide variety of cell types, including both keratinocytes and fibroblasts. Elevated PGE-2 has been associated with a wide number of inflammatory diseases, including arthritis, Alzheimers, stroke, cancer, periodontal disease, most, if not all, inflammatory skin diseases, skin aging, and of course UVR-induced sunburn (Williams and Peck 1977, Nakae et al. 2003).

In skin, elevated PGE-2 stimulates vasodilatation; induces the expression of inflammatory cytokines and chemokines; increases the chemotaxis of inflammatory cells from the circulation into the skin; and increases sensitivity of neurons to stimuli, thereby increasing neuronal substance P and calcitonin generated peptide production and secretion (Williams and Peck 1977).

IL-1 plays a key role in propagating an inflammatory response. IL-1 is produced by both keratinocytes and fibroblasts in response to a variety of external stimuli (including UVR, chemical irritants, and allergens), as well as by immune cells, specifically monocytes. IL-1 activates a wide number of inflammatory events, including stimulating the production of many cytokines and chemokines by mononuclear phagocytes, indirectly activating neutrophils through chemokines produced by IL-1–activated monocytes, and inducing endothelial cells of the vasculature to produce adhesion molecules (such as vascular cell adhesion molecules and intercellular cell adhesion molecules), which then, in turn, assist in the anchoring and chemotaxis of additional immune cells into the skin (Debets et al. 1997, Cronin et al. 1999).

Coenzyme Q10 and carotenoids produce an additive inhibitory effect on PGE-2 production in IL-1–stimulated fibroblasts. Cells were treated with IL-1 where indicated, with CoQ10 (10 μm), colorless carotenoids (total concentration of phytoene and phytofluene of 15 μg/mL), or the combination of both CoQ10 and colorless carotenoids. CoQ10 alone led to an approximately 16% inhibition of the IL-1 induction of PGE-2. The carotenoids alone inhibited the PGE-2 increase significantly (47%). The combination of CoQ10 and carotenoids inhibited PGE-2 production by 70%.

With IL-1-stimulated fibroblasts, CoQ10 was ineffective in blocking IL-1 induction of matrix metalloproteinase 1 (MMP-1) in fibroblasts. In contrast, the colorless carotenoids inhibited IL-1-induced MMP-1 by a small, although reproducible, 16%. It is interesting that the combination of CoQ10 and the colorless carotenoids decreased IL-1–induced MMP-1 significantly by almost 30%, which indicates a synergistic inhibitory effect of these two compounds (Fuller et al. 2006, von Oppen-Bezalel et al. 2006).

The other cytokine examined in this study, IL-6, also plays an important role in propagating an inflammatory response in the skin. This cytokine is produced by both keratinocytes and fibroblasts, as well as by T lymphocytes, and acts principally as an immune-cell growth factor, aiding in immune cell growth and differentiation. However, increased production of IL-6 leads to excessive growth and activation of immune cells and has been linked to arthritis and psoriasis. The surprising finding that CoQ10 and/or carotenoids can reduce UVR-induced IL-6 levels and IL-1–mediated PGE-2 production markedly suggests the value of incorporating these compounds into topical skin care products. Furthermore, their ability to inhibit the UVR and IL-1–mediated increase in MMP-1 suggests that their long-term use might help prevent the loss of collagen in the dermal matrix during the natural aging process (Fuller et al. 2006).

Phytoene and Phytofluene Skin Whitening Effects

Skin color is primarily determined by the amount of the pigment melanin present in melanocytes, cells that are present in the epidermal basal layer of the skin. Melanin production takes place in unique organelles within the melanocytes, known as melanosomes, and protects the skin from the harmful effects of UV light. The darker skin tone of certain societies results from a naturally increased production of melanin, and high production in response to the stimulus of UV light leads to the well-known tanning effect of the skin. Aging, exposure to sun, hormonal abnormalities, and various skin disorders increase the deposition of melanin pigment in skin, resulting in dark spots and freckles. Such dark spot are not only considered to be unattractive in many societies, they may also be malignant and therefore, undesired.

For those who are interested in eliminating the presence of dark spots on the skin or a lighter skin tone, whitening or bleaching compositions are useful. Many modern skin-bleaching compositions either destroy melanin (typically by destroying or disrupting melanin granules); inhibit its formation (often by inhibiting tyrosinase, a melanin biosynthetic enzyme, or by inhibiting melanocyte activity and proliferation), or both. Many of these bleaching compositions contain harsh chemicals (such as peroxides, acids or formaldehyde), or thiolated materials (such as glutathione, cysteine, mercaptosuccinic acid, mercaptodextran, and mercaptoethanol). These chemicals, in addition to having a stringent effect on skin, have an objectionable odor that makes the products containing them undesirable to a consumer.

Less stringent therapies have other disadvantages. For example, hydroquinone is used as a whitening agent, because it acts by suppressing melanocyte activity. However, hydroquinone is oxidized by air, light, and tyrosinase itself. The oxidized products of hydroquinone have been implicated in skin irritation and inflammation and in pigmentation rebound (i.e., initial lightening followed by darkening). In addition, due to these oxidation processes, the shelf life of preparations containing hydroquinone and its bioavailability upon application are relatively short.

Typical whitening agents in cosmetic formulations are kojic acid, arbutin, licorice extract, and vitamin C. These are effective tyrosinase inhibitors and anti-oxidants, but are usually not stable and are oxidized and degraded easily. Moreover, there is a high risk of skin irritation and inflammation with these compounds.

IBR, Israeli Biotechnology Research Ltd. has recently discovered that a combination of phytoene and phytofluene has a direct effect on skin pigmentation by reducing the melanin content in the skin cells (data not shown). The advantages of these compounds for skin whitening are that these are natural carotenoids with no known adverse effects either when applied to skin or when taken orally, while being highly effective. Phytoene and phytofluene at concentrations lower in order of magnitude have higher effects on melanin synthesis compared to known skin-whitening ingredients, such as arbutin (courtesy of IBR Ltd.). The phytoene and phytofluene showed a reduction of 22% of the melanin content in B16 Murine melanocytes at a concentration of 7.5 μM with no effect on cell viability. Arbutin, a known skin-whitening agent, showed an effect on melanin synthesis without inhibitory effect on cell growth at maximum concentration of 50 μM (Akiu et al. 1991).

Phytoene and Phytofluene Applications

Cosmetics and wellness nutrition are the main fields in which the colorless carotenoids, phytoene and phytofluene, were implemented and used for protection and prevention against internal and external stress and oxidative stress damage.

The benefiting activities of these components as anti-inflammatory ingredients and in reducing UV-induced oxidation and reducing damage to DNA and other biomolecules make them ideal members in products designed for protection, prevention and relief of skin problems, as well as for nutrition.

One can find the colorless carotenoids, phytoene and phytofluene, from *Dunaliella* and from tomatoes in sun protection products, moisturizers, and boosters of other biomolecular activities, such as the CoQ10 in cosmetic formulations. New cosmetic products are now being produced with the colorless carotenoids as skin whitening agents as well.

There is only one food supplement available on the market today (PhytoflORAL™), which contains phytoene, phytofluene and ξ-carotene for oral intake, makes these ingredients available in higher doses and for enrichment of ones nutrition for health protection and prevention of premature aging and photo-aging that occurs due to external (sun, oxidants, contaminants) and internal (metabolism and oxidation processes) damage. It is also being marketed now as a nutricosmetics product both to prevent skin aging and as a skin whitener.

Therapeutics

Epidemiological Studies

Numerous epidemiological studies have suggested an association between the high intake of carotenoid-rich fruits and vegetables and reduced risk of cancer (Khachick 2002. Conference Notes, Speaker: Sharoni 2003, Conference Notes, Speaker: Kucuk 2003, Conference Notes, Speaker: Levy 2003).

Skin Protection

β-carotene, phytoene, and canthaxanthin protected mice against UVB-induced skin tumors. For tumors induced by DMBA/cotton oil or DMBA/UV-B, mice receiving β-carotene showed a significant difference in tumor number, although phytoene and canthaxanthin had no effect (Mathews-Roth 1982).

Guinea pigs received phytoene by daily intraperitoneal injection for 14 days followed by UV irradiation developed significantly less erythema than animals that received the placebo (Mathews-Roth and Pathak 1975).

It is well documented that supplementation with tomato-based products increases lycopene, phytofluene, and phytoene levels in human serum and protects against UV-light-induced erythema (Aust et al. 2005). Phytofluene and phytoene may have contributed to protection. Both of these carotenoids exhibit absorption maxima at wavelengths of UV light. Absorption of UV light protects skin from photodamage and may explain the differences observed between groups.

Cell Culture

Expression of phase II detoxification enzymes is, in part, regulated by the transcription factor Nrf2 (nuclear factor E2-related factor 2). In transiently transfected cancer cells, phytoene and phytofluene had a much smaller effect than lycopene on the expression of reporter genes fused to ARE sequence (Ben-Dor et al. 2005). The acyclic carotenoids phytofluene, ζ-carotene, and lycopene reduced the human prostate cancer cell line's viability; phytoene had no effect (Nara 2001).

Levy and Sharoni have demonstrated that physiological concentration of lycopene alone (0.3 μM) or phytoene plus phytofluene, did not decrease the growth of human breast cancer cells, while physiological concentration of lycopene plus phytoene and phytofluene effectively decreased the growth of the cancer cells.

The acyclic carotenoids ζ-carotene and phytofluene and the oxidation products of lycopene inhibited the growth of HL-60 human promyelocytic cells through apoptosis (Nara 2001).

Summary

Phytoene and phytofluene, the colorless carotenoids, are in a way, newly recognized members of the carotenoids group of phytonutrient and active ingredients being used for cosmetics, as well as for nutritional and therapeutic applications.

For topical cosmetic use, the advantages of phytoene and phytofluene, the colorless carotenoids, over colored carotenoids are clear since using them allows use of effective quantities without staining the skin which will occur with very low levels of pigmented carotenoids. Having anti-oxidative capabilities, prevention and protection against UV and environmental stress, together with anti-inflammatory properties without additional color, make them natural members in topical formulations and a needed active ingredient.

Moreover, phytoene mostly is shown to be more resistant to internal degradation than lycopene by being accumulated in higher rates than the other carotenoids (Stahl et al. 2005, Porrini et al. 2005) and could potentially stabilize and protect other ingredients and carotenoids, as was demonstrated with β-carotene and CoQ10 from destruction (von Oppen-Bezalel et al. 2006).

Intervention studies and extended topical uses of phytoene and phytofluene are still to be done. Evidential results of intervention studies from combined carotenoids mixtures with and without phytoene and phytofluene, together with a list of *in vitro* studies, are supporting the significant activity of the compounds on (1) reducing inflammation and inflammatory events, (2) prevention of cancer, sunburn effects and UV erythema, and (3) oxidative stress damage caused by free radicals and UV, as well as (4) protection of DNA.

References

Akiu, S., Y. Suzuki, T. Asahara, Y. Fujinuma, and M. Fukuda. 1991. Inhibitory effect of arbutin on melanogenesis—biochemical study using cultured B16 melanoma cells. Nippon Hifuka Gakkai Zasshi. 101(6): 609-613.

Alaluf, S., U. Heinrich, W. Stahl, H. Tronnier, and S. Wiseman. 2002. Dietary carotenoids contribute to normal human skin color and UV photosensitivity. Journal of Nutrition. 132: 399-403

Aust, O., W. Stahl, H. Sies, H. Tronnier, and U. Heinrich. 2005. Supplementation with tomato-based products increases lycopene, phytofluene, and phytoene levels in human serum and protects against UV-light-induced erythema. International Journal of Vitamins and Nutrition Research. 75(1): 54-60.

Bartley, G.E. and P.A. Scolnik. 1995. Plant carotenoids: Pigments for photoprotection, attraction, and human health. Plant Cell 7: 1027-1038.

Ben-Amotz, A., J. Gressel, and M. Avron 1987. Massive accumulation of phytoene induced by norflurazon in *Dunaliella bardawil* (Chlorophyceae) prevents recovery from photoinhibition. Journal of Phycology 23: 176-181.

Ben-Amotz, A., A. Lers, and M. Avron. 1988. Stereoisomers of beta-Carotene and Phytoene in the Alga *Dunaliella bardawil.* Plant Physiology 86: 1286-1291.

Ben-Dor, A., M. Steiner, L. Gheber, M. Danilenko, N. Dubi, K. Linnewiel, A. Zick, Y. Sharoni, and J. Levy. 2005. Carotenoids activate the antioxidant response element transcription system Molecular Cancer Therapy 4: 177-186.

Bendich, A. and J.A. Olson. 1989. Biological actions of carotenoids. FASEB Journal 3: 1927-1932.

Bezalel, L., J. Paltiel, E. Soudant, and A. Ben-Amotz. 1998. Non conventional carotenoids. In Platform & Poster papers from the 20th IFSCC Congress P088.

Boileau, A.C., N.R. Merchen, K. Wasson, C.A. Atkinson, and J.W. Erdman, Jr. 1999. *Cis*-lycopene is more bioavailable that *trans*-lycopene in vitro and in vivo in lymph-cannulated ferrets. Journal of Nutrition 129: 1176-1181.

Boileau, T.W.-M., A.C. Boileau, and J.W. Erdman, Jr. 2002. Bioavailability of *all-trans* and *cis*-isomers of lycopene. Experimental Biology and Medicine (Maywood) 227: 914-919.

Britton, G. 1988. Biosynthesis of carotenoids. pp. 133-180. In Plant Pigments T.W. Goodwin (ed) Academic Press, London, UK.

Brown, B.G., X.Q. Zhao, A. Chait, L.D. Fisher, M.C. Cheung, J.S. Morse, A.A. Dowdy, E.K. Marino, E.L. Bolson, P. Alaupovic, J. Frohlich, and J.J. Albers. 2001. Simvastatin and niacin, antioxidant vitamins, or the combination for the prevention of coronary disease. New England Journal of Medicine 345(22): 1583-1592.

Burton, G.W. and K.U. Ingold. 1984. ß-Carotene: An unusual type of lipid antioxidant. Science. 224: 569-573.

Campbell, J.K., S. Zaripheh, M.A. Lila, and J.W. Erdman, Jr. 2005. Relative bioavailability of phytoene and phytofluene in male F344 rats. FASEB Journal 19: 4385 (abs.).

Cesarini, J.P., L. Michel, J.M. Maurette, H. Adhoute, and M. Bejot. 2003. Immediate effects of UV radiation on the skin: modification by an antioxidant complex containing carotenoids. Photodermatology Photoimmunology Photomedicine. 19(4): 182-189.

Clinton, S.K., C. Emenhiser, S.J. Schwartz, D.G. Bostwick, A.W. Williams, B.J. Moore, and J.W. Erdman, Jr. 1996. *Cis-trans* lycopene isomers, carotenoids, and retinol in the human prostate. Cancer Epidemiological Biomarkers and Prevention 5: 823-833.

Conference notes: Examining the Health Benefits of Lycopene from Tomatoes, April 1-2, 2003, Washington, DC, Convened by the Center for Food and Nutrition Policy at

Virginia Tech, Sponsored by the H.J. Heinz Company. "Lycopene, Skin Cancer and UV Exposure", Speaker: Dr. Yoav Sharoni.

Conference notes: Examining the Health Benefits of Lycopene from Tomatoes, April 1-2, 2003, Washington, DC, Convened by the Center for Food and Nutrition Policy at Virginia Tech, Sponsored by the H.J. Heinz Company "Evidence for Reducing the Risk of Prostate Cancer—A Clinical Trial" Speaker: Dr. Omer Kucuk.

Conference notes: Examining the Health Benefits of Lycopene from Tomatoes, April 1-2, 2003, Washington, DC, Convened by the Center for Food and Nutrition Policy at Virginia Tech, Sponsored by the H.J. Heinz Company "The Role of Lycopene and Other Tomato Carotenoids in Breast and Endometrial Cancers". Speaker: Dr. Joseph Levy.

Cronin, M.T., J.C. Dearden, G.P. Moss, and G. Murray-Dickson. 1999. Investigation of the mechanism of flux across human skin *in vitro* by quantitative structure-permeability relationships. European Journal of Pharmacological Sciences 7: 325-330.

Dagenais, G.R., R. Marchioli, S. Yusuf, and G. Tognoni. 2000. Beta-carotene, vitamin C, and vitamin E and cardiovascular diseases. Current Cardiology Reports. 2: 293-299. 1997. The IL-1 system in psoriatic skin: IL-1 antagonist sphere of influence in lesional psoriatic epidermis. R. Debets, J.P. Hegmans and P. Croughs (eds) Journal of Immunology 158: 2955-2963.

Ebenezer, W.J. and G. Pattendėn. 1993. *cis*-Stereoisomers of beta-carotene and its congeners in the alga *Dunaliella bardawil*, and their biogenetic interrelationships. Journal of the Chemical Society Perkin Trans. I: 1869-1873.

Edwards, A., B.T. Vinyard, E.R. Wiley, E.D. Brown, J.K. Collins, P. Perkins-Veazie, R.P. Baker, and B.A. Clevidence. 2003. Consumption of watermelon juice increases plasma concentrations of lycopene and beta-carotene in humans. Journal of Nutrition 133, 1043-1050.

Erdman, J.W. 2005. Promises and Perils of Lycopene/Tomato Supplementation and Cancer Prevention, How Do Nutritional and Hormonal Status Modify the Bioavailability, Uptake, and Distribution of Different Isomers of Lycopene? Journal of Nutrition 135: 2046S-2047S.

Ermakov, I.V., M.R. Ermakova, W. Gellermann, and J. Lademann 2004. Noninvasive selective detection of lycopene and beta-carotene in human skin using Raman spectroscopy. Journal of Biomedical Optics. 9(2): 332-338.

Fraser, P. D. and P.M. Bramley. 2004. The biosynthesis and nutritional uses of carotenoids. Progress in Lipid Research 43: 228-265.

Fuller, B.B., D.R. Smith, A.J. Howerton, and D. Kern. 2006. Anti-inflammatory effects of CoQ10 and colorless carotenoids. Journal of Cosmetic Dermatology. 5(1): 30-38.

Goldstein, J.L., Y.K. Ho, and S.K. Basu. 1979. Binding site on macrophages that mediates uptake and degradation of acetylated low density lipoprotein, producing massive cholesterol deposition. Proceedings of the National Academy of Sciences. USA. 76: 333-337.

Gollnick, H.P.M., W. Hopfenmüller, C. Hemmes, S.C. Chun, C. Schmid, K. Sundermeier, and H.K. Biesalski. 1996. Systemic beta-carotene plus topical UV-sunscreen are an optimal protection against harmful effects of natural UV-sunlight: Results of the Berlin-Eilath study. European Journal of Dermatology. 6: 200-205.

Haberland, M.E., D. Fong, and L. Cheng. 1988. Malondialdehyde altered protein occurs in atheroma of Watanabe heritable hyperlipidemic rabbits. Science 241: 215-218.

Halliwell, B. and J.M.C. Gutteridge. 1985. The chemistry of oxygen radicals and other oxygen-derived species. pp. 20-64. In Free Radicals in Biology and Medicine. Oxford University Press, New York, USA.

Hata, T.R., T.A. Scholz, I.V. Ermakov, R.W. McClane, F. Khachik, W. Gellermann, and L.K. Pershing. 2000. Non-invasive raman spectroscopic detection of carotenoids in human skin. Journal of Investigative Dermatology. 115(3): 441-448.

Heinecke, J.W., H. Rosen, and A. Chait. 1984. Iron and copper promote modification of low density lipoprotein by human arterial smooth muscle cells in culture. Journal of Clinical Investigation. 74: 1890-1894.

Henriksen, T., E.M. Mahoney, and D.Steinberg. 1981. Enhanced macrophage degradation of low density lipoprotein previously incubated with cultured endothelial cells: Recognition by receptors for acetylated low density lipoproteins. Proceedings of the National Academy of Sciences. USA. 78: 6499-6503.

Kaplan, L.A., J.M. Lau, and E.A. Stein. 1990. Carotenoid composition, concentrations, and relationships in various human organs. Clinical Physiology And Biochemistry. 8: 1-10.

Khachik, F., G.R. Beecher, M.B. Goli, and W.R. Lusby. 1991. Separation, identification, and quantification of carotenoids in fruits, vegetables and human plasma by high performance liquid chromatography. Pure&App/. Chemistry. 63(1): 71-80.

Khachik, F., L. Carvalho, P.S. Bernstein, G.J. Muire, D.Y. Zhao, and N.B. Katz. 2002. Chemistry, distribution, and metabolism of tomato carotenoids and their impact on human health. Experimental Biology and Medicine. 227: 845-851.

Krinski, N.I., M.M. Matthews-Roth, and R.F. Taylor (eds), 1989. In Carotenoids, Chemistry and Biology. Planum Press, New York, London.

Lee, J., S. Jiang, N. Levine, and R.R. Watson. 2000. Carotenoid supplementation reduces erythema in human skin after simulated solar radiation exposure. Proceedings of the Society of Experimental Biology and Medicine. 223: 170-174.

Linnane, A.W., C. Zhang, and N. Yarovaya. 2002. Human aging and global function of coenzyme Q10. Ann NY Academic Science. 959: 396-411.

Mathews, M.M. 1964. Protective effects of β-carotene lethal photosensitization by haematoporphyrin. Nature. 203: 1092.

Mathews-Roth, M.M. 1982. Antitumor activity of beta-carotene, canthaxanthin and phytoene. Oncology. 39(1): 33-37.

Mathews-Roth, M.M. 1987. Photoprotection by carotenoids. Federation Proceedings. 46(5): 1890-1893.

Mathews-Roth, M.M. 1990. Carotenoid functions in photoprotection and cancer prevention. Journal of Environmental Pathology Toxicology and Oncology. 10(4-5): 181-192.

Mathews-Roth, M.M. and M.A. Pathak. 1975. Phytoene as a protective agent against sunburn (>280 nm) radiation in guinea pigs. Phytochemistry and Phytobiology. 21-4: 261-26.

Mathews-Roth, M.M. and N.I. Krinsky. 1987. Carotenoids affect development of UV-B induced skin cancer. Photochemistry and Photobiology. 46(4): 507-509.

Mathews-Roth, M.M., M.A. Pathak, J.A. Parrish, T.B. Fitzpatrick, E.H. Kass, K. Toda, and W. Clemens. 1972. A clinical trial of the effects of oral beta-carotene on the responses of human skin to solar radiation. Journal of Investigative Dermatology 59: 349-353.

Müller, H., A. Bub, B. Watzl, and G. Rechkemmer. 1999. Plasma concentrations of carotenoids in healthy volunteers after intervention with carotenoid-rich foods. European Journal of Nutrition 38, 35-44.

Nakae, S., Y. Komiyama, and S. Narumi. 2003. IL-1-induced tumor necrosis factor-alpha elicits inflammatory cell infiltration in the skin by inducing IFN-gamma-inducible protein 10 in the elicitation phase of the contact hypersensitivity response. International Journal of Immunology 15: 251-260.

Nara, E., H. Hayashi, M. Kotake, K. Miyashita, and A. Nagao. 2001. Acyclic carotenoids and their oxidation mixtures inhibit the growth of HL-60 human promyelocytic leukemia cells. Nutrition Cancer. 2001; 39(2): 273-283.

Nara, E.K., M. Kushiro, H. Zhang, T. Sugawara, K. Miyashita, and A. Nagao. 2001. Carotenoids Affect Proliferation of Human Prostate Cancer Cells *Journal of Nutrition.* 131: 3303-3306.

Obermüller-Jevic, U.C., P.I. Francz, J. Frank, A. Flaccus, and H.K. Biesalski. 1999. Enhancement of the UVA induction of haem oxygenase-1 expression by beta-carotene in human skin fibroblasts. FEBS Letters. 460: 212-216.

Orset, S.C. and A.J. Young. 2000. Exposure to low irradiances favors the synthesis of 9-cis beta, beta-carotene in *Dunaliella salina* (Teod.). Plant Physiology. 122(2): 609-618.

Paetau, I., F. Khachik, E.D. Brown, G.R. Beecher, T.R. Kramer, J. Chittams, and B.A. Clevidence. 1998. Chronic ingestion of lycopene-rich tomato juice or lycopene supplements significantly increases plasma concentrations of lycopene and related tomato carotenoids in humans. American Journal of Clinical Nutrition. 68: 1187-1195.

Palozza, P. and N.I. Krinsky. 1992. beta-Carotene and alpha-tocopherol are synergistic antioxidants. Archives of Biochemistry and Biophysics. 297(1): 184-7.

Porrini, M., P. Riso, A. Brusamolino, C. Berti, S. Guarnieri, and F. Visioli. 2005. Daily intake of a formulated tomato drink affects carotenoid plasma and lymphocyte concentrations and improves cellular antioxidant protection. British Journal of Nutrition. 93(1): 93-99.

Richelle, M., K. Bortlik, S. Liardet, C. Hager, P. Lambelet, M. Baur, L.A. Applegate, and E.A. Offord. 2002. A food-based formulation provides lycopene with the same bioavailability to humans as that from tomato paste. Journal of Nutrition. 132: 404–408.

Ross, R. 1993. The pathogenesis of atherosclerosis: a perspective for the 1990s. Nature 362: 801-809.

Salles, B., C. Provot, P. Calsou, I. Hennebelle, I. Gosset, and G.J. Fournié. 1995. A chemiluminescent microplate assay to detect DNA damage induced by genotoxic treatments. Analytical Biochemistry. 232: 37-42.

Sarter, B. 2002. Coenzyme Q10 and cardiovascular disease: a review. Journal of Cardiovascular Nursing. 16: 9-20.

Satomi, Y., N. Misawa, T. Maoka, and H. Nishino. 2004. Production of phytoene, a carotenoid, and induction of connexin 26 in transgenic mice carrying the phytoene synthase gene *crtB*. Biocemistry Biophysics Research Communications. 320: 398-401.

Schickler, H., J. Paltiel, L. Bezalel, A. Shaish, A. Ben-Amotz, and E. Soudant. 2000. Colorless carotenoids from *Dunaliella sp.* exhibiting UV protection and anti-oxidative properties. Oral presentation at the Phytochemical Society of Europe (PSE) Meeting, Lisbon, Portugal.

Schmitz, H.H., C.L. Poor, R.B. Wellman, and J.W. Erdman, Jr. 1991. Concentrations of selected carotenoids and vitamin A in human liver, kidney and lung tissue. Journal of Nutrition. 121: 1613-1621.

Shaish, A., M. Avron, and A. Ben-Amotz. 1990. Effect of Inhibitors on the Formation of Stereoisomers in the Biosynthesis of ß-Carotene in *Dunaliella bardawil*. Plant and Cell Physiology. 31(5): 689-696.

Shults, C.W. 2003. Coenzyme Q10 in neurodegenerative diseases. Current Medicinal Chemistry. 10: 1917-1921.

Sies, H. and W. Stahl. 2004. Carotenoids and UV protection. Photochemical Photobiological Sciences. 3(8): 749-752.

Soudant, E., A. Shaish, M. Berkovitch, L. von Oppen-Bezalel, and A. Ben-Amotz. 2005. The Colorless Carotenoids phytoene and phytofluene. Direct and indirect protection of biological systems. p. 195 In Carotenoid Science, Vol. 9. H. Hashimoto (ed) Osaka City University. Osaka, Japan.

Stahl, W. and H. Sies. 2002. Carotenoids and Protection against Solar UV Radiation. Skin Pharmacology and Applied Skin Physiology. 15: 291-296.

Stahl, W., U. Heinrich, H. Jungmann, H. Sies, and H. Tronnier. 2000. Carotenoids and carotenoids plus vitamin E protect against ultraviolet light-induced erythema in humans. American Journal of Clinical Nutrition. 71: 795-798.

Stahl, W., U. Heinrich, S. Wiseman, O. Eichler, H. Sies, and H.Tronnier. 2001. Dietary tomato paste protects against UV-induced erythema in humans. Journal of Nutrition. 131: 1449-1451.

Stahl, W. and H. Sies. 2005. Bioactivity and bioprotective effects of natural carotenoids. Biochemistry and Biophysica Acta 1740(2): 101-107.

Stahl, W. and H. Sies. 2005. Bioactivity and bioprotective effects of natural carotenoids. Biochemistry and Biophysica Acta 1740(2): 101-107.

Taira, J., K. Mimura, T. Yoneya, A. Hagi, A. Murakami, and K. Makino. 1992. Hydroxyl radical formation by UV-irradiated epidermal cells. Journal of Biochemistry (Tokyo). 111(6): 693-695.

von Oppen-Bezalel, L., E. Lerner, D.G. Kern, B. Fuller, E. Soudant, and A. Shaish. 2006. Colorless carotenoids from IBR: phytoene and phytofluene from unicellular algae – applications in cosmetics, wellness and nutrition. Fragrance Journal. 34-3-307: 48-53.

Werman, M.J., S. Mokady, and A. Ben-Amotz. 2002. Bioavailability of the isomer mixture of phytoene and phytofluene-rich alga *Dunaliella bardawil* in rat plasma and tissues. Journal of Nutritional Biochememistry.13(10): 585-591.

Williams, T.J. and M.J. Peck. 1977. Role of prostaglandin-mediated vasodilatation in inflammation. Nature 270: 530–532.

Yamada,N., D. Kusano, and E. Kuwan. 2002. Bleaching Activity of 4-Phenyl-3-(substituted benzylthio)-4*H*-1, 2, 4-triazoles. Bioscience, Biotechnology, and Biochemistry. 66(8): 1671-1676.

Zhang, P. and S.T. Omaye. 2001. Antioxidant and prooxidant roles for beta-carotene, alpha-tocopherol and ascoebic acid in human lung cells. Toxicology in vitro 15: 13-24.

19

Biofuels from Microalgae: Review of Products, Processes and Potential, with Special Focus on *Dunaliella* sp.

Michael H. Huesemann[1] *and John R. Benemann*[2]

Abstract

There is currently great interest in using microalgae for the production of biofuels, mainly due to the fact that microalgae can produce biofuels at a much higher productivity than conventional plants and that they can be cultivated using water, in particular seawater, and land not competing for resources with conventional agriculture. However, at present such microalgae-based technologies are not yet developed and the economics of such processes are uncertain. The production of microalgae biomass has been described in many publications, therefore we limit our review here on the various types of biofuels and bioenergy that can be generated from microalgal biomass, emphasizing the genus *Dunaliella*. We review power generation by direct combustion, production of hydrogen and other fuel gases and liquids by gasification and pyrolysis, methane generation by anaerobic digestion, ethanol fermentations, and hydrogen production by dark and light-driven metabolism. We in particular discuss the production of lipids, vegetable oils and hydrocarbons, which could be converted to biodiesel. Direct combustion for power generation has two major disadvantages in that the high N-content of algal biomass causes unacceptably high NO_x emissions and losses of nitrogen fertilizer. Thus, the use of sun-dried microalgal biomass would not be cost-competitive with other solid fuels such as coal and wood. Thermochemical conversion processes such as gasification and pyrolysis have been successfully demonstrated in the laboratory but will be difficult to scale up commercially and suffer from similar, though sometimes not as stringent, limitations as combustion. Anaerobic digestion of microalgal cells yields only about 0.3 L methane per g volatile solids destroyed, about half of the maximum achievable, but yields can be increased by adding carbon rich substrates to circumvent ammonia toxicity caused by the N-rich algal biomass. Anaerobic digestion would be best suited for the treatment of algal biomass waste after value-added products have been separated. Algae can also be grown to accumulate starches or similar fermentable

[1] Pacific Northwest National Laboratory (PNNL), Marine Sciences Laboratory, 1529 West Sequim Bay Road, Sequim, Washington 98382, USA. Email: michael.huesemann@pnl.gov

[2] Benemann Associates, 3434 Tice Creek Dr. No.1, Walnut Creek, CA 94595, USA.

products, and ethanol or similar (e.g., butanol) fermentations could be applied to such biomass, but research is required on increasing solvent yields. Dark fermentation of algal biomass can also produce hydrogen, but, as with other fermentations, only at low yields. Hydrogen can also be generated by algae in the light; however; this process has not yet been demonstrated in any way that could be scaled up and, in any event, *Dunaliella*, is not known to produce hydrogen. In response to nutrient deficiency (nitrogen or silicon), some microalgae accumulate neutral lipids which, after physical extraction, could be converted, via transesterification with methanol, to biodiesel. Nitrogen-limitation does not appear to increase either cellular lipid content or lipid productivity in *Dunaliella*. Results from life cycle energy analyses indicate that cultivation of microalgal biomass in open raceway ponds has a positive energy output ratio (EOR), approaching up to 10 (i.e., the caloric energy output from the algae is 10 times greater than the fossil energy inputs), but EOR is less than 1 for biomass grown in engineered photobioreactors. Thus, from both an energetic as well as economic perspective, only open ponds systems can be considered. Significant long-term R&D will be required to make microalgal biofuels processes economically competitive. Specifically, future research should focus on (a) the improvement of biomass productivities (i.e., maximizing solar conversion efficiencies), (b) the selection and isolation of algal strains that can be mass cultured and maintained stably for long periods, (c) the production of algal biomass with a high content of lipids, carbohydrates, and co-products, at high productivity, (d) the low-cost harvesting of the biomass, and (e) the extraction and conversion processes to actually derive the biofuels. For *Dunaliella* specifically, the highest potential is in the co-production of biofuels with high-value animal feeds based on their carotenoid content.

Introduction

Given current trends, world energy use will increase five-fold by 2100 (Huesemann 2006). However, fossil fuel resources are depleting rapidly, and their combustion is the major cause for global climate change. It is thus imperative that energy demand be reduced substantially and be supplied in large part by renewable fuels, of which biofuels are among the most appealing. However, currently the production of biofuels, such as ethanol, biodiesel, or methane, from terrestrial biomass requires extremely large areas of productive land, thereby competing with food production. Replacing even the current worldwide primary energy demand of 351 EJ/yr with biofuels would require a land area comparable to the ~1.5 billion hectares used for all of agriculture (Huesemann 2006). For ethanol from corn to substitute for just the U.S. gasoline consumption (about 25% of U.S. energy demand) would require all 190 million hectares of U.S. crop land (Kheshgi et al. 2000).

Microalgae have two major advantages over higher plants with respect to biofuels production: First, biomass productivities are expected to be significantly greater for microalgae, with currently achievable productivities projected at about

70 metric tons per hectare-year of ash-free dry weight (AFDW, e.g. organic matter) (Sheehan et al. 1998). This compares to 3 t/ha·yr for soybeans, 9 t/ha·yr for corn, and 10-13 t/ha·yr for switchgrass or hybrid poplars (Perlack et al. 2005, McLaughlin et al. 2002). High productivities result in a much smaller footprint for microalgal biofuel systems, in terms of both land and water resources used (Huntley and Redalje 2007). Second, the cultivation of microalgae does not require arable land or freshwater - it can be carried out in shallow ponds on hardpan soils, using saline or brackish water. Many species of microalgae, such as *Dunaliella*, grow in seawater, allowing its utilization for CO_2-enriched air, requiring their co-location with CO_2 sources, such as flue-gases from fossil-fuel fired power plants. The combination of high biomass productivities and the lack of need for arable land and freshwater allows the large-scale production of microalgal biofuels without affecting agricultural commodities prices, thereby avoiding the ethical conflict that arises when diverting crops that are desperately needed to feed a growing world population for biofuel production.

Dunaliella is a particularly promising algae for biofuels production for several reasons: First, it is already produced commercially in Australia, India, Israel and (formerly) the United States. Second, some strains have a high content of commercially valuable natural beta-carotene. Third, it can grow in seawater and brackish water with hypersaline salt concentrations. Fourth, it can produce high levels of oil, starch, or glycerol, and also contains hydrocarbons (Avron and Ben-Amotz 1992, Ben-Amotz and Avron 1983, Oren 2005). It is used herein as a specific example of an alga suitable for biofuels production.

Biofuels from Microalgae

Like other plant biomass, there are several alternatives for using microalgae as fuels:

- drying the biomass followed by direct combustion for power generation or other thermochemical conversions (gasification, pyrolysis) to generate synthesis gas, oils, etc;
- chemical or physical separation of lipids for the production of biodiesel;
- fermentations to produce ethanol, methane or other fuels by microbial action; and
- photobiological processes for hydrogen production.

The production of biofuels using these processes (see Figure 1), including environmental and operational factors that affect their optimal performance, with particular emphasis on *Dunaliella sp.*, will be discussed next.

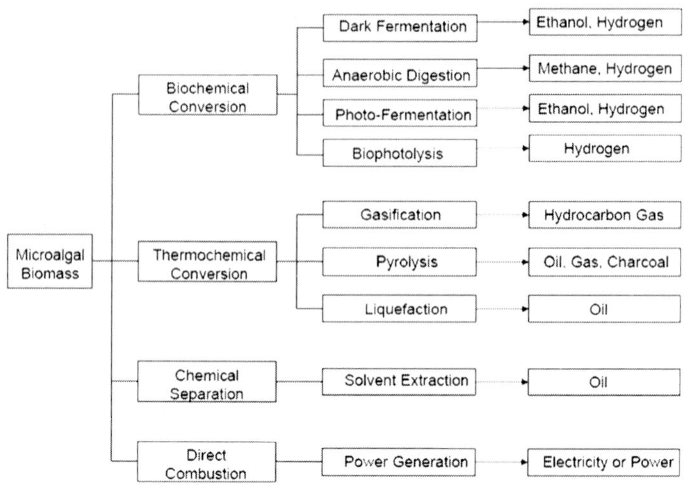

Figure 1: Overview of Microalgal Biomass Conversion Processes for the Generation of Biofuels and Bioenergy (adopted from Tsukahara and Sawayama 2005).

Direct Combustion and Thermochemical Conversions of Microalgal Biomass

Direct Combustion

The most straightforward way to generate energy from biomass is to burn it. Biomass power plants are widely used, and a few coal-fired power plants also co-fire some biomass along with coal. In Europe, in particular, combined heat and power (CHP) biomass systems are used to provide local power and meet heating needs. Direct combustion has been suggested to be applicable to microalgae biomass (Matsumoto et al. 1995, Kadam 2002). Other thermochemical biomass-to-fuels processes can also be considered such as gasification (combustion with limited O_2 to generate a low-BTU fuel gas, also called syngas) and pyrolysis (externally heating biomass to generate a mixture of gaseous and liquid fuels, with a solid fuel residue). However, combustion or other thermochemical processes first require that the algal biomass be dried. Since microalgal cells after harvest contain at least 80% water by weight, it requires more energy to evaporate the water than is contained in the biomass. Thus it is not practical to dry it using any thermal drying process relying on fossil or biomass fuel. The use of waste heat is also difficult as heat transfer would be a limiting factor. Thus only sun-drying can be considered as a low cost and practical process. It requires a lined (plastic or concrete) drying pad, with a greenhouse-type cover in areas where rainfall is likely. Assuming an 85% moisture content of the harvested microalgal biomass, an 0.8 cm/day evaporation rate, and a 25% residual water content after drying, 8 kg of water would evaporate using a 1 m^2 drying pad. Assuming a productivity of 30 g/ m^2·day of ash-free algae biomass (a reasonable average summertime productivity), only 2% of the cultivation area would have to be dedicated for drying. Thus sun-drying is a plausible, although not a costless, alternative. Such sun-drying, which involves daily spreading of the harvested algal slurry on a drying pad and subsequent collection of the dried algal 'chips', can be estimated to roughly cost in the range of $10/ton, or about $1/GJ of the algal biomass heating value, which is not a prohibitive cost.

The main challenge is that the dried algal biomass is a problematic fuel due to its high nitrogen content, which is typically about 10% of the organic, ash-free, dry weight (AFDW), about two orders of magnitude higher than the nitrogen content in woody biomass. Combustion or other thermochemical processes can cause the oxidation of the nitrogen, resulting in NO_x emissions, a major local, regional, and even global pollutant. In addition, such processes would destroy the nitrogen fertilizer contained in the algal biomass, a valuable high energy component. Indeed, most schemes for algal biomass production depend on either recycle or other reuse of the nitrogen fertilizers contained in microalgae. As discussed below, the production of lipids or starches by microalgae may be achievable by making nitrogen the limiting factor in their growth, but even then the nitrogen content of the biomass would be about 4 to 5%, still too high for allowing its thermochemical destruction and the resulting air pollution. Thus, the environmental benefits of combusting microalgae biomass compared to coal or other fossil fuels, e.g., a net reduction in CO_2 and possibly also lower SO_x emissions, must be weighed against the problems caused by the high nitrogen content of algae biomass. But perhaps the most important argument against direct combustion of algae biomass is that this competes with the lowest cost fuels available – coal and wood – which typically cost less than $5/GJ. As discussed below, the cost of producing microalgae biomass is much higher. Thus the production of gaseous and, most importantly, liquid biofuels – ethanol, biodiesel, bio-oils, – are more plausible fuel products from microalgae biomass.

One interesting method of direct combustion is the direct use of microalgae-slurries in diesel engines, where fuel value is high, i.e., greater than $20/GJ at the retail level. First suggested and tested by Jenkins (1993), dried microalgal cells, because of their small size (5 – 15 μm), are comparable to the droplet size of oil sprays used in compression-ignition engines, could be a suitable fuel for diesel engines. However, there is no report on the NO_x emissions from such a system. Scragg et al. (2003) tested this general concept by running a diesel engine on an emulsion consisting of transesterified rapeseed oil (i.e., biodiesel), a surfactant

(Triton X-100), and a 10% slurry of *Chlorella vulgaris* (3-10 μm). This fuel caused no problems, despite the slightly increased viscosity of the fuel, for operating a diesel engine. It was claimed that mixing a thick aqueous slurry of microalgae with the surfactant-amended biodiesel would avoid the energy-intensive drying of the biomass. This however ignores the simple fact that the heat of vaporization of water is a "hidden tax" on the process, thereby reducing its efficiency.

Gasification and Pyrolysis

With insufficient, or even total lack of O_2, any biomass, including algae biomass, can be converted under elevated temperature, and possibly pressure, into gaseous, liquid, and solid fuels (Tsukahara and Sawayama 2005). The composition and relative yields of the various fuel products depend on the process operating conditions: temperature, pressure, catalyst concentrations, treatment time, and O_2 supply (Miao et al. 2004). Gasification processes partially combust the biomass with limited air, providing mainly a gaseous fuel (synthesis gas) with some tar by-products, while pyrolysis processes heat the biomass in the absence of O_2, resulting in a mixture of fuel gases, oil, and chars.

Minowa et al. (1995) carried out the thermochemical liquefaction of wet *Dunaliella tertiolecta* biomass at around 300 °C and 10 MPa, and demonstrated oil yields ranging from 31 to 43%. It was found that under one specific operating condition, the resulting oil had a viscosity and caloric value similar to petroleum fuel oil. The thermochemical liquefaction process exhibited a positive energy output ratio (EOR), which is the ratio of the caloric energy output from the oil to the energy inputs needed to heat up the wet biomass to the required reaction temperature. However, the EOR values were low, with the energy input amounting to two-thirds or more of the caloric energy output from the oil, i.e., the EOR was only 1.5 or lower.

Pyrolysis studies of microalgal biomass (i.e., *Chlorella protothecoides* and *Spirulina platensis*) have evaluated oil yields as a function of temperature, heating rates, and residence times, with maximum oil yields obtained at 500 °C for 5 minutes or at 200-300 °C for 120 minutes (Peng et al. 2000, 2001, see also Miao et al. 2004). Such data should be generally applicable to all algal biomass sources, including *Dunaliella.* Ginzburg (1993) studied *Dunaliella* oil production and considered bio-oil production via pyrolysis of biomass a better option to producing algal lipids, as productivity of algae biomass high in lipids (induced by nutrient stress) is lower than for biomass high in proteins and starches (see also below).

Sawayama et al. (1999) examined the thermochemical liquefaction of wet algal biomass under nitrogen at 2-4 MPa and 300 °C, with oil yields for *Botryococcus braunii* and *Dunaliella tertiolecta* 46% and 35%, respectively. They reported highly favorable EOR values, of about 7 for *Botryococcus braunii,* not surprising considering the high hydrocarbon content of this alga, and a somewhat lower

EOR value of 3 for *Dunaliella tertiolecta*. These investigators also compared EOR values for the liquefaction of other biomass sources such as wood, sewage sludge, and kitchen garbage, and found that they are significantly lower than for the algal biomass, indicating that algae appear to be a good candidate for thermochemical liquefaction, at least from an energy balance perspective.

Minowa and Sawayama (1999) examined low temperature gasification of wet *Chlorella vulgaris* biomass at 350 °C and 18 MPa under nitrogen in the presence of a commercial nickel catalyst. Fuel gas yields increased with catalyst loading and approached the calculated equilibrium gas composition of ca. 50% CH_4, 6% H_2, and 44% CO_2. Low temperature catalytic gasification has the advantage of not requiring dried biomass as other pyrolysis processes, which translates into a more favorable EOR of 3, i.e., three units of fuel energy output per unit energy input. In addition, the ammonia resulting from the gasification of nitrogenous compounds can be recycled as fertilizer for microalgal cultivation, thereby reducing the need for energy-intensive nitrogen fertilizers (Tsukahara and Sawayama 2005). As a result of these benefits, low-temperature high pressure catalytic conversion of microalgal biomass to biogas has significantly greater EOR than incineration.

Wu et al. (1999) subjected the biomass of the marine coccolithophore, *Emiliana huxleyi*, to vacuum pyrolysis, and Miao et al. (2004) studied the fast pyrolysis of *Chlorella protothecoides* and *Microcystis aeroginosa* biomass at 500 °C. These investigators found that bio-oils derived via pyrolysis of microalgae are preferable to those derived from wood because of their lower oxygen content, higher heating value, and greater stability. Miao and Wu (2004) also compared bio-oil production via pyrolysis of *C. prototheocoides* cells grown either autotrophically (light energy and CO_2) or under heterotrophic (fed glucose in the dark) conditions. Heterotrophic cells had high lipid (55%) and low protein (10%) contents, while the converse was true for autotrophic cells (15% lipids and 53% protein). Upon fast pyrolysis at 500%, the heterotrophic cells produced 3.4 times higher oil yields (57%) than the autotrophic ones (17%). The bio-oil from the heterotrophic cells had a much lower oxygen content, higher heating value, lower density, and lower viscosity compared to oils derived from either autotrophic cells or wood. An energy analysis indicated that the fast pyrolysis of algal biomass yielded net biofuel energy since all EOR values were positive, between 10 and 2 for oil derived from the heterotrophic and autotrophic biomass.

However, it must be cautioned that these EOR calculations, and those of the prior cited studies, aside from the heat balances, did not include compression and other parasitic energy inputs used in the process, all of which could significantly change the overall energy balance results. In conclusion, thermochemical energy conversion processes, although interesting experimentally, are not likely to be the best route to biofuels from algal biomass, as the need for drying is not easily circumvented, nitrogen preservation is difficult, and, in any event, the resulting pyrolysis oils or synthesis gases would require significant upgrading prior to commercial use.

Methane Production by Anaerobic Digestion of Microalgal Biomass

Anaerobic digestion involves the microbial conversion of organic matter to biogas, a mixture of CH_4, CO_2, water vapor, and small amounts (typically <1%) H_2S, and sometimes H_2. A wide variety of biomass feedstocks and wastes have been used to generate methane by anaerobic digestion, with lignocellulosic biomass (wood, crop residues, etc.) resulting in little or no gas production while more putrecible substrates (food wastes, wastewater sludges, animal wastes, etc.) generate considerable biogas, typically expressed in terms of liters of methane per gram of volatile solids (VS) destroyed (Gunaseelan 1997). Anaerobic digestion of marine macroalgae such as *Macrocystis* (kelp) and *Sargassum* has been extensively investigated (Chynoweth et al. 2001, Chynoweth 2005), and it is generally observed that their methane yield ranged from 0.26-0.41 L CH_4/g VS, comparable to that of non-marine biomass sources. Since the anaerobic digestion of these macroalgae was carried out under saline conditions, there should be in principle no problems in converting high-salinity *Dunaliella* biomass to methane. The relatively high sulfate content of seawater will result in significant H_2S formation, but there are relatively low-cost biological (sulfide oxidizing bacteria) and chemical (iron sponge) processes for H_2S removal.

There have been, however, only relatively few published studies on the anaerobic digestion of microalgae, none involving biomass from *Dunaliella*. Golueke et al. (1957) studied biogas production from the green microalgae biomass, *Scenedesmus sp.* and *Chlorella sp.*, harvested from algal wastewater ponds. Compared to raw sewage sludge, the algae yielded about one-third lower gas production on a volatile solids input basis, about 0.25 L/g VS added, but the yields and rates increased, also relatively to sewage sludge, as the temperature was raised from 35 to 50 °C, with best yields of about 0.32 L CH_4/g VS input at 11 day retention times. The maximum destruction of algal VS was about 45%, compared to 60% for the sewage sludge VS destruction. They suggested that the relatively low digestability and thus yield of microalgal biomass was the result of cell walls resisting bacterial attack, with the cells remaining viable under anaerobic conditions, but being more readily degraded by bacteria at higher temperatures.

Uziel (1974), using sun-dried *Scenedesmus* sp. reconstituted with water, obtained essentially the same results at mesophilic temperatures. Eisenberg et al. (1979) carried out comparable experiments with dried and fresh green algal biomass and observed similar results, obtaining 0.33 L CH_4/g VS fed for both types of algae at a loading rate of 0.67 kg VS/L·d and a 30 day retention time. Chen and Oswald (1998) evaluated the influence of thermochemical pretreatment of green microalgae biomass, also harvested from sewage ponds, finding that methane production rates increased by one-third when the microalgae were pretreated at

100 °C for 8 hours at a solids concentration of 3.7%. They did not consider the energy required for the thermal pretreatment. Rigoni-Stern et al. (1990) measured a maximum biogas production rate of 0.35 L/g VS during the anaerobic digestion of a natural algal biomass bloom from the highly eutrophic Venice Lagoon in Italy, and this seems to be a generally achieved yield.

Compared to most green algae, cyanobacteria cell walls are more easily broken down by bacterial action. Samson and LeDuy (1983a) subjected *Spirulina maxima* biomass to mechanical (ultrasonic and mechanic disintegration) and thermochemical (pH and temperature) pretreatment and found that while these pretreatments increased the solubility of microalgal biomass and had a positive effect on-acid forming bacteria, they did not result in a significant increase in methane production rates compared to the untreated control biomass, suggesting that, indeed, the cyanobacteria are more readily digested than green algae, although CH_4 yields were not higher. Subsequently, Samson and LeDuy (1986) carried out a detailed parametric and kinetic study of the anaerobic digestion of *Spirulina maxima* biomass and reported a maximum CH_4 yield of 0.35 L/g VS added, equivalent to an energy conversion efficiency of about 59% (as evidenced from volatile solids destruction). This is not that different from the results obtained with green algae, although the higher yield is achieved at lower temperatures.

One possible reason for low methane yields from *Spirulina* is the high nitrogen content of microalgal biomass (C/N~6/1) which results in the accumulation of ammonia, known to inhibit the anaerobic digestion process (Parkin and Owen 1986). In order to circumvent this problem, Samson and LeDuy (1983b) tested the addition of carbon-rich wastes such as sewage sludge, peat hydrolyzate, and spent sulfite liquor to *Spirulina maxima* biomass and found that both methane yield and productivity were doubled when algal biomass was amended with 50% (wt) sewage sludge. Similarly, Yen (2004) and Yen and Brune (2007) added waste paper (50% wt) to microalgal sludge to adjust the C/N ratio to around 20-25/1 which, in turn, doubled the methane production rate from 573 mL/L day to 1170 mL/L day in an anaerobic digester operated at 35 °C and with a hydraulic retention time of 10 days. Samson and LeDuy (1986) reported energy recovery efficiencies ranging from 6 to 60% depending on digester residence time and volatile solids concentrations. Low energy conversion yields for microalgal biomass and the need for digester heating will often result in a negative energy balance for the process, making it intrinsically unviable for biofuel production.

The economics of anaerobic digestion depends on the process used, with a wide range of engineering options available, such as fixed tank, mechanically mixed, heated digesters, used for digestion of wastewater sludges, low solids 'up-flow-sludge blanket reactors', and even covered ponds with gas capture used for animal wastes treatment. The latter is the lowest cost option, but would need to be tested for algal biomass. Cost of methane produced in such systems would range from about $1 to 10/GJ of fuel, with does not include the cost of the biomass inputs.

Which of these processes would be most applicable to methane fermentations of microalgae biomass would depend on the particular circumstances, e.g., if the biomass is produced as part of a wastewater treatment process or if some higher value products (carotenoids, lipids, etc.) are first extracted prior to digestion. The latter is most applicable to *Dunaliella,* where the extraction of carotenoids, and possibly other constituents, is of commercial interest, with the residual biomass likely to be available for anaerobic digestion. It should also be noted that this green alga has only a minimal cell wall, and thus would be highly susceptible to anaerobic digestions.

Production of Ethanol and Other Solvents

There are two different processes by which ethanol can be generated by microalgae. First, by yeast fermentation of carbohydrate storage products such as starch in green algae, glycogen in cyanobacteria, or even glycerol that is accumulated at high salinities by *Dunaliella*. Second, by an endogenous 'self-fermentation' of carbohydrate storage products by algal enzymes induced in the absence of oxygen.

Hirano et al. (1997) cultured *Chlorella vulgaris* and digested the microalgal starch (37% dry wt of biomass) with alpha amylase and glucoamylase to sugars which were then fermented by the common yeast, *Saccharomyces cerevisiae*, to ethanol. Matsumoto et al. (2003) tested the saccharification of starch produced by marine microalgae using a salt-tolerant amylase from a marine bacterium since terrestrial amylases were found to be inactive in saline cell suspensions. This concept may also be applied to *Dunaliella*, which is cultured in saline media. Shirai et al. (1998) cultivated *Dunaliella* on a high-salinity (3% wt NaCl) soy sauce waste, disrupted the cells, an easy step due to the absence of rigid cell walls in *Dunaliella*, and saccharified the intracellular starch using glucoamylase produced by *Rhizopus sp*. The saccharified starch was then fermented by *S. cerevisiae*, yielding 11 mg ethanol per gram *Dunaliella* biomass (dry wt basis).

Glycerol produced by *Dunaliella* can be fermented to solvents, including ethanol. Nakas et al. (1983) found that glycerol produced by five species of *Dunaliella*, when fermented by *Chlostridium pasteurianum*, yielded n-butanol (80-100% on a molar basis), with smaller amounts of 1,3-propanediol, ethanol (<10%), and acetate. Solvent formation is 50% inhibited at 2%, and 100% at 3%, NaCl. An industrial process will therefore require reducing the salt in the algal biomass and/or isolating salt-tolerant strains of solventogenic *Chlostridia*. Von Hippel (1987) fermented glycerol from *Dunaliella salina* lysates with *Klebsiella pneumonia*, reporting ethanol yields of 0.04 to 0.49 g ethanol/g glycerol (10% - 98% of theoretical yield), depending on fermentation conditions. Ethanol production declined sharply when the NaCl concentration in the medium exceeds 1.5%.

Gfeller and Gibbs (1984) demonstrated microalgal self-fermentation of intracellular starch to formate, acetate, ethanol, glycerol, and hydrogen in *Chlamydomonas reinhardtii*, in the dark under anaerobic conditions. Akano et al. (1996) cultured *Chlamydomonas sp.* MGA 161 in an alternating cycle of aerobic photosynthesis and anaerobic dark fermentation, resulting in the accumulation of starch in the light phase and subsequent conversion to ethanol, glycerol, and acetic acid in the dark phase. Hirano et al. (1997) and Hirayama et al. (1998) screened strains of marine microalgae for starch accumulation and subsequent self-fermentation to ethanol. One strain, *Chlamydomonas sp.* (YA-SH-1), was found to have an intracellular starch content of 30%, half of which was converted to ethanol in the dark under anaerobic conditions (Hirayama et al. 1998). Hirano et al. (1997) reported the highest ethanol productivity in *Chlamydomonas reinhardtii* (UTEX2247), with ethanol concentrations reaching 1% (wt). Hon-Nami (1996) grew *Chlamydmonas perigranulata*, containing 35% starch, and demonstrated that the stored starch was converted into ethanol and 2,3-butanediol (in a ~ 2:1 ratio) using the endogenous fermentation enzymes. No self-fermentation of intracellular glycerol or carbohydrates has been reported for any *Dunaliella* species, and is not likely as these algae are not known to ferment under anaerobic conditions.

Hydrogen Production

Photobiological H_2 production has been investigated since the early 1970s as a potential source of this clean fuel (Kheshgi et al. 2005). There are three principal processes by which hydrogen can be generated by microalgae: Dark-fermentations, light-driven fermentations, also called photo-fermentations, and biophotolysis (Benemann 1996, Benemann 2000, Nath and Das 2004).

Dark fermentation involves the anaerobic conversion of microalgal-reduced substrates (e.g., starch, glycogen, glycerol, etc.) into hydrogen, solvents, and mixed acids. This process is carried out either by externally supplied anaerobic heterotrophs (e.g., chlostridia, enteric bacteria, etc.) or, in some cases, by the microalgal cell itself, via a 'self-fermentation' as in the above described ethanol self-fermentations. However, such dark anaerobic H_2 fermentations are limited by thermodynamic constraints to rather low yields, with only about 25% of the energy in the starch being converted to H_2 fuel (Hallenbeck and Benemann 2002). H_2 production by self-fermentations of intracellular starch in *Chlamydomonas* gives similar results to those with of anaerobic bacteria (Gfeller and Gibbs 1984, Klein and Betz 1978, Miura et al. 1981, 1982, Ohta et al. 1987).

The organic acids formed during dark self- or bacterial fermentations can be converted into H_2 using photosynthetic bacteria, a 'photofermentation' (Benemann 1977). Such a process was demonstrated in a small outdoor system (Akano et al. 1996), but had a very low solar energy conversion efficiency. Fermentation of starch from various algae, including *Dunaliella tertiolecta,* to lactic acid and subsequently conversion to H_2 with photosynthetic bacteria was also studied by others (Ike et al. 1997 a,b, Kawaguchi et al. 2001), yielding between 5 to 8 H_2 per mole of starch from *D. tertiolecta*, several-fold the yield that can be obtained in dark fermentations. However, the overall light-energy conversions were low. Ike et al. (1999) used a mixed halophilic culture of *Vibrio fluvialis* and *Rhodobium marinum* to generate H_2 from starch accumulated by green algae, also including *D. tertiolecta*, allowing the utlization of seawater and minimizing contamination.

H_2 can also be produced by microalgae via direct or indirect biophotolysis in which the fundamental concept is to use microalgae to catalyze the conversion of solar energy and water into H_2 fuel, with O_2 as by-product. Most current research is focusing on direct biophotolysis, in which water is split into H_2 and O_2 without intermediate CO_2 fixation. However, this reaction is limited by the strong inhibition of the process (i.e., the hydrogenase enzyme) by the simultaneously produced O_2. Even if O_2 inhibition were somehow overcome, direct biophotolysis would still suffer from the practical problems of generating explosive H_2:O_2 mixtures as well as requiring expensive photobioreactors to contain such a reaction. In indirect biophotolysis, O_2 and H_2 production take place in different stages, with intermediate fixation and release of CO_2, which would be recycled (Hallenbeck and Benemann 2002). In the first, photosynthetic O_2 evolving/CO_2 fixation stage, carbohydrates (starch or glycogen) accumulate, and these are then used in a second stage for light-driven H_2 production. However, neither biophotolysis process, or for that matter dark or photo-fermentations, provide a mechanism for practical applications.

In any case, there are no reports of H_2 generation by *Dunaliella*, either in the dark or light, despite several groups testing for it. *Dunaliella* may lack hydrogenase genes or even a fermentative pathways, which could be rather easily determined with modern tools. Information regarding the presence or absence of a hydrogenase gene should be available as soon as the currently ongoing sequencing of the *Dunaliella salina* genome has been completed by the Joint Genome Institute (JGI, 2006).

Lipid Production

In response to nutrient deficiency (nitrogen or silicon) and other environmental stressors such as extreme pH values, salinity, or heavy metal toxicity, many microalgae species accumulate neutral lipids, in particular triacylglycerols (TAGs) with can be converted to liquid biofuels such as biodiesel, a subject of great current industrial interest.

The *de-novo* synthesis of such high energy storage products is triggered, in so-called oleagineous microalgae, as in oleagineous yeast and fungi, when cellular growth is limited, but metabolic energy supply, photosynthesis for microalgae, is not. Nutrient deficiency, N for green algae and Si for diatoms, is the most common approach to increase intracellular lipid content (Benemann and Tillett 1987, Harwood and Jones 1989, Roessler 1990). It should be noted that some green algae preferentially store starch, rather than lipids, and some can store either, under different conditions. Cyanobacteria do not produce TAGs, storing polybeta-hydroxy butyrate and/or glycogen. Here we only discuss TAGs and lipids in eukaryotic algae.

Lipid accumulation by algae seems to have been first studied by Von Witsch (1948, 1953) in Germany, who reported some algae containing up to 80% TAGs. Shifrin and Chisholm (1981) surveyed the lipid content in 30 species during exponential growth and under nitrogen and silicon limited conditions and found that nitrogen deprivation on average more than doubled the total lipid content of green microalgae from about 17% to 35%. Similarly, Piorreck et al. (1984) found that nitrogen limitation in two green algae (*Chlorella vulgaris* and *Scenedesmus obliquus*) leads to an increase in neutral lipids (up to 53% of biomass), 70% of which were TAGs. Suen et al. (1987) reported that under nitrogen-deficient conditions, *Nannochloropsis sp. QII* accumulated 55% lipids, consisting of 79% TAGs, 9% polar compounds, and 2.5% hydrocarbons. They also showed that such enhanced lipid biosynthesis resulted principally from *de novo* CO_2 fixation. Many others observed similar results: McGinnis et al. (1997) found that the lipids in nitrogen-limited culture of *Chaetoceros muelleri* increased 5-7 fold compared to nitrogen-replete cultures. Wainman and Smith (1997) observed that the lipid content in phytoplankton from natural fresh and saltwater correlated with the nitrate or silicate limitation. Sheehan et al. (1998) reported that nitrogen deficiency led to increased lipid content in, for examples, *Ankistrodesmus* (from 24% to 40%), *Isochrysis* (from 7% to 26%), and *Nannochloris* (from 21% to 35%).

For lipid production, the main objective is to achieve high lipid productivity, which, for photosynthetic systems, must translate into high solar energy conversion efficiencies. However, under nutrient, e.g., nitrogen, limitation, although the fraction of cellular lipids may increase, cellular growth and photosynthesis rates decline, with the overall effect being a reduction in lipid productivity and photosynthetic efficiency. The challenge is to achieve high lipid content at high productivity and solar energy conversion efficiency.

Benemann and Tillett (1987) followed the kinetics of lipid induction in *Nannochloropsis*, shifting cells to increased light intensity upon nitrogen limitation, thus increasing lipid production enabling continued biomass growth. Takagi et al. (2000) attempted to optimize lipid production rates by inducing high lipid accumulation in *Nannochloris* through nitrogen deficiency but as soon as biomass growth slowed down, initiated an intermitting nitrate feeding schedule to enable continued biomass growth. By creating conditions under which cells grew repeatedly under chronic nitrogen deficiency, these investigators found that 10 separate consecutive additions of 0.9 mM nitrate resulted in the same final biomass concentration (2.16 g/L) than a single 9.9 mM nitrate addition, but final lipid content was much higher in the intermittently feed culture than in the one that received the same nitrate concentration as a single initial dose (i.e, 51% versus 31%, respectively).

There have been several studies on lipid production and composition in *Dunaliella*. Tornabene et al. (1980) reported a lipid content of 45-55% (of biomass) in *Dunaliella salina*, with the principal lipids being non-polar and consisting of 30% hydrocarbons, a rather high content, second only to the hydrocarbon-producing alga *Botryococcus braunii*. Pak et al. (1991, 1993) grew *D. salina* in batch and fed-batch cultures and found optimum conditions for hydrocarbon (i.e., hexane extractable lipids) production at pH of 8 and 1.4 g/L NaCl, but this amounted to only 52 mg/L. In batch light/dark cycle (12:12 hours) cultures, maximum hydrocarbon production rates were 35 mg/L·day. Similarly, Park et al. (1998) observed very low hydrocarbon productivities in *Dunaliella salina*.

Nitrogen-limitation appears to have the opposite effect in *Dunaliella* compared to other microalgae. Shifrin and Chrisholm (1981) found that the lipid content in *D. tertiolecta* actually decreased slightly under nitrogen-starved conditions. Sheehan et al. (1998) reviewed a study that noted a large drop in lipid content of *D. salina* upon nitrogen limitation, from 25% to 9%. Gordillo et al. (1998) found that total lipids per cell decreased in a nitrogen-limited culture of *D. viridis* grown in air (0.035% CO_2), but not at 1% CO_2. Recently, Weldy and Huesemann (2007) measured lipid content and productivity in nitrogen-limited (2 mM) and nitrogen-replete (20 mM) batch cultures illuminated at either 200 or 800 μmole/m²·sec. While the biomass lipid content was slightly higher in the nitrogen deficient cultures (i.e., 44% vs. 38%), total lipid content per liter culture was about three times higher in the nitrogen sufficient cultures since nitrogen deficiency greatly slowed down biomass growth (Figure 2). In summary, it appears that nitrogen-limitation in *Dunaliella* cultures does not increase either lipid content or lipid productivity.

Other factors may be more important for lipid production by this alga: An increase in CO_2 concentrations from 2% to 10% resulted in an increase in total lipid fatty acid content by one-third (i.e., from 27.5 to 37 mg/g) after one day and 170% (i.e., from 49 to 133 mg/g) after 7 days (Muradyan et al. 2004, Sergeenko et al. 2000). The effects of increasing medium salinity on lipid content in *Dunaliella* cells are ambiguous, possibly reflecting unique characteristics of specific strains, with two studies reporting decreases in lipid content (Al-Hasan et al. 1987, Sheehan et al. 1998), and two others showing the opposite (Gautam et al. 1994, Takagi et al. 2006). In the latter study, addition of 0.5 or 1 M NaCl at the mid- or end-log phase in cultures containing initially 1 M NaCl further increased the lipid content to 70%, a very high value. As before, the important question is not lipid content but productivity, which was not addressed in these studies. Finally, several studies have evaluated the effects of increasing light intensity on cellular lipid content in *Dunaliella* cultures: Gordillo et al. (1998) observed with increasing light intensity a decline in the mass of total lipids per cell in *Dunaliella*, while Weldy and Huesemann (2007) found no statistically significant change in the

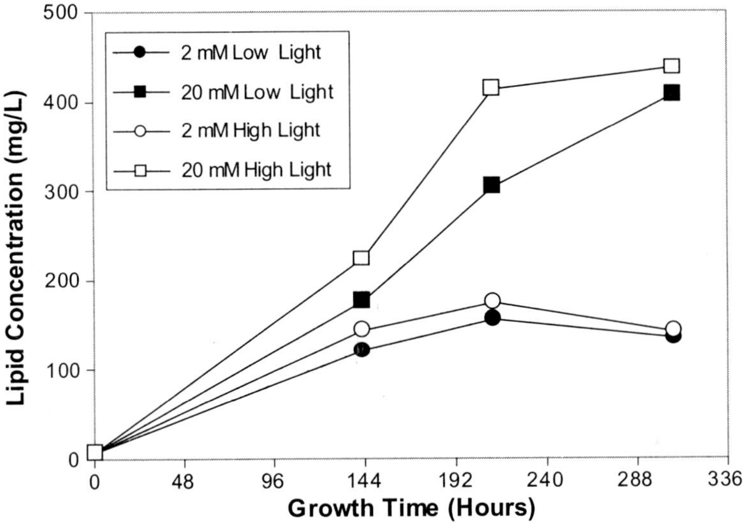

Figure 2: Lipid Production as a Function of Nitrate Concentrations and Light Intensity (Low = 200 μmole/m^2·sec and High = 800 μmole/m^2·sec) in Batch Culture of *Dunaliella salina* (Weldy and Huesemann, 2007).

lipid content of *Dunaliella salina* when the light intensity increased from 200 to 800 μmole/m^2·sec in either nitrogen-deficient (2 mM nitrate) or nitrogen-replete (20 mM nitrate) cultures (Figure 2).

The interest in algal oil production, hydrocarbons, or TAGs is in their use as vehicular fuels, specifically through transesterification of TAGS with methanol to biodiesel by a well-established industrial process (Chisti 2007, Frondel and Peters 2007, Ma and Hanna 1999, Muniyappa 1996). One issue to be considered is that the high moisture content of the algal biomass at harvest interferes with conventional hexane extraction of lipids.

Life Cycle Energy Analysis

For biofuels to reduce atmospheric CO_2 emissions, they must replace fossil fuels. To achieve this, the entire biofuels production and use cycle must have lower fossil CO_2 emissions than the fossil fuels they replace. From a practical perspective, the EOR, the biofuels energy output to fossil fuels input ratio (see above), should also be positive, at least 2, or preferably higher. In such energy analyses the entire life cycle of the fuels should be included.

Life cycle energy analyses are not simple and even for well-established processes there is much uncertainty and argument. For example, there is ongoing debate about whether ethanol from corn, as made in the U.S. at over 100 operating plants, has a positive EOR (Farrell et al. 2006, Hill et al. 2006, Wang 2005, Pimentel and Patzek 2005). The analysis and results depend on selection of system boundaries, the technology used (new plants are more energy efficient), plant size, the type of fuel it uses for process steam (e.g., coal, natural gas, or biogas from animal manures), the agronomic practices of corn farming (e.g., fertilizer use, irrigation, etc.), transportation distances, and many others. Actually, the differences between the high and low EOR estimates for corn ethanol are not as divergent as the arguments surrounding them might suggest, with most showing a very modest positive EOR of about 1.25, with some slightly higher and a few below 1 (e.g., a negative energy efficiency). Even the optimistic studies project no real reduction in CO_2 emissions, although corn ethanol does reduce oil imports (Farrell et al. 2006).

In the case of microalgae, where the technology is not yet at the point of even a pilot plant, such energy and greenhouse gas analyses must be based on a number of assumptions related to system input parameters, which then define the process that must be developed. These can be categorized as the embodied energy in the materials (steel, concrete) and other energy used for system construction, and the operating energy inputs for algal cultivation, harvesting, processing, and fuel conversion systems. Embodied energy is amortized over the expected lifetime of the plant or components. Embodied energy is often difficult to determine, but the uncertainties are generally not large enough to materially affect the overall outcome, as the operating energy inputs typically dominate. Operating energy costs include the pumping of the CO_2 to and then transfer into the algal culture, the energy required to mix the algal culture, the energy required to pump the culture to the harvesting subsystem, the energy for operating the fuel conversion plant, for pumping water to the plant, for wastewater treatment, for inoculum production, for fertilizers, salts, chemicals, and any others that may apply.

Goldman and Ryther (1977) argued that the inputs into large-scale microalgal biofuels (methane) production systems - pond mixing, algal harvesting, water pumping, CO_2 transfer, and, in particular, nitrogen fertilizer supplies - would outweigh the gains, e.g., the EOR would be negative. However, in a rebuttal to this position, Oswald and Benemann (1977) argued that fertilizer nutrients could be recycled from the anaerobic digester, as could the water after harvesting the cells. Further, they argued, that there was no need for fast mixing (which greatly increases power consumption) of such ponds, and that harvesting might be accomplished by a simple sedimentation process ('bioflocculation'). Even if not a sole solution to the energy crisis, algal biofuels could contribute to this goal.

The issue of energy inputs vs. outputs was addressed in some detail by Benemann et al. (1982), and more recently by Benemann and Oswald (1996). These studies were based on fairly detailed, though still conceptual, engineering designs and analyses. For example piping length, diameter, and pumping energy required for CO_2 fertilization were all calculated, as was the mixing energy vs.

velocity in the ponds, based on prior experimental data, among many other parameters. These engineering analyses were also based on many favorable assumptions, most importantly that it would be possible to produce algal biomass at high productivity (about 100 t/ha·yr) and with high oil content (40%) in large, earthen, unlined, paddle wheel mixed raceway ponds. They concluded that a very favorable energy output ratio (EOR) approaching 10 (e.g., fossil energy inputs about 10% of outputs), could be achieved for algal systems that produced oils. One major factor in these favorable analyses is that the residue left over after biodiesel production could be converted to methane gas by anaerobic digestion, and then used to generate all the operating power (mixing, pumping, gas transfer, etc.) required, leaving a surplus.

Minova and Sawayama (1999) carried out a life-cycle energy analysis for cultivation and harvesting of *Chlorella vulgaris* and subsequent low temperature catalytic gasification. They reported a positive energy balance, EOR 1.7, with energy production (17.8 MJ/kg dry cells) exceeding energy consumption of 10.5 MJ/kg (1.54 MJ/kg for nutrients, 2.15 MJ/kg for cultivation, 0.85 MJ/kg for harvesting and 6.0 MJ/kg for the gasification step). Sawayama et al. (1999) estimated the energy embodied in the final microalgal bio-oil to the various energy inputs for fertilizers, cultivation, harvesting, and thermochemical liquefaction for three different biofuel production systems using two different algae. They calculated an EOR of 2.5 when *Botryococcus braunii* was grown in wastewater medium, which decreased slightly, to 2.2, when nitrogen fertilizer was added. A similar system using *Dunaliella tertiolecta* had a negative EOR of 0.6 mainly due to the large amounts of energy required for the thermochemical liquefaction of *Dunaliella* biomass.

These prior studies were all based on open pond systems. Recently, Rodolfi et al. (2007) compared the energy embodied in materials and energy used for mixing, to the energy outputs of algae grown in open ponds, tubular, and flat-plate photobioreactors. They projected that the biomass production (thus biomass energy output) would be greater for the photobioreactors (1300 GJ/ha·yr for tubular and 1900 GJ/ha·yr for flat-plate designs) than for the raceway ponds (1200 GJ ha·yr). However, the energy required for mixing was much greater in the tubular and flat plate photobioreactors (i.e., 180 and 670 GJ/ha·yr, respectively) than the only 6.5 GJ/ha·yr required for the raceway pond. Similarly, the life-time amortized energy content of the materials used in the construction of the flat-bed photobioreactor was 1200 GJ ha·yr while it was only ca. 115 GJ ha·yr for the raceway pond. Thus, while the highly engineered flat-bed photobioreactor has a much greater gross biomass energy output than the simpler outdoor raceway pond (i.e., 1900 versus 1200 GJ ha·yr), the EOR for the ponds is 10 while that for the photobioreactors is 1, so the latter have a zero net energy output. As most energy inputs were not considered (e.g., fertilizers, harvesting, processing, etc.) in this preliminary analysis, it is clear that highly engineered photobioreactors would have a negative energy balance, i.e., EOR would be less than unity and close to zero, indicating that the entire process consumes rather than produces energy. Thus, given that open raceway ponds are also much more economical, as discussed next, they are the only plausible system for production of microalgae biofuels.

Economic Feasibility Analysis

Commercial Systems

The most important issue for microalgae biofuels production is the cost of producing the biomass, which would then be processed to yield the biofuels. The current state-of-the-art of algal biomass production is not very encouraging in this regard. At present, total world production for algae biomass, mainly *Spirulina*, *Chlorella* and *Dunaliella*, is only about 10,000 dry tons per year, about half of this *Spirulina*, most of the rest *Chlorella*, and only about 10% *Dunaliella*. These are not published statistics, but estimates based on personal knowledge (by JRB) of this industry. The cost of production (plant gate, prior to marketing, transportation, overhead charges) can be roughly estimated, on a dry basis, at \$5 per kg of *Spirulina*, \$10/kg of *Chlorella* and about \$20/kg of *Dunaliella*, and these vary significantly with plant location and production process specifics. *Spirulina*, used mainly as a human food supplement, is the cheapest to produce because it can be grown relatively easily in mass culture by virtue of the high bicarbonate medium used, which discourages most contaminants, and because it can be harvested with screens. *Chlorella*, also sold as a food supplement, is more costly because it requires greater control over the culture, with significant amounts of inoculum production, and the culture has to be harvested by centrifugation. *Dunaliella* is the most expensive, because it is produced under highly saline conditions (>100 g/L NaCl) to induce a maximum level of beta-carotene, the commercial product of interest, but this also greatly reduces productivity. In addition, *Dunaliella* harvesting requires expensive chemicals or centrifuges.

Spirulina is cultivated in plastic-lined, raceway ponds, of maximum size between a quarter to a third of a hectare. CO_2 is piped into the ponds and the algae culture is harvested by screens, the underflow returned to the ponds, and the biomass, after washing and further dewatering by belt press is immediately spray-dried. This process is relatively low cost, both capital and operating, and, still, the production costs of the biomass (nominally \$5/kg) exceed those required for fuel production (about \$0.25/kg, prior to conversion to a biofuel) by over an order of magnitude. As just noted, *Chlorella* or *Dunaliella* are even less promising. *Chlorella* is typically produced in circular rather than raceway ponds, often with added acetate, with production moving towards dark fermentations. *Dunaliella salina* is produced in Australia in a total of about 1000 hectares of ponds, with large individual (ca. 100 ha) unmixed, shallow ponds, where the culture naturally develops with minimal management and without CO_2 supplementation. Two smaller plants, in Israel and India, produce this alga in the same type of raceway paddle wheel mixed ponds as used for *Spirulina* production, but production costs are even higher than in the unmixed ponds. One reason is the already noted very low productivity of *Dunaliella* grown at high salt, much lower than that of *Spirulina*, whose productivity already is not very high, at about 30 mt/ha·yr of dry biomass (somewhat higher perhaps at very favorable locations, such as Hawaii). Harvesting these algae is also very expensive - the centrifuges used at one plant have a higher capital cost than the entire pond. cultivation system. It should be noted that there is essentially no information published on any of the commercial systems operating around the world, thus costs and productivities are based on informed speculation rather than published data.

Biofuel Production Systems

The question thus becomes: how to bridge the gap between the current very high cost of algal biomass production for high-value products, and the very low cost required for biofuels production. Some production cost reductions would be possible just by scaling up the process, in terms of total cultivation area and the size of individual growth ponds. For economics of scale, individual growth ponds should be well over one hectare in size, and unlined, as plastic liners are very expensive. It can be roughly estimated that a durable liner (a cheap liner is not durable) would cost roughly $100,000/hectare, installed. Assuming a relatively low 15% annual average capital charge (5% depreciation, 8% cost of capital, 2% fixed costs for insurance, taxes, etc.), and projecting a future production rate of 100 mt/ha·yr, the liner alone adds $150/ton of dry algal biomass, or over half the total cost goal for such a system. Further, the energy embedded in producing the plastic would be significant in an EOR analysis. For this reason, prior cost analyses for algal biofuels were all based on unlined open ponds (Benemann et al. 1982, Benemann and Oswald 1996, Weissman and Goebel 1987).

Indeed, large, mixed, raceway growth ponds have been used in wastewater treatment, and some large, paddle-wheel mixed raceway ponds are used at atleast one *Spirulina* plant, for the purpose of managing excess water (e.g., rain events) and for producing a lower (than human food) quality animal food product. The productivity of such unlined ponds is a major issue. Weissman and Tillett (1990) observed only a modest difference between the productivities of a lined and an unlined pond, suggesting that this may not be a limiting factor. However, currently there is no experience with very large (>1 ha), unlined open ponds for high productivity algae cultivation. Pilot-scale work is required to confirm the feasibility of such large ponds.

In this context, the productivity of algal mass cultures, present and projected, is perhaps the single major issue in this field. Enormous quantities of biofuels, mainly biodiesel, are projected to be produced with microalgae in closed photobioreactors. Most of these projections are made by newly formed companies eager to exploit the interest engendered by the problems of global warming and high oil prices. Some projections even exceed the maximum theoretical limits of photosynthesis, generally given as 10% total solar energy conversion. Even half such a solar efficiency, equivalent to about 100 mt/ha·yr of biomass with a one-third content of extractable TAGs, is already about twice the maximum currently achieved productivity. Its achievement would represent a major advance and is a central goal of R&D in this field. Such a productivity would yield about 30,000 liters of biodiesel/hectare·yr, almost 10 times that of palm oil, the currently highest oil producing crop. Of course such a projection would need to be adjusted for local conditions. It must be noted that this productivity is based on many assumptions, not only a high photosynthetic efficiency by the algae, but also the ability to maintain a stable algal culture, efficient harvest of the algal biomass, and induction of a high level of TAGs without lowering productivity. Most importantly, it is based on developing new algal strains with higher productivity (see below). Land used for harvesting ponds, pond berms, inoculum production, roads, and such, is not included. Most importantly, this productivity is a relatively long-term goal, not a present reality, as portrayed by some.

Algae Harvesting

Another major issue in costs of production is the harvesting of the algal biomass. This has been, perhaps, the major stumbling block to wider applications of microalgae, such as in wastewater treatment. The difficulty of the problem can perhaps be understood by considering that at a typical standing biomass of about 100 g AFDW/m^2 of pond (i.e., equivalent to a concentration of only 400 mg/L at 25 cm depth) and with a daily harvest rate of 30%, yielding 30 g AFDW of cells per m^2 per day, and further assuming that the algal cells at harvest contain 70% water by weight, this represents an approximately 100 micron thin layer of algal cells, each only 10-30 microns in size, that must be concentrated at least a 100 fold (e.g., to 4% solids). And this must be done daily, processing 750 m^3 of liquid per hectare, for a yield of only 30 kg of biomass. How to accomplish this task, cheaply and at high efficiency, has been a major focus of algal biomass R&D.

Chemical flocculation and centrifugation, the main techniques currently in use, are much too expensive in terms of both money and energy to contemplate for the present applications. Screening works well for filamentous algae such as *Spirulina*, and could be cheap enough (though perhaps not quite as cheap as desired), but it cannot be used for most algae, which are unicellular. And for various reasons such as relatively low productivity and the need for high bicarbonate concentrations, *Spirulina* does not currently appear as a suitable species for mass culture. Other filamentous algae could and do deserve more attention.

However, most focus, in particular for biofuels production, is on single-cell or small colonial green algae, such as *Dunaliella*, and possibly diatoms, although for the latter the silicate requirement may be difficult to meet. For such algae the only plausible low-cost harvesting method is 'bioflocculation' , the spontaneous flocculation and settling of the algal cells. Variations thereof are, for example, spontaneous floatation, which may apply to algae with a high lipid/hydrocarbon content and 'autoflocculation' in which the algae co-precipitate with carbonates and phosphates at high pH, induced when CO_2 supply to the culture is curtailed. When small algal cell aggregate into larger flocs, they settle (or float, as the case may be) much more rapidly than single cells, due to Stokes Law. Algae are well known to flocculate spontaneously in natural environments, and also in algal mass cultures, in particular in wastewater treatment systems, but the mechanisms underlying these properties are not understood and thus it has not been possible to apply such a process reliably. All the engineering/economic analyses of algal mass cultures for energy production are based on the assumption that such harvesting processes will be developed. Bioflocculation harvesting could involve a continuous settling tank or a batch flocculation-settling pond. In the latter the algae would first flocculate and then settle, with the algal concentrate collected from the bottom of the plastic-lined pond after the supernatant has been decanted. The settled algae, expected to be about 3 to 4% solids, can then be further concentrated by centrifugation, which is cost-effective even for biofuels production, as the flows handled are small. A critical design is the time required for flocculation and the settling velocity of the flocs. Unfortunately, thus far little work has been done on such a harvesting with *Dunaliella*. For this alga other harvesting alternatives have been explored, including the use of their phototactic behavior. However, these are at present also only conceptual.

Production of Biofuels from *Dunaliella*

For the large-scale production of *Dunaliella,* seawater would be used, which requires access to this resource without excessive pumping. Otherwise there is no great difference between the mass culture of this alga and others. Of course, the entire discussion in this review is based on the assumption that it is actually possible to stably cultivate these (or any) microalgae in open pond mass cultures, something that presently remains to be demonstrated at the scales and costs required for biofuels. The experience with commercial cultivation of *Dunaliella* is not applicable as these systems are based on hypersaline conditions, which although they select for an essentially monoculture of *D. salina*, also results in an overall very low productivity.

Several relatively detailed engineering design studies of algal biofuels production have been published (Benemann et al. 1982, Weissman and Goebel 1987, Benemann and Oswald 1996) which conclude that the production of biofuels using microalgae is feasible in principle, but they had to make many rather optimistic assumptions about the process details to achieve the required low costs. Major assumptions were detailed above: large, open, unlined growth ponds; stable cultivation of the algae; achievement of very high productivities with high oil content; and a low-cost and efficient harvesting process. Assuming these all are actually feasible and achievable, the other issues, such as the supply of CO_2 to the ponds, appear more readily feasible, though they also provide some challenges. Of course, producing an algal biomass high in oils (or fermentable carbohydrate, which is actually easier to produce), is only a first step, it still requires the transformation of the biomass into biofuels, which presents additional issues and costs. As pointed out above, a viable algal oil extraction remains to be developed, or at least demonstrated. It may be possible to combine algae oil recovery and a final concentration by centrifugation into one step (Benemann and Oswald 1996).

In conclusion, thus far, the production of algal biofuels is still a long-term, high risk R&D proposition. A nearer-term approach is to combine production of biofuels with higher value co-products or services, of sufficient market potential to allow production of significant quantities of biofuels. Co-production of food supplements, the main current products of microalgae, would not be suitable, as their markets are miniscule compared to those of biofuels. One attractive approach is to co-produce algal biofuels as part of wastewater treatment, but for seawater algae, such as *Dunaliella*, this option is not available. Co-production of algal biofuels with animal feeds is the more attractive alternative. Two high-value animal feed co-products of seawater microalgae algae are the carotenoids, such as lutein, used in chicken feeds, with markets of several hundred million dollars, and omega-3 fatty acids. As *Dunaliella* is not a good source of the latter, the main opportunity for this alga is to co-produce carotenoids with TAGs or carbohydrates, which would be converted to biodiesel and ethanol, respectively. For example, a 1% content of lutein, only about three-times above current levels, could justify production of *Dunaliella* biomass costing $2/kg, about 10 times the cost point for a biofuels-only process. As *Dunaliella* is already used in the commercial production of beta-carotene, this approach would appear to have merit. Of course, it will require a significant R&D effort to develop such an integrated biofuels-animal feed process.

Research Needs and Conclusions

The above review demonstrates the current early stage of development of microalgae biofuels technologies. None of the biofuels that could be derived from microalgae has been demonstrated even in the laboratory to be produced by or from microalgae in the amounts or yields that meet minimal requirements for practical development. For example, high yields of methane from algal biomass remain to be demonstrated, as does a high triglyceride content in algal biomass produced at high productivity. Hydrogen production, despite literally thousands of publications, remains a laboratory curiosity, and a viable mechanism on which to base even a plausible conceptual process remains to be demonstrated. Combustion and other thermochemical processes are problematic, due to the high nitrogen and moisture content of the algal biomass. In addition, aside from combustion, thermochemical processes, such as gasification and pyrolysis are not yet commercial even with other biomass resources (lignocellulosic biomass, etc.), raising the question of why they would be any better with algal biomass.

The greatest challenge, however, is the production of algal biomass of the required composition at low cost, a few hundred dollars per ton, or over one order of magnitude below current commercial technology. However, the intrinsic advantages of microalgae biofuels – the potential for very high productivity and no competition for agricultural crop land or water resources – justify the long-term R&D required to determine the potential of this technology. Only open raceway ponds can be considered for such applications, and, in the near-term at least, even these will have to include co-production of higher value products or services, such as specialty animal feeds and wastewater treatment.

There is no lack of research needs in the development of microalgae biofuel processes. Perhaps most important is the problem of productivity, i.e., the maximization of solar energy conversion efficiencies. Contrary to popular and even some scientific beliefs, the fundamental problem of microalgae cultures is that they are overall no more efficient at converting solar energy to biomass than higher plants. Photosynthesis is highly efficient at low light intensity, but saturates at about one-tenth of full sunlight, which means that most of the photons are not efficiently used. The solution to this problem, recognized for a long time (Benemann 1990 and citations therein), is to reduce the antenna pigment content of algae, and thus avoid their mutual shading. Work on this problem is ongoing (Huesemann et al. 2008, Melis et al. 1999, Nakajima and Ueda 1997, 2000, Nakajima et al. 2001, Polle et al. 2002, Radmer and Kok 1977, Sukenik et al. 1987) but no improved algal strain is available yet that has been demonstrated to have a higher productivity outside the laboratory. Other research needs, already mentioned above, include the selection and isolation of algal strains that can be mass cultured; the maintenance of stable cultures in ponds for long periods; the production of algal biomass with a high content of lipids, carbohydrates, and co-products, at high productivity; the low-cost harvesting of the biomass; and the extraction and conversion processes to actually derive the biofuels.

For *Dunaliella* specifically, the highest potential is in the co-production of biofuels with high-value carotenoids that are useful in animal feed production. For this, continuing research will be required, including advancing genetic tools to allow development of improved strains of this most interesting and already commercially important alga (Dunahay et al. 1992, Walker et al. 2005).

Acknowledgments

The authors thank Dr. J. Polle for his support and suggestions in the preparation of this chapter.

References

Akano, T., Y. Miura, K. Fukatsu, H. Miyasaka, Y. Ikuta, H. Matsumoto, A. Hamasaki, N. Shioji, T. Mizoguchi, K. Yagi and I. Maeda. 1996. Hydrogen production by photosynthetic microorganisms. Applied Biochemistry and Biotechnology 57/58:677-688.

Al-Hasan, R.H., M.A. Ghannoum, A.K. Sallal, K.H. Abu-Elteen and S.S. Radwan. 1987. Correlative changes of growth, pigmentation and lipid composition of *Dunaliella salina* in response to halostress. Journal of General Microbiology 133:2607-2616.

Avron, M. and A. Ben-Amotz. (eds). 1992. *Dunaliella:* physiology, biochemistry, and biotechnology. CRC Press. Boca Raton, FL, USA.

Ben-Amotz, A. and M. Avron. 1983. Accumulation of metabolites by halotolerant algae and its industrial potential. Annual Review Microbiology 37:95-119.

Benemann, J. 1996. Hydrogen biotechnology: Progress and prospects. Nature Biotechnology 14 (September):1101-1103.

Benemann, J.R. 1977. Hydrogen and methane production through microbial photosynthesis. pp. 285-298. In Living Systems as Energy Converters. R. Buvet et al. (eds). Elsevier/North-Holland Biomedical Press, Amsterdam, The Netherlands.

Benemann, J.R. 2000. Hydrogen production by microalgae. Journal of Applied Phycology 12:291-300.

Benemann, J.R. and D.M. Tillett. 1987. Effects of fluctuating environments on the selection of high yielding microalgae. In Final Report to the Solar Energy Research Institute.

Benemann, J.R. and W.J. Oswald. 1996. Systems and economic analysis of microalgae ponds for conversion of CO_2 to biomass. Final report. US DOE. http://govdocs.aquake.org/cgi/reprint/2004/915/9150050.pdf

Benemann, J.R. and R.P. Goebel, J.C. Weissman and J.C. Augenstein. 1982. Microalgae as a source of liquid fuels. Final technical report US DOE-OER. http://www.osti.gov/bridge/product.biblio.jsp?query_id=0&page=0&osti_id=6374113

Chen, P.H. and W.J. Oswald. 1998. Thermochemical treatment for algal fermentation. Environment International 24 (8):889-897.

Chisti, Y. 2007. Biodiesel from microalgae. Biotechnology Advances 25:294-306.
Chynoweth, D.P. 2005. Renewable biomethane from land and ocean energy crops and organic wastes. HortScience 40 (2):283-286.
Chynoweth, D.P., J.M. Owens and R. Legrand. 2001. Renewable methane from anaerobic digestion of biomass. Renewable Energy 22:1-8.
Dunahay, T.G., E.E. Jarvis, K.G. Zeiler, P.G. Roessler and L.M. Brown. 1992. Genetic engineering of microalgae for fuel production. Applied Biochemistry and Biotechnology 34/35:331-339.
Eisenberg, D.M., W.J. Oswald, J.R. Benemann, R.P. Goebel and T.T. Tiburzi. 1979. Methane fermentation of microalgae. In Anaerobic digestion. D.A. Stafford, B.I. Wheatley and D.E. Hughes (eds). Applied Science Publishers LTD. London, UK.
Farrell, A.E., R.J. Plevin, B.T. Turner, A.D. Jones, M. O'Hare and D.M. Kammen. 2006. Ethanol can contribute to energy and environmental goals. Science 311:506-508.
Frondel, M. and J. Peters. 2007. Biodiesel: a new oildorado? Energy Policy 35:1675-1684.
Gautam, M., T.R. Madaan and B.D. Kaushik. 1994. Lipid profile of *Dunaliella maritima* in response to salinity stress. Indian Journal of Plant Physiology 37 (4):256-258.
Gfeller, R.P. and M. Gibbs. 1984. Fermentative metabolism of *Chlamydomonas reinhardtii* - I. Analysis of fermentative products from starch in the dark and light. Plant Physiology 75:212-218.
Ginzburg, B.-Z. 1993. Liquid fuel (oil) from halophilic algae: a renewable source of non-polluting energy. Renewable Energy 3 (2/3):249-252.
Goldman, J.C. and J.H. Ryther. 1977. Mass production of algae: bioengineering aspects. In Biological Solar Energy Conversion. A. Mitsui, S. Miyachi, A.S. Pietro and S. Tamura (eds). Academic Press. New York, USA.
Golueke, C.G., W.J. Oswald and H.B. Gotaas. 1957. Anaerobic digestion of algae. Applied and Environmental Microbiology 5 (1):47-55.
Gordillo, F.J.L., M. Goutx, F.L. Figueroa and F. X. Niell. 1998. Effects of light intensity, CO_2 and nitrogen supply on lipid class composition of *Dunaliella viridis*. Journal of Applied Phycology 10:135-144.
Gunaseelan, V.N. 1997. Anaerobic digestion of biomass for methane production: a review. Biomass and Bioenergy 13 (1/2):83-114.
Hallenbeck, P.C. and J.R. Benemann. 2002. Biological hydrogen production: Fundamentals and limiting processes. International Journal of Hydrogen Energy 27:1185-1193.
Harwood, J.L. and A.L. Jones. 1989. Lipid metabolism in algae. Advances in Botanical Research 16:1-53.
Hill, J., E. Nelson, D. Tilman, S. Polasky and D. Tiffany. 2006. Environmental, ecomomic, and energetic costs and benefits of biodiesel and ethanol biofuels. PNAS. 103(30): 11206-11210.
Hirano, A., R. Ueda, S. Hirayama and Y. Ogushi. 1997. CO_2 fixation and ethanol production with microalgal photosynthesis and intracellular anaerobic fermentation. Energy 22 (2/3):137-142.
Hirayama, S., R. Ueda, Y. Ogushi, A. Hirano, Y. Samejima, K. Hon-Nami and S. Kunito. 1998. Ethanol production from carbon dioxide by fermentative microalgae. Paper read at Fourth International Conference on Carbon Dioxide Utilization, September 7-11, 1997, at Kyoto, Japan.
Hon-Nami, K. 1996. A unique feature of hydrogen recovery in endogenous starch-to-alcohol fermentation of the marine microalga, *Chlamydomonas perigranulata*. Applied Biochemistry and Biotechnology 131 (1-3):808-828.

Huesemann, M.H. 2006. Can advances in science and technology prevent global warming? A critical review of limitations and challenges. Mitigation and Adaptation Strategies for Global Change, 11:539-577.

Huesemann, M.H., T.S. Hausmann, R. Bartha, M. Aksoy, J. Polle, J.C. Weissman, and J.R. Benemann, "Biomass productivities in wild type and a new pigment mutant of Cyclotella sp. (Diatom)"., Applied Biochemistry and Biotechnology, Online First, July 3, 2008.

Huntley, M.E. and D.G. Redalje. 2007. CO_2 mitigation and renewable oil from photosynthetic microbes: A new appraisal. Mitigation and Adaptation Strategies for Global Change. 12:573-608.

Ike, A., N. Toda, K. Hirata and K. Miyamoto. 1997a. Hydrogen photoproduction from CO_2-fixing microalgal biomass: application of lactic acid fermentation by *Lactobacillus amylovorus*. Journal of Fermentation and Bioengineering 84 (5):428-433.

Ike, A., N. Toda, N. Tsuji, K. Hirata and K. Miyamoto. 1997b. Hydrogen photoproduction from CO_2-fixing microalgal biomass: application of halotolerant photosynthetic bacteria. Journal of Fermentation and Bioengineering 84 (6):606-609.

Ike, A., T. Murakawa, H. Kawaguchi, K. Hirata and K. Miyamoto. 1999. Photoproduction of hydrogen from raw starch using a halophilic bacterial community. Journal of Bioscience and Bioengineering 88 (1):72-77.

Jenkins, D.P. 1993. Diesel power generation without carbon dioxide emissions. Chemical Engineering Research and Design 71 (A3):322-333.

JGI, Joint Genome Institute, 2006. FY2006 Sequencing for the Department of Energy Microbial Genome Program, 2006. http://www.jgi.doe.gov/sequencing/DOEmicrobes2006.html

Kadam, K.L. 2002. Environmental implications of power generation via coal-microalgae cofiring. Energy 27:905-922.

Kawaguchi, H., K. Hashimoto, K. Hirata and K. Miyamoto. 2001. Hydrogen production from algal biomass by a mixed culture of *Rhodobium marinum* A-501 and *Lactobacillus amylovorus*. Journal of Bioscience and Bioengineering 91 (3):277-282.

Kheshgi, H.S., R.C. Prince, and G. Marland. 2000. The potential of biomass fuels in the context of global climate change: Focus on transportation fuels. Annual Review Energy Environment 25:199-244.

Ma, F. and M. Hanna. 1999. Biodiesel production: a review. Bioresource Technology 70:1-15.

Matsumoto, H., N. Shioji, A. Hamasaki, Y. Ikuta, Y. Fukuda, M. Sato, N. Endo and T. Tsukamoto. 1995. Carbon dioxide fixation by microalgae photosynthesis using actual flue gas discharged from a boiler. Applied Biochemistry and Biotechnology 51/52:681-692.

Matsumoto, M., H. Yokouchi, N. Suzuki, H. Ohata and T. Matsunaga. 2003. Saccharification of marine microalgae using marine bacteria for ethanol production. Applied Biochemistry and Biotechnology 105-108:247-254.

McGinnis, K.M., T.A. Dempster and M.R. Sommerfeld. 1997. Characterization of the growth and lipid content of the diatom *Chaetoceros muelleri*. Journal of Applied Phycology 9:19-24.

Miao, X. and Q. Wu. 2004. High yield bio-oil production from fast pyrolysis by metabolic controlling of Chlorella protothecoides. Journal of Biotechnology 110:85-93.

McLaughlin, S.B., D.G. De La Torre Ugarte, C.T. Garten, L.R. Lynd, M.A. Sanderson, V.R. Tolbert and D.D. Wolf. 2002. High-value renewable energy from prairie grasses. Environmental Science and Technology 36(10):2122-2129.

Melis A., J. Neidhardt and J.R. Benemann. 1999. *Dunaliella salina* (Chlorophyta) with small chlorophyll antenna sizes exhibit higher photosynthetic productivities and photon use efficiencies than normally pigmented cells. J. Appl. Phycol. 10: 515–525.

Miao, X., Q. Wu and C. Yang. 2004. Fast pyrolysis of microalgae to produce renewable fuels. Journal of Analytical and Applied Pyrolysis 71:855-863.

Minowa, T. and S. Sawayama. 1999. A novel microalgal system for energy production with nitrogen cycling. Fuel 78:1213-1215.

Minowa, T., S.-Y. Yokoyama, M. Kishimoto and T. Okakura. 1995. Oil production from algal cells of *Dunaliella tertiolecta* by direct thermochemical liquefaction. Fuel 12:1735-1738.

Miura, Y., K. Yagi, Y. Nakano and K. Miyamoto. 1981. Requirement of oxygen for dark hydrogen evolution by a green alga, *Chlamydomonas reinhardtii*. Journal of Fermentation Technology 59:441-446.

Miura, Y., K. Yagi, M. Shoga and K. Miyamoto. 1982. Hydrogen production by a green alga, *Chamydomonas reinhardtii*, in an alternating light/dark cycle. Biotechnology and Bioengineering 24:1555-1563.

Muniyappa, P.R., S.C. Brammer and H. Noureddini. 1996. Improved conversion of plant oils and animal fats into biodiesel and co-products. Bioresource Technology 56:19-24.

Muradyan, E.A., G.L. Klyachko-Gurvich, L.N. Tsoglin, T.V. Sergeyenko and N.A. Pronina. 2004. Changes in lipid metabolism during adaptation of the *Dunaliella salina* photosynthetic apparatus to high CO_2 concentration. Russian Journal of Plant Physiology 51 (1):62-72.

Nakajima, Y. and R. Ueda. 1997. Improvement of microalgal photosynthetic productivity by reducing the content of light harvesting pigment. Journal of Applied Phycology. 9:503-510.

Nakajima, Y. and R. Ueda. 2000. The effect of reducing light-harvesting pigment on marine microalgal productivity. Journal of Applied Phycology 12:285-290.

Nakajima, Y., M. Tsuzuki and R. Ueda. 2001. Improved productivity by reduction of the content of light-harvesting pigment in *Chlamydomonas perigranulata*. Journal of Applied Phycology 13:95-101.

Nakas, J.P., M. Schaedle, C.M. Parkinson, C.E. Coonley and S.W. Tanenbaum. 1983. System development for linked-fermentation production of solvents from algal biomass. Applied and Environmental Microbiology 46 (5):1017-1023.

Nath, K. and D. Das. 2004. Improvement of fermentative hydrogen production: various approaches. Applied Microbiology and Biotechnology 65:520-529.

Ohta, S., K. Miyamoto and Y. Miura. 1987. Hydrogen evolution as a consumption mode of reducing equivalents in green alga fermentation. Plant Physiology 83:1022-1026.

Oren, A. 2005. A hundred years of *Dunaliella* research. Saline Systems 1 (2):1-14.

Oswald, W.J. and J.R. Benemann. 1977. A critical analysis of bioconversion with microalgae. pp. 379-396. In Biological Solar Energy Conversion. A. Mitsui, S. Miyachi, A. San Pietro and S. Tamura (eds). Academic Press, New York, U.S.A.

Pak, J.H., S.-Y. Lee, W.C. Shin and H.-Y. Lee. 1991. Optimization of producing liquid fuel from photosynthetic algal growth. Journal of Microbiology and Biotechnology 1 (2):111-115.

Pak, J.H., S.-Y. Lee, Y.-N. Kim and H.-Y. Lee. 1993. The production of algal hydrocarbons in outdoor cultivations of *Dunaliella salina* 1650. Journal of Microbiology and Biotechnology 3 (1):46-50.

Park, D.-H., H.-W. Ruy, K.-Y. Lee, C.-H. Kang, T.-H. Kim and H.-Y. Lee. 1998. The production of hydrocarbons from photoautotrophic growth of *Dunaliella salina* 1650. Applied Biochemistry and Biotechnology 70-72:739-746.

Parkin, G.F. and W.F. Owen. 1986. Fundamentals of anaerobic digestion of wastewater sludges. Journal of Environmental Engineering 112 (5):867-920.

Peng, W., Q. Wu and P. Tu. 2000. Effects of temperature and holding time on production of renewable fuels from pyrolysis of *Chlorella protothecoides*. Journal of Applied Phycology 12:147-152.

Peng, W., Q. Wu, and P. Tu. Pyrolytic characteristics of heterotrophic *Chlorella protothecoides* for renewable bio-fuel production. Journal of Applied Phycology 13:5-12.

Peng, W., Q. Wu, P. Tu and N. Zhao. 2001. Pyrolytic characteristics of microalgae as renewable energy source determined by thermogravimetric analysis. Bioresource Technology 80:1-7.

Pimentel, D. and T.W. Patzek. 2005. Ethanol production using corn, switchgrass, and wood; Biodiesel production using soybean and sunflower. Natural Resources Research 14(1):65-76.

Piorreck, M., K.-H. Baasch and P. Pohl. 1984. Biomass production, total protein, chlorophylls, lipids and fatty acids of freshwater green and blue-green algae under different nitrogen regimes. Phytochemistry 23:207-216.

Perlack, R.D., L.L. Wright, A.F. Turhollow, R.L. Graham, B. Stokes and D.C. Erbach 2005. Biomass as feedstock for a bioenergy and bioproducts industry: The technical feasibility of a billion-ton annual supply. ORNL/TM-2005/66. www.osti.gov/bridge

Polle, J.E.W., S. Kanakagiri, E.S. Jin, T. Masuda and A. Melis. 2002. Truncated chlorophyll antenna size of the photosystems – A practical method to improve microalgal productivity and hydrogen production in mass culture. International Journal of Hydrogen Energy 27:1257-1264.

Radmer, R.J. and B. Kok. 1977. Light conversion efficiency in photosynthesis. In Photosynthesis I. A. Trebst and M. Avron. (eds) Springer Verlag. Berlin, Germany.

Rigoni-Stern, S., R. Rismondo, L. Szpyrkowicz, F. Zilio-Grandi and P.A. Vigato. 1990. Anaerobic digestion of nitrophilic algal biomass from the Venice Lagoon. Biomass 23 (3):179-199.

Roessler, P.G. 1990. Environmental control of glycerolipid metabolism in microalgae: commercial implications and future research directions. Journal of Phycology 26:393-399.

Rudolfi, L., N. Bassi, G. Padovani, G. Bonini, G.C. Zittelli, N. Biondi and M.R. Tredici. 2007. Presented at the 7th European Workshop on Biotechnology of Microalgae, Berlin, Germany, June 11-14, 2007.

Samson, R. and A. LeDuy. 1983a. Influence of mechanical and thermochemical pretreatments on anaerobic digestion of *Spirulina maxima* algal biomass. Biotechnology Letters 5 (10):671-676.

Samson, R., and A. LeDuy. 1983b. Improved performance of anaerobic digestion of *Spirulina maxima* algal biomass by addition of carbon-rich wastes. Biotechnology Letters 5 (10):677-682.

Samson, R., and A. LeDuy. 1986. Detailed study of anaerobic digestion of *Spirulina maxima* algal biomass. Biotechnology and Bioengineering 28:1014-11023.

Sawayama, S., T. Minowa and S.-Y. Yokoyama. 1999. Possibility of renewable energy production and CO_2 mitigation by thermochemical liquefaction of microalgae. Biomass and Bioenergy 17:33-39.

Scragg, A.H., J. Morrison and S.W. Shales. 2003. The use of a fuel containing Chlorella vulgaris in a diesel engine. Enzyme and Microbial Technology 33:884-889.

Sergeenko, T.V., E.A. Muradyan, N.A. Pronina, G.L. Klyachko-Gurvich, I.M. Mishina and L.N. Tsoglin. 2000. The effect of extremely high CO_2 concentration on the growth and biochemical composition of microalgae. Russian Journal of Plant Physiology 47 (5):632-638.

Sheehan, J., T. Dunahay, J. Benemann and P. Roessler. 1998. A Look Back at the U.S. Department of Energy's Aquatic Species Program - Biodiesel from Algae. Golden. National Renewable Energy Laboratory. Colorado, USA.

Shifrin, N.S. and S.W. Chisholm. 1981. Phytoplankton lipids: interspecific differences and effects of nitrate, silicate, and light-dark cycles. Journal of Phycology 17:374-384.

Shirai, F., K. Kunii, C. Sata, Y. Teramoto, E. Mizuki, S. Murao and S. Nakayama. 1998. Cultivation of microalgae in the solution from the desalting process of soy sauce waste treatment and utilization of algal biomass for ethanol fermentation. World Journal of Microbiology and Biotechnology 14:839-843.

Suen, Y., J.S. Hubbard, G. Holzer and T.G. Tornabene. 1987. Total lipid production of the green alga *Nannochloropsis sp.* Q11 under different nitrogen regimes. Journal of Phycology 23:289-296.

Sukenik, A., P.G. Falkowski, and J. Bennett. 1987. Potential enhancements of photosynthetic energy conversion in algal mass culture. Biotechnology and Bioengineering 30: 970-977.

Takagi, M., Karseno and T. Yoshida. 2006. Effect of salt concentration on intracellular accumulation of lipids and triacylglyceride in marine microalgae *Dunaliella* cells. Journal of Bioscience and Bioengineering 101 (3):223-226.

Takagi, M., K. Wanatabe, K. Yamaberi and T. Yoshida. 2000. Limited feeding of potassium nitrate for intracellular lipid and triglycericde accumulation of *Nannochloris sp.* UTEX LB1999. Applied Microbiology and Biotechnology 54:112-117.

Tornabene, T.G., G. Holzer and S.L. Peterson. 1980. Lipid profile of the halophilic alga, *Dunaliella salina*. Biochemical and Biophysical Research Communications 96 (3):1349-1356.

Tsukahara, K. and S. Sawayama. 2005. Liquid fuel production using microalgae. Journal of the Japan Petroleum Institute 48 (5):251-259.

Uziel, M. 1978. Solar energy fixation and conservation with algal-bacterial systems. Ph.D., University of California, Berkeley, California, USA.

von-Hippel, D.F. 1987. An analysis of an ethanol-producing solar-bioconversion process using the microalga *Dunaliella* sp. as the biomass crop, University of California, Berkeley. USA.

von Witsch, H. 1948. Physiologischer Zustand und Wachstum Intensitaet bei *Chlorella*. Arch. Microbiol. 14:128-141.

von Witsch, H. and R. Harder.1953. Stoffproduktion durch Gruenalgen und Diatomeen in Massenkultur. pp. 154-165. In Algal Culture: From Laboratory to Pilot Plant. J.S. Burlew (ed). Carnegie Institute of Washington, Publication No. 600.Washington, DC, USA.

Wainman, B.C. and R.E.H. Smith. 1997. Can physiochemical factors predict lipid content in phytoplankton? Freshwater Biology 38:571-579.

Walker, T.L., C. Collet and S. Purton. 2005. Algal transgenics in the genome era. J. Phycol. 41(6): 1077–1093.

Wang, M. 2005. Updated energy and greenhouse gas emission results of fuel ethanol. Proceedings of the 15th International Symposium on Alcohol Biofuels 26-28 September 2005.San Diego, California, USA.

Weissman, J.C. and R.P. Goebel. 1987. Design and analysis of microalgal open pond systems for the purpose of producing fuels: A subcontract report. US DOE. http://www.osti.gov/bridge/product.biblio.jsp?query_id=0&page=0&osti_id=6546458

Wu, Q., J. Dai, Y. Shiraiwa, G. Sheng and J. Fu. 1999. A renewable energy source - hydrocarbon gases resulting from pyrolysis of the marine nanoplanktonic alga *Emiliania huxleyi*. Journal of Applied Phycology 11:137-142.

Weldy, C.S. and M.H. Huesemann. 2007. Lipid production by Dunaliella salina in batch culture: Effects of nitrogen limitation and light intensity. Journal of Undergraduate Research, Volume 7, in press.

Weldy, C.S. and M. H. Huesemann. 2007. Lipid production by Dunaliella salina in batch culture: Effects of nitrogen limitation and light intensity. Journal of Undergraduate Research 7:115-122.

Yen, H.-W. 2004. Anaerobic bioassay of methane potential of microalgal biomass. Biosystems Engineering, Clemson University, Clemson, NC, USA.

Yen, H.-W. and D.E. Brune. 2007. Anaerobic co-digestion of algal sludge and waste paper to produce methane. Bioresource Technology 98:130-134.

20 Application of *Dunaliella* in Atherosclerosis

Aviv Shaish[1], Ayelet Harari[1], Yehuda Kamari[1], Hofit Cohen[1], Gustav Schonfeld[2] and Dror Harats[1]*

Abstract

The alga *Dunaliella bardawil* has been investigated as a protective means against atherosclerosis, mainly because of its high content of β-carotene, that potentially, can protect LDL against oxidation, which is a key event in atherogenesis. The effect of *Dunaliella* on risk factors for atherosclerosis (such as increased plasma levels of total cholesterol, LDL cholesterol, TGs, and decreased HDL cholesterol) has been investigated as well. The gathered data suggest that *Dunaliella bardawil* administration has the potential to inhibit LDL oxidation and to influence plasma TG cholesterol, and HDL levels. In studies performed in mouse models of atherosclerosis and human trials, we found that 9-cis rich powder has a more potent anti-atherogenic effect than poor 9-cis β-carotene *Dunaliella* powder. Therefore, we hypothesize that *Dunaliella* powder exerts beneficial effects on atherosclerosis, because of its high content of 9-cis β-carotene, a source for the rexinoid X receptor ligand, 9-cis retinoic acid.

The effect of *Dunaliella* on atherogenesis in patients with risk factors for atherosclerosis has not been studied yet. However, studies in mouse models of atherosclerosis and the beneficial effect on plasma lipids and HDL-cholesterol levels in humans imply that the alga has the potential to inhibit atherosclerosis formation in patients with risk factors for the disease.

Introduction

The alga *Dunaliella bardawil* has been investigated as a protective agent in atherosclerosis, because of its high content of β-carotene, which could potentially

---

[1] The Bert W. Strassburger Lipid Center, Sheba Medical Center, Tel-Hashomer, and Sackler Faculty of Medicine, Tel Aviv University Israel.

[2] 6602 Wohl Hospital, 660 S. Euclid Ave., Campus Box 8046, St. Louis, MO 63110, USA

[*] Corresponding author: Email: avivsha@netvision.net.il

reduce LDL oxidation, a key process in atherogenesis. The effect of *Dunaliella* on risk factors for atherosclerosis (such as increased plasma levels of total cholesterol, LDL cholesterol and TG and decreased HDL cholesterol) has been investigated as well. Our research group studied the hypothesis that *Dunaliella* powder exerts beneficial effects on atherosclerosis because of its high content of 9-cis β-carotene, a source of the rexinoid X receptor ligand, 9-cis retinoic acid. This chapter focuses on the effect of *Dunaliella* on the development of atherosclerosis, as well as on risk factors for the disease.

Atherosclerosis-Atherogenesis and Atherosclerosis Risk Factors

Atherosclerosis is the major cause of morbidity and mortality in developed societies. The name of the disease originated from the Greek words athero (paste) and sclerosis (hardness). The disease is characterized by the accumulation of deposits of fatty substances, cholesterol, cellular waste products, and calcium in the inner lining of large- and medium-sized arteries. The data gathered in the past two decades established the idea that atherosclerosis is an inflammatory disease; inflammatory cells, including monocytes, macrophages, and lymphocytes, play an important role throughout the developing stages of atherosclerosis. In addition, inflammatory pathways contribute to atherothrombosis formation, the late complication of the disease (Libby 2002). The atherogenic process begins when activated endothelial cells recruit monocytes and T-cells from the bloodstream into the wall of the artery. Macrophages that accumulate cholesterol and other fatty materials are transformed into foam cells (Glass and Witztum 2001). With time, these fat-laden foam cells increase both in size and number and form deposits in the wall of the artery, causing a thickening of the artery that ultimately can lead to a reduction of blood flow to the heart (heart attack) or to the brain (stroke), collectively termed 'cardiovascular diseases.'

There are numerous risk factors for atherosclerosis, among them: elevated plasma cholesterol levels, in particular, low-density lipoprotein (LDL) cholesterol; low plasma levels of high-density lipoprotein (HDL) cholesterol; elevated triglyceride (TG) levels; smoking; high blood pressure; diabetes mellitus; elevated levels of C-reactive protein (CRP) in the blood; obesity; inactive lifestyle; age; gender; and family history of cardiovascular diseases (Glass and Witztum 2001).

The next paragraphs provide more background information regarding risk factors for atherosclerosis. Administration of *Dunaliella* was anticipated to modify these risk factors, which were studied in animal models or human trials. These studies will be described and discussed in depth in the following sections.

LDL cholesterol – Cholesterol is carried in the plasma by several lipoproteins that transport dietary and endogenously-produced lipids. In humans, the majority of plasma cholesterol is carried by LDL particles. Substantial evidence has implicated increased levels of LDL-cholesterol with increased prevalence of atherosclerosis (Glass and Witztum 2001; Steinberg 2004) The hypothesis that high plasma levels of cholesterol causally contribute to atherogenesis was initially suggested by the pioneer study of Anitschkow in 1923. In his research, Anitschkow demonstrated both that feeding rabbits with purified cholesterol induced atherosclerosis and that the lesion area was proportional to the levels of blood cholesterol (Anitschkow 1913). The 'cholesterol hypothesis' is supported by several epidemiological studies, showing that coronary death rate is proportional to blood cholesterol and dietary-intervention trials (reviewed by Steinberg 2005). Moreover, in recent years, aggressive LDL-cholesterol-lowering therapy with statins has been shown to prevent cardiovascular events (Grines 2006).

Several lines of evidence suggest that LDL oxidation plays an important role in atherogenesis: *I.* Minimally oxidized LDL induces monocytes to adhere to endothelium *in vitro* (Berliner et al. 1990), which may trigger the recruitment of circulating mononuclear cells into evolving atherosclerotic lesions. *II.* Cultured macrophages rapidly degrade extensively oxidized LDL (Henriksen et al. 1981; Rajavashisth et al. 1990), which may play a role in converting intimal macrophages into lipid-laden foam cells, the pathological hallmark of atherosclerotic lesions. *III.* Lipoprotein-like particles, which appear to have been damaged oxidatively, have been isolated from atherosclerotic tissue (Daugherty et al. 1988). Moreover, protein-bound lipid oxidation products have been detected immunohistochemically in human and animal atherosclerotic lesions (Palinski et al. 1989), suggesting that oxidized LDL is present in the artery wall. *IV.* Finally, chemically unrelated antioxidants slow lesion formation in hyperlipidemic rabbits and primates (Carew et al. 1987; Bjorkhem et al. 1991; Sparrow et al. 1992), implicating oxidized lipoproteins in atherosclerotic vascular disease. Despite these supporting data for the oxidative modification hypothesis, studies in animal models showed dissociation between the effect of several antioxidants on atherosclerosis and LDL susceptibility to oxidation or aortic lipid oxidation (Witting et al. 1999; Upston et al. 2001; Stockner and Keaney 2004). Moreover, supplementing antioxidants to patients, including the synthetic, all-trans β-carotene (one of the two major β-carotene stereoisomers in *Dunaliella*), failed to affect either atherosclerosis or cardiovascular diseases in several trials (Dagenais et al. 2000). The ATBC clinical study used a combination of vitamin E and synthetic all-trans β-carotene in male smokers, in a secondary prevention strategy. However, no benefit on major coronary events was found (Rapola et al. 1996). The Heart Protection Study (HPS) found no effect from a combination of antioxidants (vitamins E and C and all-trans β-carotene) on mortality and incidence of vascular diseases. Moreover, in the HDL Atherosclerosis Treatment Study (HATS), antioxidants (vitamin C, E, all-trans β-carotene, and selenium) negated the benefit of Simvastatin/Niacin on plasma lipid profile and stenosis progression (Brown et al. 2001). These findings cast doubt on the hypothesis that the all-trans stereoisomer of β-carotene from *Dunaliella* has the potential to inhibit atherogenesis by its anti-oxidative activity, *per se*.

HDL-cholesterol – In contrast to LDL-cholesterol, plasma levels of HDL-cholesterol are inversely associated with the prevalence of atherosclerosis. The Veterans Administration HDL Intervention Trial (VA-HIT) showed that increasing HDL-cholesterol in patients treated with the fibrate Gemfibrozil resulted in significant reductions in the risk of major cardiovascular events (Rubins et al.1999). The protective effects of HDL are largely attributed to the ability of HDL particles to remove cholesterol from cells, particularly macrophages. In this process, termed 'Reverse Cholesterol Transport' (RCT), apolipoprotein A-I (apoA-I), the major lipoprotein in HDL, facilitates both the removal of excess cholesterol from peripheral tissues and its delivery to the liver. ATP-binding membrane cassette transport protein A1 (ABCA1) plays a major role in this process by translocating cholesterol from cells into the extracellular space (Ohashi et al. 2005).

Elevated TG levels – The bulk of the evidence suggests that elevated plasma levels of TG is an independent risk factor for coronary heart disease (Mooney 1999), despite some epidemiological studies that did not find an association between plasma levels of TG and cardiovascular mortality (Szapary and Rader 2004). Most of the TGs are carried by chylomicrons and their remnants and by very low density lipoproteins (VLDL). Postprandial chylomicron remnants are believed to be extremely atherogenic, and the liberation of non-esterified fatty acids from TG-rich lipoproteins may contribute to endothelial toxicity and promote atherogenesis.

Diabetes Mellitus – Epidemiological studies established the association between coronary heart disease and both type I and type II diabetes mellitus (Valeri et al. 2004). The hallmark of diabetes is impaired signaling of insulin, which results in the elevation of plasma glucose levels. The consequence of 'glucose toxicity' is accelerated atherosclerosis progression by the increase of advanced glycation end products (AGEs) (Goldin et al. 2006). In addition, diabetes mellitus often increases the conventional risk factors for atherosclerosis, namely, elevated TG, total cholesterol, and LDL cholesterol, and lowers the levels of HDL-cholesterol.

The Effect of *Dunaliella* on Risk Factors for Atherosclerosis in Animal and Human Studies

A. The Effect of *Dunaliella* on Plasma LDL Cholesterol, HDL Cholesterol, and TG Levels

Animal Studies – The effect of *Dunaliella* powder or *Dunaliella* extracts on plasma lipoprotein levels has been studied in several animal models. Itoh et al.

demonstrated that either *Dunaliella bardawil* powder rich in β-carotene, or β-carotene extracted from the alga lowered plasma lipid levels both in mice fed a high-cholesterol diet (Takahashi et al. 2000) and in a rat model of triton-induced hyperlipidemia (Itoh et al.2000). *Dunaliella bardawil* extract decreased total and LDL cholesterol levels in diet-induced hypercholesterolemia in mice. However, this treatment also reduced HDL-cholesterol and increased TG plasma levels, which are associated with increased risk for atherosclerosis (Takahashi et al. 2000).

We studied the effect of *Dunaliella* on plasma lipid and lipoprotein levels in several mouse models: human apoA-I transgenic mice (Rubin et al. 1991); the db/db diabetes model; and two atherosclerotic models, LDLR-/(Harari et al. 2008) and apoE-/- mice. Although the drug fibrate raises HDL-cholesterol in humans (see more details on fibrate action in the following paragraph), it lowers HDL-cholesterol in rodents, which carry most of their plasma cholesterol in HDL particle. Combined treatment of fibrate plus *Dunaliella bardawil* displayed a synergistic effect in human apoA-I transgenic mice (Figure 1).

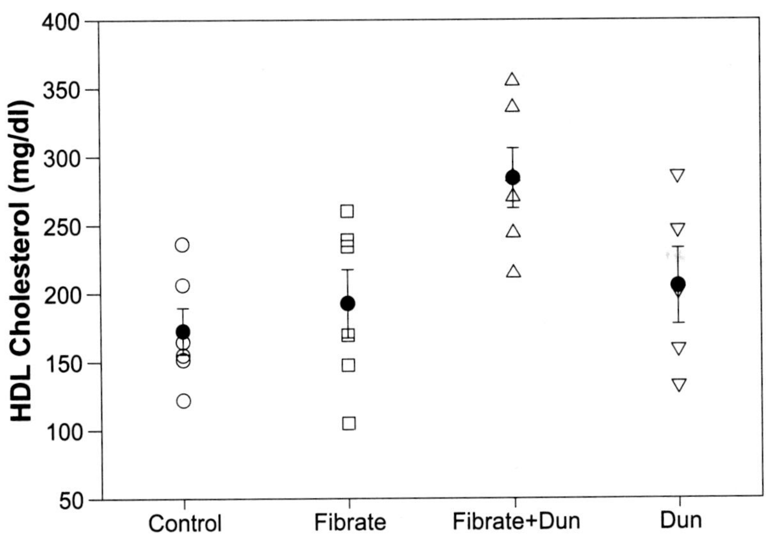

Figure 1: *Dunaliella* increased HDL-cholesterol in Human apoAI transgenic mice. Human apoA-I mice (n=24) were divided into four groups and treated for 2 wk: control group (fed by PBS), fibrate group (fed by Ciprofibrate, 30 μg/g body weight), fibrate plus *Dunaliella* (fed by *Dunaliella* powder) group and *Dunaliella* group. The results are expressed as mean ± SE. P=0.011 compared with control; P<0.044 compared with fibrate group. (with permission from Atherosclerosis (Shaish et al. 2006)).

We also studied the effect of *Dunaliella bardawil* on plasma TG levels in the diabetic db/db mouse model. Plasma TG in a control group was elevated, as expected in this mouse model (134±17 mg/dl), and TG levels in the *Dunaliella*-treated group were 95±8 mg/dl only, after 6 wk of treatment (Figure 2).

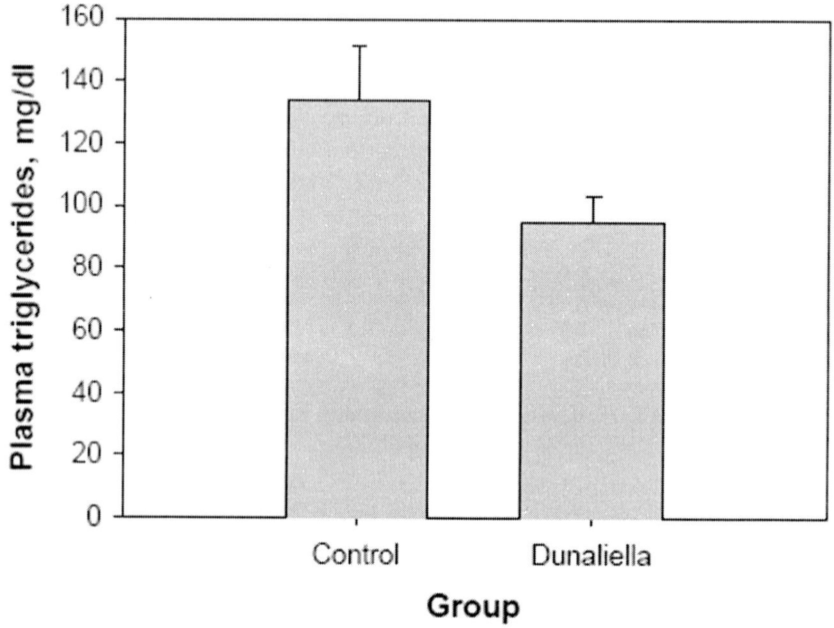

Figure 2: *Dunaliella* reduced TG levels in db/db mice.
db/db mice were treated with *Dunaliella*. The results are expressed as mean ±SE. * $P<0.05$.

We further studied the effect of *Dunaliella bardawil* on atherosclerosis in two mouse models, LDL-R-/-(Harari et al. 2008) and apoE-/- mice. The alga reduced plasma cholesterol in LDLR-/- and apoE-/- mice fed atherogenic diet (Figure 3). β-carotene-deficient algal powder had no effect on the lipid profile in LDLR-/- mice, suggesting that β-carotene, and not other components of the alga, is the active compound in the alga. We further showed that algal preparation containing a high ratio of 9-cis to all-trans (50%:50%) had a greater effect on plasma cholesterol levels in LDLR-/- mice than the algal preparation containing a lower ratio (25%:75%) (Figure 4). This finding suggests that the beneficial effects of *Dunaliella* on plasma lipids can be attributed particularly to the 9-cis β-carotene stereoisomer. In addition to the mouse models, the effects of 9-cis rich *Dunaliella bardawil* powder was studied in a rat model of fructose-induced hypertriglyceridemia. Similar to the mouse models, *Dunaliella* treatment reduced plasma TG levels (data not shown).

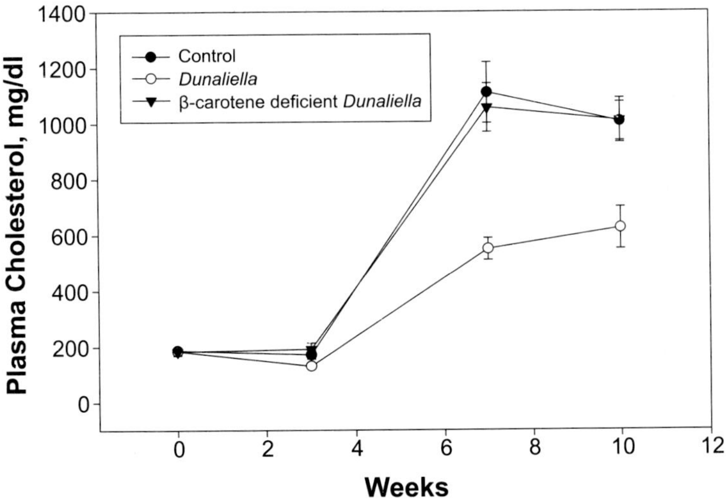

Figure 3: *Dunaliella* reduced plasma cholesterol levels in LDLR-/- mice. Mice were treated with *Dunaliella* or β-carotene-deficient *Dunaliella* powder as described in the text. Results represent the mean ± SE. Differences between groups were compared by ANOVA. $P<0.001$.

In contrast to the beneficial effect of *Dunaliella* on plasma lipid levels in rat and mouse models, the administration of 9-cis extract (isolated from the *Dunaliella bardawil*) to high-cholesterol fed, New Zealand white rabbits did not affect plasma cholesterol or TG levels. The conflicting results between the mouse and the rabbit are probably due to the much lower dose administrated to the rabbits (Shaish et al. 1995).

Human trials – Several research groups have studied the effect of 9-cis β-carotene stereoisomer *Dunaliella bardawil* powder on plasma lipoprotein levels in humans. Itoh et al. (2000) treated 32 hyperlipidemic patients with *Dunaliella bardawil* powder containing 40 mg or 120 mg of β-carotene per d, for 10 wk. The treatment decreased plasma levels of total cholesterol, LDL cholesterol, TG, and lipid peroxides. In this study, no effect on plasma levels of HDL-cholesterol was found. We studied the effect of a combined treatment of the drug fibrate and *Dunaliella bardawil* treatment on plasma HDL cholesterol and TG levels (Shaish et al. 2006). Fibrates are effective in lowering plasma TG and increasing HDL-cholesterol levels when their baseline plasma concentrations are low (Staels et al. 1998). Raising HDL-cholesterol and lowering plasma TG levels with fibrates reduced coronary events in patients with low plasma HDL-cholesterol levels (Rubins et al. 1999).

It was postulated that the effect of fibrates on HDL-cholesterol levels would be mediated by their effects on apoA-I expression, which would be controlled by the peroxisome proliferator-activated receptor-a (PPARα), a member of the PPAR

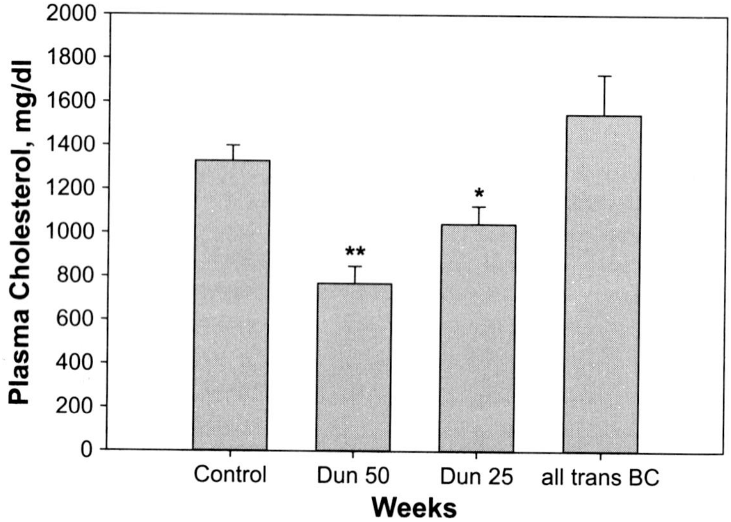

Figure 4: *Dunaliella* powder containing high 9-cis β-carotene is more efficient than low 9-cis β-carotene powder in reducing plasma cholesterol levels in LDLR-/- mice.
Mice were treated with *Dunaliella* powder as described in the text. Results represent the mean ± SE. Differences between groups were compared by ANOVA.
Dun 25=*Dunaliella* powder containing 25% 9-cis β-carotene and 75% all-trans β-carotene. Dun 50=*Dunaliella* powder containing 50% 9-cis β-carotene and 50% all-trans β-carotene.
All-trans BC = synthetic all-trans β-carotene. *p<0.05; **p<0.001.

family (Vu-Dac et al. 1994). PPARs function as ligand-dependent transcription factors that, upon activation, heterodimerize with the 9-cis retinoic acid receptor (RXR). The heterodimer binds to specific response elements, termed peroxisome proliferator response elements (PPREs), and regulates gene expression (Schoonjans et al. 1996). Several studies have shown that PPAR ligands increase the expression of ATP binding cassette transporter A1 (ABCA1), leading to enhanced reverse-cholesterol transport and elevation in plasma HDL-cholesterol levels (Chinetti et al. 2001). It should be noted that either partner, RXR or PPAR, can regulate the transcriptional activity of the DNA bound complex by interacting with its own ligand; however, co-treatment with both ligands has a synergistic effect (Kliewer et al. 1992).

The 9-cis β-carotene stereoisomer has been shown to be a precursor of 9-cis retinoic acid both *in vitro* in human intestinal mucosa (Wang et al. 1994) and *in vivo* in a ferret perfused with 9-cis β-carotene (Hebuterne et al. 1995). Moreover, β-carotene rich powder of the alga *Dunaliella bardawil* was shown as a source of retinol in a rat (Ben-Amotz et al. 1988). Therefore, 9-cis β-carotene rich powder administration has the potential to improve fibrate action via its conversion to 9-cis retinoic acid.

We performed two small-scale human trials that examined the effect of combined fibrate and *Dunaliella* on TG and HDL levels (Shaish et al. 2006). In an open-labeled trial, 20 fibrate-treated men with plasma HDL-cholesterol levels below 40 mg/dl were given *Dunaliella* capsules, providing 60 mg β-carotene per day, containing all-trans and 9-cis β-carotene (1:1 ratio; weight/weight). In a second study, 22 fibrate-treated patients participated in a double-blind, placebo controlled trial. Eleven patients were treated with *Dunaliella* capsules, and 11 patients were treated with β-carotene-deficient *Dunaliella* capsules. Six wk of the combined treatment increased plasma HDL-cholesterol by 24.5% and 12.7% in the first and second trials respectively (P=0.002 and 0.012) (Figure 5a,b). An inverse correlation was found between baseline dietary vitamin A intake, as calculated from a dietary interview, and delta HDL (R=-0.467, P=0.03). A trend toward inverse correlation was also detected between baseline dietary β-carotene intake and delta HDL (R=-0.395, P=0.08). In contrast to the first trial, TG levels decreased significantly (P=0.017) by 26.7% (Figure 5b). Levels of apoC-III and apoC-II, and the TG/apoC-III and apoC-II/apoC-III ratio were not affected by fibrate plus *Dunaliella* treatment (data not shown). In all measured parameters within the placebo group, no significant changes were observed.

The Effect of *Dunaliella* on LDL Oxidation

The effect of all-trans β-carotene, one of the two main stereoisomers found in *Dunaliella bardawil* on LDL oxidation, has been investigated in several studies by using a synthetic all-trans stereoisomer. However, the data regarding the ability of 9-cis β-carotene to protect LDL against oxidation are limited. β-carotene quenches singlet oxygen and free radicals rapidly and exhibits good radical-trapping antioxidant properties at low partial pressure of oxygen (Burton and Ingold 1984). β-carotene also acts synergistically with α-tocopherol as a radical-trapping antioxidant in membranes. The reports of the ability of β-carotene to protect LDL against oxidation are controversial: Jialal et al. 1991. reported that all-trans β-carotene inhibits LDL oxidation both in a cell-free system (using copper as an oxidant) and in cellular systems. Supplementation of LDL *in vitro* with β-carotene protected against copper, AAPH or γ-radiolysis-induced oxidation, measured as)mobility of LDL by laser doppler electrophoresis (Packer, 1993; Arrio et al. 1993; Glass and Witzum 2001) and protected LDL against oxidation by endothelial cells in culture (Dugas et al. 1998). In other work, the addition of all-trans β-carotene to Lp(a) *in vitro* partially protected Lp(a) against oxidation by copper (Naruszewicz et al. 1992). In contrast to other studies research found that β-carotene had no effect on LDL

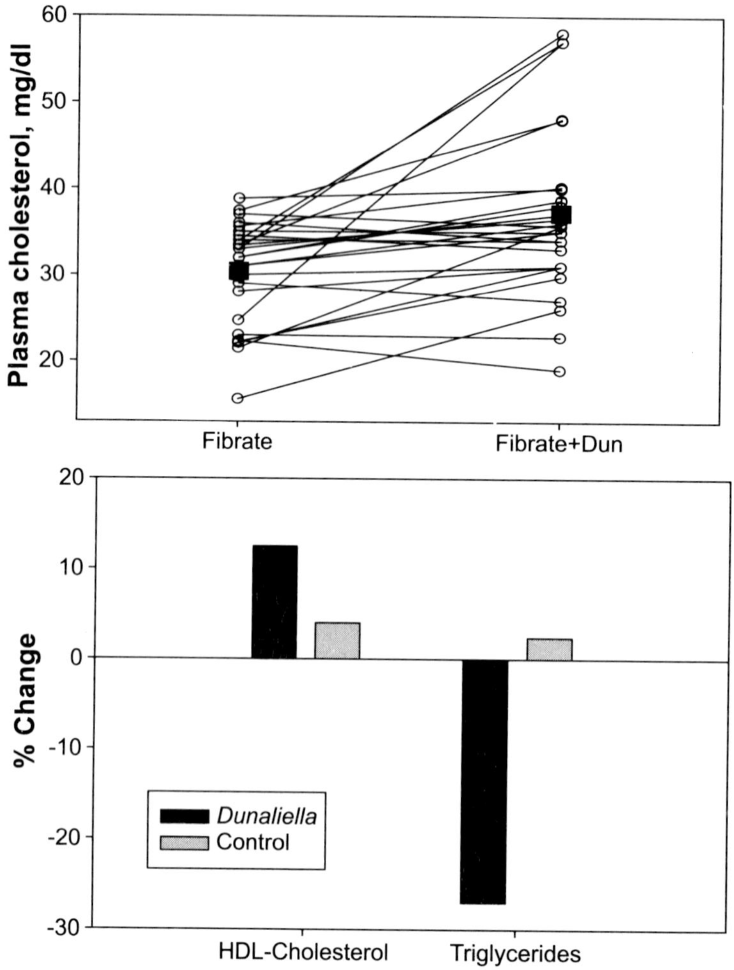

Figure 5: *Dunaliella* increased plasma HDL-cholesterol and decreased plasma TG levels in patients with dyslipidemia.
a: Plasma HDL-cholesterol levels (n=20) at baseline (Fibrate= fibrate treatment only), and after combined treatment (Fibrate+DUN= fibrate plus *Dunaliella*).
b: Plasma HDL-cholesterol and triglycerides levels in the placebo treated group (n=11) and in *Dunaliella* treated group (n=11). (with permission from Atherosclerosis (Shaish et al. 2006)).

oxidation induced by copper *in vitro.* (Reaven et al. 1993; Morel et al. 1983, van Hinsbergh etal. 1986) and *ex vivo* (28-30). β-carotene did not protect against metal ion-dependent and independent oxidation of LDL *ex vivo*, measured as the lag phase of diene oxidation formation (Frei and Gaaziano 1993). Moreover, one study (Bowen and Omaye 1998) showed that in copper-mediated oxidation, increased LDL β-carotene levels may cancel the protective effect of a-tocopherol.

The effect of natural cis stereoisomers of β-carotene on lipoprotein oxidation is not clear. Such isomers include 9-cis, which is found in plant organs that are exposed to sunlight and in *Dunaliella bardawil* that accumulates massive amounts of this stereoisomer (Ben-Amotz et al. 1988).

In Vitro Studies

Lavy et al. (1993) loaded isolated human lipoproteins with all-trans or a mixture of 9-cis and all-trans β-carotene from the alga *Dunaliella bardawil*. The incubation of whole plasma with the β-carotene stereoisomers resulted in substantial β-carotene binding to LDL and VLDL. Lipid peroxidation of VLDL and LDL was inhibited by all-trans and *Dunaliella*-derived β-carotene significantly. However, the all-trans was twice as effective as *Dunaliella* β-carotene.

We compared the effects of all-trans and 9-cis, the two β-carotene stereoisomers present in the highest amount in *Dunaliella bardawil*, against oxidation of lipoprotein core lipids. Human LDL was loaded with the two stereoisomers using a modified method for microsome enrichment with antioxidants (Palozza et al. 2006). We found that all-trans and 9-cis stereoisomers of β-carotene protect the core lipids of LDL against oxidation by water soluble peroxyl radicals initiated by AAPH (Table 1; Table 2).

In contrast to the protection against AAPH oxidation, we did not detect any protective effect of β-carotene against copper-induced lipoprotein oxidation. Although both copper and AAPH are water-soluble radical generators, the initiation of radical production is different. AAPH is a compound that thermally decomposes and generates peroxyl radicals at a constant rate. The mechanism by which copper initiates LDL oxidation is not fully understood. Copper binds to LDL, which has been shown to contain binding sites crucial for lipid peroxidation initiation (Wagner and Heinecke 1997; Roland et al. 2001). Copper and other metal ions may initiate oxidation by generating alkoxy and peroxy lipids from preformed lipid hydroperoxides. Frei and Gaziano (1993) found no correlation between the length of the lag phase of LDL exposed to copper or AAPH. Following β-carotene supplementation *in vivo*, the lag phase of LDL oxidation by copper was shorter ($P<0.01$), and there was a trend with AAPH toward a longer lag phase ($P=0.07$) (Gaziano et al. 1995; Reaven et al. 1994) on the other hand, found no effect of β-carotene on AAPH-induced oxidation. However, in that study, a very high level of 10mM AAPH was used (Reaven et al. 1994). When 0.1mM AAPH was used, β-carotene protected the lipoprotein (Table 2). The physiological significance of β-carotene protective effect against cholesterol

Table 1: *Dunaliella* 9-cis and synthetic all-trans β-carotene protected LDL against AAPH-induced oxidation. LDL was oxidized with 1 mM AAPH for 5 h and MDA equivalents formation was measured. Data represent the mean±SD of three experiments. Differences between groups were compared by Mann-Whitney test. *p<0.05.

Antioxidant	**nmol MDA**
Control	13.8±2.3
α-tocopherol	2.3±1.2*
all-trans β-carotene	7.9±4.4*
9-cis β-carotene	5.1±3.3*

Table 2: *Dunaliella* 9-cis and synthetic all-trans β-carotene protected cholesteryl esters in LDL against AAPH-induced oxidation. Human LDL was loaded with antioxidants. The LDL was oxidized with 1mM AAPH at 37°C for 6 h. PLOOH and CEOOH were extracted and reduced to alcohols with 1.2mM triphenylphosphine. CEOH was measured by reverse phase C18 column. Elution was performed isocratically at 1 ml/min, with acetonitrile:2-propanol:H_2O (44:54:2 v/v/v) as a solvent. PLOH was measured by a silica column. Elution was performed isocratically at 1 ml/min. With acetonitrile:tert butanol:H_2O (55:30:15 v/v/v) as a solvent. Results represent the mean±SD. Differences between groups were compared by Mann-Whitney test. *P<0.05 compared with the control group.

Antioxidant	**PLOOH**	**CEOOH**
	μg/mg protein	μg/mg protein
Control	30.3±13.2	1171±729
α-tocopherol	15.3±6.0*	518±345*
all-trans β-carotene	29.3±6.9	487±220*
9-cis β-carotene	23.3±11.6	628±332*

ester oxidation is unknown. Although epidemiological studies have shown that a high intake of β-carotene is associated with a decreased risk for coronary heart disease (Gaziano 1996; Steinberg 2005), it is not clear whether β-carotene has an anti-atherogenic effect by protecting LDL against oxidation. In humans, in contrast to rabbits or mice, LDL contains β-carotene, and its level is controlled by β-carotene intake in the diet (Gaziano et al. 1995) and by other factors, such as smoking (Dietrich et al. 2003). In the experiment described above, very high concentrations of β-carotene were used to demonstrate its protective effect, and it is not known whether the lower levels of β-carotene in LDL *in vivo*, achieved by dietary supplementation, can provide similar protection against LDL oxidation in the arterial wall.

Animal Studies

The potential for *Dunaliella* to act as an antioxidant *in vivo* was first studied in rats (Levin et al. 1997). The rats were fed a diet containing fresh or oxidized soybean oil supplemented with synthetic all-trans β-carotene or *Dunaliella* extract containing 75% 9-cis β-carotene. Both carotene sources inhibited erythrocyte peroxidation similarly. However, the authors suggested that the observation that 9-cis β-carotene degraded faster than the all-trans isomer may indicate that the 9-cis isomer is a more efficient antioxidant *in vivo*. Administration of 9-cis extract, isolated from the alga *Dunaliella bardawil*, to New Zealand white rabbits that were fed a high-cholesterol diet did not affect *ex vivo* LDL oxidation. Similar results were obtained with the synthetic all-trans β-carotene. These results were not unexpected, because β-carotene is not accumulated in the plasma and in LDL in rabbits (Shaish et al. 1995).

Human Trials

To study the anti-oxidation activity of β-carotene in LDL, Levy et al. (Levy and Aviram 1995) LDL, Levy et al. 1995 compared the effect of synthetic all-trans β-carotene with *Dunaliella bardawil* powder containing a mixture of 9-cis and all-trans stereoisomers. Administration of β-carotene from the two sources to healthy volunteers, for 14 d, resulted in elevation of plasma and LDL β-carotene levels. However, although the elevation in all-trans group was 3.6 fold and the elevation in *Dunaliella* group was only 1.8, both treatments protected LDL against copper-induced oxidation similarly, measured by thiobarbituric acid-reactive substances accumulation. A further study of this group (Levy et al. 2000) demonstrated that *Dunaliella bardawil* administered to patients with diabetes mellitus protected LDL against copper-induced LDL oxidation *ex vivo*. In that trial, 20 patients were treated with *Dunaliella bardawil* powder that provided 60 mg β-carotene per d for 3 wk. LDL isolated from the diabetic patients was more susceptible to copper-induced oxidation than LDL from normal volunteers. *Dunaliella bardawil* powder administration protected the LDL particles as measured by three parameters: lag phase of LDL oxidation, malondialdehyde formation, and lipid peroxides accumulation.

The Effect of *Dunaliella* on Atherosclerosis in Animal Models

To test the effect of *Dunaliella* on atherosclerosis, we used male apoE-deficient mice. This mouse model develops atherosclerotic lesions similar to human atherosclerotic plaques. Mice were fed with a standard chow diet supplemented with 8% *Dunaliella* powder for 6 wk. A control group received the chow diet alone. Atherosclerosis, determined by the lesion area at the aortic sinus, was significantly lower in the *Dunaliella*-treated mice compared with the control group of untreated mice (Figure 6). Similar results were obtained in LDL receptor -/- mice (Harari et al. 2008).

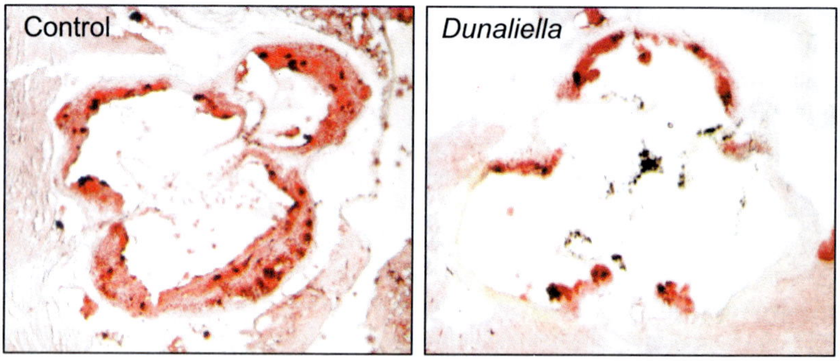

Figure 6: *Dunaliella* reduced atherosclerotic lesion area in apoE-/- mice. Mice were fed *Dunaliella* powder or β-carotene-deficient powder (*Dunaliella*-BC).

The effect of the 9-cis β-carotene stereoisomer, isolated from the alga *Dunaliella bardawil* on atherosclerosis, was studied also in New Zealand. white rabbits, fed a high-cholesterol diet. In this study, 9-cis failed to inhibit atherogenesis, but synthetic all-trans reduced atherogenesis significantly. We assumed that the use of very low levels of the 9-cis stereoisomer and its relatively fast oxidation in the food (less than 0.01%) were insufficient to inhibit atherogenesis in that study.

Summary

The following findings reflect the data accumulated from the experiments performed, to date, to study the effect of *Dunaliella* on atherosclerosis or atherosclerosis-related risk factors:

- **Plasma total cholesterol levels:** *Dunaliella bardawil* powder inhibited plasma total cholesterol elevation in LDL-R-/- mice that were fed with a Western diet.
- **Plasma TG levels:** *Dunaliella bardawil* powder reduced plasma TG levels in rats and in db/db mice. Moreover, a combined treatment of the drug Fibrate plus *Dunaliella* reduced TG in hyperlipidemic patients.
- **Plasma HDL-cholesterol levels:** *Dunaliella bardawil* powder increased plasma HDL-cholesterol levels in a mouse model, and a combined treatment of the drug Fibrate plus *Dunaliella* increased plasma HDL-cholesterol in patients with low levels of HDL-cholesterol.
- **LDL oxidation:** β-carotene derived from *Dunaliella* inhibited LDL oxidation *in vitro* and *ex vivo* in healthy volunteers and in diabetic patients treated with *Dunaliella bardawil* powder. Regardless of these results, the effect of β-carotene on LDL oxidation seems irrelevant to atherogenesis, because antioxidant supplementation (including the synthetic, all-trans β-carotene) to patients failed to affect either atherosclerosis or cardiovascular diseases in several trials.
- **Atherogenesis:** 9-cis β-carotene-rich *Dunaliella bardawil* powder reduced atherogenesis in both LDLR-/- and apoE-/- mouse models.

Conclusions

The amassed data suggest that *Dunaliella bardawil* administration has the potential to manipulate several risk factors associated with atherosclerosis, including increased plasma TG levels, and low plasma HDL levels. We hypothesize that the beneficial effects can be attributed mainly to the 9-cis β-carotene stereoisomer and not to the protective anti-oxidant properties of all-trans β-carotene or 9-cis β-carotene. The findings that 9-cis-rich *Dunaliella* powder augmented the activity of PPARα ligand point out that 9-cis β-carotene may act as the precursor of 9-cis retinoic acid, the natural ligand of the nuclear receptor RXR (see Figure 7). Activation of RXR by a diet enriched with RXR ligand precursor has the potential to augment numerous metabolic pathways.

The effect of *Dunaliella* on atherogenesis in patients has not been studied yet. However, encouraging results in mouse models of atherosclerosis and the beneficial effects on plasma lipids and HDL-cholesterol levels in humans suggest that the alga has the potential to inhibit atherosclerosis progression in patients.

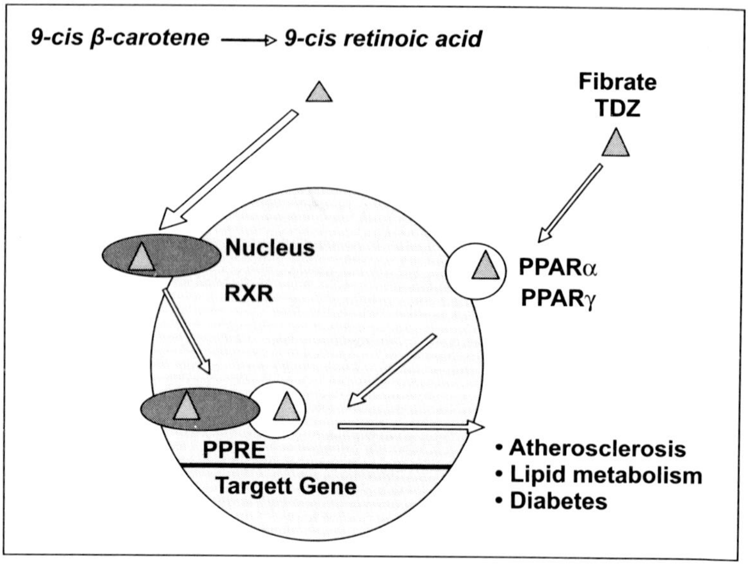

Figure 7: 9-cis β-carotene *from Dunaliella* - a precursor for RXR ligand augments the effect of PPAR ligands. PPARs function as ligand-dependent transcription factors, which upon activation heterodimerize with the 9-cis retinoic acid receptor (RXR). The heterodimer binds to specific response elements, termed peroxisome proliferator response elements (PPREs), and regulates genes involved in atherosclerosis and diabetes. We hypothesize that 9-cis β-carotene from *Dunaliella* is converted to 9-cis retinoic acid and, therefore, augments the effect of PPAR-α and PPAR-γ ligand.

Acknowledgments

This work was supported by the Nikken Sohonsha Corporation, Gifu, Japan; Ministry of Health, Chief Scientist's Office, Jerusalem, Grant no. 4055; Green Foundation, Tel-Aviv University. A complete credit is given to the editor of Atherosclerosis, Elsevier Science Ireland Ltd. for the permission to replicate Figures 1 and 5.

References

Anitschkow, N.N. 1913. Ueberexperimentelle choleserinstaetose und ihre bedeutung fur die entstehung einiger pathologischer prozesse. Zentralbl. Allg. Pathol. 24: 1-9.

Arrio, B., D. Bonnefont-Rousselot, J.D. Catudioc and L. Packer. 1993. Electrophoretic mobility changes of oxidized human low density lipoprotein measured by laser Doppler electrophoresis. Biochem Mol Biol Int 30: 1101-1114.

Ben-Amotz, A., A. Lers and M. Avron. 1988. Stereoisomers of beta-Carotene and Phytoene in the Alga Dunaliella bardawil. Plant Physiol 86: 1286-1291.

Ben-Amotz, A., S. Mokady and M. Avron. 1988. The beta-carotene-rich alga *Dunaliella bardawil* as a source of retinol in a rat diet. Br J Nutr 59: 443-449.

Berliner, J. A., M.C. Territo, A. Sevanian, S. Ramin, J.A. Kim, A. Bamshad, M. Esterson and A.M. Fogelman. 1990. Minimally modified low density lipoprotein stimulates monocyte endothelial interactions. J Clin Invest 85: 1260-1266.

Bjorkhem, I., A. Henriksson-Freyschuss, O. Breuer, U. Diczfalusy, L. Berglund and P. Henriksson, 1991. The antioxidant butylated hydroxytoluene protects against atherosclerosis. Arterioscler Thromb 11: 15-22.

Bowen, H.T. and S.T. Omaye. 1998. Oxidative changes associated with beta-carotene and alpha-tocopherol enrichment of human low-density lipoproteins. J Am Coll Nutr 17: 171-179.

Brown, B.G., X.Q. Zhao, A. Chait, L. D. Fisher, M. C. Cheung, J. S. Morse, A. A. Dowdy, E.K. Marino, E.L. Bolson et al. 2001. Simvastatin and niacin, antioxidant vitamins, or the combination for the prevention of coronary disease. N Engl J Med 345: 1583-1592.

Burton, G.W. and K.U. Ingold. 1984. beta-Carotene: an unusual type of lipid antioxidant. Science 224: 569-573.

Carew, T.E., D.C. Schwenke and D. Steinberg. 1987. Antiatherogenic effect of probucol unrelated to its hypocholesterolemic effect: evidence that antioxidants in vivo can selectively inhibit low density lipoprotein degradation in macrophage-rich fatty streaks and slow the progression of atherosclerosis in the Watanabe heritable hyperlipidemic rabbit. Proc Natl Acad Sci USA 84: 7725-7729.

Chinetti, G., S. Lestavel, V. Bocher, A.T. Remaley, B. Neve, I.P. Torra, E. Teissier, A. Minnich, M. Jaye et al. 2001. PPAR-alpha and PPAR-gamma activators induce cholesterol removal from human macrophage foam cells through stimulation of the ABCA1 pathway. Nat Med 7: 53-58.

Dagenais, G.R., R. Marchioli, S. Yusuf and G. Tognoni. 2000. Beta-carotene, vitamin C, and vitamin E and cardiovascular diseases. Curr Cardiol Rep 2: 293-299.

Daugherty, A., B.S. Zweifel, B.E. Sobel and G. Schonfeld. 1988. Isolation of low density lipoprotein from atherosclerotic vascular tissue of Watanabe heritable hyperlipidemic rabbits. Arteriosclerosis 8: 768-777.

Dietrich, M., G. Block, E.P. Norkus, M. Hudes, M.G. Traber, C.E. Cross and L.Packer. 2003. Smoking and exposure to environmental tobacco smoke decrease some plasma antioxidants and increase gamma-tocopherol in vivo after adjustment for dietary antioxidant intakes. Am J Clin Nutr 77: 160-166.

Dugas, T.R., D.W. Morel and E.H. Harrison. 1998. Impact of LDL carotenoid and alpha-tocopherol content on LDL oxidation by endothelial cells in culture. J Lipid Res 39: 999-1007.

Frei, B. and J.M. Gaziano. 1993. Content of antioxidants, preformed lipid hydroperoxides, and cholesterol as predictors of the susceptibility of human LDL to metal ion-dependent and -independent oxidation. J Lipid Res 34: 2135-2145.

Gaziano, J.M. 1996. Randomized trials of dietary antioxidants in cardiovascular disease prevention and treatment. J Cardiovasc Risk 3: 368-371.

Gaziano, J.M., A. Hatta, M. Flynn, E.J. Johnson, N.I. Krinsky, P.M. Ridker, C.H. Hennekens and B. Frei. 1995. Supplementation with beta-carotene in vivo and in vitro does not inhibit low density lipoprotein oxidation. Atherosclerosis 112: 187-195.

Glass, C.K. and J.L. Witztum. 2001. Atherosclerosis. the road ahead. Cell 104: 503-516.

Goldin, A., J.A. Beckman, A.M. Schmidt and M.A. Creager. 2006. Advanced glycation end products: sparking the development of diabetic vascular injury. Circulation 114: 597-605.

Grines, C.L. 2006. The role of statins in reversing atherosclerosis: what the latest regression studies show. J Interv Cardiol 19: 3-9.

Harari, A., O. Harats, D. Mark, H. Cohen, I. Barshack, Y. Kamari, A. Gonch, Y. Gerber, A. Ben-Amotz, A. Shaish. 2008. α 9-cis-β-carotene-enriched diet inhabits atherosclerosis and fatty liver formation in LDL receptor knockout mice. J. Nutr. 138.

Hebuterne, X., X.D. Wang, E.J. Johnson, N.I. Krinsky and R.M. Russell. 1995. Intestinal absorption and metabolism of 9-cis-beta-carotene in vivo: biosynthesis of 9-cis-retinoic acid. J Lipid Res 36: 1264-1273.

Henriksen, T., E.M. Mahoney and D. Steinberg. 1981. Enhanced macrophage degradation of low density lipoprotein previously incubated with cultured endothelial cells: recognition by receptors for acetylated low density lipoproteins. Proc Natl Acad Sci U S A 78: 6499-6503.

Itoh, Y.H., S. R. Noguchi, R. Kurabe, T. Kajiwara, T. Nitta M. Ayakawa, Y.N. Miyata. 2000. Inhibitory effect of Dunaliella beta-carotene extracted from *Dunaliella bardawil* on plasma lipid in rats. Aicho Ika Daigaku Igakkai Zasshi 28: 263-271.

Jialal, I., E.P. Norkus, L. Cristol and S.M. Grundy. 1991. beta-Carotene inhibits the oxidative modification of low-density lipoprotein. Biochim Biophys Acta 1086: 134-138.

Kliewer, S. A., K. Umesono, D.J. Noonan, R.A. Heyman and R.M. Evans. 1992. Convergence of 9-cis retinoic acid and peroxisome proliferator signalling pathways through heterodimer formation of their receptors. Nature 358: 771-774.

Lavy, A., A. Ben Amotz and M. Aviram. 1993. Preferential inhibition of LDL oxidation by the all-trans isomer of beta-carotene in comparison with 9-cis beta-carotene. Eur J Clin Chem Clin Biochem 31: 83-90.

Levin, G., M. Yeshurun and S. Mokady. 1997. In vivo antiperoxidative effect of 9-cis beta-carotene compared with that of the all-trans isomer. Nutr Cancer 27: 293-297.

Levy, Y., B.-A. A. and M. Aviram. 1995. Effect of dietary supplementation of different β-carotene isomers on lipoprotein oxidative modification. J Nutr Med 6: 13-22.

Levy, Y., H. Zaltsberg, A. Ben-Amotz, Y. Kanter and M. Aviram. 2000. Dietary supplementation of a natural isomer mixture of beta-carotene inhibits oxidation of LDL derived from patients with diabetes mellitus. Ann Nutr Metab 44: 54-60.

Libby, P. 2002. Inflammation in atherosclerosis. Nature 420: 868-874.

Mooney, A. 1999. Treating patients with hypertriglyceridaemia saves lives: triglyceride revisited. Curr Med Res Opin 15: 65-77.

Morel, D.W., J.R. Hessler and G.M. Chisolm. 1983. Low density lipoprotein cytotoxicity induced by free radical peroxidation of lipid. J Lipid Res 24: 1070-1076.

Naruszewicz, M., E. Selinger and J. Davignon. 1992. Oxidative modification of lipoprotein(a) and the effect of beta-carotene. Metabolism 41: 1215-1224.

Ohashi, R., H. Mu, X. Wang, Q. Yao and C. Chen. 2005. Reverse cholesterol transport and cholesterol efflux in atherosclerosis. Qjm 98: 845-856.

Packer, L. 1993. Antioxidant action of carotenoids in vitro and in vivo and protection against oxidation of human low-density lipoproteins. Ann N Y Acad Sci 691: 48-60.

Palinski, W., M.E. Rosenfeld, S. Yla-Herttuala, G.C. Gurtner, S.S. Socher, S.W. Butler, S. Parthasarathy, T.E. Carew, D. Steinberg, and J.L. Witztum. 1989. Low density lipoprotein undergoes oxidative modification in vivo. Proc Natl Acad Sci USA 86: 1372-1376.

Palozza, P. and N.I. Krinsky. 1992. beta-Carotene and alpha-tocopherol are synergistic antioxidants. Arch Biochem Biophys 297: 184-187.

Palozza, P., R. Muzzalupo, S. Trombino, A. Valdannini and N. Picci. 2006. Solubilization and stabilization of beta-carotene in niosomes: delivery to cultured cells. Chem Phys Lipids 139: 32-42.

Rajavashisth, T. B., A. Andalibi, M.C. Territo, J.A. Berliner, M. Navab, A.M. Fogelman and A.J. Lusis. 1990. Induction of endothelial cell expression of granulocyte and macrophage colony-stimulating factors by modified low-density lipoproteins. Nature 344: 254-257.

Rapola, J.M., J. Virtamo, J.K. Haukka, O.P. Heinonen, D. Albanes, P.R. Taylor and J.K.Huttunen. 1996. Effect of vitamin E and beta carotene on the incidence of angina pectoris. A randomized, double-blind, controlled trial. Jama 275: 693-698.

Reaven, P. D., A. Khouw, W.F. Beltz, S. Parthasarathy and J.L. Witztum. 1993. Effect of dietary antioxidant combinations in humans. Protection of LDL by vitamin E but not by beta-carotene. Arterioscler Thromb 13: 590-600.

Reaven, P.D., E. Ferguson, M. Navab and F.L. Powell. 1994 Susceptibility of human LDL to oxidative modification. Effects of variations in beta-carotene concentration and oxygen tension. Arterioscler Thromb 14: 1162-1169.

Roland, A., R.A. Patterson and D.S. Leake. 2001. Measurement of copper-binding sites on low density lipoprotein. Arterioscler Thromb Vasc Biol 21: 594-602.

Rubin, E.M., B.Y. Ishida, S.M. Clift and R.M. Krauss. 1991. Expression of human apolipoprotein A-I in transgenic mice results in reduced plasma levels of murine apolipoprotein A-I and the appearance of two new high density lipoprotein size subclasses. Proc Natl Acad Sci USA 88: 434-438.

Rubins, H.B., S.J. Robins, D. Collins, C.L. Fye, J.W. Anderson, M.B. Elam, F.H. Faas, E. Linares, E.J. Schaefer et al. 1999. Gemfibrozil for the secondary prevention of coronary heart disease in men with low levels of high-density lipoprotein cholesterol. Veterans Affairs High-Density Lipoprotein Cholesterol Intervention Trial Study Group. N Engl J Med 341: 410-418.

Schoonjans, K., B. Staels and J. Auwerx. 1996. Role of the peroxisome proliferator-activated receptor (PPAR) in mediating the effects of fibrates and fatty acids on gene expression. J Lipid Res 37: 907-925.

Shaish, A., A. Daugherty, F. O'Sullivan, G. Schonfeld and J.W. Heinecke. 1995. Beta-carotene inhibits atherosclerosis in hypercholesterolemic rabbits. J Clin Invest 96: 2075-2082.

Shaish, A., A. Harari, L. Hananshvili, H. Cohen, R. Bitzur, T. Luvish, E. Ulman, M. Golan, A. Ben-Amotz et al. 2006. 9-cis beta-carotene-rich powder of the alga *Dunaliella bardawil* increases plasma HDL-cholesterol in fibrate-treated patients. Atherosclerosis. 189: 215-221.

Sparrow, C.P., T.W. Doebber, J. Olszewski, M.S. Wu, J.Ventre, K.A. Stevens and Y.S. Chao. 1992. Low density lipoprotein is protected from oxidation and the progression of atherosclerosis is slowed in cholesterol-fed rabbits by the antioxidant N,N'-diphenyl-phenylenediamine. J Clin Invest 89: 1885-1891.

Staels, B., J. Dallongeville, J. Auwerx, K. Schoonjans, E. Leitersdorf and J.C.Fruchart. 1998. Mechanism of action of fibrates on lipid and lipoprotein metabolism. Circulation 98: 2088-2093.

Steinberg, D. 2004. Thematic review series: the pathogenesis of atherosclerosis. An interpretive history of the cholesterol controversy: part I. J Lipid Res 45: 1583-1593.

Steinberg, D. 2005 Thematic review series: the pathogenesis of atherosclerosis. An interpretive history of the cholesterol controversy: part II: the early evidence linking hypercholesterolemia to coronary disease in humans. J Lipid Res 46: 179-190.

Stocker, R. and J.F. Keaney Jr. 2004. Role of oxidative modifications in atherosclerosis. Physiol Rev 84: 1381-1478.

Szapary, P.O. and D.J. Rader. 2004. The triglyceride-high-density lipoprotein axis: an important target of therapy? Am Heart J 148: 211-221.

Takahashi, H.I., H. Youko, T. Shiomi, Y. Ayakawa, N. Miyata. 2000. Decrease of the plasma cholesterol level by administration of *Dunaliella* extract in exogenous hyperchloesterolemic mice. Aicho Ika Daigaku Igakkai Zasshi 28: 249-255.

Upston, J.M., P.K. Witting, A.J. Brown, R. Stocker and J.F. Keaney Jr. 2001. Effect of vitamin E on aortic lipid oxidation and intimal proliferation after arterial injury in cholesterol-fed rabbits. Free Radic Biol Med 31: 1245-1253.

Valeri, C., P. Pozzilli and D. Leslie. 2004. Glucose control in diabetes. Diabetes Metab Res Rev 20 Suppl 2: S1-8.

van Hinsbergh, V. W., M. Scheffer, L. Havekes and H.J. Kempen. 1986. Role of endothelial cells and their products in the modification of low-density lipoproteins. Biochim Biophys Acta 878: 49-64.

Vu-Dac, N., K. Schoonjans, B. Laine, J.C. Fruchart, J. Auwerx and B. Staels. 1994. Negative regulation of the human apolipoprotein A-I promoter by fibrates can be attenuated by the interaction of the peroxisome proliferator-activated receptor with its response element. J Biol Chem 269: 31012-31018.

Wagner, P. and J.W. Heinecke 1997. Copper ions promote peroxidation of low density lipoprotein lipid by binding to histidine residues of apolipoprotein B100, but they are reduced at other sites on LDL. Arterioscler Thromb Vasc Biol 17: 3338-3346.

Wang, X.D., N.I. Krinsky, P.N. Benotti and R.M. Russell. 1994. Biosynthesis of 9-cis-retinoic acid from 9-cis-beta-carotene in human intestinal mucosa in vitro. Arch Biochem Biophys 313: 150-155.

Witting, P., K. Pettersson, A.M. Ostlund-Lindqvist., C. Westerlund, M. Wagberg and R. Stocker. 1999. Dissociation of atherogenesis from aortic accumulation of lipid hydro (pero)xides in Watanabe heritable hyperlipidemic rabbits. J Clin Invest 104: 213-220.

YH, I. 2000. Decrease of bloodlipid levels in hyperlipidemia by administration of Dunaliella powder (dry powder of *Dunaliella bardawil*). Nippon Rinsho Seiri Gakkai Zasshi 30: 217-226.

21 Utility of *Dunaliella* in Ecotoxicity Testing

Marie E. DeLorenzo

Abstract

Aquatic ecosystems are susceptible to anthropogenic contaminants such as pesticides, metals, pharmaceuticals, and persistent organic chemicals. The field of ecotoxicology focuses on determining the risk of such pollutants to non-target organisms. The inclusion of algal species in ecotoxicity testing provides an assessment of impacts to primary production and phototrophic biomass that are the base of the aquatic food web. *Dunaliella* has been used as a representative marine and estuarine algal test species. This chapter describes toxicity testing methods and results achieved with a variety of chemical compounds using *Dunaliella tertiolecta*. *D. tertiolecta* is an excellent toxicity test organism due to its ease of culture in the laboratory, size and morphology, rapid growth rate, and consistent response to environmental contaminants. To provide data on both structural and functional effects, a number of different test endpoints may be employed such as cell density, growth rate, primary productivity, chlorophyll, and cellular biovolume. The results generated have applications in environmental risk assessment and environmental management, particularly when combined with toxicity testing using other endpoints and species.

Introduction

Algae play important roles in aquatic ecosystems. They provide food for higher trophic levels, produce oxygen, and cycle nutrients. They typically have a large surface area to volume ratio, which gives them significant capacity for both uptake and interaction with chemicals in the environment. Other properties that make

---

NOAA, National Ocean Service, 219 Fort Johnson Rd., Charleston, SC 29412, Email: Marie.DeLorenzo@noaa.gov

algae valuable test species are their ubiquitous distribution, rapid turnover rate, ease of collection, and their ability to be cultured in the laboratory.

Detrimental effects on these primary producers may serve as an early indicator of contaminant stress in the ecosystem. Some United States federal regulations that require phytotoxicity data include the Clean Water Act, Federal Insecticide, Fungicide and Rodenticide Act, Superfund Act, Resource Conservation and Recovery Act, Toxic Substances Control Act, Federal Food Drug and Cosmetic Act, and the Marine Protection, Research and Sanctuaries Act (Lewis 1995). The species recommended by the U.S. Environmental Protection Agency (U.S. EPA) for freshwater green algae toxicity testing are *Selenastrum capricornutum* (*Pseudokirchneriella subcapitata*), *Scenedesmus subspicatus, Scenedesmus quadricada*, and *Chlorella vulgaris* (U.S. EPA 1974). The cyanobacteria species recommended are *Microcystis aeruginosa* and *Anabaena flos-aquae*, and the recommended diatom species are *Cyclotella* spp., *Navicula pelliculosa, Nitzschia* sp., and *Synedra* sp. (U.S. EPA 1974). Recommended marine microalgal toxicity test species are *Dunaliella tertiolecta* (green algae), *Skeletonema costatum, Thalassiosira pseudonana* (diatoms), and *Champia parvula* (red algae) (U.S. EPA 1974). *Dunaliella tertiolecta* is also recommended for use in toxicity assessments by the American Society for Testing and Materials (ASTM) (ASTM 1996), and the American Public Health Association (APHA), the American Water Works Association, and the Water Pollution Control Federation (APHA et al. 1989).

Dunaliella tertiolecta is a planktonic, euryhaline species in the class Chlorophyceae (green algae), order Dunaliellales, and family Dunaliellaceae (Borowitzka and Siva 2007). *D. tertiolecta* has been used as a toxicity test organism for assessing the effects of pollutants on marine and estuarine primary producers. A number of toxicity studies have also used the hypersaline species *Dunaliella bioculata* (Borowitzka and Siva 2007) for toxicity assessments (Heldal et al. 1984, Felix et al. 1988, Kusk and Nyholm 1992, Krishnaswamy-Chang 1997).

Phytotoxicity is defined as the toxicity of environmental pollutants based on the growth and survival of plants. Two other terms, algicidal and algistatic, are used to describe algal responses to environmental contaminants. Algicidal denotes a chemical that is lethal to the plant population. This is used to describe cellular necrosis, or death of the algal cell. Algistatic chemicals completely inhibit cell growth but allow the test species to resume growth when resuspended in clean medium. Certain chemicals can be stimulatory to algal growth, while others are inhibitory. Typically, inhibition is concentration dependent. Certain chemicals, however, are stimulatory at low levels and then toxic at higher levels, producing a dose-response curve that is described as hormesis. Calabrese and Baldwin (2003) described the hormetic response as a common, but usually overlooked phenomenon. Rarely, the opposite effect may occur with low levels causing inhibition, and higher levels causing stimulation. Figure 1 shows some different dose-response curves observed.

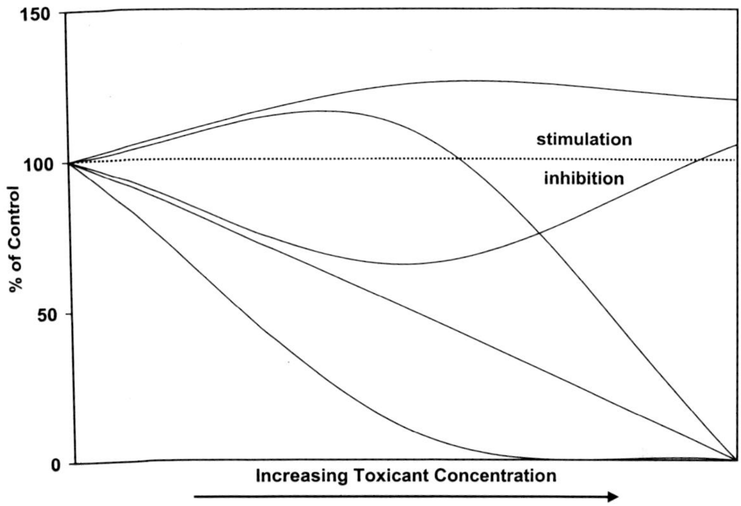

Figure 1: Representative algal responses to contaminant exposure (reproduced from Lewis 1995).

Once the chemical response has been measured, the test results are used to generate a toxicity value. These are either the ICx or ECx (estimated inhibitory or effective concentration of a chemical or effluent that reduces the measured parameter x percent relative to the control population) or the SC_{20} (the test concentration that stimulates algal growth 20% above that of the control population). The most common EC value used is the EC_{50}, which represents a median effect level, or a 50% reduction of the response variable compared with controls that lack the chemical. These standardized values are useful when comparing responses of different algal species to the same chemical and when comparing responses of the same species to different chemicals. Several agencies have published recommended guidelines for phytotoxicity testing, including ASTM (1996), the International Organization for Standardization (ISO) (1987), U.S. EPA (1985), and the Organization for Economic Cooperation and Development (OECD) (1985). These prescribed methods are similar in most aspects, but vary in the length of exposure (72 h for ISO and OECD; 96 h for ASTM and U.S. EPA), illumination level (30-90 μE/m^2/s for ASTM, 120 μE/m^2/s for ISO and OECD, and 300 μE/m^2/s for U.S. EPA), and number of replicates (3 replicates for ASTM and U.S. EPA; 3 replicates with 6 replicates in controls for ISO and OECD). The initial cell density for all tests is designed to maintain the cells in log phase growth throughout the exposure. Algal test protocols with *D. tertiolecta* are described below. The 96 h test period is considered a chronic exposure because the rapid growth rate of *D. tertiolecta* allows the response of several generations to be assessed during that time.

Methods

Algal Culture

Parent cultures of *D. tertiolecta* can be obtained from source laboratories (e.g. the University of Texas Culture Collection) and then sterile-transferred into F/2 marine media (Guillard 1983). Recommended culture conditions for *D. tertiolecta* are 25°C under cool-white fluorescent lighting (86± 8.6 $\mu E/m^2/s$) with a 16:8 h light:dark photoperiod and continual mixing using an orbital shaker (DeLorenzo and Serrano 2003). Cultures should be sterile-transferred as needed to maintain log phase growth.

Toxicity Testing

Standard 96 h static algal toxicity bioassay protocols (ASTM 1996) can be employed to determine the effective pesticide concentrations that reduce population growth rate by 50% (EC_{50}). Analytical grade standards (>96% purity) of each test compound should be used to prepare stock solutions. Organic compounds may be dissolved in 100% acetone and the doses then administered to obtain an equivalent acetone concentration in each treatment and control. Acetone concentrations $\leq 0.4\%$ have been verified to have no significant effect on *D. tertiolecta* growth compared to non-acetone controls (M.E. DeLorenzo, unpublished data). Inorganics may be dissolved in F/2 media. Once in solution, the chemical may become less bioavailable to the algae due to binding to the container, or with components of the media. It is, therefore, recommended to quantify the chemical concentrations in the algal exposure.

The series of concentrations tested should include a control and five equally spaced treatment concentrations. Three replicate glass test tubes are suitable for each treatment, each containing 25 mL of media. Each tube is inoculated from a sterile culture flask to provide an initial cell density of approximately 50,000 cells/mL. Testing is conducted under culture conditions described above, and tubes are repositioned within the environmental chamber each day to minimize possible spatial differences in illumination and temperature on growth rate.

This method has also been adapted to a microplate technique (Hall and Golding 1998). This scaled-down version uses a 96-well microplate and a 72 h exposure. The advantage is the use of smaller volumes and increased speed of testing. There is potential for chemical binding to the plastic microplate, which may affect bioavailability of the chemical to the algae and the assay results. Use of glass microplates, although more costly than plastic, would be advisable for the testing of hydrophobic compounds.

Test Endpoints

Cell Density

Cell density of the algal population can be measured via direct counts. A minimum of 18 grids or 400 cells are counted from 100 μL aliquots on an improved Neubauer hemacytometer at approximately the same time each day (0, 24, 48, 72, and 96 h). Two homogenized samples are counted per tube per time period. Direct cell counts have the advantage of allowing the observer to notice changes in algal cell morphology.

An estimate of cell density can also be determined using *in vivo* fluorescence. This method uses a fluorometer to measure the fluorescent signal from the cells at approximately the same time each day (0, 24, 48, 72, and 96 h). It should be noted, that while this method typically has very good correlation with direct counts (Figure 2), certain chemicals may interfere with the fluorescent signal (e.g. color from the antibiotic oxytetracycline) and bacterial contamination of the samples may also give erroneous readings (M.E. DeLorenzo, personal observation). Other methods for estimating cell density include electronic particle counters (e.g. Coulter counter), or spectrophotometric measurements (Pennington and

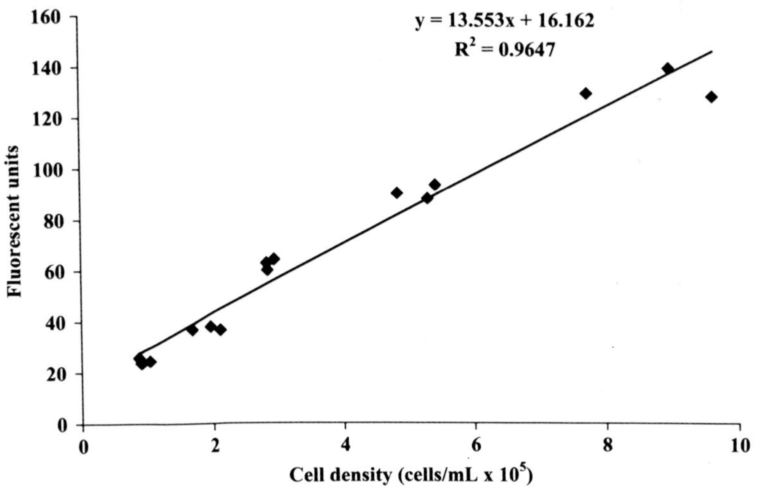

Figure 2: Relationship between direct cell counts and fluorescent readings as estimates of *D. tertiolecta* population density.

Scott 2001). It is recommended that the accuracy of these estimated cell density measures be initially verified by correlations with direct cell counts, particularly for new test chemicals.

Growth Rate

The cell count data or estimated cell density data (e.g. *in vivo* fluorescence) at 96 h can be used as the toxicity test endpoint, or the daily measurements can be used to generate a population growth rate. The population growth rate (divisions/day) is determined by log transforming ($\log_{10}$ + 1) the cell count data and calculating the slope of the converted values over time Sorokin, 1973. An average growth rate determined for *D. tertiolecta* under control conditions is 0.75 divisions/day (± standard error (SE) = 0.05) (DeLorenzo et al. 2004).

Biomass

Extracted chlorophyll *a* is used as a standard estimate of algal biomass and can be employed as a toxicity test endpoint. Described here is a fluorometric method adapted from Glover and Morris (1979). Ten mL of algal sample are filtered onto Whatman Type GF/F filters. The filters are placed in glass vials with one mL of a saturated magnesium carbonate solution and frozen at -20°C until extraction. The samples are allowed to thaw, and nine mL of 100% acetone are added to each vial. The vials are shaken, refrigerated overnight, shaken again the next day, and refrigerated once more overnight. After this 48 h extraction period, the samples are brought to room temperature in the dark, and fluorescence is measured using a fluorometer. Correction for phaeophytin is not needed if using a Turner 10-AU fluorometer with narrow band width optical chracteristics (Welshmeyer et al. 1994). Using a chlorophyll *a* standard curve, raw fluorescence data are converted to chlorophyll *a* concentration (µg/L). These values can be expressed on a per cell basis using the cell density determined for each sample. An average chlorophyll *a* concentration reported for *D. tertiolecta* under control conditions is 9.83×10^{-3} µg/cell (± SE = 6.46×10^{-4}) (DeLorenzo et al. 2004).

Primary Productivity

Total phototrophic carbon assimilation can be used to assess contaminant effects on the rate of primary production. Following the methods of Li et al. (1980), five mL of algal sample are spiked with ^{14}C-labeled sodium bicarbonate (50 µCi/mL) and incubated under culture conditions. After 24 h, samples are filtered onto cellulose nitrate filters, rinsed with media, placed in scintillation vials, and fumed overnight with 10% hydrochloric acid to remove inorganic carbon. Scintillation cocktail is then added, and after a 48 h stabilization period, radioactivity is measured using a liquid scintillation counter. Use of internal standards and correction for quenching is recommended. The photosynthetic carbon assimilation rate can be expressed as mg C/L/hour or using cell density data, as mg C assimilated/cell/hour. An average primary productivity rate reported for *D. tertiolecta* under control conditions was 1.03×10^{-6} mg C assimilated/cell/hour (± SE =1.23×10^{-7}) (DeLorenzo et al. 2004).

Cellular Characteristics

Changes in cellular biovolume and surface area can be useful estimates of toxicity, particularly for algistatic compounds. Twenty-five cells from each treatment and control are randomly selected to determine the mean cell volume and surface area. Length and width measurements are estimated using an ocular micrometer. Biovolume and surface area calculations are performed as described by Hillebrand et al. (1999). Average cellular biovolume (using the equation for volume of a cylinder) and surface area values determined for *D. tertiolecta* under control conditions were 396 μm^3 (± SE = 43.42) and 181.33 μm^2 (± SE = 15.37), respectively (Weiner and DeLorenzo 2004).

An average cellular dry mass can also be calculated as a test endpoint. Using a method modified from Tang et al. (1998), 40 mL of algal cell suspension from treatment and control samples are filtered onto pre-weighed Whatman GF/F filters. Filters are dried at 105°C for 12 h and weighed again. Three replicate samples are measured for each treatment. An average dry weight determined for *D. tertiolecta* under control conditions was 7.61 ng/cell (± SE = 0.16) (Weiner and DeLorenzo 2004).

Physiological Endpoints

Cellular lipid and protein content can be determined with spectrophotometric methods (Holland and Gabbott 1971). For lipid determination, 15 mL of sample is centrifuged (700 g for 10 min). A chloroform/methanol extraction is performed on the resulting pellet, and absorbance is measured at 375 nm. For protein determination, five mL of sample is centrifuged (447 g for 10 min), and the pellet is resuspended in sodium hydroxide, heated, and mixed with binding reagent. Absorbance is then measured at 595 nm. Spectrophotometric absorbances are converted to lipid and protein concentrations (µg/L) using standard curves created with tripalmitin (lipid assay) and bovine serum albumin (protein assay). Average concentrations determined for *D. tertiolecta* under control conditions were 0.174 ng protein/cell (± SE = 0.010) and 0.155 ng lipid/cell (± SE = 0.015) (DeLorenzo et al. 2004).

Morris et al. (1981) indicated that the measurement of ^{14}C accumulated in various photosynthetic end products (polysaccharides, lipids, proteins, and amino acids) can be used to assess the physiological state of organisms or populations.

This differentiation of incorporated ^{14}C into the main macromolecular pools is an extension of the classical ^{14}C-labeled sodium bicarbonate method for measuring productivity (Li et al. 1980), and it allows not only the quantity but also the quality of the synthesized organic substances to be evaluated. The extraction method measures newly incorporated carbon, thus only active microalgae are examined, without the interference of detritus or other living particles. A solvent fractionation scheme is used to separate microalgal macromolecules (DiTullio 1993). Radioactivity of the fractions is measured using liquid scintillation and expressed as a percentage of total carbon assimilated. Using this method, *D. tertiolecta* was found to incorporate most of its photosynthetic carbon into proteins (54%, ± SE = 7.39), followed by low molecular weight compounds such as amino acids (28%, ± SE = 9.33), lipids (16.5%, ± SE = 4.19), and polysaccharides (1.5%, ± SE = 0.17) (Weiner et al. 2007).

Peterson and Stauber (1996) describe use of the β-D-galactosidase enzyme as a sub-acute endpoint with *D. tertiolecta* in a three hour bioassay. The benefit of this endpoint is that toxicity results can be achieved with a shorter exposure. Toxicity is measured as a reduction in fluorescence relative to the controls.

Statistics and Analysis

Various computer programs for calculating the EC_{50}/IC_{50} and confidence limits are available (e.g. the linear interpolation method for sublethal toxicity (Norberg-King 1993), ToxcalcTN version 5.0 from Tidepool Scientific Software (Hall and Golding 1998), or non-linear regression in SAS by fitting the logistic equation to the data. Statistical differences among treatments may be determined using analysis of variance (ANOVA) or the non-parametric Kruskal-Wallace analysis where appropriate. Dunnett's procedure for multiple comparisons (or the non-parametric equivalent) may be used to determine which treatments differ significantly from the control (Zar 1999). The highest concentration tested that resulted in an effect not significantly different from the control is defined as the no observed effect concentration (NOEC), and the lowest concentration tested that resulted in an effect significantly different from the control is defined as the lowest observed effect concentration (LOEC).

Mixtures

Chemicals rarely exist individually in the environment. Characterizing the combined effect of multiple contaminants is an important issue in ecotoxicology. Chemicals in mixture may exhibit effects that are the sum of the individual contaminant toxicities (additive), are greater than expected (synergistic) toxicity, or are less than expected (antagonistic) toxicity.

One method of assessment is to conduct simple binary mixture tests using proportions of the respective EC_{50} values. Mixture toxicity can be determined according to Marking (1977), using the equation:

$$S = (A_m/A_i) + (B_m / B_i)$$

Where, S = sum of biological activity or TTU (total toxicant unit); $A_m = EC_{50}$ for compound A in mixture; $A_i = EC_{50}$ for compound A individually; $B_m = EC_{50}$ for compound B in mixture; and $B_i = EC_{50}$ for compound B individually.

S values are then used to calculate an Additive Index. If $S \leq 1.0$ then the Additive Index = (1/S)-1.0. If $S \geq 1.0$ then the Additive Index = S (-1) + 1.0. An additive index value less than zero denotes antagonistic toxicity. An index value greater than zero denotes synergistic toxicity. An index with confidence limits overlapping zero indicates the mixture has additive toxicity. The confidence limits surrounding the additive index are used to determine a significant deviation from zero. The additive index confidence limits are calculated using the upper and lower 95% confidence interval values from the test compounds' EC_{50} values and the additive index equation.

Mixture toxicity may also be assessed using the Modified Toxic Unit approach (Marking 1985), where the response of the mixture is compared to a predicted response based on toxic units. The predicted response of the model is complete additivity. The percent effect (based on the control values) of each mixture treatment is calculated and graphed as a dose response curve. A 50% effect of the mixture is predicted to occur at 1 toxic unit (TU), which is the treatment where the individual compounds are present at one half of their individual EC_{50} values. When a 50% effect occurs at less than 1 TU, the mixture is considered to be greater than additive, or synergistic. When a 50% effect occurs at greater than 1 TU, the mixture is considered to be less than additive, or antagonistic. The Additive Index and Modified Toxic Unit approaches assume that the mixture compounds have similar modes of action (Faust et al. 2003). If the individual compounds have completely different modes of toxic action and are predicted to act independently in mixture, the Response Addition model may be applied (Faust et al. 2003). This model uses the equation E(cmix) = 1 - (1-E(c1))(1-E(c2)), where E(c1) and E(c2) denote the percent effect caused by the individual constituents c1 and c2, and E(cmix) is the total effect of the mixture.

Results and Discussion

Of the chemicals tested with *D. tertiolecta*, the largest data set available is for the triazine herbicide atrazine. Atrazine is a potent photosystem II inhibitor. Specifically, atrazine competes with plastoquinone II (QB) at its binding site on the D1 protein, blocking electron transport from photosystem II (Fuerst and Norman 1991). The interruption of photosynthesis also leads to oxidative stress and photooxidative damage of lipid and chlorophyll molecules.

This algistatic compound has been well studied, and a range of toxicity values reported for marine algal species is 20 μg/L to 600 μg/L (Huber 1993, Solomon et al. 1996). *D. tertiolecta* responses to atrazine are in the lower end of this range. DeLorenzo et al. (2004) compared the sensitivity of different test endpoints (total lipid content, chlorophyll *a*, primary productivity, cell density, cellular biovolume, and growth rate) with *D. tertiolecta* in response to atrazine exposure (Table 1). All of the 95% confidence intervals overlapped for the 96 h EC_{50} values generated (Table 1), indicating that, for these endpoints and this compound, there were no significant differences in sensitivity. In this same study, protein was a significantly less sensitive endpoint, and EC_{50} values could not be generated at the atrazine concentrations tested. Total protein concentration was significantly reduced relative to controls at ≥ 50 μg/L atrazine. Variability was highest for lipid measurements, yielding a wide confidence interval (Table 1).

When examining the effects of atrazine on a per cell basis, it was observed that total protein concentration of *D. tertiolecta* cells increased with atrazine exposure relative to controls (Table 2) (DeLorenzo et al. 2004). To differentiate effects on total protein from those on protein synthesis, subsequent analyses using the biochemical fractionation technique (DiTullio 1993) revealed the percent of low molecular weight macromolecules such as amino acids increased, while the percent of protein per cell decreased (Weiner et al. 2007). This indicates that the cells were unable to complete metabolic pathways required to synthesize proteins.

Another triazine herbicide, Irgarol 1051, was 100 times more toxic to *D. tertiolecta* than atrazine. Irgarol is used as an additive in antifoulant paints. It yielded a 96 h growth rate EC_{50} value of 0.7 mg/L. Irgarol and atrazine exhibited additive toxicity when tested in mixture (Table 3) (DeLorenzo and Serrano 2006). Other herbicides tested with *D. tertiolecta* included 2,4-D, diquat, and deethylatrazine (a metabolite of atrazine). These compounds were all less toxic to *D. tertiolecta*, with EC_{50} values greater than 1000 mg/L (Table 2).

Felix et al. (1988) tested multiple herbicides with the hypersaline *Dunaliella* species *D. bioculata*. DeLorenzo and Serrano (2006) reported a 96 h growth rate EC_{50} of 45,470 μg/L for 2,4-D with *D. tertiolecta*, whereas Felix et al. (1988) reported a 48 h growth rate EC_{50} of 22,104 μg/L using the species *D. bioculata*. *D. bioculata* had a 48 h growth rate EC_{50} of 216 μg/L for atrazine, 270 μg/L for alachlor, 16,909 μg/L for glysophate, and 257 for paraquat dichloride (Felix et al. 1988).

The fungicide chlorothalonil has been tested with *D. tertiolecta*, yielding toxicity similar to that of atrazine. Chlorothalonil had a 96 h growth rate EC_{50} value of 64 μg/L, whereas its primary degradation product, hydroxychlorothalonil, was relatively non-toxic (Table 2). Chlorothalonil was observed to cause cell lysis in *D. tertiolecta*, thus it would be considered an algicidal compound (DeLorenzo

Table 1: Summary of atrazine effects on *D. tertiolecta*: EC_{50} values (μg/L) with 95% confidence intervals, lowest observed effect concentrations (LOEC) (μg/L), and no observed effect concentrations (NOEC) (μg/L).

	Cell Density	**Growth Rate**	**Primary Productivity**	**Total Biovolume**	**Chlorophyll-α**	**Total Lipid**
EC_{50}= 95% CI=	66.4 (59.6-87.9)	69.4 (66.0-81.7)	66.8 (60.3-82.1)	68.7 (56.2-90.8)	65.0 (61.3-73.4)	47.9 (4.5-70.1)
LOEC= NOEC=	251 2.5	50 25	25 12.5	50 25	25 12.5	25 12.5

Table 2: Summary of pesticide effects on *D. tertiolecta* cell density.

Chemical	96 h EC_{50} (μg/L)	95% CI (μg/L)	Reference
irgarol	0.7 growth rate	0.7 - 0.8	DeLorenzo and Serrano (2006)
2,4-D	45,470 growth rate	44,330-46,680	DeLorenzo and Serrano (2006)
diquat	2,249 growth rate1,139 cell density	2,147-2,347 923-1,300	DeLorenzo (unpublished data)
deethylatrazine	>1000 growth rate	not determined	DeLorenzo (unpublished data)
chlorothalonil	64 growth rate	62-65	DeLorenzo and Serrano (2003)
hydroxychlorothalonil	54,290 growth rate	46,910-59,910	DeLorenzo (unpublished data)
fipronil	631 biovolume >1000 cell density >1000 growth rate	not determined	Overmyer et al. (2007)
endosulfan	194 growth rate 135 cell density	175-213 117- 156	DeLorenzo (unpublished data)
chlorpyrifos	769 growth rate	727 - 814	DeLorenzo and Serrano (2003)
permethrin	>500 growth rate	not determined	DeLorenzo (unpublished data)
Scourge® (active ingredients are resmethrin and piperonyl butoxide)	104 cell density	8-247	DeLorenzo (unpublished data)

and Serrano 2006). Chlorothalonil and irgarol exhibited synergistic toxicity (approximately 1.5 times greater toxicity) when tested in mixture (Table 3). In addition, chlorothalonil and atrazine were approximately twice as toxic in mixture than individually (Table 4). These results suggest potential for even greater toxicity to occur where irgarol is in mixture with both atrazine and chlorothalonil.

Table 3: Effects of binary mixtures of irgarol and three other pesticides on *D. tertiolecta* population growth rate (DeLorenzo and Serrano 2006). Each mixture tested consisted of equal proportions of the individual EC_{50} values.

irgarol / atrazine mixture	individual 96 h EC_{50} (95% CI)	mixture 96 h EC_{50} (95% CI)
irgarol atrazine	0.7 μg/L (0.7-0.8) 69 μg/L (66-82)	0.5 μg/L (0.5-0.5) 23 μg/L (21-25)
additive index value (range)	0.002 (-0.004...0.076); additivity	
irgarol / chlorothalonil mixture	individual 96 h EC_{50} (95% CI)	mixture 96 h EC_{50} (95% CI)
irgarol chlorothalonil	0.7 μg/L (0.7-0.8) 64 μg/L (62-65)	0.3 μg/L (0.3-0.3) 16 μg/L (16-18)
additive index value (range)	0.51 (0.47...0.53); synergism	
irgarol / 2,4-D mixture	individual 96 h EC_{50} (95% CI)	mixture 96 h EC_{50} (95% CI)
irgarol 2,4-D	0.7 μg/L (0.7-0.8) 45,470 μg/L (44,330-46,680)	0.9 μg/L (0.8-1.0) 48,170 μg/L (41,860-52,940)
additive index value (range)	-1.24 (-1.10...-1.30); antagonism	

Several insecticides have been tested with *D. tertiolecta*, and these compounds are usually less toxic. The organophosphate insecticide chlorpyrifos yielded a 96 h growth rate EC_{50} value of 769 μg/L (DeLorenzo and Serrano 2003) and demonstrated additive toxicity when tested in mixture with atrazine (Table 4). For the phenylpyrazole insecticide fipronil and the pyrethroid insecticide permethrin, 96 h growth rate EC_{50} values were >1000 μg/L and >500 μg/L, respectively (Table 2). Walsh et al. (1977) reported a 7 d EC_{50} of 580 μg/L for the organochlorine insecticide kepone. *D. tertiolecta* cells have also been shown to bioaccumulate mirex and biotransform DDT and naphthalene (Kobayashi and Rittman 1982). In *Dunaliella*, flagellar motility is controlled by calcium ions. Marano et al. (1988) showed that exposure to the insecticide lindane, which disrupts sodium channels, inhibited *D. bioculata* cellular velocity and caused erratic movements.

Table 4: Toxicity of two mixtures (atrazine / chlorpyrifos and atrazine / chlorothalonil) to *D. tertiolecta* based on population growth rate (DeLorenzo and Serrano 2003). Atrazine and chlorpyrifos were tested in a 1:16 mixture (5.88% atrazine, 94.11% chlorpyrifos). Atrazine and chlorothalonil were tested in a 1:1 mixture (50% atrazine, 50% chlorothalonil).

atrazine/chlorpyrifos mixture	individual 96 h EC_{50} (95% CI)	mixture 96 h EC_{50} (95% CI)
atrazine chlorpyrifos	69 μg/L (66-82) 769 μg/L (727-814)	27 μg/L (25-29) 438 μg/L (397-463)
additive index value (range)	(-0.07…0.04…0.08); additive	
atrazine / chlorothalonil mixture	individual 96 h EC_{50} (95% CI)	mixture 96 h EC_{50} (95% CI)
atrazine chlorothalonil	69 μg/L (66-82) 64 μg/L (62-65)	18 μg/L (17-19) 18 μg/L (17-19)
additive index value (range)	(0.72…0.83…0.95); synergistic	

Aromatic hydrocarbons such as benzene, toluene, and xylene were found to be stimulatory to *D. tertiolecta* growth at low mg/L concentrations (Dunstan et al. 1975). Wiegman et al. (2001) examined the toxicity of azaarenes, a family of N-heterocyclic polycyclic aromatic hydrocarbons, to *D. tertiolecta*. The 72 h growth rate EC_{50} values for five azaarene isomers ranged from 22 μg/L to 73,700 μg/L. In general, toxicity increased with increasing number of rings, with toxicity attributed to narcotic and photoenhanced mechanisms.

A number of metals have been tested with *D. tertiolecta*. Studies include copper (Abalde et al. 1995), lead and aluminum (Türker Saçan et al. 2007), cadmium, zinc, silver, and mercury (Fisher et al. 1984). *D. tertiolecta* has been described as metal resistant relative to other algal species, and metals have been shown to accumulate in *D. tertiolecta* cells (Fisher et al. 1984).

An emerging area of toxicological concern is the presence of numerous pharmaceutical and personal care products (PPCPs) in aquatic environments worldwide (Zuccato et al. 2000). DeLorenzo and Fleming (2007) examined the effects of six PPCP compounds (simvastatin, clofibric acid, triclosan, fluoxetine, diclofenac, and carbamazepine) on *D. tertiolecta*. The 96 h EC_{50} values for triclosan, fluoxetine, simvastatin, diclofenac, and clofibric acid were 3.5 μg/L, 170 μg/L, 22,800 μg/L, 185,690 μg/L, and 224,180 μg/L, respectively. Of the PPCPs tested, only the antimicrobial compound triclosan yielded toxicity at typical marine concentrations. Concentrations of many compounds, however, are usually higher in freshwater environments, particularly downstream of wastewater treatment plants.

Both mixtures, simvastatin-clofibric acid and fluoxetine-triclosan, demonstrated additive toxicity (DeLorenzo and Fleming 2007).

Conclusions

While no single species can be expected to represent the response of a given ecosystem to stress, *Dunaliella* has significant applications in aquatic toxicity testing and risk assessment. As a native species to marine environments and a primary producer, *Dunaliella* is a model test species. This short term but chronic toxicity bioassay is repeatable, inexpensive, and uses small volumes of test chemical. There are test data available for a considerable number of contaminants. Studies of the effects of temperature, salinity, and nutrient manipulations on contaminant toxicity are also possible with *D. tertiolecta*. The use of *D. tertiolecta*, together with a suite of well-defined toxicity tests where a variety of endpoints and species are employed, can achieve a holistic picture of toxic impacts.

Acknowledgements

I would like to thank Dr. Mike Fulton and personnel of the National Ocean Service, Center for Coastal Environmental Health and Biomolecular Research Charleston Laboratory, Marine Ecotoxicology Branch for funding, assistance with laboratory bioassays, and review of this document. The National Ocean Service (NOS) does not approve, recommend, or endorse any proprietary product or material mentioned in this publication.

References

Abalde, J., A. Cid, S. Reiriz, E. Torres and C. Herrero. 1995. Response of the marine microalga *Dunaliella tertiolecta* (Chlorophyceae) to copper toxicity in short time experiments. Bull. Environ. Contam. Toxicol. 54(2): 317-324.

[APHA] American Public Health Association, American Water Works Association, and Water Pollution Control Federation. 1989. Toxicity testing with phytoplankton. Standard Methods for Examination of Water and Wastewater, 17th Ed, Suppl. Washington, DC.USA

[ASTM] American Society for Testing and Materials. 1996. Standard guide for conducting static 96-h toxicity tests with microalgae. Vol. 11.05, ASTM, West Conshohocken, PA. USA.

Borowitzka, M.A. and C.J. Siva. 2007. The taxonomy of the genus *Dunaliella* (Chlorophyta, Dunaliellales) with emphasis on the marine and halophilic species. J. Appl. Phycol. 567-590.

Calabrese, E.J. and L.A. Baldwin. 1993. Performing Ecological Risk Assessments. Lewis Publishers, Chelsea, MI. 288 pp.
DeLorenzo, M.E. and L. Serrano. 2003. Individual and mixture toxicity of three pesticides; atrazine, chlorpyrifos and chlorothalonil, to the marine phytoplankton species *Dunaliella tertiolecta*. J. Environ. Sci. Health, Part B. 38(5): 529-538
DeLorenzo, M.E. and L. Serrano. 2006. Mixture toxicity of the antifouling compound irgarol to the marine phytoplankton species *Dunaliella tertiolecta*. J. Environ. Sci. Health, Part B. 41(8): 1349-1360
DeLorenzo, M.E. and J. Fleming. 2008. Individual and mixture eff ects of selected pharmaceuticals on the marine phytoplankton species *Dunaliella tertiolecta*. Arch. Environ. Chem. Toxicol. 54:203-210.
DeLorenzo, M.E., M. Leatherbury, J.A. Weiner, A.J. Lewitus and M.H. Fulton. 2004. Physiological factors contributing to the species-specific sensitivity of four estuarine microalgal species exposed to the herbicide atrazine. J. Aquat. Ecosys. Health Manag. 7(1): 137-146.
DiTullio, G.R. 1993. Incorporation of C-14 into protein as an estimate of phytoplankton N-assimilation and relative growth rate. pp. 573-578. In Handbook of Methods in Aquatic Microbial Ecology. P.F. Kemp, B.F. Sherr and J.J. Cole (eds) CRC Press, Boca Raton, FL, USA.
Dunstan, W.M., L.P. Atkinson and J. Natoli. 1975. Stimulation and inhibition of phytoplankton growth by low molecular weight hydrocarbons. Mar. Biol. 31(4): 305-310.
Faust, M., R. Altenburger, T. Backhaus, H. Blanck, W. Boedeker, P. Gramatica, V. Hamer, M. Scholze, M. Vighi and L.H. Grimme. 2003. Joint algal toxicity of 16 dissimilarly acting chemicals is predictable by the concept of independent action. Aquat. Toxicol. 63: 43-63.
Felix, H.R., R. Chollet and J. Harr. 1988. Use of the cell wall-less alga *Dunaliella bioculata* in herbicide screening tests. Ann. Appl. Biol. 113(1): 55-60.
Fisher, N.S., M. Bohe and J.-L. Teyssie. 1984. Accumulation and toxicity of Cd, Zn, Ag, and Hg in four marine phytoplankters. Mar. Ecol. Prog. Series. 18: 201-213.
Fuerst, E.P. and M.A. Norman. 1991. Interactions of herbicides with photosynthetic electron transport. Weed Sci. 39: 458-464.
Glover, H.E. and I. Morris. 1979. Photosynthetic carboxylating enzymes in marine phytoplankton. Limnol. Oceanogr. 23: 510-519.
Guillard, R.R.L. 1983. Culture of phytoplankton for feeding marine invertebrates. pp. 29-60. In Culture of Marine Invertebrates.C.J. Berg (ed.) Hutchinson Ross Publishing Company, New York, USA.
Hall, J.A. and L.A. Golding. 1998. Marine Algae (*Dunaliella tertiolecta)* Chronic Toxicity Test Protocol. Appendix to: Standard methods for whole effluent toxicity testing: development and application. Report no. MFE80205. NIWA report for the Ministry for the Environment, Wellington, New Zealand.
Heldal, M., S. Norland, T. Lien, G. Knutsen, K. Tjessem and A. Aarberg. 1984. Toxic responses of the green alga *Dunaliella bioculata* (Chlorophycea, Volvocales) to selected oxidised hydrocarbons. Environ. Pollut. Ser. A Ecol. Biol. 34: 119-132.
Hillebrand, H., C. Durselen, D. Kirschtel, U. Pollingher and T. Zohary. 1999. Biovolume calculation for pelagic and benthic microalgae. J. Phycol. 35: 403-424.

Holland, D.L. and P.A. Gabbott. 1971. A micro-analytical scheme for the determination of protein, carbohydrate, lipid, and RNA levels in marine invertebrate larvae. J. Mar. Biol. Assoc. U.K. 51: 659-668.

Huber, W. 1993. Ecotoxicological relevance of atrazine in aquatic systems. Environ. Toxicol. Chem. 12: 1865-1881.

Kobayashi, H. and B.E. Rittman. 1982. Microbial removal of hazardous organic compounds. Environ. Sci. Technol. 16: 170A-183A.

Krishnaswamy-Chang, S. 1997. Biflagellated unicellular algae as detectors of pesticide contamination in water. Environ. Monit. Assess. 44(1-3): 487-502.

Kusk, K.O. and N. Nyholm. 1992. Toxic effects of chlorinated organic compounds and potassium dichromate on growth rate and photosynthesis of marine phytoplankton. Chemosphere. 25(6): 875-886.

Lewis, M.A. 1995. Algae and vascular plant tests. pp. 135-169. In Fundamentals of Aquatic Toxicology. G.M. Rand and S.R. Petrocelli (eds) Hemisphere Publishing Co., New York, USA.

Li, W.K.W., H.E. Glover and I. Morris. 1980. Physiology of carbon assimilation by *Oscillatoria theibautti* in the Caribbean Sea. Limnol. Oceanogr. 25: 447-456.

Marking, L.L. 1977. Methods for assessing additive toxicity of chemical mixtures. pp. 99-108. In Aquatic Toxicology and Hazard Evaluation. F.L. Mayer and J.L. Hamelink (eds) ASTM STP 634, American Society for Testing and Materials, West Conshohocken, PA, USA.

Marking, L.L. 1985. Toxicity of chemical mixtures. pp. 164-176. In Fundamentals of Aquatic Toxicology. G.M. Rand and S.R. Petrocelli (eds) Hemisphere Publishing Co., New York, USA.

Marano, F., S. Krishnaswamy, C. Bettrencourt, D. Schoevaert, J.M. Provost and B. Volochine. 1988. Control of ciliary beat by calcium: the effects of Lindane, a potent insecticide. Biol. Cell. 63: 143-150.

Morris, A., E. Smith and H.E. Glover. 1981. Products of photosynthesis in phytoplankton off the Orinoco River and in the Caribbean Sea. Limnol. Oceanogr. 26: 1034-1044.

Norberg-King, T.J. 1993. A linear interpolation method for sublethal toxicity: The inhibition concentration (ICP) approach. Technical Report 03-93. U.S. Environmental Protection Agency, Duluth, MN, USA.

[OECD] Organization for Economic Cooperation and Development. 1985. Algal growth inhibition test. Test Guideline No. 201. OECD Guidelines for Testing of Chemicals. Paris.

Overmyer, J.P., D.R. Rouse, J.K. Avants, A.W. Garrison, M.E. DeLorenzo, K.W. Chung, P.B. Key, W.A. Wilson and M.C. Black. 2007. Toxicity of fipronil and its enantiomers to marine and freshwater non-targets. J. Environ. Sci. Health, Part B. 42(5): 471-480.

Pennington, P.L. and G.I. Scott. 2001. Toxicity of atrazine to the estuarine phytoplankter *Pavlova* sp. (Prymnesiophyceae): increased sensitivity after long-term, low-level exposure. Environ. Toxicol. Chem. 20(10): 2237-2242.

Peterson, S.M. and J.L. Stauber. 1996. New algal enzyme bioassay for the rapid assessment of aquatic toxicity. Bull. Environ. Contam. Toxicol. 56: 750-757.

Solomon, K.R., D.B. Baker, P. Richards, K.R. Dixon, S.J. Klaine, T.W. LaPoint, R.J. Kendall, C.P. Weisskopf, J.M. Giddings, J.P. Geisy, L.W. Hall and W.M. Williams. 1996. Ecological risk assessment of atrazine in North American surface waters. Environ. Toxicol. Chem. 15(1): 31-76.

Sorokin, C. 1973. Dry weight, packed cell volume and optical density. pp. 321-343. In Handbook of Phycological Methods: Culture Methods and Growth Measurements. J. Stein (ed.) Cambridge University Press, New York, USA.

Tang, J., K.D. Hoagland and B.D. Siegfried. 1998. Uptake and bioconcentration of atrazine by selected freshwater algae. Environ. Toxicol. Chem. 17(6): 1085-1090.

Türker Saçan, M., F. Oztay and S. Bolkent. 2007. Exposure of *Dunaliella tertiolecta* to Lead and Aluminum: Toxicity and Effects on Ultrastructure. J. Biol. Trace Element Res. (in press).

[U.S. EPA] U.S. Environmental Protection Agency. 1974. Marine algal assay procedure bottle test: Eutrophication and Lake Restoration Branch National Environmental Research Center, Corvallis, OR, USA.

[U.S. EPA] U.S. Environmental Protection Agency. 1985. Toxic substances control act test guidelines; final rules. Fed Regist 50: 797.1050, 797.1075, and 797.1060. (Also Technical Support Document).

Walsh, G.E., K. Ainsworth and A.J. Wilson. 1977. Toxicity and uptake of kepone in marine unicellular algae. *Chesapeake Sci.* 18(2): 222-223.

Weiner, J.A., M.E. DeLorenzo and M.H. Fulton. 2007. Atrazine induced species-specific alterations in the subcellular content of microalgal cells. Pest. Biochem. Physiol. 87: 47-53.

Weiner, J.A. and M.E. DeLorenzo. 2004. Relationship between uptake capacity and differential toxicity of the herbicide atrazine in selected microalgal species. Aquat. Toxicol. 68: 121-128.

Welschmeyer, N.A. 1994. Fluorometric Analysis of chlorophyll a in the presence of chlorophyll b and pheopigments. Limnol. Oceanogr. 39(8): 1985-1992.

Wiegman, S., P.L Van Vlaardingen, E.A.J. Bleeker, P. De Voogt and M.H.S. Kraak. 2001. Phototoxicity of azaarene isomers to the marine flagellate *Dunaliella tertiolecta*. Environ. Toxicol. Chem. 20(7): 1544–1550.

Zar, J.H. 1999. Biostatistical Analysis. 4th Ed.. Prentice Hall, Upper Saddle River, NJ, USA.

Zuccato, E., D. Calamari, M. Natangelo and R. Fanelli. 2000. Presence of therapeutic drugs in the environment. The Lancet. 355: 1789-1790.

Glossary

2D: Two Dimensional

α: initial slope or photosynthetic efficiency.

Acetyl CoA carboxylase: A carboxylase that converts one molecule of HCO_3^- and one molecule of acetyl CoA into a molecule of malonyl CoA. Malonyl CoA is the substrate for chain elongation of fatty acids by addition of acetyl units with the release of CO_2.

Adaptation: A process occurring over generations that involve changes in the genetic makeup of cells and selection ultimately helping offspring of organisms to survive in changing habitats.

Aerobic: Requiring oxygen; processes that use oxygen.

AFDW: Ash-free dry weight. The dry weight of the combustible (at 550° C) organic matter contained in biomass.

Alga: A non-taxonomic term describing a diverse group of oxygen evolving photosynthetic organisms lacking true stems, roots, or leaves.

Algicidal: Describes a chemical that is lethal to the plant population.

Algistatic: Describes a chemical that completely inhibits cell growth but test species can resume growth when resuspended in clean medium.

AMT/Mep protein family of transporters: Ammonium transporter/methylamine permease proteins (often more fully referred to as ammonium transporter/methylamine permease/Rhesus [Amt/Mep/Rh] proteins) involved in ammonia transport in bacteria and eucaryotic organisms.

Anaerobic digestion: The microbial conversion of organic matter to biogas, a mixture of methane, carbon dioxide, water vapor, and small amounts of hydrogen sulfide, and sometimes hydrogen.

Anaplerotic Carboxylase: A carboxylase that is involved in an anaplerotic pathway (qv).

Anaplerotic pathway: A metabolic pathway that functions to replenish the intermediate pools of a pathway from which carbon skeletons are withdrawn for biosynthesis. An example is the role of (C_3+C_1) carboxylases in replenishing oxaloacetate to the tricarboxylic acid cycle when C_5 and C_4 compounds are removed for bisynthesis.

Antagonistic: Chemicals exhibit less than additive toxicity in mixture.

Antioxidant: A reactive compound that delays or prevents reactions with oxygen.

Anterior: Refers to the front part of a cell where the flagella are located.

Anthropogenic: Effects, processes, or materials that are derived from human activities aplanospore a non-motile spore (single cell reproductive unit)

Aplanospore: A non-flagellated spore.

Arylsulphatase activity: Aryl sulphatases are a diverse group of proteins with a range of substrate specificities and physicochemical properties. They are responsible for the release of sulphate from arylsulphate esters.

Asexual reproduction: A process involving mitosis in which a cell first doubles its DNA content and then divides into two identical daughter cells.

Athalassic: Waters of elevated salinity that have not had a connection with the sea in the geologically recent past or where connection to the sea was lost and the seawater evaporated to dryness before being reinstated as an aquatic environment.

Atherogenesis: The process of accumulating fatty substances in the inner lining of arteries.

Atherosclerosis: A disease characterized by the accumulation of deposits of fatty substances, cholesterol, cellular waste products, and calcium in the inner lining of large- and medium-sized arteries.

Axenic culture: An axenic culture contains only one species; free from all contaminants and other organisms including bacteria.

Batch culture: A closed system culture produced by inoculating a sterile medium. In advanced growth stages the algae get self limited for nutrients, light and carbon dioxide.

Betacarotene (α carotene): A plant pigment belonging to the family of isoprenoids that are made up of C5 isoprene units. Derived from phytoene through a sequence of dehydrogenation, isomerization, and cyclization reactions.

Bioassay: experiment conducted to measure the effects of a substance on a living organism.

Bioavailability: Fraction of a contaminant that is free to be taken into the organism; e.g. binding of the chemical to sediments may make it less bioavailable to pelagic organisms.

Biofouling: Describes the process of development of a biofilm on surfaces of materials.

Biofuel: A fuel made from a renewable biological source.

Biomass: Quantitative amount of living material; e.g. chlorophyll measurement is an estimate of phytoplankton biomass.

Biovolume: Estimate of microalgal cell volume using microscopic measurement of cell linear dimensions, assuming geometric shapes.

BN: Blue-Native (gels).

Bootstrap support: Statistic measurement to support phylogenetic reconstruction. The bootstrap analysis involves the resampling of nucleotide from an alignment, with replacement, so as to generate a number of pseudoreplicate data sets of the same length as the original alignment. The optimal topology is then estimated from each pseudoreplicate.

BWF Biological Weighting Function: A mathematical correction factor de-signed to account for the effect of levels of different wavelengths of radiation (largely UV-B) on specific physiological processes under specific conditions for specified species.

Carotenes: Non-polar polyene carotenoids (simple hydrocarbons) consisting of only carbon and hydrogen atoms.

Carotenoids: Yellow, orange and red lipid-soluble polyene pigments present in all photosynthetic tissues; also present in some bacteria, fungi and animals; may additionally contain oxygen (xanthophylls).

C14, C16 etc: Fatty acids designated by chain length, digit following :, number of double bonds.

C_3 carbon reduction cycle (Calvin cycle): The sequence of reactions that is employed by autotrophs in the net assimilation of inorganic carbon into 3 carbon sugar phosphates.

C_3+C_1 carboxylase: A carboxylase that adds a C_1 compound (CO_2 or HCO_3^-) to a C_3 compound (the monocarboxylic acid anions phosphoenolpyruvate, pyruvate or propionyl CoA) to produce a C_4 compound. Examples of such carboxylases are phosphoenolpyruvate carboxylase and phosphoenolpyruvate carboxykinase (can also function in reverse as a (C_4-C_1) decarboxylase (qv)), pyruvate carboxylase, and propionyl CoA carboxylase. Some of these enzymes can operate in anaplerotic pathways (qv) and in the variant of a CCM termed C_4 photosynthesis.

C_4-C_1 decarboxylase: A decarboxylase that converts a C_4 compound (the dicarboxylic acid anions oxaloacetate or malate) into CO_2 and a monocarboxylic C_3 compound. Examples of such decarboxylases are phosphoenolpyruvate carboxy-kinas (can also function as a craboxylase), NAD malic enzyme and NADP malic enzyme. These enzymes can function in the variant of a CCM (qv) termed C_4 photosynthesis.

CA: Carbonic anhydrase.

CAB: Chlorophyll *a*/*b* binding protein.

Carbon Concentrating Mechanism (CCM): Any process that results in the occurrence of a higher steady-state concentration of CO_2 which is available to RUBISCO (qv) in steady-state photosynthesis than the concentration in the bulk medium. CCMs always involve an energy input.

Carbonic anhydrase: An enzyme that catalyses the equilibrium between CO_2 and HCO_3^-

Carcinogenic: Molecules that produce cancer or promote cancer growth.

Cardiovascular: Having to do with the heart and the blood vessels.

Carotenogenesis: The process of developmentally or stress induced over-accumulation of carotenoids in cells leading to production of carotene.

Carotenoid: A group of yellow to red pigments derived from a C40 backbone that is made from the building blocks isopentenyl pyrophosphate and dimethylallyl phyrophosphate.

CCM: The CO_2 concentrating mechanism, employed by algae and cyanobacteria that involves the active transport of inorganic carbon and its concentration at the active site of RUBISCO, increasing the carboxylation activity of that enzyme and supporessing photorespiration.

CCMs Carbon Concentrating Mechanisms: Membrane bound pumps that actively transport inorganic carbon (DIC) into algal cells for photosynthetic fixation by RUBISCO.

cDNA : Complementary DNA that is reverse-transcribed from messenger RNA by the enzyme reverse transcriptase.

Cellulose: A complex carbohydrate molecule found in the cell wall of plants that is made up of the basic building block glucose.

Chemostat: A continuous culture system where the level of biomass is adjusted by an inflow of fresh sterile medium; the growth rate is equal to the dilution rate. When the algal cell density of the culture is held constant the culture is designated as a steady state culture.

Chemotroph : Organism that obtains energy from inorganic molecules.

Chlorophyceae: A taxonomic term describing the class of green algae.

Chlorophylls: Green lipid-soluble photosynthetic pigments of algae and higher plants, essential in photosynthesis, consisting of closed tetrapyrrole rings with magnesium.

Chloroplast: An organelle (part of a plant or algal cell), containing the thylakoid membranes, pigments and enzymes necessary for photosynthesis

Chlorophyta: A taxonomic division containing the green algae which are characterized by having photosynthetic pigments similar to those in higher plants.

Cholesterol: A fat soluble organic C27 molecule made by cells which belongs to the class of sterols.

Chylomicron: A microscopic globule made up of proteins and lipids. Located in the blood and lymphatic fluids.

cis/trans isomers: Carotenoids occur in both cis and trans forms; cis isomers which are easily formed in solution from the naturally occurring straight trans forms have the main polyene chain bent due to the rotation of a double bond.

Cladistic analysis: A method of grouping organisms by common ancestor.

Classification: Describing a process of categorizing organisms into groups of the same type.

Clockwise (CW) basal body configuration: Basal bodies of the two flagella are slightly displaced clockwise when the zooid is seen from the anterior end.

Clonal culture: Cultures established from a single cell and maintained by vegetative reproduction.

Clone: A cell or group of cells genetically identical derived from one single cell.

CM: Confocal Microscope.

Compensatory base pair change (CBC): A phenomenon where one side base of a pair changes, the opposite side base also changes, compensating to conserve that base pair in a double strand helix.

Conjugation: A processes in which two cells first establish direct contact followed by gene transfer.

Continuous culture: An open culture system where the inflow of medium is at the same rate as that flows out. The chemostat and turbidostat cultures are examples of continuous cultures.

CP43': Chloropyll-binding Protein induced in PS-I in cyanobacteria by iron deprivation.

Cryptochromes: Protein pigments that are blue/UV-A sensitive and can influence physiological responses. They show significant structural homology with DNA repairing photolyase enzymes, but do not repair DNA.

Cyst: Resting cell surrounded by a thick cell wall.

D1, D2: The two primary structural proteins at the reaction center of photosystem II.

dCA I: *Dunaliella* carbonic anhydrase I

dCA II: *Dunaliella* carbonic anhydrase II

dCAI, dCAII: *Dunaliella* plasma membrane Carbonic Anhydrases.

D-Fox: *Dunaliella* multicopper Ferroxidase.

DIC: Dissolved Inorganic Carbon. The forms of inorganic carbon (carbon dioxide, bicarbonate and carbonate) that can be found dissolved in water.

DTf: *Dunaliella* Transferrin.

EC_{50}: Effective concentration of a chemical or effluent that reduces the measured parameter 50 percent relative to the control population.

Ecotoxicology: Study of the adverse or toxic effects of chemicals and other anthropogenic materials on plants and animals.

Ek: The light level at which the linear part of the P-E curve intercepts a plateau (light saturation index.

ELIPS Early Light-Induced Proteins: Proteins that are rapidly induced by light-stress. They appear to bind pigments and offer a photo-protective function.

Endosymbiont: An alga that lives within the tissues of another alga (the host), the association acting to the mutual benefit of both partners.

Endosymbiosis: The ancestral incorporation of a minor alga (procaryotic or eucaryotic) into a larger host alga, the association behaving as a single organism. Transfer of genes from endosymbiont to host usually occurs.

Endpoint: Measurement taken at the end of a toxicant exposure (e.g. mortality, percent hatch, cellular biomarker level).

E_{opt}: Optimal light intensity or the light level that maximizes photosynthesis under given nutrient temperature conditions.

EOR: Energy output ratio, i.e., the ratio of the caloric energy output obtained from the microalgal fuel and the fossil energy inputs to produce the fuel. Only processes with EOR>1 have a net positive energy balance.

E_s: The light level that saturates photosynthesis (in the absence of photoinhibition).

EST: Expressed sequence tag is a short cDNA sequence corresponding to part of the mRNA molecule. ESTs can be used to identify genes that are expressed at any particular condition.

Estuarine: Of the estuary, habitat where tidal mixing of fresh and salt water occurs.

Eukaryote (eucaryote): An organism with membrane-bound organelles (e.g. nucleus, mitochondria, golgi apparatus and a 9 + 2 flagellum structure), as distinct from a more primitive prokaryotic organism, lacking membrane-bound organelles.

Euryhaline: Waters of a wide range of salinities.

Euryhaline organism: An organism capable of withstanding a wide range of environmental salinities.

Extinction coefficient: A constant value used for the quantitation of a highly purified pigment from its light absorption at a given wavelength, measured in a cell of 1 cm thickness.

Excitation pressure: Refers to the relative oxidation reduction (redox) state of the primary electron acceptor of photosystem II, Q_A, thus it is as an estimate of the redox state of PSII itself. Excitation pressure can be measured non-invasively in intact cells by the fluorescence parameter 1-qP.

Extremophile: An organism that likes to live in an extreme environment.

Eyespot: Confined area(s) containing pigments within cells of certain algae. Usually the eyespot pigments are orange or red carotenoids.

FAE1: *Dunaliella* Fatty Acid Elongase.

Fdx: Flvodoxin.

Fermentation: Catabolic reactions producing ATP in which organic compounds serve as both primary electron donor and ultimate electron acceptor.

Fermentor: A closed system for heterotrophic growth of algae using organic carbon source such as glucose, glycerol etc.

Flagellum: A 'tail' of a cell (long hair-like extension) used for locomotion.

Fluorometric: Measurements based on the intensity and wavelength distribution of emission spectrum after excitation by a certain spectrum of light.

Fluorometry: An analytical method for the measurement of the fluorescence of a compound induced by ultraviolet light.

Fv/Fm: The ratio of variable fluorescence to maximal fluorescence. It is a sensitive indicator of the efficiency of light driven electron transfer out of photosystem II.

Gamete: A reproductive cell containing a haploid set of chromosomes.

Gametogenesis: The process of development of gametes.

Gasification: A process that converts organic materials into carbon monoxide and hydrogen (synthesis or syngas) at high temperatures with a controlled amount of oxygen.

Germination: The process where a spore or zygote begins to grow and develop, often occurring after some time of dormancy.

Glycocalyx: A gelatinous extra-cellular covering of cells. It is made up of polysaccharides and/or glycoproteins that coat a cell.

Glycoprotein: A molecule consisting of a protein linked with carbohydrates (= sugar molecules).

Growth rate: Describes the increase of growth in a culture of cells. It is generally measured as an increase in the number of cells, biomass, or optical density.

Habitat: The type of location with resources where an organism or population of organisms occurs.

Halobacteria: Type of bacteria capable of living only in hypersaline water bodies.

Halotolerant: Being tolerant to high range of salinities.

Herbicide: A chemical substance that is used to inhibit growth or kill plants.

Heterothallic: The condition of gametes from one organism being self-sterile, thus requiring a partner for sexual reproduction.

Homeostasis: Refers to an organism maintaining a constant internal environment.

Homothallic: The condition of gametes from one organism being able to reproduce sexually. Only one clone is required for sexual reproduction; self-compatible.

HSP: Heat Shock Protein

Hypersaline: Waters that contain higher concentrations of dissolved mineral salts than seawater.

I_{50}: Inhibitor concentration causing 50 % inhibition of a given activity.

IEF: Iso-Electric Focussing.

In vivo: Experimentation done in or on a whole, living organism; e.g. in vivo fluorescence is an estimate of phytoplankton biomass using the whole, living cell, whereas extracted chlorophyll a is an estimate of phytoplankton biomass from cellular extracts.

Inoculum: Material containing organisms to begin new cultures and usually consists of actively dividing or algal cells in exponential phase.

Intron: A non-coding region within an eukaryotic gene which is transcribed and then removed in the process of RNA maturation; resides between exons.

Isogamete: A gamete from a species where all gametes look alike.

Isolate: Strain of a single cell.

Isolation: Separation of a single cell, usually done with a micro syringe or a Pasteur pipette.

Isomerization: The process of converting one structural isomer into another structural isomer which involves changes in the organization of atoms.

Isomers (Stereoisomers): Molecules that have the same number and types of atoms, but having different geometric structures.

Isoprenoid: A compound derived from the isoprene building bocks isopentenyl pyrophosphate and dimethylallyl phyrophosphate.

ITS region: Region of the nuclear ribosomal cistron which includes the Internal Transcribed Spacer 1, 5.8S rRNA gene and Internal Transcribed Spacer 2.

ITS-2 secondary structure: the general three-dimensional form of a segment of RNA, in this case, the Internal Transcribed Spacer 2. RNA secondary structure is generally divided into helixes (contiguous base pairs) and loops (unpaired nucleotides surrounded by helixes).

Kcs: â-ketoacyl CoA synthase

Lamella: A structure resembling a thin plate within a chloroplast, often extending the whole length of the chloroplast.

LC: Liquid Chromatography.

LCHII: The major light harvesting pigment-protein complex that surrounds photosystem II.

LCHII proteins are encoded by a small family of nuclear genes the expression of which has been shown to be response to the redox state of the photosynthetic electron transport chain.

Life cycle: The progression of an organism through a series of developmental changes.

Lipid: A group of organic molecules that are insoluble in water (also called fats). Examples are oils, waxes, sterols, and triglycerides.

Lipid globule: A cellular compartment made up of a small spherical mass which contains a lipid mono-layer. Examples for lipid globules are oleosomes (located in the cytoplasma) and plastoglobules (located in plastids).

Lipoprotein: A group of proteins that have conjugated lipids.

LOEC: Lowest observable effect concentration—lowest concentration tested that resulted in an effect significantly different from the control.

MALDI-TOF: Matrix Assisted Laser Desorption/Ionization-Time Of Flight

MAP kinase: Mitogen Activated Protein kinase

Maximum Likelihood (ML): A method for the inference of phylogeny. It evaluates a hypothesis about evolutionary history in terms of the probability that the proposed model and the hypothesized history would give rise to the observed data set. The supposition is that a history with a higher probability of reaching the observed state is preferred to a history with a lower probability. The method searches for the tree with the highest probability or likelihood.

Maximum Parsimony (MP): A method for the inference of phylogeny based on cladistics; it establishes groups based on their shares, derived attributes (synapomorphies). Organisms that share common ancestors are grouped in clades. It minimizes the total number of evolutionary steps required to explain a given set of data, or in other words by minimizing the total tree length.

Media: To grow marine micro algae seawater or artificial seawater is enriched with nutrients. sterilized and seeded with an algal cell.

Metabolic engineering: A procedure that uses recombinant DNA techniques to purposely modify the metabolism of an organism so that it will produce specific desired molecules.

Metabolite: A chemical molecule made by chemical reactions in cells.

Microalga: An alga of microscopic size.

Mitochondrion: A cytoplasmic organelle, whose major function is the generation of ATP (adenosine triphosphate) by respiration.

Monophyletic lineage: A group of related species descended from a single ancestral form.

MS BLAST: MS-driven Basic Local Alignment Search Tool.

MS: Mass Spectrometry.

Mycosporines: A family of water soluble amino acid derivatives found in algae. They increase photoprotection by harmlessly absorbing UV radiation before it can hit a sensitive target in the cell.

Nanoplankton: Unicellular microalgae in the size range >2 – 20 μm

NOEC: No observable effect concentration—highest concentration tested that resulted in an effect not significantly different from the control.

NPQ Non-Photochemical Quenching: An indicator of the rate of thermal dissipation of excess potentially damaging radiation absorbed by the photosystems. The xanthophyll cycle is part of this form of photoprotection.

Nuclear genome: The DNA within the cell nucleus organelle a membrane-delimited structure within a cell (e.g. nucleus, chloroplast).

Osmoregulation: Control and maintenance of the osmotic balance within a cell.

p130b: *Dunaliella* iron-deprivation induced glycoprotein.

P680: A special pair of chlorophyll a molecules located in the center of photosystem II reaction enter. Its function is to perform the primary charge separation with pheophytin upon receiving excitation energy.

Pallmeloid: A stage of cells that are non-motile and embedded in a gelatinous matrix.

PAR Photosynthetically Active Radiation: The visible portion of the spectrum (400–700 nm) that drives photosynthesis.

Paratransgenesis: In this strategy, commensal or symbiotic bacteria found at mucosal sites of pathogen transmission are isolated and genetically altered to elaborate immune peptides or engineered single chain antibody fragments (scFv) that neutralize infectious agents. The transgenic bacteria are then delivered back to mucosal sites where disease transmission occurs. This strategy was initially developed to combat transmission of the Chagas disease parasite, Trypanosoma cruzi, by reduviid bug vectors.

Pasteurization: Sea water is heated at 95–100 ^{0} C for one hour.

periplasmic: Located outside the plasmalemma.

Peroxiredoxin: A family of non-heme peroxide scavenging enzymes. Their electron donor is typically reduced thioredoxin.

Pesticide: Chemical designed to kill and or inhibit pest organisms; specifically, herbicides target pest plants, insecticides target pest insects, and fungicides target pest fungi.

Phosphorylation: An addition of a phosphate group to a protein or other organic biomolecules.

Photoacclimation: Changes to molecular, physiological and biochemical components of a cells in response to changes to the redox state of the photosynthetic electron transport chain.

Photobioreactor: In general terms the description for a cultivation system for growth of phototroph organisms.

Photodamage: A deleterious process altering structure and function of molecules which is induced by absorption of photons by molecules. Protein damage occurred as a consequence of exposure to excessive irradiance. In chloroplast, photodamage is commonly referred to as damage to D1 protein of photosystem II.

Photoinhibition: A condition when the rate of photodamage exceeds the capacity of photoprotection and of the repair process, causing the decline of the overall photosynthesis rate.

Photorespiratory carbon oxidation cycle (PCOC): A pathway by which two molecules of the 2-phosphoglycolate produced in the oxygenase activity of RUBISCO are converted into one molecule of glycerladehyde-3-phosphate and one molecule of CO_2. The pathway involves recycling of amino-groups, oxidation reactions, and the input of reductant and ATP.

Photostasis: Refers to the balance between energy absorption and energy utilization in photosynthetic organisms. Changes in light intensity, temperature and nutrient status, for example, can disrupt photostasis. Acclimation responses by the photosynthetic organism are an attempt to regain photostasis.

Photosynthetic carbon reduction cycle (PCRC): A pathway involving Rubisco (qv) as the sole carboxylase which reduces CO_2 to the oxidation-reduction level of carbohydrate as glyceraldehyse-3-phosphate.

Phototroph: Organism that obtains its energy from light.

Phototropin: A flavoprotein pigment that is an auto-kinase activated by blue/UV-A radiation and signals many developmental stages in algae and higher plants.

Phycoplast: Array of microtubules oriented parallel to the plane of cytokinesis, through which the developing wall is formed; present in chlorophycean green algae.

Phytoene: The first C40 hydrocarbon molecule in the biosynthesis of carotenoids. Derived from the conjugation of two geranylgeranylpyrophosphate molecules.

Phytofluene: Derived by dehydrogenation from the molecule phytoene.

Phytoplankton: Photosynthetic microalgae (usually 2 to 200μm), living free, or suspended (swimming feebly) in the water column.

Phytotoxicity: Toxicity of environmental pollutants determined on the growth and survival of plants plastid genome the DNA within the chloroplast (plastid) of a eukaryotic plant or algal cell, separate from the chromosomal DNA of the nucleus.

Plastoquinone (PQ): A quinone molecule which is found within the thylakoid membrane and transfers electrons from PSII to the cytochrome b_6f complex and concurrently moves protons from the stromal to the luminal side of the thylakoid membrane. PQ exists as a pool of molecules and the relative redox state of this pool of molecules A pool of is believed to act as a sensor which reflects the overall redox state of the electron transport chain.

Posterior: Refers to the back part of a cell opposite to the location of the flagella.

PPCP: Pharmaceuticals and personal care products, a group of emerging environmental contaminants.

PPFD Photosynthetic Photon Flux Density: The rate of flow of photons (400–700 nm) through a standard area. Usually expressed as ìmol photons per second through an area of 1 meter squared. A mole of photons was formerly called an 'Einstein'.

Primary carotenoid: A carotenoid that is structural/functional component of the photosynthetic apparatus.

Primary carotenoids: Carotenoids produced under normal conditions

Primary productivity: Rate of production of organic compounds from carbon dioxide, principally through the process of photosynthesis.

Prokaryote: A primitive organism lacking membrane-bound organelles (e.g. bacteria/ blue-green algae)

Productivity: The rate of output (for example oxygen during photosynthesis) per given time.

PS-I: Photosystem I

PSII repair process: A process that photosynthetic organisms bring back activity of photosystem II that is lost as a result of photodamage. The process involves selective removal of damaged D1 protein and replacement with a de novo synthesized copy.

Pyrenoid: A subcompartment of the chloroplast stroma of many algae and hornworts that have a CCM. The pyrenoid contains most of the cell complement of RUBISCO (qv). It is usually a small, dense, proteinaceous body (sometimes crystalline) connected to the chloroplast of some eukaryotic algae; often surrounded by storage carbohydrate (e.g.starch).

Pyrolysis: The chemical decomposition of organic materials by heating in the absence of oxygen or any other reagents, except possibly steam.

Raceway pond: A shallow pond in the form of a raceway which is mixed by paddle wheels.

Random Amplified Polymorphic DNA (RAPD): Technique used for detecting genomic polymorphisms at a number of different loci using a single oligonucleotide of arbitrary sequence as PCR primer.

rbcL and rbcS: Genes which encode the large and small subunits of the Ribulose 1,5- Bisphosphate Carboxylase/Oxigenase enzyme (RUBISCO).

Reciprocity: Assumes that only the rate times the exposure duration (i.e. total accumulated dose of radiation exposure) determines the biological effect, independent of the rate or intensity of the exposure.

Restriction Fragment Length Polymorphism (RFLP): Molecular analysis based on the property of endonucleases to cut a DNA fragment at specific sites (recognition sequence or restriction sites). The restriction fragments are then separated according to length by agarose gel electrophoresis.

Reverse-Transcription PCR: The polymerase chain reaction reverse transcribing messenger RNA into complementary DNA.

Ribulose bisphosphate carboxylase-oxygenase (RUBISCO): A carboxylase that catalyses the conversion of one molecule of ribulose-1,5-bisphosphate and one molecule of each of CO_2 and H_2O into two molecules of 3-phosphoglycerate. The enzyme also catalyses a competing oxygenase reaction; here one molecule of ribulose bisphosphate and one of O_2 is converted into one molecule of 3-phosphoglycerate and one molecule of 2-phosphoglycolate. It is the enzyme employed by nearly all autotrophs for the assimilation of CO_2 into organic carbon.

ROS Reactive Oxygen Species: Partially reduced oxygen molecules including superoxide, hydrogen and lipid peroxides, hydroxyl radical and singlet excited oxygen.

RUBISCO Ribulose 1,5 bis phosphate carboxylase/oxygenase: The photosynthetic enzyme which fixes carbon dioxide, by covalently bonding it to another carbon atom.

SDS-PAGE: Sodium Dodecyl Sulfate Polycarylamide Gel Electrophoresis.

Secondary carotenoid: A carotenoid that serves a function different than structure and function in the photosynthetic apparatus. Many secondary carotenoids are produced by organisms in response to environmental stress. A class of highly unsaturated pigment molecules that belongs to the group of Isoprenoids (= Terpenes) and that is not directly involved in photosynthesis.

Secondary carotenoids carotenoids produced under abnormal conditions (e.g. lack of nitrogen, high irradiance)

Secondary endosymbiosis: Endosymbiosis occurring a second time within the same host organism; the second endosymbiont may be of different origin from the first.

Selectable marker: Refers to a gene whose expression within a cell allows selection for a specific trait conferred by that gene.

Semicontinuous batch culture: Before the nutrient exhaustion, at the end of its exponential phase, 95% of the culture is poured out and fresh medium is added to the culture. If the culture is diluted each day with a portion of fresh medium it is known as semicontinuous culture.

Sexual reproduction: The process of offspring generation that involves union of gametes.

Signal transduction: A cascade of cellular processes which begin with interaction of a signal with a receptor thus causing alterations in activity of intermediate messengers and ultimately leading to changes in metabolism.

Single chain antibodies (scFv): In the scFv fragment the V_H and V_L domains are joined with a hydrophilic and flexible peptide linker, which improves expression and folding efficiency. It has a high affinity for its antigen and can be expressed in a variety of hosts. Usually linkers of about 15 amino acids are used, of which the $(Gly_4Ser)_3$ linker has been used most frequently.

Singlet oxygen: Oxygen molecules in which there are no unpaired electrons.

Sodium symport: The co-transport of Na^+ with other materials in the same direction across the membrane.

Species: A taxonomic unit which today has several different approaches to its definition. Example for the biological species concept—Group of organisms whose members can interbreed.

Spectrophotometric: Measurements based on the absorbance of light at a particular wavelength.

Spectrophotometry: The measurement of photometric intensity of each wavelength (or colour) in the optical spectrum of a compound.

Sporopollenin: An organic UV absorbing polymer found in the cell walls of some UV-tolerant algae.

State transitions: Refers to the movement of a population of LHCII complexes between PSII and PSI to ensure a balance of excitation energy reaching both photosystems.

Sterilization Procedure: Used to remove or kill all microorganisms and can be achieved by heating, autoclaving (125 ^{0} C, 15–25 min), filtration (0.2 μm) or by exposure to ultraviolet radiation. Sterility of the medium or culture has to be established by the standard procedure of spreading cells on seawater agar plates.

Stereoisomers: Molecules with an identical chemical composition, but having a different arrangement of the atoms in space

Strain: A group of organisms within a species associated to a geographic location.

Symplesiomorphy: A character shared by a number of groups, but inherited from ancestors older than the last common ancestor.

Synergistic: Chemicals exhibit greater than additive toxicity in mixture.

System I and System II fibers: Two main types of fibrous flagellar roots present in the flagellar apparatus of green algae: System I fibers, cross-striated bundles of 2 nm filaments which are associated with flagellar root microtubules; System II fibers, contractile bundles of 4-8 nm filaments which are often cross-striated.

Taxon: Any group of organisms to which any rank of taxonomic name can be applied (e.g. genus, family, class or division).

Tertiary endosymbiosis: Endosymbiosis that is occuring a third time within the same host organism, by a different endosymbiont.

Triplet chlorophyll: A state in which there are two unpaired electrons.

Toxic unit: Treatment where the individual compounds are present at one half of their individual EC_{50} values.

TTf: *Dunaliella* Triplicated Transferrin-like protein.

Turbidostat: When the culture is diluted and maintained by an inflow of fresh medium to set threshold turbidity.

Tthylakoid: A flattened, sac-like, membrane structure within the chloroplast, containing the photosynthetic pigments

Unialgal culture: Culture contains only one species of alga but may contain contaminants such as bacteria.

Vibriosis: Infections caused by *Vibrio* spp. *Vibrio* spp. are maost often considered opportunistic pathogens in shrimp, but primary disease caused by highly virulent strains has also been reported. Main *Vibrio* spp. which causes vibriosis in fish and shell fish are *V harveyi*, *V parahaemalyticus* and *V alginolyticus*.

VS: Volatile Solids. The mass of solids in water or sludges that is lost on ignition of the dried solids at 550° C.

WSSV, White spot syndrome virus: WSSV is an invertrabrate virus belongs to the nimaviridae family and is extremely virulent. Infection of peneaid shrimp can result in up to 100% mortality within 3 to 7 days.

Xanthophyll: Consists of three carotenoids, violaxanthin, antheraxanthin and zeaxanthin which are interconverted by enzymes which are sensitive to the pH of the lumen. While under low light conditions violaxanthin is the dominant member of the cycle and aids light harvesting, under high excitation pressure, antheraxanthin and notably zeaxanthin dominant which play a role in energy dissipation. The carotenoids in addition to the hydrocarbon backbone, also contain oxygen atom(s) within the molecule. These are polar oxygenated carotenoids in the chloroplasts of all plant and algal cell which act as accessory light-harvesting pigments, or have a photoprotective function.

Xanthophyll Cycle: A photoprotective cycle of three interconvertable xanthophyll carotenoids (violaxanthin, antheraxanthin and zeaxanthin) found in the antenna of the photosystems in green algae and land plants. In a high light environment, zeathanthin predominates and causes increased conversion of excess energy to heat (see NPQ). In low light violaxanthin predominates, and less energy is lost as heat. In chlorophyll c containing algae an analogous cycle is found with different xanthophylls (i.e. diadinaxanthin and diatoxanthin).

Index

C

D

E

I

K

L

M

N

O

P

Q

R

U

V

W

X

Y

Z

About the Contributors

Dr. Maria Barbosa obtained her university degree in Food Engineering at Catholic University, Portugal in 1997 and a PhD in Nutrition, Food technology and Biotechnology at Wageningen University, The Netherlands in 2003. During the PhD she specialised in Bioprocess Engineering, she worked on the development, optimization and scale-up of photobioreactors. In 2004 she became a post doc at the Swiss Federal Institute of Technology (ETH) where she worked in the field of systems biology. In 2005 she moved to the Institute of Experimental and Technological Biology (IBET), in Oeiras, Portugal to work in the animal cell technology group on the development of metabolic models. In 2007 she went back to the Bioprocess Engineering group at Wageningen University where she worked as a post doc on the development of an economical and technical feasibility model for the production of biofuels from marine biomass.

Dr. John Beardall was educated in the UK and moved to Australia in 1982. Starting out as a microbiologist (BSc Hons in 1973) he developed an interest in microalgal physiology and was awarded his PhD (University of London, 1976) for work on photosynthetic biochemistry and physiology of phytoplankton. He has worked on various aspects of this area for the last 30 years. He is currently Professor of Biology at Monash University. Professor Beardall is an Associate Editor for *Phycologia* and plays a significant role on a number of national and international professional bodies. He is author/co-author of over 90 journal articles, 15 book chapters and one text book on plant biochemistry. He has supervised 11 PhD and 2 MSc students and currently runs a busy laboratory, with students working on topics ranging from primary productivity and carbon flow in Cambodian lakes to the biochemistry of photosynthesis and biospectroscopy

Dr. Ami Ben-Amotz is Emeritus Professor of Marine Phycology at the National Institute of Oceanography (NIO) in Haifa, Israel. Ami Ben-Amotz received his MSc degree at the Hebrew University of Jerusalem and the PhD at Weitzman Institute of Science (WIS), Israel, both on studies on the alga *Dunaliella*. After post doctorate studies at Brandeis University, USA, he returned to Israel and initiated research academic activity at the NIO and at the WIS to study the biology, physiology, biochemistry and biotechnology of *Dunaliella* in long collaboration with the late Prof. M. Avron of the WIS. The fruitful cooperation led the way to establishment of the commercial *Dunaliella* production plant in Eilat, known today as Nature Beta Technologies Ltd., (NBT) Israel, a subsidiary of Nikken Sohonsha Co., Japan. Along his long career with *Dunaliella*, Prof. Ben-Amotz served as Head of the Department of Marine Biology at NIO and Head of the *Dunaliella* Section at the WIS, Chief Scientist of NBT and recently as Chief Scientist of Nikken Sohonsha Co., Japan. Prof. Ben-Amotz was nominated as President of the 8th International Marine Biotechnology Conference which was held successfully in Eilat in March 2007. Prof. Ben-Amotz has more than 130 publications and supervised dozens of students mostly on aspects related to *Dunaliella* and natural carotenoids.

Dr. John Benneman received his higher education at the University of California Berkeley in Chemistry (BS) and Biochemistry (PhD), working on biological nitrogen fixation. After three years as a Postdoctoral Fellows at U.C. San Diego, where he studied biological nitrogen fixation and hydrogen production, he rejoined the University of California Berkeley, working at the Sanitary Engineering Research Laboratory as an independent investigator on microalgae for wastewater treatment, biofuels and fertilizer production. He supervised the thesis work of five PhD students at U.C. Berkeley, in the departments of Civil Engineering, Biophysics and Plant Physiology. In the early 1980s he started a small biotechnology company and a few years later joined the Georgia Institute of Technology as Associate Professor. By the late 1980s he returned to California where he has been a full time consultant, and, from 1990 to 2000, also a part-time researcher at U.C. Berkeley (Departments of Civil Engineering and Plant and Microbial Biology). His research and consulting work span across disciplines and institutions, from the fundamental problems of photosynthesis and biological hydrogen production to the engineering of wastewater treatment ponds and landfills designed to produce renewable fuels and to abate greenhouse gas emissions.

Dr. Hofit Cohen is a senior physician and researcher of atherosclerosis at the Bert W. Strassburger Lipid Center, of Sheba Medical Center, Israel. Dr. Cohen received her MD degree from the Hebrew University, Hadassah Medical School, Jerusalem, Israel in 1992.She is Israeli Board certified in Endocrinology and Internal Medicine. A lecturer at the Sackler Faculty of Medicine Tel-Aviv University, Tel-Aviv, Israel since 2000. Dr. Cohen is a member of the Israeli Medical Association-Society for Research, Prevention and Treatment of Atherosclerosis and Israeli Endocrine Society. Dr. Cohen has published more than 10 papers and has written several chapters in endocrinology text books. Her current research interest is in the prevention and treatment of atherosclerosis and diabetes mellitus, with emphasis on the use of *Dunaliella* as a therapeutic tool in these metabolic disorders.

Dr. Marie E. DeLorenzo is a research ecologist with the National Oceanographic and Atmospheric Administration's National Ocean Service laboratory in Charleston, SC. Dr. DeLorenzo received a BS in Environmental Resource Management with a minor in Marine Science from Penn State University, as well as a master's degree in Ecology. She earned her PhD at Clemson University in Environmental Toxicology. Dr. DeLorenzo serves on the graduate faculty at the College of Charleston in the Marine Biology master's program and the Environmental Studies master's program. She is also a member of the graduate faculty at the Medical University of South Carolina in the Marine Biomedicine and Environmental Sciences PhD program.

Dr. DeLorenzo's research interests include environmental toxicology, physiological ecology, food web dynamics, effects and mechanisms of pesticide toxicity, eutrophication, nutrient cycling, and coastal resource management. She has contributed to the environmental risk assessment of individual pesticides and pharmaceuticals, as well as mixtures of chemicals. Her research has focused largely on the ecotoxicology of estuarine phytoplankton and invertebrate species, and she has published numerous peer-review journal articles on this subject.

Dr. DeLorenzo has served as president of the Southeastern Estuarine Research Society and as board member of the Estuarine Research Federation. She has also served as president of the Carolinas Chapter of the Society of Environmental Toxicology and Chemistry. Other committees served include the College of Charleston Marine Biology Program Graduate Council, the Coastal Pesticide Advisory Committee, and the Society of Environmental Toxicology and Chemistry's Pharmaceuticals Advisory Group.

Dr. Pedro Duarte was born in Lisbon, Portugal, in 1964, obtained his degree in Biology at the University of Lisbon and his PhD in Environmental Sciences at the New University of Lisbon, in 1995. Presently he works as a senior professor at University Fernando Pessoa, Oporto, Portugal. He has been involved in several European projects doing research in the field of Ecological Modeling applied to environmental problems such as benthic and pelagic primary production and carrying capacity estimation for aquaculture. He is author/co-author of 33 papers in journals (28 in international and 5 in national journals), 4 book chapters and 10 papers in conference proceedings. Presently he supervises three PhD and one MS scholars.

Dr. Ravi Durvasula is an Associate Professor of Medicine and Infectious Diseases at University of New Mexico School of Medicine in Albuquerque, USA. He also serves as Chief of Medicine at the New Mexico VA Health Care System and co-Director of the International Medicine program at UNM. Dr. Durvasula obtained his MD at McGill University in Montreal, Canada and trained in infectious diseases at Yale University School of Medicine. He was a research fellow of the Howard Hughes Medical Institute at Yale and served on faculty at Yale for five years. Dr. Durvasula's lab is engaged in several research programs aimed at paratransgenic control of infectious diseases and has been funded by the National Institutes of Health, US Department of Agriculture and Howard Hughes Medical Institute.

Dr. Einar Skarstad Egeland started his research on algal carotenoids as an MSc student, which was followed up as a PhD student under the supervision of Prof. Synnøve Liaaen-Jensen, Trondheim. His doctorate degree was on "algal carotenoids and chemosystematics" which led to the identification of eleven novel carotenoids from the Prasinophyceae. It also included suggestion of the biosynthetic pathways of the carotenoids in prasinophytes, comparison of the occurrence of marker carotenoids with algal systematics based on morphological details, together with study of carotenoids from other algal classes and unidentified algae.

After some years with temporary positions as a teacher at university/college and researcher on various projects (phytoplankton pigments, seaweed pigments, carotenoids and fatty acids in farmed fish), he was employed as an associate professor at the Bodø University College in 2002, where he is responsible for the chemistry courses. He is a member of the novel Marine Ecology Research Group, but is also involved in projects linked to seafood quality and aquaculture.

Dr. Mario Giordano did his undergraduate training in Biological Sciences at the University of Messina, followed by a PhD in Environmental Marine Sciences at the University of Genoa, and postodctoral work at the University of Florida and Sheffield University. He joined the Università Politecnica delle Marche (then University of Ancona) in 1994 and is currently Associate Professor in the Department of Marine Science. He has worked extensively on the physiology of /*Dunaliella*/species, especially regarding the acquisition of inorganic carbon and its interactions with the availability of other nutrients such as S an N, and has published over 40 papers on this topic. He has supervised 15 Masters and 5 doctoral students. His current research interests include the strategies for resource allocation in response to environmental changes, the regulation of inorganic carbon acquisition by environmental factors (including rising CO_2 levels), the use of FTIR spectroscopy for investigating algal composition and the use of algae for biodiesel production.

Dr. Patricia I. Gómez holds a PhD in Biological Sciences is a Full Time Assistant Professor at the Botany Departament, University of Concepción, Chile. Dr. Gómez has received training on advanced molecular biology techniques in Spain, Israel and USA. She has published 12 papers on Microalgal Biotechnology and Molecular Systematics which are her main research interest areas. Dr. Gómez has supervised two Master and three PhD students.

Dr. Mariela A Gonzalez is a PhD in Biology is a Full Time Professor at the Botany Department of the University of Concepción, Chile. She has published 40 papers dealing with taxonomy of marine benthic Cyanobacteria, freshwater microalgae (excl. Diatoms) and marine reds epiphytes. Within the last 12 years she had included molecular tools in her research, focused mainly in the genus Dunaliella. Dr. González has supervised three Master and two PhD students

Ayelet Harari, RD, MSc is a researcher and dietician at the Bert W. Strassburger Lipid Center, of Sheba Medical Center, Israel. Mrs. Harari received her B.Sc. and M.Sc. from the Hebrew University of Jerusalem, Israel, in 1983 and 2002 respectively. She is currently a PhD student in Tel-Aviv University, School of Medicine.Mrs. Harari is a member of the Israeli Medical Association-Society for Research, Prevention and Treatment of Atherosclerosis. Her current research interest is in the prevention and treatment of atherosclerosis, with emphasis on the use of 9-cis β-carotene as a therapeutic means in this disease.

Dr. Dror Harats is a professor and the head of The Bert W. Strassburger Lipid Center and Chairman of the IRB committee of the Sheba Medical Center, is a graduate of the Hadassah Medical School at the Hebrew University, Jerusalem. He completed his residency in Internal Medicine at Hadassah and fellowship in Pulmonary and molecular biology at the University of California, San Francisco (UCSF). For the past 20 years Prof. Harats has been involved in Research in Lipid Metabolism, Atherosclerosis and Vascular Biology. He was one of the pioneers who discovered the role of the immune system in atherosclerosis and invented a new genetic tool for the treatment of angiogenesis, a process that plays a major role in cancer and cardiovascular disorders. One of his current researches is the prevention and treatment of atherosclerosis, with emphasis on the use of *Dunaliella* as a therapeutic means in this disease. His research was published in more than 100 papers and chapters in books, and rewarded him with numerous prizes and grants in the field of atherosclerosis and cancer. Prof. Harats is a professor of medicine in the department of biochemistry at the Sackler Faculty of Medicine, Tel-Aviv University.

Dr. Ivy Hurwitz graduated from University of Kansas Medical Center. She did her post-doctoral research at Childrens' Mercy Hospital in Kansas City, MO and at the University of Pennsylvania, PA. Ivy is currently working on developing paratransgenic strategies for controlling human diseases.

Dr. Michael Huesemann stationed at the Pacific Northwest National Laboratory (PNNL), Richland, Washington, has conducted both experimental and theoretical research in environmental and marine biotechnology for more than fifteen years. Dr. Huesemann currently is or has been the principal investigator on numerous major U.S. Department of Energy funded research projects focusing on diverse areas such as photosynthetic hydrogen production, optimization of microalgal lipid and hydrocarbon production, biofixation of carbon dioxide from flue gases by marine microalgae, the effects of ocean carbon sequestration on nitrogen cycling, and hydrocarbon bioavailability in aged petroleum contaminated soils undergoing bioremediation treatment. He is also the principal investigator on an Office of Naval Research funded project on *in-situ* phytoremediation of PAH and PCB contaminated marine sediments with sea-grasses. In addition, Dr. Huesemann has published journal articles on the metabolic regulation of solvent production in anaerobic fermentations, the modeling of the leaching kinetics of hydrocarbons in aged soils, statistical soil sampling, critical analyses of mitigation responses to global climate change, environmental policy analysis, sustainable development, and professional ethics. He currently also serves as editorial board member of *Soil and Sediment Contamination: An International Journal, and Progress in Industrial Ecology*.

Dr. Marcel Janssen obtained an MSc in Environmental Technology at Wageningen University in The Netherlands. He obtained his PhD degree at the same university in 2002. During this PhD he studied microalgal photosynthesis under medium duration (seconds scale) light-dark cycles. His topic of research in the years after was the application of phototrophic micro-organisms in biotechnology in general and their cultivation in photobioreactors. Up to today he held three different postdoctoral positions, one at the Federal Institute of Technology Lausanne (Switzerland) and two at Wageningen University. Topics of research were the photoheterotrophic production of hydrogen, the development of panel photobioreactors for microalgal cultivation and the use of biocalorimetry to monitor microalgal growth. Currently he is leading a group of three PhD students within a project on carotenoid production by *Dunaliella salina* and he is co-supervising five other PhD students. The focus of research has shifted to photosynthesis in a wider sense including metabolic flux modeling.

Dr. Shirley Jeffrey MSc (Sydney), PhD (King's College, London), Chief Research Scientist and Honorary Fellow, CSIRO Marine and Atmospheric Research, Hobart, Tasmania, Australia, has always concentrated on the biology of marine phytoplankton (microalgae), and applications of microalgal culture and nutrition to Australian aquaculture. Her major focus has been on marine photosynthetic pigments as a potential measure of phytoplankton biomass. New families of marine chlorophylls (especially Chls c) and carotenoids were discovered, including new chemotaxonomic markers for phytoplankton groups, useful for oceanography. The Algal Culture Collection, which was developed, now comprises over 800 strains of Australian and international species. Career highlights include a study of symbiotic dinoflagellates on the Great Barrier Reef (R.V.Alpha Helix, 1966); the crystallization of chlorophylls c1 and c2 for extinction coefficient measurements (1969-72); the publication (as Senior Editor) of a SCOR - UNESCO Monograph on Phytoplankton Pigment Methods in Oceanography (1999), a study of UV protective pigments in microalgae (1999); and an intense interest in ocean colour applications.

Dr Jeffrey became a Fellow, Australian Academy of Science in 1991; a member of the Order of Australia (AM) in 1993; and a Foreign Associate of the US National Academy of Sciences in 2000. Other awards include the Gilbert Morgan Smith medal from the US National Academy of Sciences (2000), and the Shinkishi Hatai medal from the 21st Pacific Science Congress (2007).

Dr. Leland Jhanke holds a BA in Philosophy (Philosophy of Science) and MS and PhD degrees in Botany (Plant Physiology) from the University of Minnesota. Dr. Jahnke has previously been a member of the faculty at Central Washington University and currently is at the University of New Hampshire where he is Professor of Plant Biology. He has been Visiting Professor of Biology at the University of Essex, U.K. (1987 and 1996). Dr. Jahnke's research interests concern physical and chemical stressors that produce reactive oxygen species in photosynthetic micro-algae.

Dr. EonSeon Jin is Associate Professor in the Department of Life Science at Hanyang University in Korea. After receiving her Diploma in Biology from the Yonsei University in Korea, she continued to study and she received PhD degree at University of Minnesota, USA in major of plant physiology. Her postdoctoral research was performed from 1995 to 2000 at the Yonsei University and from 2000 to 2002 at the University of California at Berkeley, respectively. At Hanyang University, Professor Jin is studying how environmental stress influence physiology on the alga *Dunaliella* and Antarctic algae. She is also interested in applied microalgal research using green alga *Dunaliella salina* and *Haematococcus* in regard to improving high-value pigment production. She generated three EST data sets from *Dunaliella* sp., *Haematococcus pluvialis* and *Chaetoceros neogracile* to submit to NCBI. She published more than 20 papers in scientific journals.

Dr. Yehuda Kamari is a physician and researcher in the fields of atherosclerosis and hypertension at The Bert W. Strassburger Lipid Center, Hypertension Unit and Internal Medicine D, of The Chaim Sheba Medical Center, affiliated with the Sackler Faculty of Medicine, Tel Aviv University, Israel. Dr. Kamari received his MD with distinction (clinical years) from Tel Aviv University, Israel, in 1998. He specialized in internal medicine (1998-2002), and then worked as a senior physician (2002-2003) at the Department of Internal Medicine D, Sheba Medical Center. He started a PhD program in 2004, at The Bert W. Strassburger Lipid Center, Sheba Medical Center, Tel Hashomer, Sackler School of Medicine, Tel-Aviv University studying "The role of IL-1 in atherosclerosis". Dr. Kamari is a member of the Israeli Medical Association-Society for Research, Prevention and Treatment of Atherosclerosis and the Israeli Society of Hypertension. Dr. Kamari has published over 10 research papers. His current research interest is in the prevention and treatment of atherosclerosis and the metabolic syndrome, with emphasis on the role of inflammation in the pathophysiology of these conditions, and the use of *Dunaliella* as a therapeutic means in these conditions.

Dr. Adriana Katz has been studying *Dunaliella* for the last 25 years, in collaboration with the late Prof. Mordhay Avron, with Prof. Ami Ben-Amotz and with Prof. Uri Pick. Her special interests are in the mechanism of β-carotene induction, salinity tolerance, the mechanism of Na+ homeostasis and utilization of proteomics to identify salt-regulated proteins. Her pioneering proteomic work, done in collaboration with Dr. A. Shevchenko at the MPI in Dresden, provided the first comprehensive assessment how *Dunaliella* responds to high salinity.

Mr. Scott Matthews is currently a medical student at University of New Mexico. He obtained a Masters in Public Health degree from Yale and has background experience as a polymer chemist at APS. Scott's research interests are in paratransgenic control of vector-borne disease and probabilistic modeling of infectious disease processes.

Dr. Denis Maxwell received his doctorate in the Department of Plant Sciences at the University of Western Ontario in 1995. His thesis under the supervision of Norman Hüner focused on the role of the redox state of photosynthetic electron transport in photoacclimation in green algae. His post-doctoral training took place at the Department of Energy- Plant Research Laboratory at Michigan State University where he studied the role of the alternative oxidase in the laboratory of Lee McIntosh. After taking up a faculty position at the University of New Brunswick in Canada he moved in 2003 to the Department of Biology at The University of Western Ontario. His current research focuses on the role of the mitochondrion in intracellular stress sensing and signalling.

Dr. Anastasios Melis received his bachelor's degree from University of Athens, Greece, in 1970. In 1975, he obtained his PhD in Biological Sciences from Florida State University, USA. After several years of postdoctoral experience, Dr. Melis joined the University of California at Berkeley as an Assistant Professor in 1981. Since 1886, he has been Professor of Enzymology at the Department of Plant and Microbial Biology. Dr. Melis's laboratory is interested in hydrogen fuel production by a process of microalgal photosynthesis and photosystem-II damage and repair cycle in chloroplasts. In the latter area of research, *Dunaliella salina*, was employed as a model organism. He is a well-recognized scientist in the photosynthesis research field. Throughout his scientific career, Dr. Melis has supervised a lot of doctoral students and postdoctoral fellows.

Dr. Liki von Oppen-Bezalel is the VP for Business Development and Marketing of IBR Ltd. and a consultant to biotechnology companies and owner of Israel-Germany Bio-Tech Consulting. Dr. von Oppen-Bezalel received her BSc and MSc with distinction from the Hebrew University (1990) and Tel-Aviv University (1992) respectively. Dr. von Oppen-Bezalel completed her doctoral work in NCTR, the National Center of Toxicological Research, US FDA and graduated from the Hebrew University of Jerusalem, Israel in microbiology and biotechnology. Immediately after, she managed the Research and Development of IBR Ltd. followed with more than ten years in the biotechnology industry in Israel and Germany, where she managed leading positions in R&D and business development. Dr. von Oppen-Bezalel is an inventor on a number of patents as an outcome of her work in the biotechnological industry and received international awards for IBR Ltd. for its products.

Dr. Yakov Paz completed his PhD thesis in the group of UP at the Weizmann Institute of Science. His work centered on the mechanism of iron transport and internalization in *D. salina*. Dr. Paz identified two novel proteins that are induced in iron deprivation, showed that iron is internalized into acidic vacuoles and proposed a new mechanism for iron binding and internalization.

Dr. Uri Pick is professor and a research group leader at the WIS. His major interests include stress resistance mechanisms, photosynthesis and ion transport. His research in *Dunaliella* focused on salinity tolerance, Na+ and pH homeostasis, P-type ATPases, β-carotene accumulation, iron deprivation and cold acclimation. He supervised 15 PhD and MSc students on different projects concerning *Dunaliella*.

Dr. Jürgen Polle is Associate Professor in the Department of Biology at Brooklyn College of the City University of New York. After receiving a Diploma in Financial Sciences and working as a tax auditor, he changed careers and received his Diploma and his Doctorate degrees in Biology from the Georg-August University in Goettingen, Germany. Dr. Polle is an alumnus of the German National Merit Foundation. His postdoctoral research was performed from 1997 to 2002 at the University of California at Berkeley. At Brooklyn College, Dr. Polle's focus on the alga *Dunaliella* resulted in creation of the *Dunaliella* Culture Collection at Brooklyn College (DCCBC) with the webpage www.dunaliella.org. He is interested in basic and applied photosynthesis research using as one example the alga *Dunaliella salina* as a model organism. Dr. Polle is a member of the consortium responsible for the genome sequencing of *Dunaliella salina*. Dr. Polle's applied research centers on improving productivity of microalgae and he published about 15 peer-reviewed papers in scientific journals. Currently, he mentors several graduate students in his laboratory.

Prof. Dr. Song Qin, Research Professor and Assistant Director, Key Institute of Oceanology, Chinese Academy of Sciences, is President, Asia-Pacific Society of Applied Phycology (APSAP) and Past President, Asia-Pacific Society for Marine Biotechnology (APMBC). He is Conference Director, International Marine

Biotechnology Association (IMBA) and Council Member of International Society of Applied Phycology (ISAP). Prof. Dr. Song Qin, whose interest is molecular genetics and genetic engineering of eucaryotic marine algae, received his D.Sc. on Marine Biology at the Institute of Oceanology by finishing a joint doctoral training program with University of California, San Diego, and performed his post-doctoral researches in the Osaka National Research Institute, Japan and at the Georg-August University in Goettingen, Germany. He edited four books and two English Conference Proceedings and published over 100 research papers in related fields.

Dr. John Raven received his BA and PhD from the University of Cambridge UK, and has an Honorary PhD from the University of Umeå, Sweden. Since 1971 he has been at the University of Dundee UK where he is now Boyd Baxter Professor of Biology in the College of Life Sciences. Professor Raven has published over 300 peer-reviewed articles on algal and plant physiology and ecophysiology as well as other biological topics, and has authored one book and co-authored another. He is a Fellow of the Royal Society of Edinburgh and a Fellow of the Royal Society of London He holds an award of excellence from the Phycological Society of America, an Honorary Life Membership of the British Phycological Society, and is a corresponding member of the Australian Society of Plant Sciences as well as a Thompson-ISI Frequently Cited Scientist in Animal and Plant Biology.

Ms. Priya Sampath-Wiley holds a BS in Environmental and Forest Biology from SUNY College of Environmental Science and Forestry and a MS degree in Plant Biology from the University of New Hampshire. She is currently pursuing a PhD degree in Plant Biology at the University of New Hampshire under the guidance of Dr. Leland Jahnke.

Dr. Gustav Schonfeld, the Samuel E. Schechter Professor, led the Division of Atherosclerosis, Nutrition and Lipid Research at Washington University from 1972 through 2002. Dr. Schonfeld earned a BA, (1956) and MD (1960) from Washington University. Dr. Schonfeld spent two years as a research flight medical officer with the U.S. Air Force School of Aerospace Medicine and two years at MIT as associate professor of nutrition. After serving as acting chair of the Department of Preventive Medicine for three years, he was named the Kountz Professor of Medicine in 1987. From 1996 to 1999, he served as Adolphus Busch Professor, Chair of the Department of Internal Medicine, and Physician in-chief at Barnes-Jewish Hospital. He became the Samuel E. Schechter Professor of Medicine in 2001. In 1995, he received an Alumni/Faculty Award from the Washington University Medical Center Alumni Association and in 2006 a Special Award of the American Heart Association. Internationally know for his research on heart disease prevention and cholesterol and for his expertise on lipid metabolism, Dr. Schonfeld has performed many dietary and drug studies in patients with various forms of dyslipidemia. Currently he employs modern genetic and metabolic techniques to examine the low-cholesterol syndromes in humans and in engineered and congenic mice.

Dr. Aviv Shaish PhD is a researcher of atherosclerosis at the Bert W. Strassburger Lipid Center, of Sheba Medical Center, Israel, Lecturer in "Achva Academic College, Under the Academic Auspices of Ben-Gurion University" and consultant to biotechnology companies. Dr. Shaish received his B.Sc. and M.Sc. with distinction from Ben-Gurion University of Beer-Sheva, Israel, in 1985 and 1987 respectively. He obtained his Ph.D. degree in 1993 from the Biochemistry department, the Weizmann Institute of Science. After his post-doctoral work at Washington University, School of Medicine, Saint-Louis, MO, he was appointed senior researcher at Sheba Medical Center, in 1993. Dr. Shaish is a member of the Israeli Medical Association-Society for Research, Prevention and Treatment of Atherosclerosis.Dr. Shaish has published more than 50 papers. His current research interest is in the prevention and treatment of atherosclerosis, with emphasis on the use of *Dunaliella* as a therapeutic means in this disease.

Dr. Subba Rao, D.V. received PhD for research on phytoplankton ecology in relation to monsoons, currents, upwelling, and on primary organic production in the Bay of Bengal. His research and teaching activities took him to CSIRO, Australia. The Johns Hopkins University, Baltimore, Bedford Institute of Oceanography and Kuwait Institute for Scientific Research. He has supervised several graduate, post graduate and post doctoral scholars. Based on data collected on natural populations and algal cultures he is author/co-author of over hundred papers on phytoplankton ecology, physiological ecology, primary production, picoplankton, red tides, phycotoxin episodes, ballast water introductions of exotic organisms, tsunamis and marine life, and impact of mega engineering projects on marine environment. He currently also serves as editor of journal of Oceanography. Dr. Subba Rao edited two books "Pelagic Ecology Methodology" (Balkema, 2002) and "Algal Cultures, Analogues of Blooms and Applications" (Science Publishers, 2006). He is Emeritus Scientist at The Bedford Institute of Oceanography.

Mr. Bobban Subhadra graduated with a MS degree from Central Institute of Fisheries Education (ICAR), Mumbai. He also received a M.S degree in Fish Nutrition at University of Arkansas, Pine Bluff and currently pursuing his Ph.D in Biomedical Sciences at University of New Mexico, Albuquerque. His main research interest involves paratransgenic control of marine fish and shell fish pathogens and novel biotechnological approaches for fish nutrition.

Dr. Duc Tran received his BS degree in Biology in Vietnam in 1998, followed by a Masters degree in Ecology and Environment in 2002. He finished his PhD work in biology at St. Johns University in New York, USA in spring 2008. For the past decade his research concentrated on algae including studies of taxonomy, morphology, biodiversity, and molecular physiology. His PhD studies focussed on secondary carotenoid biosynthesis in the alga *Dunaliella*. Duc Tran is interested in studying algae for both fundamental and applied research with algae used as feed, food, waste water treatment and toxicology.

Dr. Andrea White holds a BA in Environmental Science from New England College, and MS and PhD degrees in Plant Biology from the University of New Hampshire. As a graduate student, Dr. White won teaching awards for her work with Introduction to Plant Biology, ultimately serving as TA coordinator for the course. Graduate research publications center on antioxidant biochemistry and the accumulation of β-carotene in *Dunaliella* sp., particularly in relation to ultraviolet exposure. Following matriculation, Dr. White became a Postdoctoral Research Associate at the University of California, Berkeley, publishing on the biochemistry of hydrogen production in *Chlamydomonas reinhardtii*, as well as examining hydrocarbon accumulation in *Botryococcus braunii*. After serving as adjunct teaching faculty at the California State University, Somona, Dr. White accepted a position within industry and now serves as the chief scientific advisor for The Linus Group, a full-service marketing strategy and communications firm.

Dr. Kittisak Yokthongwattana is currently a lecturer at the Department of Biochemistry, Faculty of Science, Mahidol University in Bangkok, Thailand. He graduated BS (Biology) from Rensselaer Polytechnic Institute, New York, USA, in 1996. In 2003, he received his PhD in Agricultural and Environmental Chemistry from the University of California at Berkeley, under a supervision of Professor Anastasios Melis. Since his doctoral research, Dr. Yokthongwattana has been working on adaptation of plant chloroplast to excessive irradiance, using unicellular green algae, including *Dunaliella salina*, as model organisms. He is also interested in studying phylogeny, physiology and biochemistry of *Dunaliella* sp. locally isolated from salt farms in Thailand. Dr. Yokthongwattana also has several doctoral and master's degree students working under his supervision.

Dr. René H. Wijffels is Professor and head of the Bioprocess Engineering Group at Wageningen University. This group develops new biotechnological processes for manufacturing of healthy food ingredients, biofuels, biopolymers, pharmaceuticals, and biopesticides. He holds an MSc in the field of Environmental Technology and a PhD in Bioprocess Engineering, both obtained at Wageningen University. Since 1991 involved in this group in education for the curriculum biotechnology and research in environmental biotechnology and later marine biotechnology. In marine biotechnology research programs have been developed in the field of photobioreactor design for production of microalgal products and cultivation of sponges for production of marine pharmaceuticals. In 2005 he obtained the VICI-grant from the Netherlands Organization of Scientific Research on the topic 'Photosynthetic Cell Factories' With a VICI grant senior researchers are offered the possibility to setup and lead their own research teams. Since 2007 he is chairman of the Bioprocess Engineering Group. Rene is vice president of the Dutch Association of Biotechnology and board member of the International Society of Applied Phycology, European Society of Marine Biotechnology and the editorial boards of the journals Marine Biotechnology and Microbial Biotechnology.

Dr. Ada Zamir was born and raised in Tel Aviv, Israel. MSc received from Hebrew University, Jerusalem and DSc from Technion, Israel Institute of Technology. Haifa. Post-doctoral training, Cornell University (contributed to the 1968 Nobel-prize-winning, first sequence determination of a nucleic acid). Since 1964 she has been a researcher at the Weizmann Institute of Science with intermittent stays at Northwestern University, Cornell University, Whitehead Institute MIT and Cambridge University, UK. Research interests: Structure and function of ribosomes; heterologous expression and assembly of nitrogenase components; mechanisms involving lipids, pigments and proteins implicated in the adaptation to light and salt stresses in *Dunaliella*; molecular basis for the salt adaptation of *Dunaliella* proteins. Academic positions: Chair, Scientific Council; Chair, Biochemistry Dept; Dean, Biophysics-Biochemistry Faculty. Member, National council of higher education. Numerous publications in the fields indicated. Supervised over 35 PhD and MSc students and Post-doctoral fellows.

1
2
3
4
5
6
7
8